AF598506

Desert Sediments: Ancient and Modern

Geological Society Special Publications

Series Editor K. COE

GEOLOGICAL SOCIETY SPECIAL PUBLICATION NO 35

Desert Sediments: Ancient and Modern

EDITED BY

L. E. FROSTICK
Department of Geology
Royal Holloway and Bedford New College
University of London

I. REID
Department of Geography
Birkbeck College
University of London

1987

Published for
The Geological Society by
Blackwell Scientific Publications
OXFORD LONDON EDINBURGH
BOSTON PALO ALTO MELBOURNE

Published for
The Geological Society by
Blackwell Scientific Publications
Osney Mead, Oxford OX2 0EL
(*Orders*: Tel. 0865 240201)
8 John Street, London WC1 2ES
23 Ainslie Place, Edinburgh EH3 6AJ
52 Beacon Place, Boston Massachusetts 02108, USA
667 Lytton Avenue, Palo Alto California 94301, USA
107 Barry Street, Carlton, Victoria 3053, Australia

First published 1987

Typeset, printed and bound in Great Britain by
William Clowes Limited, Beccles and London

DISTRIBUTORS
USA and Canada
Blackwell Scientific Publications Inc.
PO Box 50009, Palo Alto
California 94303
(*Orders*: Tel. (415) 965-4081)

Australia
Blackwell Scientific Publications
(Australia) Pty Ltd.
107 Barry Street, Carlton
Victoria 3053
(*Orders*: Tel. (03) 347 0300)

British Library Cataloguing in Publication Data

Desert sediments: ancient and modern.—(Geological Society special publication, ISSN 0305-8719; no. 35).
1. Sedimentation and deposition 2. Deserts
I. Frostick, L. E. II. Reid, Ian
III. Geological Society special publication, ISSN 0305-8719 IV. Series
551.3′04 QE571

ISBN 0-632-01905-0

Library of Congress Cataloging-in-Publication Data

Desert sediments.
(Geological Society special publication; no. 35)
1. Sediments (Geology) 2. Arid regions.
I. Frostick, L. E. II. Reid, Ian. III. Geological Society of London. IV. Series.
QE471.2.D475 1987 551.3′04 87-11686

ISBN 0-632-01905-0

Contents

Dune dynamics and deposits

Grain size, process and dune environment

CHEMICAL SEDIMENTS

REMOTE SENSING OF DESERT SEDIMENTS

List of contributors

P. Abell
Department of Chemistry
University of Rhode Island
Kingston
Rhode Island 02881
USA

R. Amit
Institute of Earth Sciences
The Hebrew University of Jerusalem
Jerusalem 91904
Israel

M. M. Ashour
Department of Geography
University of Qatar
Doha
PO Box 2713
Qatar

C. S. Breed
US Geological Survey
Flagstaff
Arizona 86001
USA

R. A. Carruthers
Badley, Ashton and Associates Ltd
Aveland House
Queen Street
Spilsby
Lincs PE23 5JX
UK

L. B. Clemmensen
Institute of General Geology
Øster Voldgade 10
DK-1350
Copenhagen
Denmark

P. A. Davis
US Geological Survey
Flagstaff
Arizona 86001
USA

A. J. Edwards
Department of Geography
The Queen's University
Belfast BT7 1NN
Northern Ireland
UK

N. R. J. Fieller
Department of Probability and Statistics
University of Sheffield
Sheffield S3 7RH
UK

E. C. Flenley
Department of Probability and Statistics
University of Sheffield
Sheffield S3 7RH
UK

L. E. Frostick
Department of Geology
Royal Holloway and Bedford New College
University of London
Egham Hill
Egham
Surrey TW20 0EX
UK

R. Gerson
Institute of Earth Sciences
The Hebrew University of Jerusalem
Jerusalem 91904
Israel

D. D. Gilbertson
Department of Archaeology and Prehistory
University of Sheffield
Sheffield S10 2TN
UK

K. W. Glennie
Shell UK Ltd
Shell-Mex House
The Strand
London WC2R 0DX
UK

S. Grossman
Institute of Earth Sciences
The Hebrew University of Jerusalem
Jerusalem 91904
Israel

A. Gunatilaka
Department of Geology
Kuwait University
Box 5969
Kuwait 13060

A. M. Harvey
Department of Geography
University of Liverpool
PO Box 147
Liverpool L69 3BX
UK

M. A. Hassan
Department of Physical Geography
Institute of Earth Sciences
The Hebrew University of Jerusalem
Jerusalem 91094
Israel

D. A. Hendry
Department of Geography
King's College
University of London
London WC2R 2LS
UK

A. R. Jones
NERC
Institute of Terrestrial Ecology
Penrhos Road
Bangor
Gwynedd
Wales
UK

S. A. W. Kay
Department of Geography
University College
University of London
Gower Street
London WC1E 6BT
UK

J. Lekach
Department of Physical Geography
Institute of Earth Sciences
The Hebrew University of Jerusalem
Jerusalem 91094
Israel

I. Livingstone
School of Environmental Studies
Gloucestershire College of Arts and Technology
Oxstalls
Gloucester GL2 9HW
UK

J. J. McAlister
Department of Geography
The Queen's University
Belfast BT7 1NN
Northern Ireland
UK

J. F. McCauley
US Geological Survey
Flagstaff
Arizona 86001
USA

J. K. Maizels
Department of Geography
University of Aberdeen
St Mary's
High Street
Old Aberdeen AB9 2UF
Scotland
UK

A. C. Millington
Department of Geography
University of Reading
Whiteknights
Reading
Berkshire RG6 2AH
UK

C. W. Mitchell
Department of Geography
University of Reading
Whiteknights
Reading
Berkshire RG6 2AH
UK

S. Mwango
Department of Geology
Kuwait University
Box 5969
Kuwait 13060

H. Olsen
Geological Survey of Greenland
Øster Voldgade 10
DK-1350
Copenhagen K
Denmark

K. Pye
Department of Earth Sciences
University of Cambridge
Downing Street
Cambridge CB2 3EQ
UK

N. Quarmby
NERC
Unit for Thematic Information and Services
University of Reading
Whiteknights
Reading RG6 2AB
Berkshire
UK

I. Reid
Department of Geography
Birkbeck College
University of London
Malet Street
London WC1E 7HX
UK

C. Roberts
Department of Geography
College of St Paul and St Mary
The Park
Cheltenham
Gloucester GL50 2RH
UK

A. P. Schick
Department of Physical Geography
The Hebrew University of Jerusalem
Jerusalem 91094
Israel

J. J. Smith
Department of Geography
The Queen's University
Belfast BT7 1NN
Northern Ireland
UK

R. J. Suttill
South Australian Oil and Gas Corporation Pty Ltd
Adelaide
South Australia 5000
Australia

D. S. G. Thomas
Department of Geography
University of Sheffield
Sheffield S10 2TN
UK

J. R. G. Townshend
NERC
Unit for Thematic Information and Services
University of Reading
Whiteknights
Reading RG6 2AB
Berks
UK

H. Tsoar
Department of Geography
Ben Gurion University of the Negev
Beer Sheva 84120
PO Box 653
Israel

H. Vine
2 Swale Close
Oadby
Leicester LE2 4GF
UK

A. Warren
University College
University of London
Gower Street
London WC1E 6BT
UK

W. B. Whalley
Department of Geography
The Queen's University
Belfast BT7 1NN
Northern Ireland
UK

E. K. Wild
Department of Geology
University of Bristol
Bristol BS8 1RJ
Avon
UK

B. P. J. Williams
Department of Geology
University of Bristol
Bristol BS8 1RJ
Avon
UK

M. A. J. Williams
Department of Geography
Monash University
Clayton
Victoria
Australia 3168

Preface

Deserts provide an excellent subject for interchange between Earth scientists of differing sub-disciplines. The aridity of climate and the sparseness of plants and animals, especially Man, encourage surface processes that produce sedimentary deposits sufficiently distinctive to be recognized unequivocally both throughout the passage of time and from one planet to another. Because of this, and because Lyell's dictum of uniformitarianism is probably most appropriate in desert settings where the impingement of evolving life forms has been least, an exchange between scientists concerned with the living deserts of Earth and beyond, on the one hand, and with ancient Earth counterparts, on the other, is extremely fruitful. Lessons are learned on all sides.

This was the spirit that encouraged 185 participants to gather at a Special Scientific Meeting of the Geological Society of London in May 1986. It is also the underlying rationale for this Special Publication which incorporates a substantial proportion of the contributions made at that Meeting.

This was not the first time that desert sediments have brought students of the ancient and modern together (see, e.g. McKee, E. D. (ed.) 1979. *Global Sand Seas. US Geological Survey Professional Paper 1052*; Brookfield, M. E. & Ahlbrandt, T. S. (eds) 1983. *Eolian Sediments and Processes. Developments in Sedimentology*, 38). However, previously there has been a tendency to emphasize the process, form and structure of wind-blown sand. This reinforces the popular view of deserts as areas of marching dunes and of little else. In fact, wind-formed dunes cover only a fraction of the Earth's present deserts, and there is no reason to suspect that the same is not true of bygone ages. Because of this, both the Meeting at the Geological Society and this volume have attempted to match studies of aeolian dunes with those of riverine and chemical deposits. As a result seven papers deal with the fluvial domain and two with chemical deposits (Fig. 1). Even the papers that champion the wind as an agent of erosion (twelve in number) highlight its ability to redistribute material of grain sizes both smaller and larger than those which bank characteristically into dune forms. Hence, four papers deal with the dynamics and accumulation of dust, and one compares sheet accumulations on Earth and Mars that contain a substantial gravel component.

However, any imbalance in treatment of the various process domains (wind, water, etc.) inevitably reflects the range of high quality papers

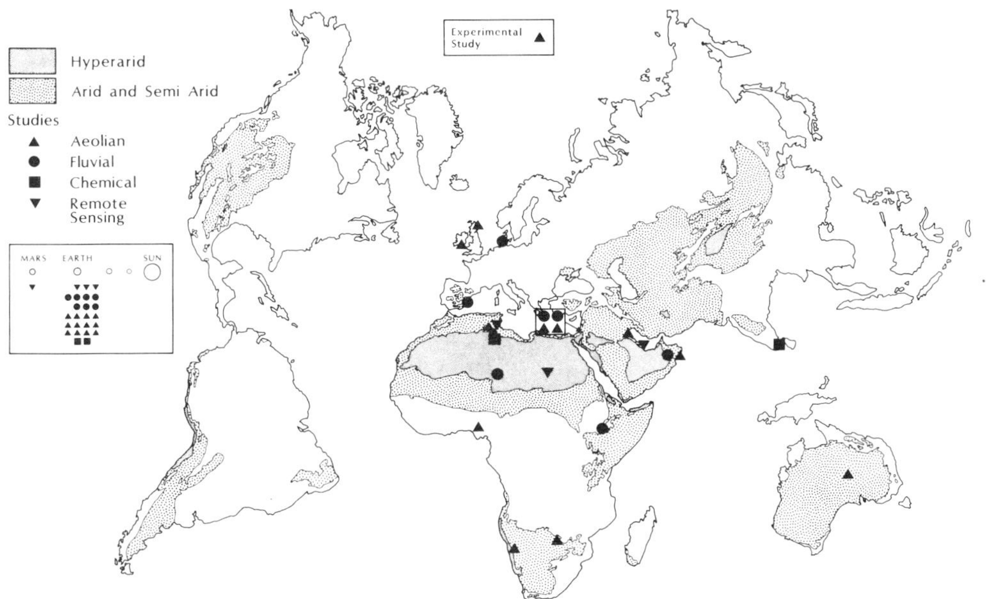

FIG. 1. Distribution of the Earth's hot deserts (after UNESCO 1977) and the location of studies in this volume.

proffered for inclusion. It also reflects the fascinations of those Earth scientists who conduct their research in deserts, ancient or modern. Thus, twenty-two papers seek to describe the processes and forms of clastic sediments, while only two directly address the problems of chemical sediments, and only three tackle weathering processes and diagenesis.

After an introduction, the volume is arranged so that fluviatile sediments are covered first. This recognizes the significance of water in shaping the desert landscape. Here there are two papers whose setting is modern, and five which deal with ancient counterparts. The volume *then* moves to aeolian sediments (seven modern; five ancient) if only because wind-blown materials often find their ultimate source in fluvial deposits. Two papers deal with contrasting chemical processes and deposits of modern age, before the volume moves to consider deserts at synoptic scale with the application of remote sensing techniques (three papers, all dealing with modern settings).

LYNNE FROSTICK
IAN REID
London, 1987

Acknowledgments

We have many people to thank. In the first instance, the permanent staff of the Geological Society and our own research officers played a vital part in the success of the Meeting that spawned this volume. We particularly mention Richard Bateman and Carolyne Symmonds, Jerry Jarvis and Angela Clark. We are indebted to the Geological Society for the generous provision of funds which allowed the participation of so many from beyond British shores. As for the production of this Special Publication, we thank the authors of each contribution for meeting each successive deadline. However, we especially thank our referees for their unsparing hard work. They were:

J. A. Allen, P. A. Allen, J. R. L. Allen, V. R. Baker, M. J. Barnsley, A. C. Brayshaw, M. E. Brookfield, J. A. Catt, C. P. Chalmers, W. G. Chaloner, L. Clemmensen, P. J. Curran, G. Evans, P. Friend, R. A. M. Gardner, R. Gerson, D. D. Gilbertson, K. W. Glennie, A. S. Goudie, A. T. Grove, A. M. Harvey, G. E. Hollis, D. K. C. Jones, P. Jungerius, D. J. Kinsman, N. Lancaster, M. M. Mainguet, J. Maizels, L. McFadden, J. McManus, A. C. Millington, G. Nanson, W. Nemec, K. Pye, H. G. Reading, C. C. Reeve, K. S. Richards, J. Rose, A. P. Schick, H. Scoging, R. C. Selley, I. J. Smalley, R. P. Steele, M. M. Sweeting, D. B. Thompson, J.-J. Tiercelin, H. Tsoar, I. Tunbridge, B. R. Turner, C. Vita-Finzi, A. Warren, R. J. Wasson, I. West, M. A. J. Williams, M. J. Wilson, D. H. Yaalon.

INTRODUCTION

Desert sediments: ancient and modern

K. W. Glennie

Tropical deserts have existed sporodically on our planet from the Precambrian to the Present, and seem not to have been a permanent feature of its surface (Glennie 1987). Depending on their definition, tropical deserts and semi-deserts currently occupy between approximately one fifth and one third of the Earth's land surface, of which only about 4% is covered by that popular concept of what a desert consists of—sand dunes. The remaining area comprises barren rock (both hill and plateau) with a variable cover of sediments transported by ephemeral streams (wadis, arroyos) to form the deposits at the terminal points of these streams. Varying with the ratio of water supply to the annual potential rate of evaporation, those terminal areas may be occupied by desert lakes that are generally of a temporary nature (permanent only if the water is provided from beyond the margins of the desert or is fed by groundwater), and become more saline as they become desiccated. The end product of such a situation is a salina or sabkha (area of sand, silt or clay, commonly encrusted with halite).

There is no universally accepted definition of a desert. In its simplest form it can be defined as a barren tract of land over which rainfall is too limited or spasmodic to support vegetation adequately. Very few desert areas are completely devoid of vegetation, and many areas that fall within a desert in terms of average annual rainfall may have an even though sparse cover of plants that have adapted to the relatively arid environment in which they live. In this context, some writers define deserts as areas that have an average upper limit of 250 mm of annual rainfall, even though it may all fall in one storm and rainfall may not recur for several years. Perhaps more important than the rainfall itself is the ratio between it and the potential rate of evaporation—the aridity or desiccation factor. This is about 1:10 in some parts of the Australian Desert and up to 1:500 in areas of the Sahara (Cooke & Warren 1973).

Tropical deserts exist either because the winds that cross their surfaces have a low relative humidity or there are no hills to push the winds up to colder cloud-forming altitudes; the Southwest Monsoon gives torrential rain over the Western Ghats of peninsular India but little over the arid low-lying Rajasthan Desert further to the N. Drawn from the high-pressure area of the Mediterranean Sea, the North African trade winds blow south and southwestward under clear skies, warming as they cross the Sahara, and so becoming capable of absorbing more moisture without condensation. By contrast, the southern hemisphere coastal deserts of the Namib and Atacama are often shrouded in mist but rarely experience rain.

Cold deserts also occur in areas of low precipitation but lack a protective cover of vegetation because the moisture necessary for growth is frozen for a large part of the year. Allied to the low temperature, peri-glacial dune fields exist because of adequate supplies of fluvio-glacial sediments that are unprotected by vegetation and have a grain size that is suitable for deflation once the summer melt-floods have ceased.

Dunes are formed because strong dry winds can transport sand-size particles as a saltation load within a few metres of the land surface (sand storm). Because winds can drive sand over a hard immobile surface faster than over one of loose sand (Bagnold 1941), sand tends to accumulate in areas that are already sand covered. These sand patches grow in size whenever the velocity of the wind lessens and is no longer capable of keeping the grains in motion; dunes eventually result. Silt and clay-size particles, however, can be carried in suspension (dust storm) and by this means may be transported right out of the desert to be deposited elsewhere, as a marine horizon in the ocean or on land as loess.

A cover of vegetation on the land surface tends to inhibit the aeolian transport of sand. Some ancient dune sequences that were deposited before higher plants began to colonize extensive areas of land during the Devonian (*eg* Precambrian dunes; Ross 1983) may reflect the ease with which sand could be moved by the wind in the absence of vegetation rather than any long-term aridity. The Devonian Old Red Sandstone sequences of Britain, for instance, are dominated by fluvial and lacustrine sediment that must indicate a fairly high annual rainfall, and yet dune sands have accumulated locally (Mykura 1983; Carruthers, this volume).

To confirm a true desert origin for ancient sequences, additional evidence of aridity is needed. In hot deserts this is most easily recog-

From Frostick, L. & Reid, I. (eds), 1987, *Desert Sediments: Ancient and Modern*, Geological Society Special Publication No. 35, pp. 1–4.

nized in the form of evaporites—gypsum crystals (or their pseudomorphs), anhydrite or halite. If the preserved ancient dune sands formed part of an originally larger sand sea, that direct evidence of aridity may be buried or could have been removed by later erosion. In such a case, recourse must be made to the climatic context of such dune sands as interpreted from the rocks above and below.

An understanding of the processes and climatic conditions that give rise to deserts is the outcome of studies that had their beginnings in the latter part of the 19th Century (*eg* Blandford 1877; Walther 1888; Buhse 1892; Hedin 1896; Cornish 1897; McGee 1897; and in the classic book on deserts by Walther 1900). The range of desert areas and topics studied by those early workers may be gleaned from the titles of their papers.

Several important papers that added to our understanding of various aspects of desert sedimentation, both ancient and modern, were published between the two World Wars (*eg* Ball 1927; Blackwelder 1928, 1933; Bagnold 1931; McKee 1934; Madigan 1936; Shotton 1937; Reiche 1938). Bagnold (1941) brought our understanding of dune-sand movement a great step forward with his classic book *The Physics of Blown Sand and Desert Dunes*; a combination of accurate observation during travels in the Egyptian Desert, combined with experimental work while on leave from the army, made this a unique contribution to geological science.

Since World War II there has been an ever-increasing volume of publications on deserts of both a popular and a scientific nature. This was a natural outcome of the increasing ease and dependability of motor transport in the desert, and its efficient use in soft sand using driving techniques that Bagnold helped to pioneer both before and during the war in North Africa. The North African campaigns did much to remove the mystery of the desert without reducing the need to respect its harsh environment. Luck plays no part in desert travel; there is a fair chance that the careless will not survive extended journeys, especially during the hottest season when recorded temperatures in parts of North Africa and Arabia, for instance, often exceed 50°C.

Aerial photography also improved under the influence of military needs. Post-war photogeological mapping of parts of North Africa and Arabia as part of the search for oil not only resulted in high-quality structural maps, but also permitted a three-dimensional study of desert morphology, including dune forms and their distribution, on a scale that hitherto had been impossible. The scale increased to global dimensions with the advent of space travel, and with the development of remote-imaging systems carried on spacecraft. We can now compare the presumed aeolian sediments of some of the planets with those of the Earth (see *eg* Greeley & Iverson 1985).

Scientific progress depends not only on advances in equipment but especially on the scientists themselves; these are men, and increasingly also women, who have both the curiosity and the drive to find out for themselves. Since World War II there has been an ever-increasing volume of publications on deserts. The outstanding contributor of that period was E. D. McKee, whose papers on deserts (and other geological topics) span more than 50 years. He will always be remembered for conceiving, contributing to and editing that excellent volume *Global Sand Seas* (McKee 1979).

Most writers have, pragmatically, concentrated on one aspect or another of deserts without viewing them as a whole. My own involvement was a direct outcome of the commercial importance of ancient desert sandstones as reservoirs for hydrocarbons. When the size of the giant Dutch Groningen gas field (Stäuble & Milius 1970) was realized in 1963, the reservoir was believed to have been of desert (probably dune) origin, but as no one within Shell had first-hand knowledge of desert sediments, they could not be sure. I was instructed to study deserts to find out what influenced their development in order to assist prediction in ancient sequences and to find criteria for identifying ancient desert strata. A lot of hydrocarbon exploration involves an understanding of time-dependent changes in palaeogeography. With respect to dunes, it was clear that they should also be studied in the context of deserts as a whole (Glennie 1970). When the lessons of modern deserts were applied to the early Permian desert sequences of NW Europe (Glennie 1972), the overall palaeogeographic distribution of the Rotliegend and associated rock sequences became of prime importance for predicting the presence of possible hydrocarbon-bearing reservoirs.

The discovery of oil and gas in a large number of structures beneath both the North Sea and some adjacent land areas led inevitably to attempts to discover the reason why the reservoir sands were better in some areas than in others. Diagenetic processes had played a major role in damaging the primary reservoir properties of the desert sands, whether as early diagenesis, such as occurs with the near-surface formation of calcretes or the cementation of wadi sands, or as the burial-related diagenetic growth of pore-filling illite and other minerals (Glennie *et al.* 1978; Hancock & Taylor 1978). The economic importance of the Rotliegend and Triassic (*eg* van Veen 1975; Colter & Ebbern 1978) sandstones of NW Europe as a reservoir for gas has meant that considerable scientific effort has gone into understanding the reasons for disappointment in well productivity.

It is logical that the more we know both about the basic causes of deserts, and about the range of sediments found in them and their ancient counterparts, the greater will be the variety of new related studies. We see this in the spread of subjects discussed in this volume. The much neglected topic of desert fluvial sediments is considered in seven papers, which include a Triassic example from Denmark (Olsen). Aeolian sediments are not neglected, with dust-size particles forming the subject of four papers and dunes of another seven. Of these, dune and fluvial sequences of Devonian age in Ireland (Carruthers), Permo–Carboniferous paraglacial aeolianites in Australia (Williams, Wild & Suttill) and possible Triassic star dunes in Scotland (Clemmensen) illustrate the range of ancient desert sequences covered. Some of the more unusual features of deserts are illustrated in the chemical precipitates associated with spring water in a highly evaporative environment in Tunisia (Roberts & Mitchell), and in the diagenetic replacement of plant roots by silica and calcium carbonate in tropical dune sands in India (Hendry).

Hyper-arid environments on Earth and Mars are compared by using satellite-borne remote-sensing tools (Breed & McCauley). The use of such equipment to assist with the mapping of surface deposits in Qatar (Ashour) and for monitoring the processes of sediment transport in Tunisia (Millington *et al.*) brings the volume well and truly into the space age.

References

Bagnold, R. A. 1931. Journeys in the Libyan Desert. *Geographical Journal* **78**, **13–39**, 524–535.

—— 1941. *The Physics of Blown Sand and Desert Dunes*. Chapman and Hall, 265 pp.

Ball, J. 1927. The problems of the Libyan Desert. *Geographical Journal* **70**, 209–244.

Blackwelder, E. 1928. Mudflow as a geological agent in semi-arid mountains. *Geological Society of America Bulletin* **39**, 465–484.

—— 1933. The insolation hypothesis of rock weathering. *American Journal of Science, 5th Series* **26** (152), 97–113.

Blandford, W. T. 1877. Geological notes on the Great Indian Desert between Sind and Rajputana. *Indian Geological Survey Records* **10**, 10–21.

Buhse, F. A. 1892. Die grosse Persische Salzwüste und ihre Umgebung. *Deutches Rundschau Geographie Statistik* **15**, 49–59.

Colter, V. S. & Ebbern, J. 1978. The petrography and reservoir properties of some Triassic sandstones of the Northern Irish Sea Basin. *Journal of the Geological Society, London* **135** (1), 57–62.

Cooke, R. U. & Warren, A. 1973. *Geomorphology in Deserts*. Batsford, London, 374 pp.

Cornish, V. 1897. On the formation of sand dunes. *Geographical Journal* **9**, 278–309.

Glennie, K. W. 1970. *Desert Sedimentary Environments*. Developments in Sedimentology **14**, Elsevier, Amsterdam, 222 pp.

—— 1972. Permian Rotliegendes of North-West Europe interpreted in light of modern desert sedimentation studies. *American Association of Petroleum Geologists Bulletin* **56** (b), 1048–1071.

—— 1987. Desert sedimentary environments, present and past: a summary. *Sedimentary Geology* **50**, 135–165.

——, Mudd, G. C. & Nagtegaal, P. J. C. 1978. Depositional environment and diagenesis of Permian Rotliegendes sandstones in Leman Bank and Sole Pit areas of the U.K. Southern North Sea. *Journal of the Geological Society, London* **135** (1), 25–34.

Greeley, R. & Iverson, J. D. 1985. *Wind as a Geological Agent on Earth, Mars, Venus and Titan*. Cambridge University Press, 333 pp.

Hancock, N. J. & Taylor, A. M. 1978. Possible causes of Rotliegend Sandstone diagenesis in northern West Germany. *Journal of the Geological Society, London* **135** (1), 35–40.

Hedin, Sven 1896. A journey through the Takla–Makan Desert, Chinese Turkestan. *Geographical Journal* **8**, 264–278.

Madigan, C. T. 1936. The Australian sand-ridge deserts. *Geographical Review* **26** (2), 205–227.

McGee, W. J. 1897. Sheet-flood erosion. *Geological Society of America Bulletin* **8**, 87–112.

McKee, E. D. 1934. An investigation of the light-coloured cross-bedded sandstones of Canyon de Chelly, Arizona. *American Journal of Science, 5th Series* **28** (165), 219–233.

—— (ed.) 1979. *A study of Global Sand Seas*. Geological Survey Professional Paper 1052, United States Government Printing Office, Washington, 429 pp.

Mykura, W. 1983. Old Red Sandstone. *In*: Graig, G. Y. (ed.) *The Geology of Scotland* (2nd edition). Oliver and Boyd, Edinburgh, 205–251.

Reiche, P. 1938. An analysis of cross-lamination—the Coconino Sandstone. *Journal of Geology* **46** (7), 905–932.

Ross, G. M. 1983. Bigbear erg: a Proterozoic intermontane eolian sand sea in the Hornby Bay Group, Northwest Territories, Canada. *In*: Brookfield,

M. E. & AHLBRANDT, T. A. (eds) *Eolian Sediments and Processes*. Elsevier, Amsterdam, 483–519.

SHOTTON, F. W. 1937. The Lower Bunter Sandstones of north Worcestershire and East Shropshire. *Geological Magazine* **74**, 534–553.

STÄUBLE, A. J. & MILIUS, G. 1970. Geology of Groningen gas field. *American Association of Petroleum Geologists Memoir* **14**, 359–369.

VEEN, F. R. VAN 1975. Geology of the Leman Gas Field. *In*: WOODLAND, A. W. (ed.) *Petroleum and the Continental Shelf of North-West Europe*. Applied Science Publishers, 223–233.

WALTHER, J. 1888. Über Egebnisse einer Forschungsreise auf der sinaihalbinsel und in der arabischen Wüste. *Verhandl. Ges. Erdkunde* **15**, 244–255.

—— 1900. *Das Gesetz der Wüstenbildung in Gegenwart und Vorzeit*. Reimer, Berlin, 175 pp.

K. W. GLENNIE, Shell UK Ltd, Shell-Mex House, Strand, London WC2R 0DX. *Present address*: 4 Morven Way, Ballater, Aberdeenshire AB3 5SF, UK.

FLUVIAL SEDIMENTS

Process and form

Vertical exchange of coarse bedload in desert streams

A. P. Schick, J. Lekach & M. A. Hassan

SUMMARY: The tracing of coarse particles, transported by flood events in two ephemeral streams in the Negev Desert, Israel, reveals that a size-independent, highly skewed longitudinal dispersion pattern is accompanied by vertical distribution throughout the scour layer. An analysis of the intermittent burial and resurfacing of these particles from one event to the next leads to the formulation of a concept of equilibrium in vertical exchange. This equilibrium requires that, for the 'dominant' flood event (frequency 1–2 years), the number of previously buried particles exposed by that event is balanced by an identical number of previously surficial particles being buried. Such an equilibrium can be maintained only if the scour layer reaches sufficiently deep into the 'sub-armour' layer to entrain a requisite number of particles of armour-layer size. The resulting functional relationship between (1) the size of coarse particles in relation to the matrix; (2) the thickness of the armour and 'sub-armour' layers and the distribution of coarse particles; and (3) aspects of the flow regime, enables the derivation of hydrological information from studies of the three-dimensional distribution of coarse particles in modern channel bed sediments of arid streams. It also helps to explain the vertical distribution of pebbles in ancient continental clastic sediments.

Fluvial processes in arid streams operate over brief periods whose aggregated duration is less than 1% of the total time. Yet, in these short bursts of activity, large amounts of sediment are eroded, transported and deposited. Much of this sediment is transported as bedload. Due to the high energy of desert floods, particles ranging in size up to large boulders are known to be moved long distances in relatively small ephemeral channels.

During floods, parts of the beds of ephemeral, coarse-grained streams are scoured, and the material is transported. Particles deposited during flood events in scoured locations become covered by later deposits, and may arrive at positions several grains below the new bed surface. Thus the active alluvial layer of ephemeral channels serves as a dynamic storage body whose components change their relative positions over time. However, viewed as a whole, the cross-section of this body of sediment, multiplied by the mean transport velocity of its components, is equal to the long-term bedload discharge of the stream (Emmett *et al.* 1983).

Many rivers exhibit a vertical concentration of coarse particles at the surface of the channel bed, termed an 'armour' (Williams & Wolman 1984; Sutherland, in press). The term armour may have several meanings. It was initially used to describe the situation below a dam when new discharges changed the character of the downstream channel and long-duration medium flows winnowed away all the fines. However, a coarse surface layer is found in streambeds even after repeated floods which are known to have affected a scour layer several diameters in depth. Hence another process, one inherent in the flow such as a dispersive stress, which tends to push large particles upwards to the level of zero stress, must take place. As used in this paper, the term armour has no genetic implication.

Desert streams in general seem to have less well developed armour layers than perennial ones. This may be due to the predominance of short and violent flood events which are characteristic of arid areas. Such floods cause the disruption of the armour at a rate faster than it can form during the relatively short and less effective periods of hydrograph recession. By contrast the relatively constant low and medium flows characteristic of perennial rivers provide plenty of time for armour formation. Desert channel beds with a cobble–pebble–sand size distribution generally show some relative concentration of coarse particles in the topmost ('armour') layer, but these same sizes are also represented in the 'sub-armour' layer. However in many perennial streams the transition between the 'armour' and 'sub-armour' layers is very abrupt; few of the largest particles that dominate the armour are found within the sub-armour layer (Andrews & Parker, in press), though other configurations have also been reported (*eg* Reid & Frostick 1986, Fig. 1*C*).

It is known from experiments with the tracing of labelled coarse particles in ephemeral streams that many of them disappear after one or more events. Some of them are re-exposed by later flood events (Leopold *et al.* 1966). This vertical exchange of coarse particles between the top and lower parts of the scour layer may be dependent on the magnitude of the flood; but, if the degree

From FROSTICK, L. & REID, I. (eds), 1987, *Desert Sediments: Ancient and Modern,* Geological Society Special Publication No. 35, pp. 7–16.

of 'armouring' is quasi-constant, a state of equilibrium must exist. This means that the number of particles resurfacing after a given event from a previously buried position equals the number of particles that were exposed prior to the event but have subsequently been found missing and have therefore become buried.

The rate of vertical exchange is not directly dependent on the degree of 'armouring'. As an example, a degree of 'armouring' of 2.0—meaning that the volumetric concentration of the coarse (*eg* $D > 32$ mm) armour particles in the top layer is twice that in the 'sub-armour' layer—can be maintained over time under various rates of exchange. As an extreme case, the entire top layer of particles can be exchanged with a 'sub-armour' layer which is double its thickness. Conversely, if a given event barely exceeds the entrainment threshold of the surface particles, only a few per cent of the coarse 'armour' population may be exchanged. In well-armoured streambeds, such as those found in many non-arid environments, the degree of armouring may be of the order of 10 or more. This means that, in order to effect even a small rate of vertical exchange, the scour layer for a given event must extend deeply into the 'sub-armour' layer, though the process by which this is accomplished is not well understood (compare Andrews & Parker, in press). The thickness of the scour layer may thus be a parameter which is functionally tied into the relation between the degree of streambed armouring and the rate of vertical exchange.

The purpose of this paper is to present data on the longitudinal and vertical dispersion of coarse particles in ephemeral streams and to analyse these data in the light of the above considerations. Both the degree of vertical exchange and the depth of the scour layer have a diagnostic value in terms of the environmental conditions which cause them. Given an indication of the longitudinal distribution of coarse labelled particles and determinations of the degree of 'armouring', it may be possible to estimate the magnitude and frequency of floods in ephemeral streams. The concepts developed in this paper may also be of use to students of ancient fluviatile strata which include coarse clasts in a sand-granule matrix.

Experimental sites

The experiments and observations on which this paper is based were conducted at two sites in the Negev Desert, southern Israel. The first site, Nahal Yael, is a small, 2 km long high-relief catchment located in the extreme arid southern tip of the Negev (Fig. 1). The site has been amply described in the literature (*eg* Gerson & Inbar 1974; Schick 1977). Mean annual rainfall is 30 mm. Flood events are spaced at intervals of up to several years, but some of these are very powerful in terms of sediment transport. The bedload tracing programme whose results are discussed below was started in 1965 and is still continuing. It is concentrated over a 1.0 km long channel reach which is 3–10 m wide and has a longitudinal slope of 0.05. The maximum recorded peak flows are 1.1 $m^3 s^{-1}$ for the upstream end of the study reach and 2.75 $m^3 s^{-1}$ for the lower end. The channel bed is alluvial throughout its length, with depth to bedrock varying from 0.5 to 4.0 m. The steep banks are either bedrock (mostly schists) or footslopes of alluvial–colluvial terraces which are the prime source of the coarse bed material presently transported by the stream. The channel bed is an irregular assemblage of straight reaches with little internal differentiation, braiding sections, and occasional gravel or pebble bars flanking inner channels that consist of finer, sand-granule material.

The Nahal Hebron site is located near Tel Shoqet, 13 km N of Beer Sheba (Fig. 1). At this point the 3–5 m wide, entrenched and relatively straight channel drains 250 km^2 of semi-arid, predominantly limestone terrain into the loess-covered Beer Sheba basin. The 150 m long injection reach used for this study consists of irregular low-relief pebble-cobble bars separated by granule-pebble pools. The slope is 0.016. At its lower end the reach merges into an irregularly spaced, entrenched meandering pattern. Flow is ephemeral, with 2–6 events per year. Flow measurements were not made prior to 1982. Indirect estimates, based on inference from observations and general models (Ben Zvi & Cohen 1975) yield a value of 7.5–15 $m^3 s^{-1}$ for the mean annual flood. The highest peak flow observed since the inception of this study (Hassan 1983) in 1982 was 33 $m^3 s^{-1}$. The recurrence interval of such a flow is estimated at 5–8 years.

Downstream dispersal of sediment

The material scoured and moved by flood events in ephemeral streams includes a coarse fraction, whose constituent particles move intermittently and at a speed lower than that of the water (Leopold *et al.* 1966). Although the entire volume of scoured material may move only a limited distance downstream during individual events, it includes components that travel widely varying distances. In one experiment in Nahal Yael, based on 56 particles ranging in weight from 117 to 1346 g, less than 10% of the particles moved

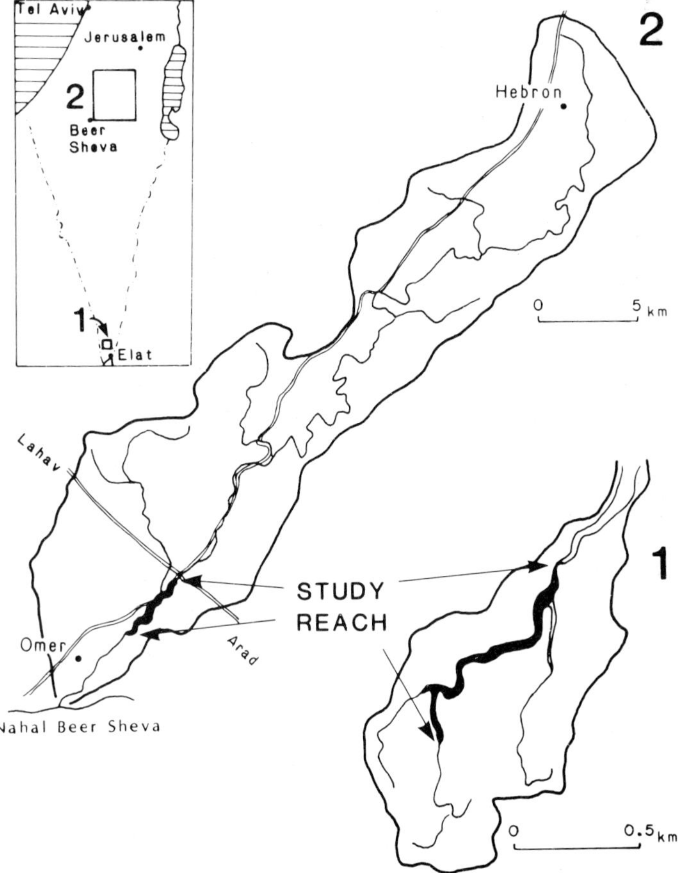

FIG. 1. Location map, Nahal Yael (1) and Nahal Hebron (2), Israel.

beyond the estimated mean distance of transport for the scour layer in two events (Schick *et al.*, in press, Fig. 5). The distance of transport does not correlate with size: a large particle, if entrained, has a similar probability of moving a given distance as a small particle.

The downstream dispersal pattern for the two events analysed (Fig. 2, A, B), is markedly similar. In terms of peak discharge, the second flood (24.05.68) was 50–100% larger than the first (25.04.68), a difference reflected in the two dispersal curves. If ranked by mobility, the transport distances exceeded for any given percentage of particle population are about double for the second event. From curve C (Fig. 2) it is evident that the extreme skewness of the frequency distribution of longitudinal dispersion characteristic for single flood events is smoothed by considering two consecutive events together. The probability of a single particle being entrained and transported over long distances in two consecutive events is small. In fact, if the probability of each particle becoming entrained and transported for a long distance is less than 10% and random, as the data on Fig. 2 seem to suggest, then the probability of two consecutive long trajectories occurring for a given individual particle is the square of the single-event probability, *ie* less than 1%.

The probability of effective re-entrainment must depend on the local sedimentary environment of the particle (*eg* Brayshaw 1985). Some are left stranded on a relatively high bar surface or in the stoss of a bush or a boulder; others may be shielded by adjacent particles (Leopold *et al.* 1966; Laronne & Carson 1976; Leopold & Emmett 1981; Brayshaw *et al.* 1983). The probability of re-entrainment is also affected by intermediate burial within the scour layer, which, as suggested by curve D (Fig. 2), is associated with particularly efficient longitudinal displacement of about one-third of the particles.

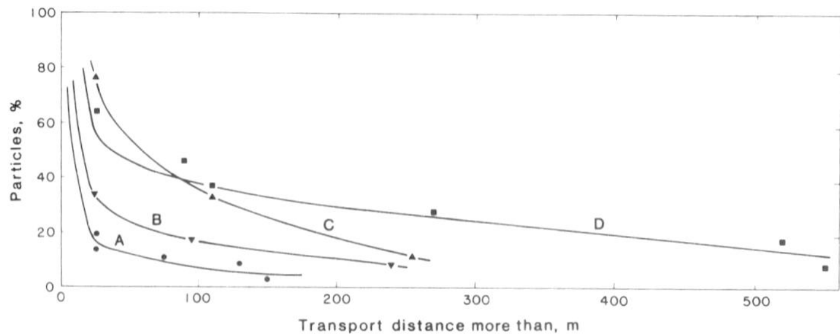

FIG. 2. Longitudinal transport distance of pebbles with weights ranging from 117 to 1346 g in Nahal Yael. Curve A—for flood event 3 (25.04.68); curve B—for event 4 (24.05.68); curve C—for both events together, based only on those particles exposed before, between, and after the two events; curve D—for both events, based on particles not exposed between the two events.

The degree of longitudinal 'mixing' of the coarser fraction within the scour layer of an ephemeral stream is considerable. The clear discrepancy in distance of transport of the most mobile 40% of the particles in curves C and D (Fig. 2) suggests that vertical and horizontal mixing of the coarse fraction are closely related.

It may be argued that rates of initial entrainment of the labelled particles, as originally placed on the channel bed surface, are higher than for those in a similar group of particles which had reached their locations by natural processes. Another possible objection to the data presented relates to the special locale of the standard injection site of the Nahal Yael tracing programme. This site is located at the upstream end of a section of the channel with a continuous alluvial bed and is immediately downstream of a series of step pools which ends in a steep chute-like channel produced as a result of the weathering of a soft dyke.

In an attempt to assess the impact of these possible problems, another analysis was undertaken. Five consecutive flood events were analysed, and only those particles were considered whose starting position was at least 20 m downstream of the point of original emplacement. Moreover, only those particles whose position was exposed and recorded before and after any two consecutive events in the series were included. In this way, it is possible to be certain that all two-step movements started from a point of fluvial deposition as well as ensuring that the process recorded was unaffected by any peculiarities in channel morphology.

The distances of particle movement so determined were divided into two groups: those that were found exposed between the two events, and those that had been assumed to have been buried by the first of the flood events in any pair. Although the data are not numerous, especially for the group of particles whose burial was not repeated, it is clear that the transport efficiency of the 'buried/exposed' particles is much higher than that of the 'exposed/exposed' ones (Fig. 3). The median distance of transport for the 'buried/exposed' particles (curve B) is, for two events, about three times that of the 'exposed/exposed' particles (curve A)—83 versus 28 m. For particles with trajectories of over 100 m, the difference between the two groups is not clear, but for shorter travelled ones, representing the majority of particles, the patterns are distinct. As a rule, long transport distances are closely related to particle burial and to the scour which must be associated with it.

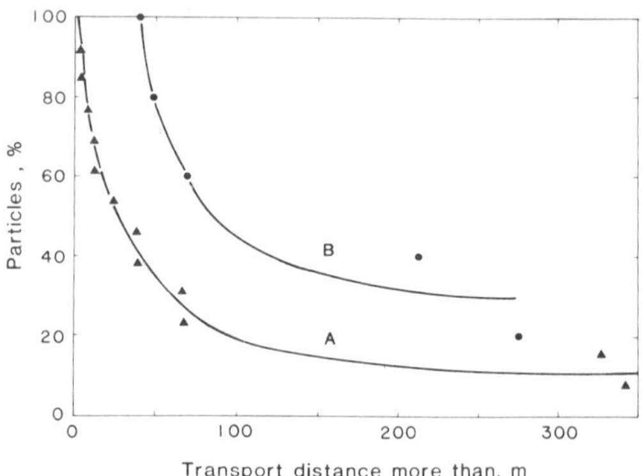

FIG. 3. Transport distance of particles in Nahal Yael during two consecutive events out of a series of five events: curve A—particles exposed before the first event, exposed after the first event, and exposed again after the second event; curve B—particles exposed before the first event, buried by the first event, and re-exposed by the second event. Period of monitoring extended over flood events 10, 12, 15, 16, and 17 (from 12.11.73 to 22.03.85) at Nahal Yael. The 256 seeded particles weighed between 400 and 10 000 g.

Vertical displacement

The event of 30.10.81 in Nahal Yael—a short, sharply peaked flood with a peak discharge frequency of 5–7 years—resulted in the burial of over one half of an individually labelled pebble–cobble sample of 256 particles (Fig. 4). The next flood, which followed more than three years later, was much smaller. It ranked only 13th in the series of 17 events monitored in Nahal Yael over a period of 21 years. Flow at the upstream end of the study reach infiltrated completely into the first 500 m of the channel bed and the flow recorded at the downstream end was derived entirely from local sources. Although this event did not cause much longitudinal movement, the deposition associated with the transmission losses in the upper part of the reach led to further burial comparable in relative terms to that associated with the previous flood: 63 out of 120 surface particles (Fig. 4). At the same time and despite the low discharge of the event of 22.03.85, 24 of the particles buried by the previous event were re-exposed. As a result, the population of particles at the surface after the second event numbered just under one-third of the original seeding total; two-thirds of these particles (*ie* 57% or 22% of the source population) did not undergo a burial stage.

The tracing method employed is visual, and so the depth of burial cannot be detected and is not easily estimated. Seventy-five per cent of the particles ranged in weight from 1000 to 3000 g, equivalent to a range in *b*-axis dimension of 88 to 127 mm. Assuming that the buried particles are deposited with their *a–b* plane parallel to the channel bed surface, and that the length of the *c*-axis is, on average, 70% of the *b*-axis, then the bulk of the traccd population can be buried completely only if the base of the particles is covered by at least 6–9 cm. This value, although a minimum one, is roughly half an order of magnitude higher than the one predicted from the relationship between depth of scour, as determined from chain measurements, and discharge per unit width of channel in Arroyo de los Frijoles, New Mexico (Leopold *et al.* 1966, fig. 159). The Arroyo de los Frijoles function, is, in metric units,

$$Y = 0.10\,q \qquad (1)$$

where Y is average scour across the channel bed, in m, and q is discharge per unit width in $m^3s^{-1}m^{-1}$. The discrepancy between the minimum Nahal Yael values as inferred from the burial data and the values predicted by the Arroyo de los Frijoles function (4.5 to 6.0 cm) may be due to the fact that the Arroyo de los Frijoles scour data were averages based on 4–10 scour chains located across the stream section, whilst buried pebbles and cobbles might be expected to favour specific transport pathways, and hence also deposition sites, in the deeper parts of the scour. However, the scatter on the original Arroyo de los Frijoles graph is high, especially where discharge is low (0.025 to 0.1 $m^3s^{-1}m^{-1}$). This range of discharge is comparable with the conditions at the upstream end of the Nahal Yael experimental reach and the observed variation may be considered to fall within the Arroyo de los Frijoles range.

Depth of burial

Direct measurements of the depth of burial of coarse particles were made in Nahal Hebron using a magnetic detector to locate previously magnetized particles after fluvial dispersion (Hassan *et al.* 1984). An average retrieval rate of 96.5% per event has been achieved. As has been described in more detail in Schick *et al.* (in press),

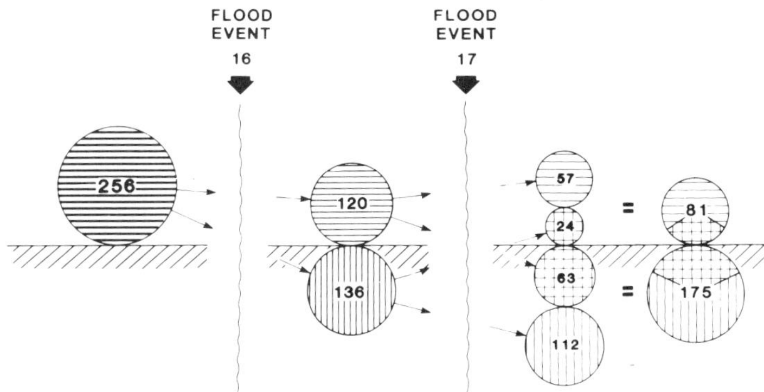

FIG. 4. Vertical displacement of pebbles and cobbles in Nahal Yael caused by flood events 16 (30.10.81) and 17 (22.03.85).

150 out of an original population of 282 magnetized pebbles were located below the surface following dispersal by two consecutive flow events, and their vertical distribution was determined (Fig. 5). Depth of burial, measured to the particle base, ranged from 5 to 42 cm, with a median value of 19 cm.

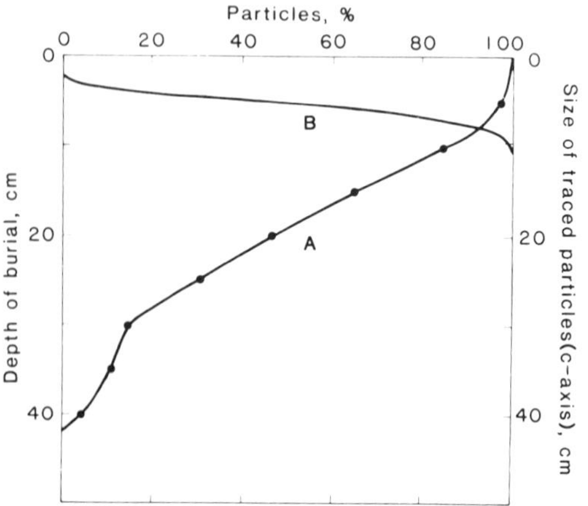

FIG. 5. Size and burial depth for magnetized pebbles weighing between 199 and 3200 g in Nahal Hebron: curve A—depth of burial following flood event 2 (23.1.1983), peak discharge 8 $m^3 s^{-1}$ per m of channel width; curve B—size distribution (*c*-axis) of the labelled particles.

Referring again to the Arroyo de los Frijoles curve (Leopold *et al.* 1966, fig. 159) as a standard for comparison, and using a peak discharge value of 8 $m^3 s^{-1} m^{-1}$ for the event in Nahal Hebron, the predicted depth of scour is 25 cm. This value is 5–6 cm higher than the actual scour depth as indicated by either the mean or the median of the vertical distribution of buried particles (Fig. 5). However, again the values fall within the overall range of the Arroyo de los Frijoles data. In reaching this conclusion we assume that burial of large clasts proceeds more or less continuously from the time of maximum scour to the time of the full redeposition of the scour layer.

The particles used for the Nahal Hebron tracing experiment range in weight from 199 to 3200 g. Their *c*-axes range from 23 to 107 mm, with a median of 54 mm. Nearly 60% of the particles were buried to a depth 3–7 times their diameter. The smallest 10% were buried to a mean depth of 17.2 cm—slightly but not significantly less than the mean value for the largest 10%, which was 18.5 cm. Thus, the dispersive stress effect (Bagnold 1968), if important, must be blurred by other factors that favour the deposition of large particles towards the base of the scour layer.

The fairly even pattern of burial with depth (Fig. 5) falls off below the depth of 30 cm, so that fewer particles per unit volume occur below 30 cm. Noting that a smooth continuation of the curve to a maximum depth of about 34 cm would bring the mean or median values closer to those predicted from the Frijoles curve, a separate explanation for the deepest 15% may be required. The reduced concentration of particles in this depth range may be the result of localized deep scour holes which attract particles during the flow, but which are difficult to detect from scour chain data on which the Frijoles function is based. More data are needed to substantiate this hypothesis.

Vertical exchange

In any coarse particle tracing experiment in a cobble–pebble–sand ephemeral stream, driven by a series of discrete flow events of equal size and form, it is predicted that the first few floods would generate the sequence shown in Fig. 4 in which the second event buries previously exposed particles, while at the same time previously buried particles become exposed. It is reasonable to presume that continuation of the experiment would result in an overall decrease in the difference between the number of particles buried and exposed and that there would eventually be some equilibrium between burial and exhumation.

The vertical exchange rate can be defined as the exposure rate divided by the burial rate. The exposure rate is the ratio between the number of particles exposed from buried locations to the total number buried prior to a given event. The burial rate is, likewise, the ratio between the number of particles buried from exposed locations and the total number situated on the surface prior to the flow event.

If long-term equilibrium exists, the vertical exchange rate will be equal to the total number of particles buried divided by the total number exposed. By way of an example, assume a before-event and after-event state wherein 20% of the particles are on the surface and 80% are buried. Assume further that the burial rate is 40%, *ie* eight particles out of every 100. To maintain equilibrium, this must be balanced by the exposure of 10% of the buried particle population.

In field conditions, the tracing of particles labelled by paint alone is unsuitable to test this hypothesis of equilibrium. Attrition and fading lead to loss of identity over time, and there is a need for an enormous number of particles. These limitations are overcome by the use of the magnetic tracing method which has yielded a

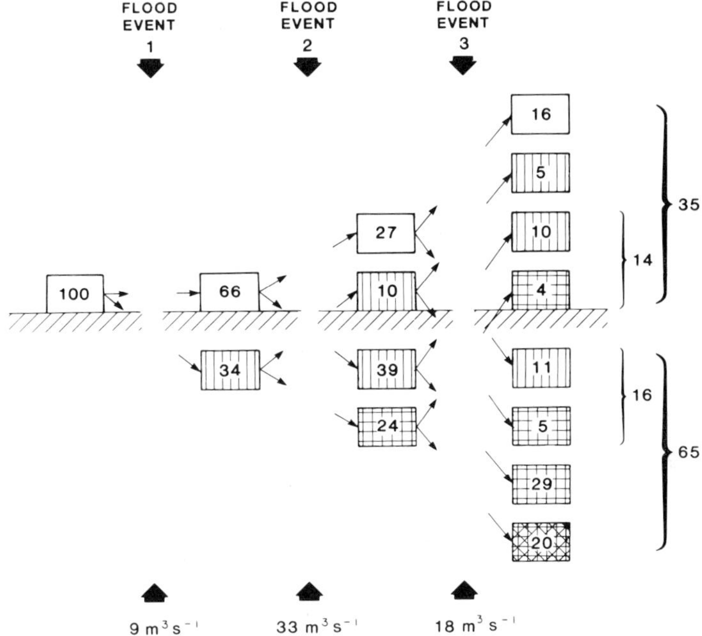

FIG. 6. Vertical exchange of coarse particles within the scour layer as a result of flood events in Nahal Hebron. Numbers are per cent of the original 282 particles, whose weight ranged between 199 and 3200 g, and which were placed on the channel bed. In this analysis, burial was defined as a state in which at least 50% of the particle was covered.

recovery rate of more than 90% per flood event (Hassan *et al.* 1984). The results using this method over three events in Nahal Hebron are shown in Fig. 6. The first event—a minor flow exceeded nearly every year—caused the burial of about one third of the original surface population. The second event—a five-year flood—buried over one half of the particle population still remaining on the surface after event 1; the counterbalancing effect of exposure was relatively small, and amounted to only one-tenth of the original population. For every four surface particles which were buried as a result of event 2, only one was exposed.

Event 3, with a peak discharge frequency of about 1.5 years, comes close to fulfilling the condition of exchange equilibrium: 14% exposures are balanced by 16% burials. This approximate equality is also expressed in the constancy of the population exposed (37% before event 3, 35% after it) and buried (63% before, 65% after). Although it seems unlikely for such an equilibrium to be attained after only three events, it is nevertheless possible that event 3 was, by chance, close to the 'dominant' event (Wolman & Miller 1960) in terms of the behaviour of the coarse particles. Such an equilibrium might even be considered a norm that is characteristic of the environmental conditions of Nahal Hebron.

High magnitude–low frequency events can be expected to result in an exchange rate higher than that of the 'dominant' event, and low magnitude–high frequency events should have a lower exchange rate. This expected relation is confirmed by comparing the results for the three Nahal Hebron events (Fig. 7). The smallest event ($Q_p = 9$ m^3 s^{-1}) buried 34% of the exposed population; the intermediate event (18 m^3 s^{-1}) buried 43%; and the five-year event (33 m^3 s^{-1}) buried 59% (Fig. 7, A). At the same time, the interme-

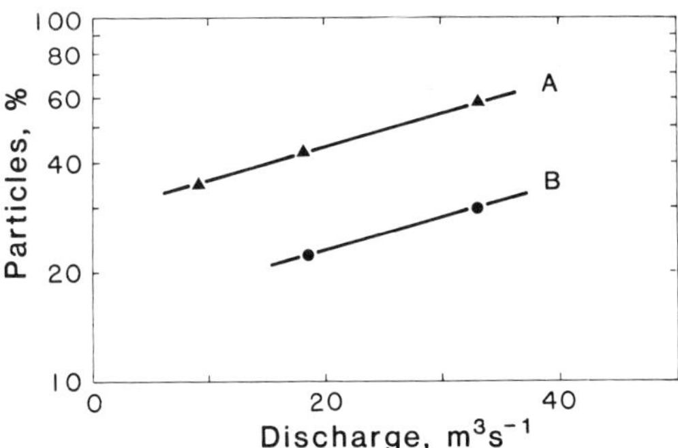

FIG. 7. Rate of vertical exchange of coarse particles within the scour layer of Nahal Hebron as a function of peak discharge. Curve A—percentage of exposed particles buried; curve B—percentage of buried particles exposed.

diate event exposed 22% of the buried population, and the five-year event exposed 29% (Fig. 7, B). The smallest event was the first in the series which started with a 100% surface population, and therefore its exposure rate is unknown.

Another factor to be considered in the context of the vertical exchange of coarse particles is the size distribution of the scour layer and more especially the relation between the coarse fraction represented by the traced particles and the predominantly sandy matrix. The traced particles, all originally native to the top layer of the channel material at the experimental site, range in size (*b*-axis) between 48 and 182 mm, with a mean of 93.5 mm. Comparing this with the surface and subsurface layers (Fig. 8), 74% of the surface layer is composed of sizes similar to those of the tracers (26% is finer); and only 21% of the 'sub-layer' is composed of these sizes (79% is finer). The concentration of coarse (>48 mm) particles in the surface layer is about 3.5 times higher than in the 'sub-layer'.

If it is true that a 'dominant' event buries roughly 40% of the coarse top layer particles in Nahal Hebron, then compensation for the vertical exchange that takes place must come from a 'sub-layer' which is scoured to a depth of 14–17.5 cm. For the five-year event, which was not an equilibrium event in the sense of vertical exchange, the depth of scour, as indicated by the exposure rate (29%), is 15–16 cm. These figures compare well with the mean scour depth indicated by the depth distribution of buried particles (Fig. 5, A), which shows a median of 19 cm.

The total depth of scour must include the top

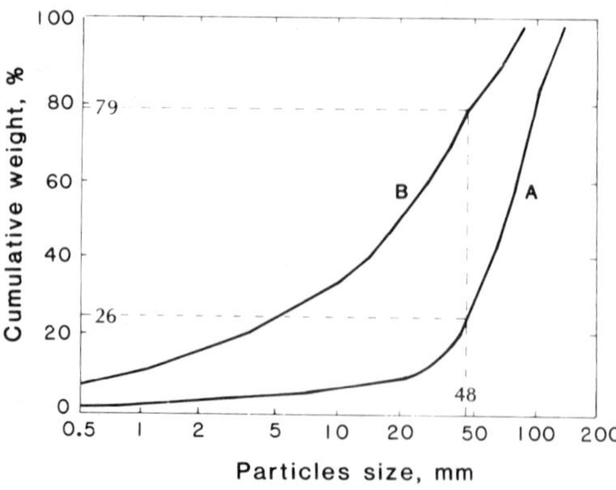

FIG. 8. Size distribution of the surface layer (A) and the 'sub-layer' (B), Nahal Hebron. Curves are composites of three sites in different sedimentary localities within the experimental reach. Depth of the surface layer was 4–5 cm. The bottom of the 'sub-layer' as sampled here was 14–30 cm below the channel bed surface. The median size of the tracer population was 48 mm.

layer, which has a thickness of 4–5 cm. So, the value indicated by the exchange method is actually 18–22 cm. On the other hand, using the magnetic method, the depth of buried particles relates to the bottom of the particle. As the mean size of the *c*-axis of the traced population is 5.5 cm (Fig. 5), these two corrections offset one another.

The close correspondence of the depth of scour values, as determined by two completely independent methods (direct determination of the depth of burial and consideration of vertical exchange in an anisotropic bed) lends support to the validity of the vertical exchange concept as outlined above.

Applications to ancient sediments

The concept of vertical exchange in the fluvial transport of arid-zone gravel-bed streams is important in several fields of hydrology, sediment transport and process geomorphology. Here, only its potential applicability to the interpretation of ancient deposits will be discussed.

Although reworking is occasionally mentioned as a factor influencing the characteristics of a gravelly sedimentary deposit, its role in modifying the deposit is seldom recognized. Fluviatile deposits are invariably the result of long-term aggradation that involves relatively thin individual net material additions, each caused by a single flood event. Many events cause scour and fill to depths which may exceed the net aggradation by an order of magnitude; these are deep enough to bury even the largest entrained particle.

Some of the coarse particles deposited in the deepest parts of the vertical exchange layer will have reached their final resting place. Other coarse particles, in shallower positions, are within reach of the vertical exchange process of subsequent flood events and are, therefore, candidates for additional 'reworking'. Thus, many of the coarse particles found in ancient pebbly sands and conglomerates may be interpreted as clasts which have escaped subsequent re-entrainment because they lie comparatively deep in the exchange layer.

Diagrams of bed thickness versus maximum particle size have been used to interpret conditions of deposition of ancient conglomerates (Steel 1974). If maximum particle size can be used as a surrogate discharge parameter, and bed thickness as a surrogate for depth of scour (and subsequent fill), then palaeoflow conditions can be interpreted in the light of work on modern desert streams. The values of scour depth and particle size reported in this paper for Nahal Yael and Nahal Hebron indeed fit reasonably well into the scattergram of bed thickness/ maximum particle size of Steel & Thompson (1983, Fig. 7) as well as into the spectrum of data which relates flow and scour depth in the Arroyo de los Frijoles function (Leopold *et al.* 1966).

In analysing ancient gravel beds, an environmentally significant parameter may be pebble depth below a given datum rather than its height above a certain base. If it can be assumed that the finer, upper part of an individual sedimentary unit such as that described, for example, by Steel & Thompson (1983, Figs 5 and 6) reflects exhumation of large clasts not balanced by burial, then the coarse layer on top of this unit may be an approximation of the channel bed surface as it existed at the time when vertical exchange affected the coarser layer below. Such an interpretation is economical in its assumptions and may lead to estimates of peak discharge at the time of deposition based both on particle size and scour depth.

Differences between depth of scour derived from the Arroyo de los Frijoles function (based on scour chain data) and those approximated from the depth distribution of the Nahal Hebron magnetized pebbles (Fig. 5) suggest that the deposition of pebbles under conditions of maximum scour (perhaps occurring simultaneously with peak flows) favour deep locations, perhaps scour holes. Although the apparent change at the depth of 30 cm in Fig. 5 does not support this view, it is probable that the 7% of the magnetized population which was not relocated after the floods was buried beyond the sensitivity limit of the detector, which was at the time 60 cm. If correct, then pebble concentrations in such scour holes should be relatively immune from further reworking and should therefore be identifiable in the sedimentary record of ancient deposits.

When the sedimentary record includes both the armour and sub-armour layers of any bed, evaluations of the vertical exchange in relation to the depth of scour can be made. The relatively poor development of an armour and a high concentration of coarse particles in the sub-armour indicate a high vertical exchange rate and may be interpreted as resulting from a flashy, typically arid-zone flow regime. Conversely, a well-developed armour and a low concentration of coarse particles in the sub-armour indicate a low vertical exchange rate, interpreted as a characteristic of perennial flow regime.

Conclusions

Desert floods transport pebbles in such a way that a few are moved great distances while the majority are only slightly moved. The pebbles that undergo long flood trajectories appear to be selected at random in successive floods, resulting in an overall lessening of differences in distances travelled between pebbles as the flood series lengthens.

For reasons not completely understood, pebbles that undergo burial travel significantly longer distances than those that do not. At the end of a flood event, a large proportion of an originally surficial pebble population ends up buried. After a number of consecutive floods, a tendency for equilibrium in vertical pebble exchange develops, wherein the number of coarse particles surfacing from buried positions equals the number being buried from surface positions. This equilibrium is achieved by the medium-size, 'dominant' flood magnitude, *ie* that with a frequency of *c.* 1.5 years. Events of higher or lower magnitude produce orderly deviations from equilibrium, but they cancel each other over time.

An equilibrium vertical exchange rate requires scour to the depth necessary for the entrainment and exposure of the number of buried pebbles needed to offset the number of surface pebbles buried by the event. Scour depths predicted from these considerations correlate well with experimental data from Israel and with the Arroyo de los Frijoles study on fluvial scour in New Mexico.

The concept of equilibrium vertical exchange and its relation to depth of scour and concentration of coarse particles both in the fluvial armour and sub-armour layers is important in understanding the geomorphological and sedimentological processes of arid-zone streams. As an interpretative tool, it has potential for the evaluation of (1) hydrology of modern, ungauged streams from data collectable over a period of a few years; and (2) environmental conditions of deposition in ancient conglomerates.

ACKNOWLEDGMENTS: This work is based in part on the ongoing PhD thesis programmes of Judith Lekach and Marwan A. Hassan. The Nahal Yael and Nahal Hebron research projects were funded in recent years by grants from the Hebrew University Authority for Research & Development, the Israel Academy of Sciences and Humanities, and the Israel Water Commission's Division of Soil Conservation and Drainage. We would also like to thank Lynne Frostick and Ian Reid for inviting Asher Schick to attend the 1986 London Geological Society Special Meeting on Desert Sediments—Ancient and Modern, which served as a catalyst for the final preparation of this paper; Luna Leopold, Ian Reid, Dan Yaalon, Moshe Inbar and an anonymous reviewer for their constructive comments; and Michal Kidron for drawing the figures.

References

ANDREWS, E. D. & PARKER, G. 1987. The coarse surface layer as a response to gravel mobility. *In:* THORNE, C., BATHURST, J. & HEY, R. (eds) *Sediment Transport in Gravel-bed Rivers.* John Wiley & Sons, in press.

BAGNOLD, R. A. 1968. Deposition in the process of hydraulic transport. *Sedimentology* **10**, 45–56.

BEN-ZVI, A. & COHEN, O. 1975. Frequency and magnitude of flows in the Negev. *Catena* **2**, 193–199.

BRAYSHAW, A. C. 1985. Bed microtopography and entrainment thresholds in gravel bed rivers. *Geological Society of America Bulletin* **96**, 218–223.

——, FROSTICK, L. E. & REID, I. 1983. The hydrodynamics of particle clusters and sediment entrainment in coarse alluvial channels. *Sedimentology* **30**, 137–143.

EMMETT, W. W., LEOPOLD, L. B. & MYRICK, R. M. 1983. *Some characteristics of fluvial processes in rivers.* 2nd International Symposium on River Sedimentation, Nanjing, China, 730–754.

GERSON, R. & INBAR, M. 1974. The field study program of the Jerusalem–Elat Symposium, 1974. Reviews and summaries of Israeli research projects. *Zeitschrift fur Geomorphology Supplementband* **20**, 1–40.

HASSAN, M. A. 1983. *Transport and dispersion of coarse bed material, Nahal Hebron.* MSc thesis, Hebrew University, Jerusalem.

——, SCHICK, A. P. & LARONNE, J. B. 1984. The recovery of flood-dispersed coarse sediment particles: a three-dimensional magnetic tracing method. *In:* SCHICK, A. P. (ed.) *Channel processes—water, sediment, catchment controls.* Catena Supplement 5, Braunschweig, 153–162.

LARONNE, J. B. & CARSON, M. A. 1976. Interrelationship between morphology and bed material transport for a small gravel-bed channel. *Sedimentology* **23**, 67–85.

LEOPOLD, L. B. & EMMETT, W. W. 1981. Some observations on movement of cobbles on a streambed. *In:* Erosion and Sediment Transport Measurements, Proceedings, Florence Symposium, June 1981, Late Papers—Poster Session, International Association for Scientific Hydrology, 49–59.

——, —— & MYRICK, R. M. 1966. Channel and hillslope processes in a semiarid area, New Mexico. *United States Geological Survey Professional Paper* 352-G.

REID, I. & FROSTICK, L. E. 1986. Dynamics of bedload transport in Turkey Brook, a coarse-grained alluvial channel. *Earth Surface Processes & Landforms* **11**, 143–155.

SCHICK, A. P. 1977. A tentative sediment budget for an extremely arid watershed in the Southern Negev. *In:* DOEHRING, D. O. (ed.) *Geomorphology in Arid Regions. Proceedings of the 8th Geomorphological Symposium*, Binghamton, NY, 139–163.

SCHICK, A. P., LEKACH, J. & HASSAN, M. A. 1987. Bedload transport in desert floods—observations in the Negev. *In:* THORNE, C., BATHURST, J. & HEY, R. (eds) *Sediment Transport in Gravel-bed Rivers.* John Wiley & Sons, in press.

STEEL, R. J. 1974. New Red Sandstone floodplain and piedmont sedimentation in the Hebridean province, Scotland. *Journal of Sedimentary Petrology* **44**, 336–357.

—— & THOMPSON, D. B. 1983. Structures and textures in Triassic braided stream conglomerates ('Bunter' Pebble Beds) in the Sherwood Sandstone Group, North Staffordshire, England. *Sedimentology* **30**, 341–367.

SUTHERLAND, A. J. 1987. Static armour layers by selective erosion. *In:* THORNE, C., BATHURST, J. & HEY, R. (eds) *Sediment Transport in Gravel-bed Rivers.* John Wiley & Sons, in press.

WILLIAMS, G. P. & WOLMAN, M. G. 1984. Downstream effect of dams on alluvial rivers. *United States Geological Survey Professional Paper 1286.*

WOLMAN, M. G. & MILLER, J. P. 1960. Magnitude and frequency of forces in geomorphic processes. *Journal of Geology* **68**, 54–74.

A. P. SCHICK, J. LEKACH & M. A. HASSAN, Department of Physical Geography, Institute of Earth Sciences, The Hebrew University of Jerusalem, 91904 Jerusalem, Israel.

Fluviatile deposits and morphology of alluvial surfaces as indicators of Quaternary environmental changes in the southern Negev, Israel

S. Grossman & R. Gerson

SUMMARY: Quaternary environmental changes in the southern Negev are recorded in the alluvial deposits and the geometry of depositional surfaces in the Timna Valley. There are some distinct differences between the alluvial deposits of the Holocene to modern flood plains and earlier (Pleistocene) ones. The Holocene to modern alluvium is characterized by sand, fine- to medium-sized gravel and very coarse gravel in large gravel bars. The sediments indicate environments ranging between extremely arid and semi-arid dominated by floods of low to moderate magnitude but with occasional short large floods. The Pleistocene sediments include: (1) debris-flow deposits indicating talus activity under a moderately arid to semi-arid climate; (2) alternating fine-grained and gravelly layers reflecting floods of low to moderate magnitudes. The nature of the soils supports these conclusions. Thick and argillic Reg soils are typical of the Pleistocene surfaces whereas shallow loamy ones are characteristic of the Holocene terraces. The alluvial surfaces usually converge downstream. The intersection points migrate upstream, following stages in the recession of a major escarpment. Broad pedimented surfaces and bajadas are characteristic of the Pleistocene period, whereas long narrow terraces of gentle declivity were formed during the Holocene. The latter suggest depletion of the available debris and in the effectiveness of large floods.

By analysing the sedimentary characteristics of alluvial deposits, it is possible to interpret both the depositional environment and the nature of the watershed. The weathering regime of the hillslopes, the availability of debris and the flow regime of the rivers can all be inferred from sediment structure, stratification, size and sorting. The environment may be interpreted by noting the presence or absence of colluvium. Identification of phases of aggradation and degradation may help define the relationship between the hydrological character of a drainage basin and the availability of sediment for transport and deposition.

The objective of this paper is to offer an explanation for the changing morphology and sedimentary facies of Quaternary alluvial surfaces in drainage basins of the southern Negev in terms of environmental changes in spite of the paucity of experimental data.

Environmental setting

The Timna Valley is located in the southern Negev, some 30 km N of the Gulf of Elat (Aqaba). It is a broad valley, drained by two major streams, Nahal (wadi) Timna and Nahal Nehushtan, which both flow into the Arava Rift Valley (Fig. 1). The catchment area of each of these streams is about 60 km^2.

The climate at present is hot and arid. Arid to semi-arid environments in general may be roughly subdivided into four regimes, according to the mean annual precipitation: (1) semi-arid, 250–400 mm a^{-1}; (2) moderately arid, 150–250 mm a^{-1}; (3) arid, 80–150 mm a^{-1}; (4) extremely arid or hyperarid <80 mm a^{-1}. Palaeoclimates are also believed to be definable within these categories. The modern climate falls into the extremely arid category, with a mean rainfall of 30 mm a^{-1}; an average summer (August) temperature of 32°–34°C; an average winter (January) temperature of 16°–18°C; and a potential evapotranspiration of 260 cm a^{-1} (*Atlas of Israel* 1985).

The geology of the area

There are three major rock types in this area (Fig. 2): (1) igneous rocks of Precambrian age; (2) ('Nubian') sandstones of Palaeozoic to Mid-Mesozoic age; (3) carbonate rocks (limestones, dolomites, chalks, marls) of late Mesozoic age.

In the Timna and Nehushtan basins the cliffs at the top of the bounding escarpment consist mostly of hard limestone and dolomite. The middle and lower parts of the bounding escarpment and the hills within the basins are formed of softer sandstones while coarse crystalline igneous rocks are exposed in the downstream parts of both basins (Fig. 3).

The region consists of two major structural units; the Arava Valley (a segment of the Dead Sea Rift); and the rift margin blocks, greatly

From FROSTICK, L. & REID, I. (eds), 1987, *Desert Sediments: Ancient and Modern*, Geological Society Special Publication No. 35, pp. 17–29.

FIG. 1. An aerial view of the Timna Valley. (*a*) Pleistocene alluvial surfaces. (*b*) Holocene alluvial surfaces. (*c*) Modern flood plains. (*d*) Relics of talus aprons. (*e*) The alluvial fan of Nahal Timna. (*f*) The main escarpment.

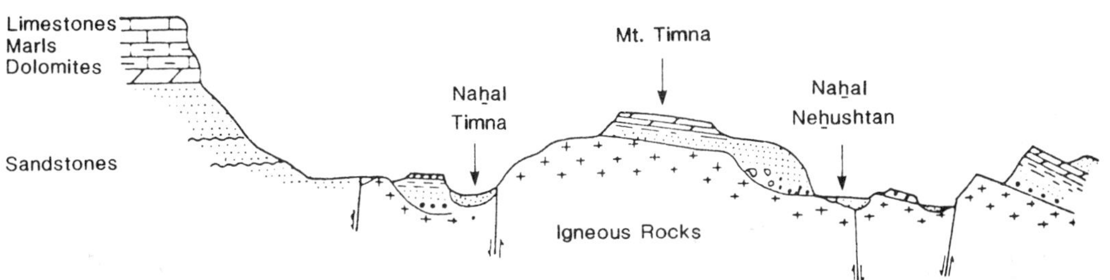

FIG. 2. A geological cross-section through the valleys of Nahal Timna and Nahal Nehushtan (after Garfunkel 1978; not to scale). The distances between the northern and southern escarpments is 8 km; the height of the northern escarpment is about 400 m.

FIG. 3. The main escarpment of Timna Valley. (*A*) Limestones, dolomites and marls. (*B*) 'Nubian' sandstones. (*C*) Relics of talus aprons. (*D*) Relics of the old terrace group.

elevated along the NNE–SSW trending Arava Valley border faults and divided into separate blocks by E–W trending faults.

The geomorphic evolution of the region

Several major stages of geomorphic evolution have been identified. Firstly, after recession of the Eocene seas an extensive erosion surface developed across the Jordan/Negev/Sinai region (Quennell 1958, 1959). Shallow streams carried gravel from E to W across the southern Negev (Garfunkel 1970, 1978). Several saddles, some of which still contain conglomerates, are found crossing the drainage divide of the Timna Valley (Gerson 1981; Gerson *et al.* 1985). Then major rifting took place during the Neogene (Freund *et al.* 1968; Freund & Garfunkel 1976). While a sinistral movement of more than 100 km occurred along the rift (Quennell 1959; Freund *et al.* 1970), at least 10 stages of rift down faulting took place in the Timna area (Gerson *et al.* 1985). Pedimented surfaces and straths truncating both igneous and sedimentary rocks alike were cut by the eastward-flowing Nahal Timna and Nahal Nehushtan. No diagnostic deposits or soils have been found on most of the surfaces. During the Quaternary a flight of talus and alluvial surfaces were developed (Fig. 1). These surfaces formed along the strike valleys in the Timna area at the contact between the erosion-resistant igneous rocks or limestones and the softer sandstones. Thus, fluvial systems of appreciable catchment size have developed in an environment of diverse lithology, changing base level (due to faulting) and climatic fluctuations.

There are no absolute dates for the alluvial deposits of the southern Negev. However studies of dated terrace sequences in the Dead Sea area (Amit & Gerson 1986) and in the Zin Valley (Goldberg 1976; Goldberg & Brimer 1983) may serve as a reference for a general subdivision of the alluvial surfaces of the Negev. Inferences of age are based on relative position within the series of terraces but evidence comes only from segments where surfaces are not converging downstream. Surface features such as bar and swale morphology; degree of desert pavement evolution (Bull 1974; Gerson 1981; Mayer *et al.* 1984; Ponti 1985); degree of chemical weathering of surficial rocks; and Reg soil properties (Danin *et al.* 1982; Amit & Gerson 1986) have been used as diagnostic parameters. The characteristics of the alluvial deposits are shown in Table 1. From these data it is possible to subdivide the alluvial surfaces into 3 groups: an upper terrace group of Pleistocene age; a lower terrace group of Holocene age; and the modern flood plain.

Landscape morphology and reconstruction of alluvial surfaces

Both the planform and the longitudinal profiles of the various surfaces in the area have been analysed. The combination of both aspects may yield additional information concerning environmental changes through time when supplemented by studies of the alluvial sediments.

In planform, the highest surfaces are usually broad plains. Although the youngest surfaces in the old group are inset into, and below, the older ones, they are still broad. Their width ranges from 300 to 800 m close to the main escarpment. The remnants of these high surfaces are relics of extensive pediments at the foot of a retreating escarpment which was often mantled by continuous talus aprons. In the Timna Valley such surfaces became narrow 1–2 km downstream of the scarps that feed them, only to widen into alluvial fans at the margins of the Arava Valley Rift. Two broad transport segments are typical of the high surfaces: one at the foot of the main escarpment, and the other at the margins of the terminal depositional basin in the lower part of the fluvial system.

The lower surfaces are narrow all along the valley course, from the escarpments in the headwater reaches to the apices of the alluvial fans at the rift margin. Their width does not exceed 200 m. The same is true of the present-day flood plains which are not wider than 180 m and in most cases are not actively widening. These narrow flood plains are associated with escarpments which are free of talus and often assume their characteristic slope and width right at the foot of the escarpment (Fig. 1). They widen appreciably only in their downstream fans.

Plotting the longitudinal profiles of the various surfaces within three categories—high (Pleistocene), low (Holocene) and present-day—yields the following results (Fig. 4):

(1) The high surfaces are convergent with the present-day stream channel in the downstream direction.
(2) The points of intersection of the high surfaces with the present-day stream channel are located 2–3 km W of the Western Arava Valley margin. Excluded are the three oldest surfaces, faulted *at* the rift margin (see point 4, below).
(3) There is a migration upvalley of the intersection points with time; the older the surface

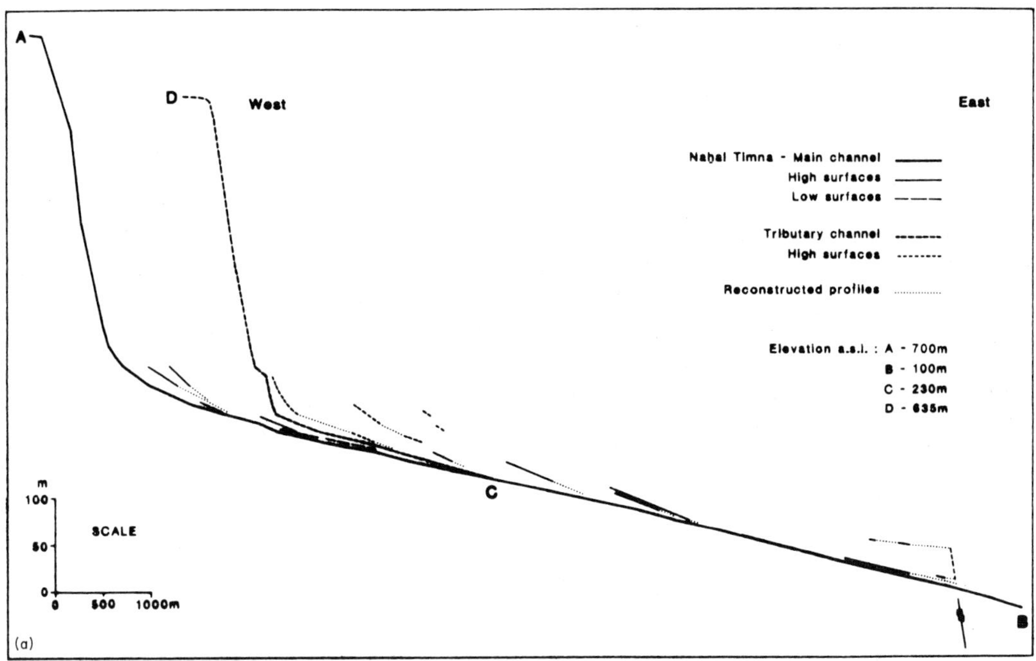

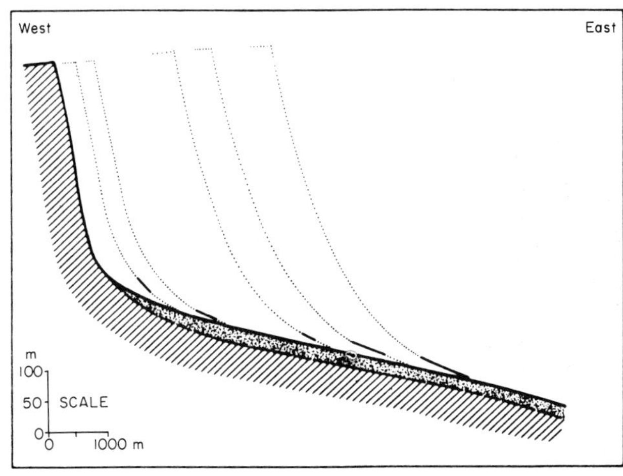

FIG. 4. Longitudinal profiles of different surfaces in the Timna Valley. Note the 10:1 vertical exaggeration. (*a*) Longitudinal profiles of Nahal (Wadi) Timna, a tributary, and their older surfaces. (*b*) Reconstructed positions of the escarpment related to old alluvial surfaces along Nahal Timna.

the further downvalley is the intersection point.

(4) There are three alluvial surfaces of the highest group which are displaced by the rift border faults. These are the oldest surfaces among the ones discussed here. Their reconstructed profiles diverge downstream *ie* eastward.

(5) The low terraces are sub-parallel to the present-day flood plain along most of their profile; they converge downstream with the present-day flood plain, 1–2 km W of the Arava Valley bounding fault zone.

(6) The modern channels are still actively downcutting in bedrock at a distance of 0.5–1.0 km from the escarpment drainage divide. Further downstream there are no visible rock outcrops and the longitudinal profile is linear.

(7) During the period of formation of the flight of alluvial surfaces in the Timna Valley, the bounding scarp has retreated some 2–3 km.

The alluvial deposits

The deposits of the alluvial surfaces of the Timna area and similar terrains in the southern Negev

TABLE 1.

Deposits and surficial features	Position in the sequence					
	Upper terrace group (Pleistocene)		Lower terrace group (Holocene)		Active flood plain	
	Sedimentary characteristics	Frequency of occurrence	Sedimentary characteristics	Frequency of occurrence	Sedimentary characteristics	Frequency of occurrence
Gravel beds*	—Pebbles: well sorted; well stratified; clast supported, frequent imbrication; void filling by sand, sometimes red.	VF	—Pebbles: well sorted; well stratified; clast supported with no matrix (sieve deposit).	VF	—Boulders, cobbles, pebbles: poorly sorted; bar morphology, clast supported; intensively weathered and patinated.	F
	—Pebbles: poorly sorted; poorly stratified; mixed with granules and sand (as matrix).	F	—Pebbles: moderately sorted; well stratified; some fine-grained sediment.	VF	—Pebbles, cobbles: poorly sorted; scattered, with sand matrix, in swales.	F
	—Cobbles: well sorted; well stratified; sometimes imbricated.	I	—Cobbles: poorly sorted; well stratified; mixed with small pebbles and coarse sand.	F		
	—Cobbles: poorly sorted; poorly stratified; discontinuous with lenses of red sand.	I	—Pebbles: poorly sorted; well stratified; mixed with fine-grained sediment.	F		
			—Boulders, cobbles, pebbles: poorly sorted; a single gravel-bar layer; no (washed away?) matrix; at the top of the sedimentary section.	F		
			—Pebbles, some cobbles: poorly sorted; poorly stratified, mixed with fine-grained sediment.	I		
			—Small pebbles: well sorted; poorly stratified; sometimes in lenses, encompassed with fine sediment.	R		
Sand and silt layers†	—Massive: with some scattered pebbles.	VF	—Massive: with some small pebbles and granules.	VF	—Moderately to poorly sorted: laminar stratification, in swales.	VF
	—Massive: with granules; usually ≤10 cm thick, between layers of gravel.	VF	—Massive: floating scattered cobbles and boulders.	F	—Moderately to poorly sorted; massive, in bars.	F
			—Moderately stratified; no coarser fractions.	I	—Sand and granules in swales and bars.	F

	—Massive: ≤1 m thick.	F			
	—Massive: with poorly sorted pebbles and granules; often reddish colour; cemented by silt and clay.	I			
	—Massive: medium to fine sand and silt; with thin layers of small pebbles.	I			
	—Massive: with scattered floating cobbles and boulders; reddish colour.	I			
Debris flows	—Large boulders; very poorly sorted; massive; discontinuous structures; clasts floating in high content of fine-grained sediment; clast lithology-limestone and sandstone.	VF	—Small boulders and cobbles; poorly sorted; massive, in lenses; moderate content of fine-grained sediment; clast lithology-limestone.	I	—No debris flow deposits.
Buried pedogenic horizons	—Thin, sometimes discontinuous buried B horizons of Reg soils; abundant fines; gypsum and salt nodules; reddish colour.	F	—Thin layers of sand with some fines; some gypsum and salt nodules; possibly buried BC of Reg soils.	I	
Surficial characteristics and morphology	—Well developed, smooth desert pavement. —Some scattered boulders. —Advanced weathering of exposed boulders on treads. —Large clasts veneered with varnish. —Eroded risers, with high concentration of coarse gravel. —Well-developed Reg soil profile, with petrogypsic and petrosalic horizon.		—Poorly developed desert pavement. —Well-preserved bar and swale morphology, frequently composed of boulders and cobbles. —Low to moderate degree of weathering of exposed gravel. —Poor varnish development. —Well-preserved risers. —Young Reg soil profile.		—Braided pattern. —No signs of weathering. —No varnish development. —No soil development.

*Gravel clasts are mainly limestone and dolomite.
†The sand fraction is usually moderately well sorted.
Frequency of occurrence: VF = very frequent; F = frequent; I = infrequent; R = rare.

are diverse, due to two factors. Firstly, source rocks greatly influence sediment texture. Coarse gravel is delivered from the hard carbonate rocks of the upper cliffs in the bounding escarpment and from the igneous rocks of Mount Timna, whereas sand comes from the abundant sandstone exposures along the headwater escarpments and on the hills within the valley. Granules are weathered from the hard rocks and coarse, gritty sandstones. Silt and small amounts of clay are derived from the matrix binding the sandstones, from mudstones as well as from the weathering of various rocks and directly from airborne dust. Secondly, the geomorphic regime is related to the climate which controls the mode and rate of weathering, soil formation, mass movement and type of stream flow. Modes of erosion, transport and deposition largely determine the texture and structure of the deposits along the fluvial system. In spite of the variations inevitably resulting from the interplay of these two factors, the scheme of subdivision on the basis of age, suggested above appears to be realistic.

Table 1 and Fig. 5 show the characteristic textures and structures of selected surfaces subdivided according to age. Table 1 may be summarized as follows.

Sand is relatively abundant in the old terraces whereas gravel is very common in the young terraces. Sand is the major constituent of the

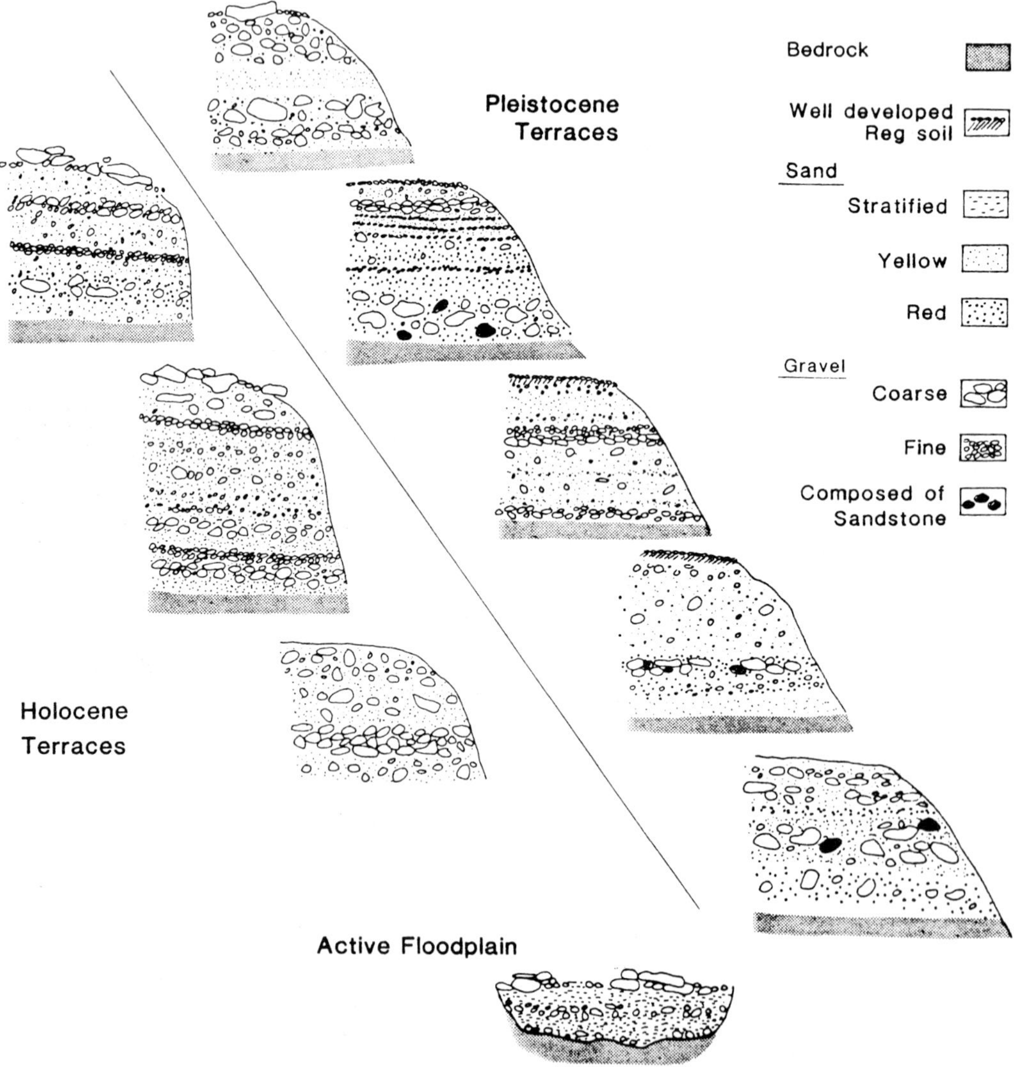

FIG. 5. Characteristic sedimentary sections in the different groups of alluvial terraces.

sediments encountered in the present-day flood plain. The sand beds in both terrace groups are massive, whereas the sandy sediments in the present-day ephemeral channels are often laminated. The upper terraces frequently contain reddened sands with relatively more silt and clay. Such a combination is less common in the lower terrace group; yellowish sand and a lower fines content are typical of the Holocene terraces. However, red sand with silt often appears as a matrix in the debris flow structures.

The coarse cobbles and pebbles of the old terrace group display poorer sorting and stratification than in the young terrace group. Small- to medium-sized pebbles are stratified in gravelly beds in the lower terraces, while in the upper terraces there is less stratification within the gravelly beds; these layers usually alternate with sandy ones. Densely packed pebbles with no fine-grained matrix (sieve deposits) are common in both terrace groups.

Debris-flow deposits frequently occur in the old terraces. They are characterized by coarse, poorly sorted gravel, including large boulders 1–1.5 m in diameter. Debris-flow deposits are rarely encountered in the young terrace group and have never been observed in the active flood plains. Concentration of very coarse gravel (large cobbles to large boulders) in large gravel bars is common at the top of the sections of young terraces, as well as in the elevated relic gravel bars of the present-day flood plains (Fig. 6).

Sections cut in the old terraces are very similar. Beds comprised of coarse sediments alternate with beds of abundant fine material. Furthermore, there is an abrupt change between beds of different texture within any particular section (Fig. 7). These phenomena are absent in the younger terraces, which often consist of single, homogeneous, units (Fig. 8).

Discussion and conclusions

Several conclusions can be drawn concerning the depositional environments and possible palaeoclimatic significance of sediments in the Timna Valley and similar watersheds in the southern Negev. Firstly, the debris-flow deposits found in the upper terraces suggest a moderately arid to semi-arid climate (Gerson 1982*a*). Secondly, thick units composed of beds of poorly sorted and poorly stratified cobbles and pebbles alternating with thin layers of sand are found mostly in the high terraces. It is proposed that they represent periods when floods were of moderate intensity and short duration, and that the climate was arid. Thirdly, thick units of sand and grit of massive structure are observed in the old terraces and were deposited by low-magnitude floods of long

FIG. 6. Large gravel bars preserved on a Holocene terrace.

FIG. 7. A section in a Pleistocene terrace. (A) Debris flow deposits. (B) Layered fluviatile sediments.

FIG. 8. A typical section of fluviatile deposits in a Holocene terrace.

duration, possibly under a moderately arid to semi-arid climate whereas the thin beds (5–30 cm thick) of sand and grit, found in the young terraces, imply floods of low to medium intensity which would occur under an extremely arid to arid climate. Moderately sorted and stratified pebbles and cobbles in extensive units of variable thickness encountered in the low terraces suggest moderate flows of medium to long duration which are characteristic of an arid climate. In addition the large gravel bars composed of poorly sorted cobbles and boulders were deposited during high-magnitude floods. These are units found at the top of the low terrace sections, and as relics in the present-day flood plain. They suggest an extremely arid to arid climate. In the modern channels, units of laminated sand and grit alternating with thin beds of small pebbles are found. These accumulate under conditions where the flow is of low magnitude and short duration, and where the climate is extremely arid.

In general the lithofacies which point to moderately arid to semi-arid palaeoclimates are characteristic of the old terrace group. Floods of low to moderate magnitude and long duration, and intense debris-flow activity, are typical. This conclusion is supported by several characteristics of the soils on the old terraces. These soils are thick (to a depth of 1–1.5 m) and well developed. They contain large amounts of silt and clay in their B horizon, and salinity is high to the bottom of the C horizon. Sprinkler experiments carried out by Greenbaum (1986) have led to the conclusion that large amounts of rainfall per storm or frequently occurring rainstorms are the conditions required to develop such Reg soil profiles. Buried argillic palaeosols in the terraces of the old group and reddened beds (Walker 1967) again point to a slightly wetter climatic regime. A detailed examination of Holocene Reg soils (Amit & Gerson 1986) has yielded different results. Shallow soil profiles (to a depth of 0.3–0.5 m) and poorly developed thin cambic B horizons have developed under extremely arid conditions. Such a climate has prevailed during most of the Holocene (Horowitz 1979; Gat & Magaritz 1980; Begin *et al.* 1985).

The sedimentary sections and the surficial patterns of the deposition in the study area may be explained by the following sequence of Quaternary environmental changes. During relatively wet climatic modes there was intensive talus activity along the main escarpment. Debris-flows transported coarse gravel mixed with fines onto the flood plains. Such deposits are often found in the lower parts of the sedimentary section of the old terraces. The transition to a drier climate was accompanied by stripping of the talus aprons and deposition of sediments of mixed facies—debris-flows and fluviatile deposits. Stream power was still insufficient to move all the sediment contributed to the channels from the main escarpment and the hills within the basin. The upper portions of the sedimentary section are mostly layered fluviatile deposits of various textures.

During a continuous arid to extremely arid climatic regime, gullying and erosion denuded much of the talus aprons; bare rock exposures were relatively widespread and the sources for sediment were depleted. High stream power and the small amount of sediment transported led to intense incision within the flood plain (Bull 1979). Several metres of downcutting within the alluvial plains, and escarpment recession of several tens of metres resulted from a long period in which the climatic regime was arid to extremely arid. An extended period of aridity may have led to extensive exposure of sandstones along the main escarpment as the talus aprons were denuded away. Sand was the major type of sediment transported during such a period.

The Holocene appears to have been rather unique. The onset of aridity was probably very rapid (Neev & Emery 1967; Begin *et al.* 1974; Danin *et al.* 1982; Gerson 1982*a*). The lower terrace group contains mostly fluviatile sediments exhibiting features consistent with overall net incision, and rather infrequent large floods depositing coarse gravel. Most of the Holocene deposits indicate events of a similar nature to those in Pleistocene deposits. However, in the latter there is a distinct alternation between beds of differing sedimentary texture, with abrupt transitions between units.

The Holocene terraces conform to some extent to the model of episodic erosion proposed by Schumm (1975) and by Womack & Schumm (1977). They seem to present a classic case of a discontinuously downcutting stream, eroding and re-depositing during an extended period of entrenchment. However, it is possible to subdivide the fluvial regimes of the Holocene into three types, occurring in the following sequence:

(1) Periods of net incision during which stream power is higher than the critical power for sediment transport (Bull 1979) and high-magnitude floods are dominant (Schick 1974). Downcutting of 1–4 m characterizes this regime.
(2) A depositional regime, in which gravel of fine to medium grain size is deposited by flows of low to moderate magnitude. Terrace fills of 1–3 m are diagnostic of this type of regime.
(3) Floods of high magnitude that are loaded

with the coarse gravelly sediments which have accumulated in the headwaters during periods of type 2 regime. Considerable quantities of large cobbles and boulders are transported and deposited in single layers as gravel bars by such extreme floods.

Cycles of this type were repeated four times during the Holocene in the Timna and Avrona Valleys. During the same period of time, 15–19 cycles of erosion and deposition have occurred along the Dead Sea coast, where fan terraces were formed by streams discharging from the less arid terrains of the northeastern Negev and Judean Desert (Bowman 1975).

Only very small relics of the older surfaces have been preserved through the long period since their formation. On the other hand, a flight of 3–5 terraces of Holocene age is still well preserved 4–6 m above the modern flood plain. Lateral widening of this flood plain at the expense of the bordering terraces may eventually leave behind only one surface, which may be falsely interpreted as representing the only depositional cycle which has occurred during the Holocene. Any comparison of the long Pleistocene with the very short Holocene should be made cautiously. The evidence of the latter is still well preserved, while the testimony left by the former is fragmentary.

Incision and deposition are the main means of response to environmental changes in the upper reaches of the Timna and Nehustan valleys. In the lower segments of the streams there were practically no changes in the position of the flood plains for long periods of time. There, the streams have generally modified their width in response to the environmental changes that have so obviously affected morphology and sediment type in the upstream segments.

One characteristic which does not appear to be very sensitive to environmental changes in the Timna Valley is the position of the apices of the alluvial fans. They have been located for long periods of time along the margin of the Arava Valley, with only very slow migration upstream. Tectonic stability along the mountain front in the Timna segment of the Arava Valley, and the negligible effect of the increase in catchment area with time on the position of the alluvial fans are the reasons for this situation.

The concept of converging fan terraces in the downstream direction is well known (Beaty 1961; Bull 1964; Hooke 1972). It usually occurs in the depositional basin at the downstream end of the fluvial system. However, in the case of Nahal Timna, Nahal Nehushtan and Nahal Avrona (some 15 km to the S of Timna Valley), the convergence of old terrace segments with those of younger ones with far gentler gradients (Fig. 4) may be best explained by a model of talus–bajada surfaces representing stages in the retreat of an escarpment (Gerson 1982*b*). The reconstruction of the alluvial terraces and the associated talus relics (Fig. 4) allows the identification of earlier positions of the escarpment bounding the drainage basin. Such a model explains the 10–20 m difference in elevation between successive alluvial surfaces in the upstream segments of the valley.

ACKNOWLEDGMENTS: We thank Michael Kidron for drawing the text figures; we are grateful to Martin Williams and two anonymous reviewers for their comments and suggestions.

References

AMIT, R. & GERSON, R. 1986. The evolution of Holocene Reg (gravelly) soils in deserts—an example from the Dead Sea region. *Catena* **13**, No. 1, 59–79.

Atlas of Israel 1985. Survey of Israel, Ministry of Labor, Jerusalem, and Elsevier Co, Amsterdam.

BEATY, C. B. 1961. Topographic effects of faulting, Death Valley, California. *Annals, American Association of Geographers* **51**, 234–240.

BEGIN, Z. B., BROECKER, W., BUCHBINDER, B., DRUCKMAN, Y., KAUFMAN, A., MAGARITZ, M. & NEEV, D. 1985. Dead Sea and Lake Lisan levels in the last 30,000 years. *Geological Survey of Israel, Preliminary Report*, 18 pp.

——, EHRLICH, A. & NATHAN, Y. 1974. Lake Lisan, the Pleistocene precursor of the Dead Sea. *Geological Survey of Israel, Bulletin* **63**, 30 pp.

BOWMAN, D. 1975. *River terraces in the Dead Sea area—morphology and genesis.* PhD thesis (in Hebrew with English summary), The Hebrew University, Jerusalem.

BULL, W. B. 1964. History and causes of channel trenching in western Fresno County, California. *American Journal of Science* **262**, 249–258.

—— 1974. *Geomorphic tectonic analysis of the Vidal region. Vidal Nuclear Generating Station.* Prepared for Southern California Edison Company, Woodward-McNeill & Associates, Appendix 2.5b, 66 pp.

—— 1979. Threshold of critical power in streams. *Geological Society of America Bulletin* **90**, No. 6, 453–464.

DANIN, A., GERSON, R., MARTON, K. & GARTY, Y. 1982. Patterns of limestone and dolomite weathering by lichens and blue-green algae and their paleoclimatic significance. *Palaeogeography, Palaeoclimatology, Palaeoecology* **37**, 221–233.

FREUND, R. & GARFUNKEL, Z. 1976. *Guidebook to excursion along the Dead Sea Rift*. Mimeograph, Department of Geology, The Hebrew University, Jerusalem, 27 pp.

——, ——, ZAK, I., GOLDBERG, M., DERIN, B. & WEISSBROD, T. 1970. The shear along the Dead Sea Rift. *Philosophical Transactions of the Royal Society of London, Series A* **276**, 107–130.

——, ZAK, I. & GARFUNKEL, Z. 1968. Age and rate of the sinistral movement along the Dead Sea Rift. *Nature* **220**, 253–255.

GARFUNKEL, Z. 1970. *The tectonics of the western margins of the southern Arava*. PhD thesis (in Hebrew with English summary), The Hebrew University, Jerusalem, 204 pp.

—— 1978. The Negev, regional synthesis of sedimentary basins. *Guidebook to excursion A2, 10th International Congress on Sedimentology*, Jerusalem, 35–110.

GAT, J. R. & MAGARITZ, M. 1980. Climatic variation in the eastern Mediterranean Sea area. *Naturwissenschaften* **67**, 80–87.

GERSON, R. 1981. Geomorphic aspects of the Elat Mountains. *In:* DAN, J., GERSON, R., KOYUMDJISKY, H. & YAALON, D. H. (eds) *Aridic soils of Israel. International Conference on Aridic soils, Israel*, 279–294.

—— 1982*a*. The Middle East: Landforms of a planetary desert through environmental changes. *Striae* **17**, 52–78.

—— 1982*b*. Talus relicts in deserts: A key to major climatic fluctuations. *Israel Journal of Earth Sciences* **31**, 123–132.

——, GROSSMAN, S. & BOWMAN, D. 1985. Stages in the creation of a large rift valley—geomorphic evolution along the southern Dead Sea Rift. *In:* MORISAWA, M. & HACK, J. T. (eds) *Tectonic geomorphology*. The Binghamton Symposia in Geomorphology, International Series, No. **15**, 53–73.

GOLDBERG, P. 1976. Upper Pleistocene geology of the Avdat/Aqev area. *In*: MARKS, A. E. (ed.) *Prehistory and Paleoenvironments in the Central Negev, Israel, Vol. 1*. Southern Methodist University Press, Dallas, 25–51.

—— & BRIMER, B. 1983. Late Pleistocene geomorphic surfaces and environmental history of Avdat/Havarim area, N. Zin. *In:* MARKS, A. E. (ed) *Prehistory and Paleoenvironment in the Central Negev, Israel, Vol. III*. Southern Methodist University Press, Dallas, 1–13.

GREENBAUM, N. 1986. *Infiltration and runoff in an extremely arid climate: infiltration experiments on small plots in the southern Arava Valley*. MSc thesis (in Hebrew), The Hebrew University, Jerusalem, 206 pp.

HOOKE, R. L. 1972. Geomorphic evidence for late Wisconsin and Holocene tectonic deformation, Death Valley, California. *Geological Society of America Bulletin* **83**, 2073–2097.

HOROWITZ, A. 1979. *The Quaternary of Israel*. Academic Press, 394 pp.

MAYER, L., GERSON, R. & BULL, W. B. 1984. Alluvial gravel production and deposition: a useful indicator of Quaternary climatic changes in deserts. *Catena Supplement* **5**, 137–151.

NEEV, D. & EMERY, R. O. 1967. The Dead Sea: depositional processes and environments of evaporites. *Geological Survey of Israel, Bulletin No.* **41**, 147 pp.

PONTI, D. J. 1985. The Quaternary alluvial sequence of the Antelope Valley, California. *Geological Society of America, Special Paper* **203**, 79–96.

QUENNELL, A. M. 1958. The structure and geomorphic evolution of the Dead Sea Rift, with discussion. *Quarterly Journal of the Geological Society of London* **114**, 1–24.

—— 1959. *Tectonics of the Dead Sea*. 20th International Geological Congress, Mexico, Association de Servicios Geologicos Africanos, 385–405.

SCHUMM, S. A. 1975. Episodic erosion: a modification of the geomorphic cycle. *In*: NELHORN, W. N. & FLEMAL, R. C. (eds) *Theories of Landform Development. Proceedings of the 6th Annual Geomorphology Symposium, Binghamton*, 85–70.

SCHICK, A. P. 1974. Formation and obliteration of desert stream terraces—a conceptual analysis. *Zeitschrift für Geomorphologie, Supplement* **21**, 88–105.

WALKER, R. T. 1967. Formation of red beds in modern and ancient deserts. *Geological Society of America Bulletin* **78**, 353–368.

WOMACK, W. R. & SCHUMM, S. A. 1977. An example of episodic erosion. *Geology* **5**, 72–76.

S. GROSSMAN & R. GERSON, Institute of Earth Sciences, Givat Ram Campus, The Hebrew University of Jerusalem, Jerusalem 91904, Israel.

Plio–Pleistocene raised channel systems of the western Sharqiya (Wahiba), Oman

J. K. Maizels

SUMMARY: Extensive alluvial fans bounding the western edge of the Sharqiya (or Wahiba) Sands are characterized by complex palaeochannel systems now forming a series of superimposed gravel ridges. The oldest, most extensive fan system is crossed by numerous sinuous, superimposed palaeochannel courses comprising highly weathered, chert-rich gravels cemented by clear crystalline calcite. Many of these old fan sediments have been chemically altered to depths of over 200 m to form a pink dolomitic clayey deposit here termed barzamanite. The formation of this calcrete-like deposit was probably associated with rising water tables during long-term fan growth. The younger, more limited fan deposit, by contrast, comprises thin, coarse-grained, broad spreads and terrace veneers of weakly cemented ophiolitic gravels. The fans have been subject to entrenchment and deflation, resulting in extensive land-surface lowering and the exhumation of buried palaeochannels. Although no dates are yet available, the older channel systems are likely to date from more humid phases during the Pliocene–Early Pleistocene, whereas the younger, terraced deposits, and the period of subsequent channel exhumation, may date from semi-arid periods of lower sea level during the Pleistocene.

The Eastern Oman Mountains are flanked on their southern margins by alluvial fan deposits, extending for distances of over 200 km, and bounding the western edge of the Wahiba sand sea and its associated aeolianite. The former alluvial stream deposits now form complex systems of 'raised' or upstanding, sinuous, superimposed linear ridges and broad gravel sheets, rising 10–20 m above the surrounding plains, producing an extensive area of inverted drainage (Figs 1 and 2).

This study aimed, firstly, to develop a model of palaeoenvironmental change and landscape development for the area of 'raised' channels. Secondly, the study aimed to identify the likely palaeohydrologic and palaeoclimatic conditions associated with periods of channel formation, fan aggradation and erosion. The final objective was to provide an overall chronological framework for these palaeoenvironmental changes in relation to the Wahiba and to the Arabian peninsula during the Quaternary.

Previous research

No detailed field investigation has previously been made into the exhumed palaeochannel systems W of the Wahiba, although there are a number of descriptions of similar raised channel deposits recorded elsewhere. Workers in Arabia and North Africa have introduced a number of different terms to describe such channels. Miller (1937) described the channels as 'suspendritic drainage lines' that stood out in 'bas-relief' above the surrounding plains in eastern Saudi Arabia. Knetsch (1954) used the unfortunate term 'pseudo-esker' to describe the sinuous gravel ridge features of the Dakka Basin in Saudi Arabia. Other workers have described these 'gravel-capped ridges' in the western Transvaal (King 1942) as 'gravel trains' (Holme 1960; Brown 1960 in Beydoun 1980), as 'perched wadis' or 'wadi ridges' in Egyptian Nubia (Butzer & Hansen 1968), as 'suspenparallel drainage' in W Texas and New Mexico (Reeves 1983), and as 'raised channels' (Warren *et al.* 1985). Less ambiguous definitions have been provided for the Omani palaeochannel systems by Glennie (1970) who describes them as 'ridges of exhumed Pleistocene (?) wadi gravels', and by Beydoun (1980) as 'exhumed or fossil river channel systems'. In this paper the terms 'raised channels', 'exhumed channels' and 'palaeochannels' are adopted.

There is general agreement in these studies that the palaeochannels have been exhumed through differential deflation of fine-grained, poorly cemented interfluve sediments, while the well-cemented coarse-grained channel sediments have remained more resistant to deflation and hence have been preserved as upstanding ridges. Most of these workers consider that channel development occurred during a more humid climatic phase, with higher stream discharges associated with alternating wet and dry periods, or seasonal flows (Miller 1937; Butzer & Hansen 1968; Glennie 1970; Beydoun 1980). However, since channel-bed cementation appears to be occurring in the present-day arid environments

From FROSTICK, L. & REID, I. (eds), 1987, *Desert Sediments: Ancient and Modern*, Geological Society Special Publication No. 35, pp. 31–50.

FIG. 1. Oblique aerial view of raised channel deposits overlying pink barzamanite. View looking SSW from Barzaman area.

FIG. 2. Ground view of thin Fan II gravels overlying weathered gravels and barzamanite of Fan I deposits. View looking N towards Barzaman and Jebel Madar.

of Oman, the link between climatic humidity and channel cementation remains uncertain (Gass pers. comm.).

However, there has been little consideration of the rates and patterns of fan development, nor any palaeohydrologic analysis of former alluvial environments, nor any consideration of the conditions either for channel exhumation or for fan entrenchment. No detailed morphological, stratigraphical or sedimentological analyses have previously been undertaken. The dates of channel formation are uncertain. Beydoun (1980) suggests that they may date from two wetter phases, identified by McClure (1976) in the Rub' al Khali, which occurred between 36 and 17 ka, and between 9 and 6 ka. However, Kassler (1973), Hötzl *et al.* (1978*a* and *b*) and Anton (1984) consider the vast sheets and ridges of alluvial gravels in central and eastern Saudi Arabia and Qatar to date from late Pliocene/Early Pleistocene sub-humid periods associated with high sea-levels.

Study area

Fieldwork was concentrated within the area of extensive alluvial fans W of the Wahiba Sands, which exhibits the most distinctive and complex systems of superimposed gravel ridges. The study area (Fig. 3) extends for *c.* 220 km from the foot of the Eastern Oman Mountains in the N towards the coast in the S, and for *c.* 50 km W–E, the western limit being marked in part by Wadi Halfayn. The eastern limit of the alluvial fan systems is marked in the N by contact with the Sands, and in the S by the abrupt western edge of the extensive aeolianite deposits that underlie the Sands (Gardner 1986).

Methods of study

The areal extent of the different alluvial fan systems was determined from *Landsat* imagery and partial air photo cover (at scales of 1 : 100 000, and *c.* 1:30 000). The thickness of the deposits was estimated from borehole records provided by the Public Authority for Water Resources (PAWR), Sultanate of Oman, as well as from field observations.

The surface morphology of the alluvial fans, the palaeochannels and the terraces was determined initially from mapping from the air photographs. Overall gradients were estimated very approximately from spot heights printed on the 1 : 100 000 topographic maps, although these were subject to regional height errors of up to 5 m. Ground surveys of the longitudinal- and cross-profiles of the channels and terraces were undertaken in two selected areas.

Sedimentary structures and clast fabric were recorded as palaeoflow indicators; coarse sediments were sampled for analysis of clast shape, size and composition as indicators of source materials, distances of transport, and post-depositional diagenetic alteration; and the mean intermediate diameter of the ten largest clasts was determined at each site for use in palaeohydraulic analysis. Samples of cemented sands and gravels were collected for petrographic and geochemical analysis. Lithological analyses and analysis of surface weathering characteristics were based on 100 surface clasts in the size range 16–32 mm from randomly located 0.5 × 0.5 m quadrats at over 300 sites.

Results

This paper summarizes the results of the field programme and preliminary analyses of the sedimentology, lithology and geochemistry and palaeohydraulics of the raised channel deposits.

Alluvial fan morphology

Two main fan systems have been identified from morphological, stratigraphical and lithological criteria (Fig. 3):

Fan system I is the oldest and most extensive, stretching southwards for *c.* 120 km from the foot of the Eastern Oman Mountains to the margins of the aeolianite outcrops in the S. PAWR borehole records (Aubel 1983; Jones 1986) suggest that buried gravels extend to depths of up to 285 m in some places, although these gravels also include the continental alluvial sediments of the Mio–Pliocene Upper Fars Group. This fan system slopes gently towards the S and SE with gradients decreasing from *c.* 0.0030 m m^{-1} W of Barzaman (Fig. 3) to only *c.* 0.0018 m m^{-1} in the SE, near the confluence of the Wadis Andam and Matam.

Fan System II forms a narrower, more recent channel system confined to the northerly, proximal parts of the present Wadi Andam piedmont zone (Fig. 3). The deposits are relatively thin, averaging only *c.* 8 m in thickness in the N (Jones 1986), thinning to only 2 or 3 m downchannel. The original channel gravels have been completely stripped off in some proximal areas, leaving undulating ridges of Hawasina chert bedrock. Many of the ridges of Fan System II rise 20–30 m above the adjacent wadi floors (Fig. 2). Proximal gradients are significantly

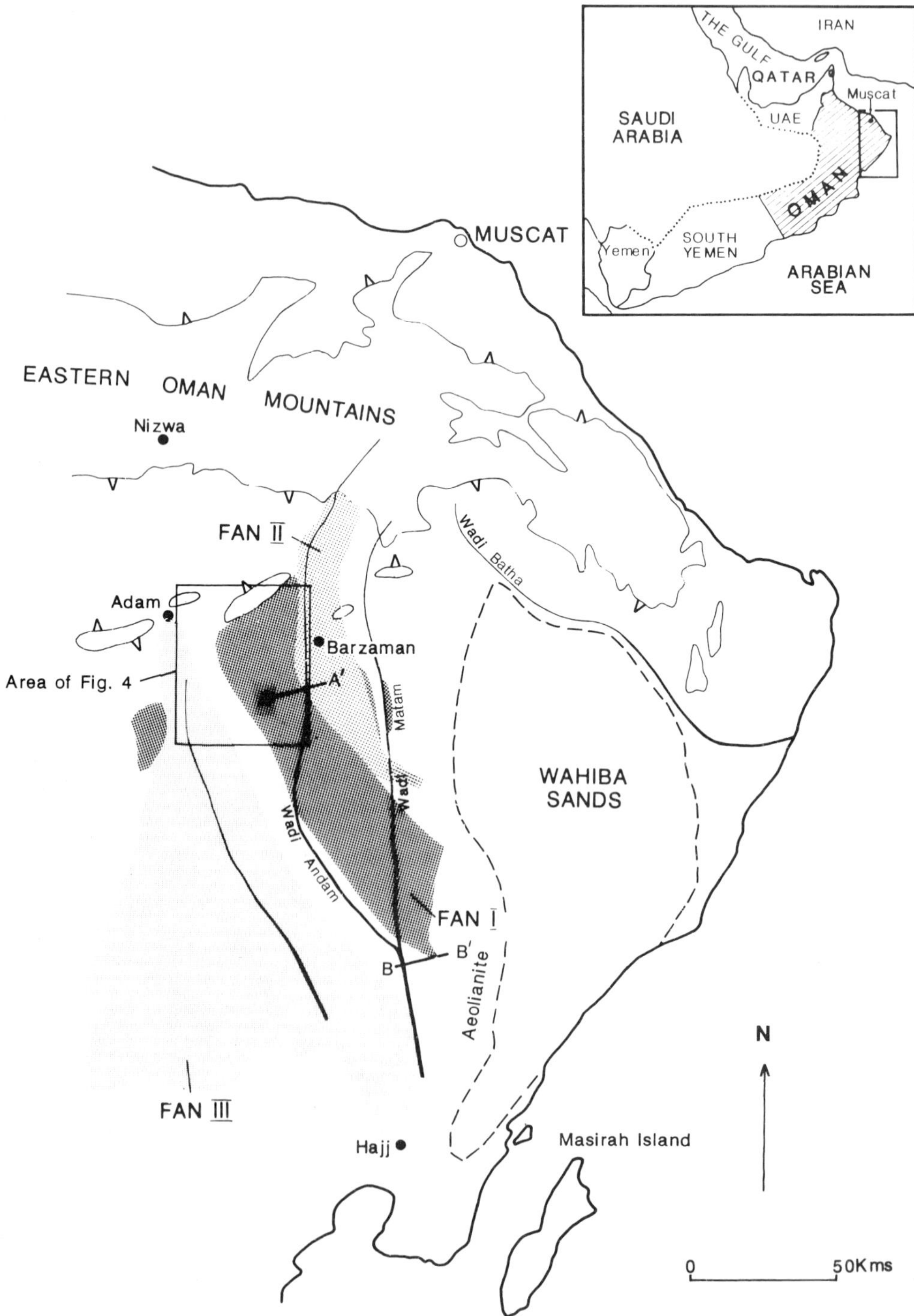

FIG. 3. Extent of main alluvial fans W of the Sharqiya (Wahiba) sands. A–A′ and B–B′ indicate locations of cross-sections in Fig. 8.

higher than on Fan System I, decreasing from *c.* 0.0046 m m^{-1} to *c.* 0.0028 m m^{-1} in the Barzaman area.

A third fan system was also identified to the W of Fan systems I and II; it extends for over 200 km from the mountain edge south south-eastwards towards the coast. Apart from its low-gradient (*c.* 0.0005 m m^{-1}) distal channel systems, this fan lies beyond the study area.

Morphology of palaeochannel systems

The surface of *Fan I* is dominated by numerous sinuous low-relief (usually up to 5 m), linear gravel ridges exhibiting markedly undulating crestlines. Sinuosity often exceeds 1.5, where sinuosity is defined as the ratio of actual channel length to channel axis length. Many of these channels intersect or cross over one another, often with successive channels forming a series of superimposed channel deposits. The older channel courses in a superimposed sequence appear to have been significantly more sinuous than the younger channels (see below).

Many areas of Fan System I exhibit older, meandering channels overlain by straight channels (see Fig. 4). The oldest courses include channels with sinuosities exceeding 1.7, whilst younger courses exhibit sinuosities averaging only about 1.03. Meander wavelengths are highly variable, but preliminary measurements from the older sinuous palaeochannels indicate an average wavelength of about 2.1 km (Table 1).

Preserved channel widths are also highly variable, being represented by ridges ranging from only 20 m across, to broad flat-topped ridges up to 500 m across, and averaging about 120 m in width. Many of the original channel deposits have been dissected or truncated by later wadi or aeolian erosive activity, thereby reducing the width of preserved channel deposits. The extent of this modification appears, however, to have been relatively insignificant since numerous long and sinous channel courses can still be so clearly identified. In addition, if it is argued that channel-floor sediments have maximum preservation potential because of maximum cementation and maximum particle sizes, it seems likely that the preserved ridge widths are not significantly smaller than those of the former main channels.

The oldest channels often appear to be the widest, and exhibit similar widths in proximal and distal zones. By contrast, the youngest channels are relatively narrow (see Figs 4 and 5), and also appear to decrease in width in the downfan direction. It seems likely that these channels accommodated progressively smaller volumes of water downstream, as a result of transmission losses and evaporation.

Fan System II is characterized by high-relief (often over 10 m), low-sinuosity channels that rarely intersect or superimpose on one another, and which overlie Fan System I channel deposits (see Fig. 8*b*). The Fan II deposits include linear ridges, extensive gravel spreads, and sequences of terrace levels bounding the main wadi courses. Sinuosities average only about 1.03, while widths are relatively high, averaging over 700 m (Table 1), reflecting either formerly broad braided or sheet flood channels. Fan II palaeochannel widths decrease significantly downstream, with many channel courses tapering off from widths of over 400 m and disappearing within 6 km downstream. The Fan II channels largely disappear towards the SE as they approach the boundary of the Sands (Fig. 3).

Sedimentology of Fan System I

(a) Facies types

Four main facies types were identified in the Fan I deposits.

Facies type 1 comprises massive, poorly imbricated, clast-supported cobble gravels, with well rounded clasts (*eg* see Fig. 6) up to 20 cm in diameter. These gravels are interbedded with sediments of *facies type 2*, represented by finer grained sand and gravel units up to 3 m thick, which exhibit distinctive foreset and trough cross-bedded structures. *Facies type 3* comprises lenses of thin horizontally bedded, medium sand, and is of relatively minor significance. All these sands and gravels are cemented by clear crystalline calcite (see Figs 6 and 9) forming the matrix of the gravels and occurring as sparry calcite and crystalline crusts around individual clasts. Many of the fine matrix materials appear to have been replaced by the carbonate cement. *Facies type 4* is the most extensive. It generally underlies the sand and gravel facies types 1–3, but may be interbedded with some of these deposits. This facies comprises a whitish-pink, massive, indurated, fine-grained, dolomitic and clay-rich rock, here termed 'barzamanite' (Fig. 10). Preliminary X-ray diffraction analyses of barzamanite samples from a single site (see Fig. 6) indicate that the oldest material (sample 18.5) is largely dolomitic (Fig. 7). The younger deposits, by contrast, are rich in clay minerals, particularly illite, montmorillonite (samples 18.1, 18.3, 18.4, 18.6 matrix; see Fig. 7) and possibly palygorskite, together with small amounts of authigenic silica (see Watts 1980).

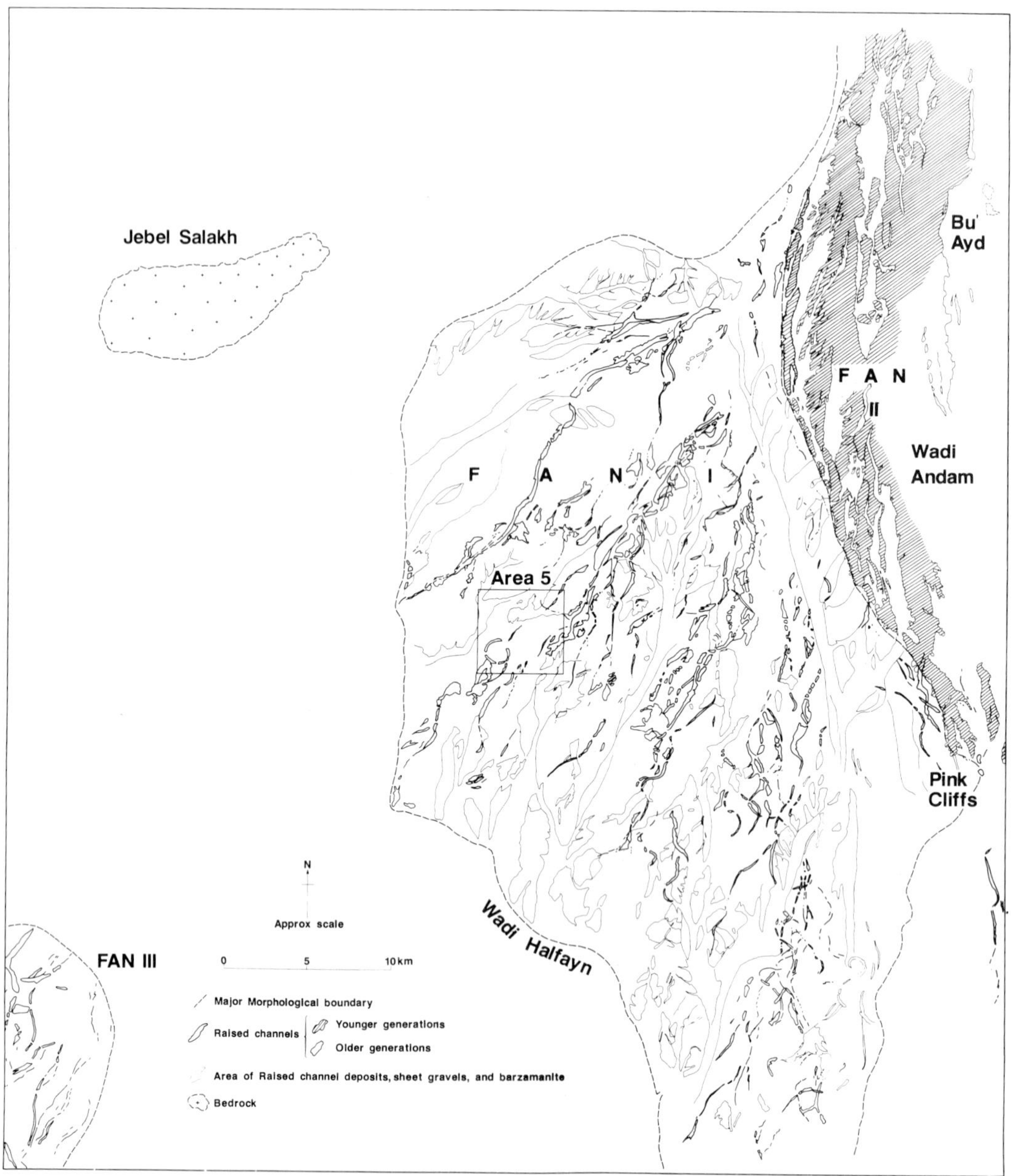

FIG. 4. Raised channel systems in the western area of Fan Systems I and II. Location shown in Fig. 3. Based on interpretation of 1:100 000 air photo mosaic.

Barzamanite often exhibits some sub-horizontal banding, a nodular structure; 'ghosting' which appears to reflect the positions of former clasts; and pockets and channels of gravels in different states of diagenesis. It often contains some small, isolated matrix-supported unaltered chert pebbles (Fig. 10). The pseudo-bedding of the barzamanite is often most distinctive towards the top of the profiles, where the barzamanite becomes increasingly nodular. At a number of sites the pseudo-bedding is deformed and exhibits either gentle large-scale folding or tight small-scale folds. Many features of the barzamanite facies resemble the characteristics of calcrete formation (Goudie 1983), although much of the barzamanite is dolomitic. Barzamanite deposits have also been recorded from further N in parts of Dubai (Bush pers. comm.).

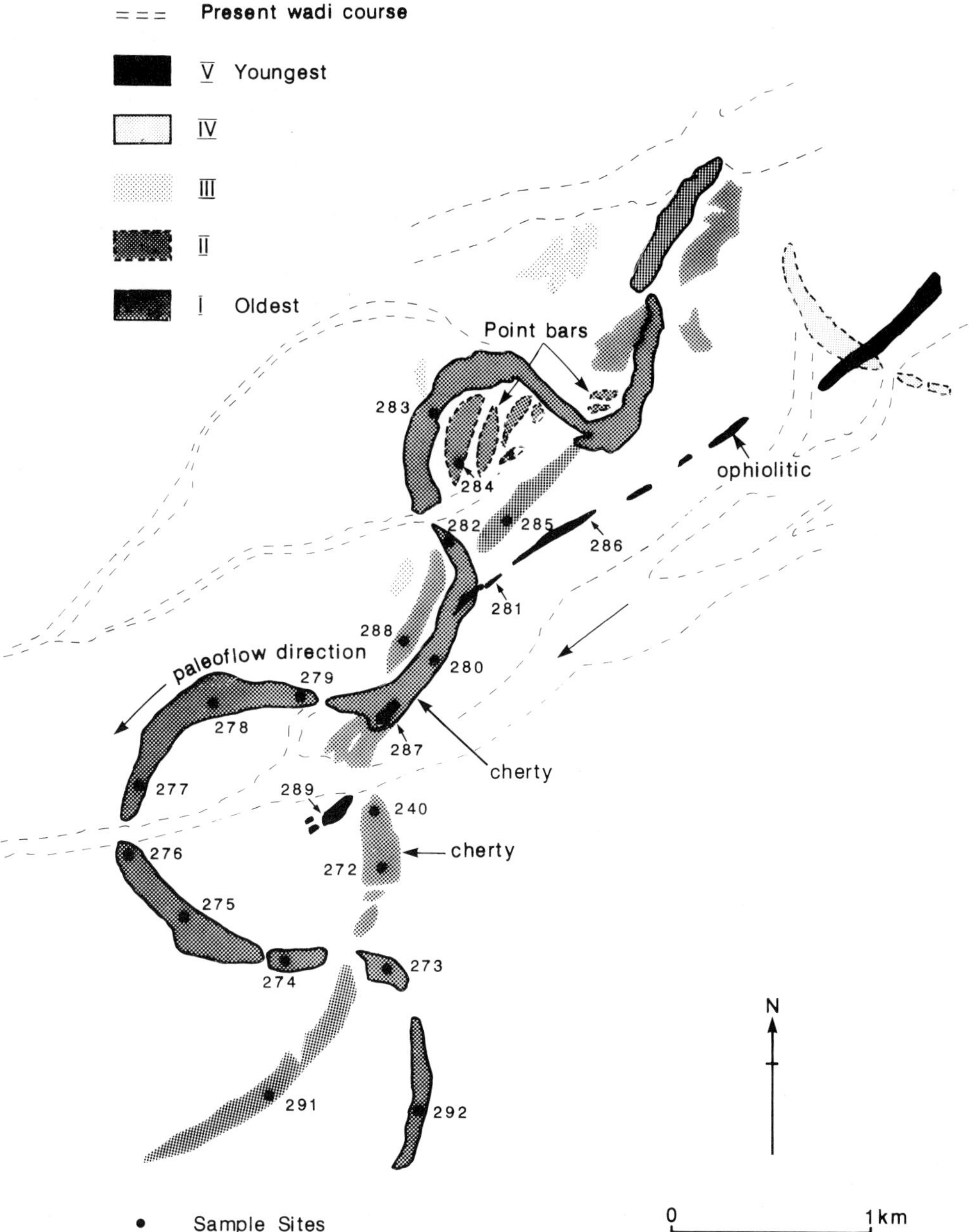

FIG. 5. Channel pattern and lithology changes associated with successive generations of palaeochannels on Fan I. Location shown in Fig. 4. Mapping based on 1:30 000 air photos, BKS Surveys Ltd, OM 85/49, 148 & 149.

TABLE 1. *Estimated palaeoflow parameters for selected raised channels of Fan Systems I and II, western Sharqiya, Wahiba, Oman*

Palaeoflow parameter	Equation no. (see text)	Fan I: Area 5 Channel generation (see Fig. 5)					Fan II: Pink Cliffs (see Fig. 4)	
		I (oldest)	II	III	IV	V (youngest)	High ridge (oldest)	Low ridge (youngest)
Gradient (regional)*				0.0044			0.0037	
Channel width m		108±35	103±34	80±52	69±11	49±18	733±351	1525±318
Max. clast size cm		11.44±2.12	12.07±2.08	–	–	21.72±2.69	15.14±2.71	15.37±4.73
Channel sinuosity		1.05	1.78	–	–	1.03	1.06	1.01
Meander wavelength† m		(2700)	2070±525	(3000)	(2220)	(4650)	–	–
Radius of curvature† m		–	450	–	–	–	–	–
Flow depth m	2	2.4	2.5			4.6	3.78	3.84
Width/depth ratio		45	41			11	194	397
Cross-sectional area m^2		259	258			225	2771	5856
n_s	3	0.0272	0.0274			0.0302	0.0285	0.0285
n_L	4	0.0344	0.0347			0.0383	0.0357	0.0358
n_J	5	0.0354	0.0351			0.0319	0.0308	0.0307
f	6	0.0428	0.0428			0.0421	0.0405	0.0405
u_s m s^{-1}	3+8	4.38	4.50			6.04	5.19	5.23
u_L	4+8	3.46	3.56			4.77	4.14	4.17
u_J	5+8	3.36	3.52			5.72	4.79	4.86
u_f	6+9	4.40	4.52			6.12	5.21	5.25
u_c	10	1.81	1.86			2.47	2.07	2.09
Fr_s	1+3+8+11	0.90	0.91			0.90	0.73	0.73
Fr_L	1+4+8+11	0.71	0.72			0.71	0.46	0.46
Fr_J	1+5+8+11	0.69	0.71			0.85	0.62	0.63
Fr_f	1+6+9+11	0.90	0.91			0.91	0.73	0.73
Fr_c	1+10+11	0.37	0.38			0.37	0.34	0.34
Q_s	3+8+12	1137‡	1176‡			1350‡	14390	30624
Q_L	4+8+12	898	929			1066	11477	24424
Q_J	5+8+12	873	918			1278‡	13292	28425
Q_f	6+9+12	1142‡	1181‡			1368‡	14440	30717
Q_c	10+12	470	485			553	5752	12237
$Q_{CA}{}^{+}$	13	(2117)	1406	(2490)	(1566)	(4890)	–	–
Q_{SCH}	14	326	325	–	–	248	2490	4866
Q_{WW}	15	440	414	298	246	157	5304	13748
$Q_{WR}{}^{+}$	16	–	1284	–	–	–	–	–

Notes
* Measures of local channel gradients too unreliable;
† These measurements and computations only possible where well-preserved sinuous channels are present (unreliable values given in brackets).
‡ Discharge estimates associated with Froude numbers between 0.75 and 1.0 (see text for explanation).
Subscripts denote use of equations as follows: s—Strickler (1923); L—Limerinos (1970); J—Jarrett (1984); f—Hey (1979); c—Costa (1983).

FIG. 6 (*opposite*). Stratigraphy and sedimentology of Fan I and Fan II deposits, site 18 near Pink Cliffs, Wadi Andam. A major unconformity separates the lower Fan I deposits from the upper Fan II deposits. The main sedimentary units are:

Fan II *Unit*	*Description*
10	Surface lag of black, varnished, ophiolite-rich pebbles < 16 cm dia., Cailleux roundness (R) = 225 ± 161.
9	Trough cross-bedded gravels with isolated sand lenses; unaltered pebbles cemented by opaque clayey-calcite; palaeocurrent vector 125°; R = 321 ± 133.
8	Horizontally bedded fine sands.
7b	Alternating sand and cobble beds.
7a	Large cross-bedded sand units, coarsening upwards; R = 449 ± 186.
6	Planar and trough cross-bedded sands, fining upwards from a basal cobble layer; palaeocurrent vector 135°–146°.
----	Unconformity

Fan I *Unit*	*Description*
5	Rubbly, structureless, poorly cemented barzamanite, with partially weathered ultrabasic clasts < 10 cm dia.; R = 308 ± 131.
4	Pseudo-bedded barzamanite, with a few scattered, bleached chert clasts < 1 cm dia.
3	Poorly cemented weathered gravels, bleached cherts and ophiolites < 8 cm dia. in sandy matrix; R = 309 ± 132.
2	Trough cross-bedded sands and gravels, with cross-bed units < 1.6 m thick; included cobbles < 20 cm dia., in friable, white calcite cement; numerous weathered basic rocks; R = 343 ± 144 and 335 ± 159.
1	Barzamanite with partially weathered bleached basic clasts.

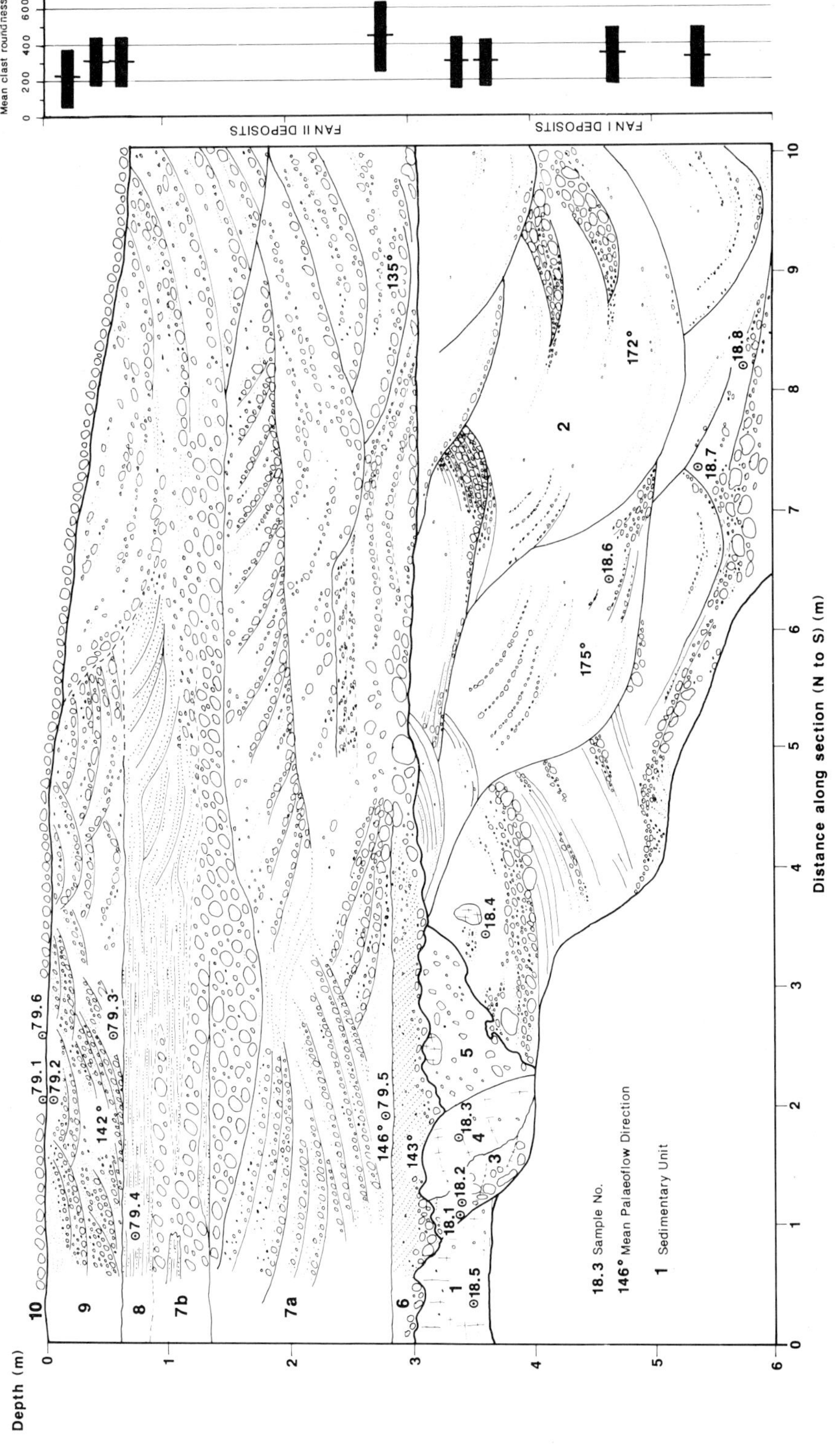

Mean clast roundness
FAN II DEPOSITS
FAN I DEPOSITS
Depth (m)
Distance along section (N to S) (m)
18.3 Sample No.
146° Mean Palaeoflow Direction
1 Sedimentary Unit

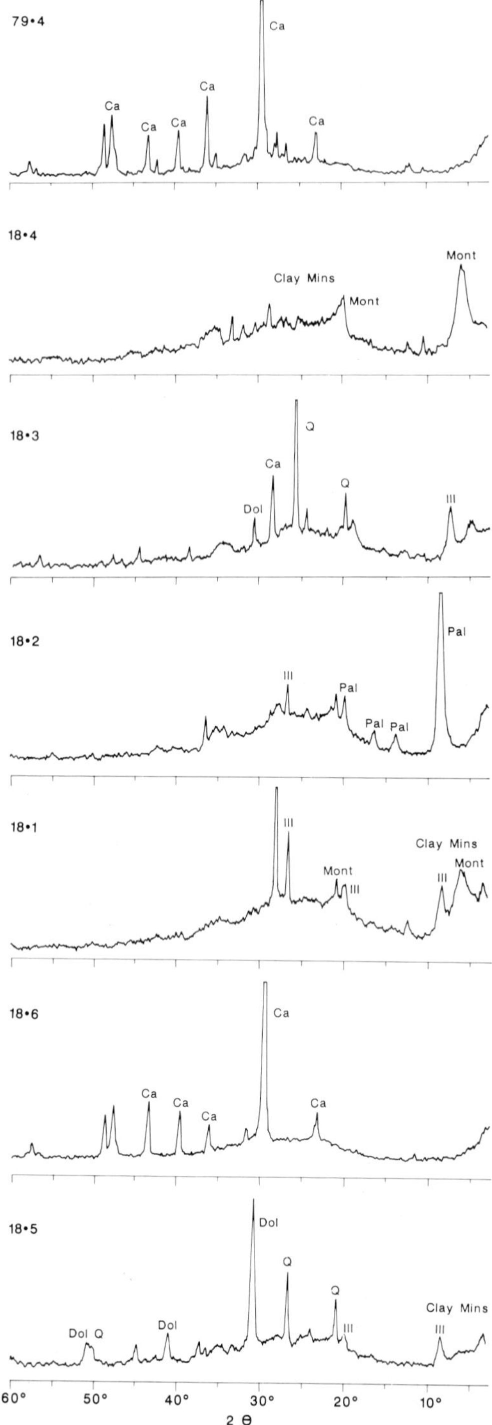

FIG. 7. X-ray diffraction traces for samples from site 18, Wadi Andam. Sample locations 79.4–18.5 shown on Fig. 6. Ca = calcite, Dol = dolomite, Ill = illite, Mont = montmorillonite, Pal = palygorskite, and Q = quartz.

(b) Sediment size characteristics

The gravels exhibit a wide range of particle sizes, normally forming a polymodal size distribution and hence a poorly sorted deposit. Maximum intermediate particle diameters exhibit a marked decrease over the 120 km of fluvial transport from *c.* 20 cm near the upland edge to only 5 cm downstream, representing an overall rate of size change of approximately 1.3 mm km^{-1}.

(c) Lithological composition

The gravels exhibit a wide range of lithologies, reflecting the heterogeneous nature of the source materials. The major constituents of the clastic fraction are: (i) *ophiolites*, which include serpentines, peridotites, gabbros and basalts, derived from the Semail Nappe outcrops in the Eastern Oman Mountains; (ii) *cherts*, derived from extensive Hawasina outcrops which bound the southern margins of the Oman Mountains. Cherts underlie most of the northern raised channel deposits, providing an abundant local supply of resistant materials for fluvial transport; and (iii) *limestones*, derived from a variety of sources of different ages, ranging from the older up-domed massifs (Permian–late Cretaceous), to the foothills of Hawasina turbidite limestones and 'exotic' marble (*eg* Hopson *et al.* 1981). A wide variety of other igneous, metamorphic and sedimentary rocks, largely derived from the Eastern Mountains, also occur within the gravels.

Significant differences in lithological composition of the gravels occur between channel deposits of different relative age. These differences reflect variations not so much in source of materials, but more in progressive post-depositional chemical alteration. The lowermost channel deposits, both in exposed sections and in superimposed palaeochannel sequences (Figs 5 and 6), exhibit the highest concentrations of cherts, usually occurring as unaltered or only partially altered, rounded and subrounded clasts. The ultramafic constituents, by contrast, appear to be most readily altered, and form much lower concentrations in the apparently oldest deposits. The barzamanite facies is often mottled by reddish-brown outlines or 'ghosts', possibly representing the diagenesis of former peridotite clasts to clay-rich materials (*eg* see Glennie *et al.* 1974, fig. 5.5*b*). The limestones also appear to exhibit progressive alteration and subsequent decreases in relative concentration.

(d) Surface weathering characteristics

Gravels exposed on channel surfaces exhibit evidence of desert varnish, case hardening and weathering rind development, of ventifacting

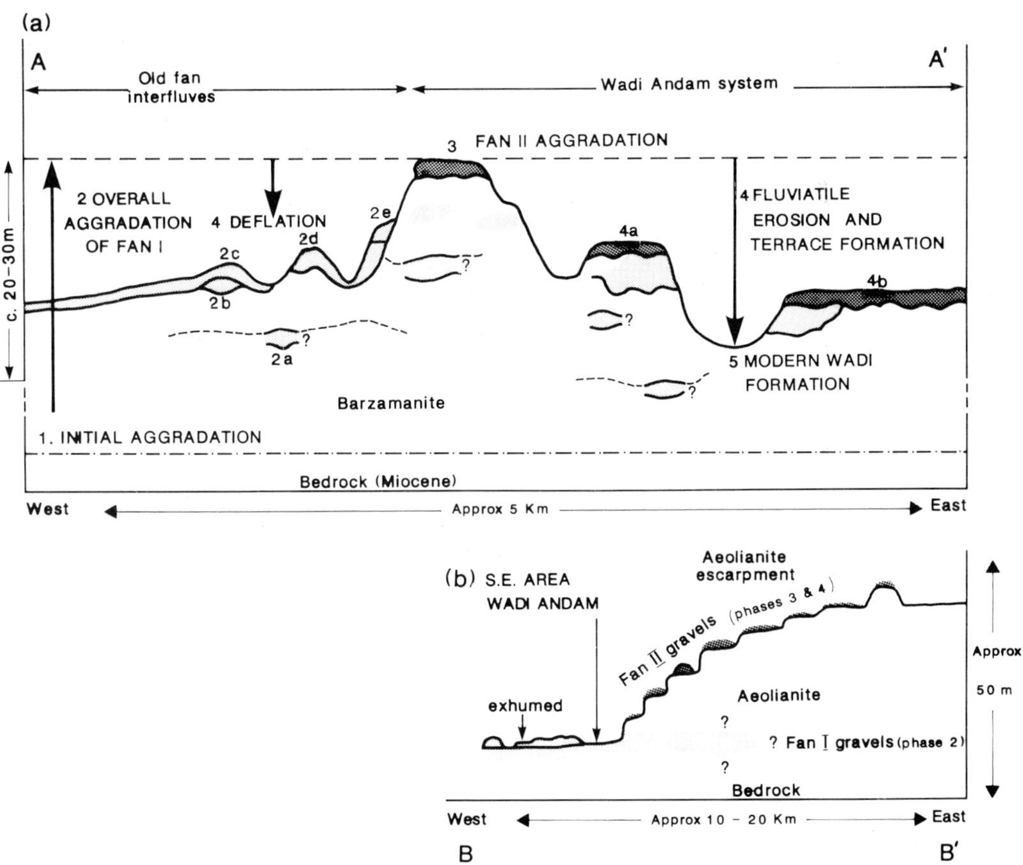

FIG. 8. Simplified model of landscape evolution of the alluvial fans and raised channel systems. (*a*) Cross-section A–A′ (see Fig. 3) across Wadi Andam at Pink Cliffs; (*b*) Cross-section B–B′ (see Fig. 3) SE of Wadi Andam across aeolianite escarpment. Phases of landscape evolution are denoted by numerals (see text for details).

and faceting, and of solution rilling and pitting, the latter particularly characteristic of limestone clasts. These clast surface features reflect long periods of subaerial weathering processes, including solution, dew etching, and aeolian sand abrasion. The gravels commonly form a layer of scattering only one particle thick (see McClure 1978) over barzamanite deposits, suggesting extensive removal of the former channel gravels by decomposition, solution, mechanical breakdown and deflation of the more basic and ultramafic lithologies. Many varnished surface gravels are underlain by 10–20 cm of clast-free silty sands and up to 50 cm of gypsiferous displacement and replacement growth (*eg* see Watson 1983).

Sedimentology of Fan System II

(a) Facies types

Two main facies types characterize the Fan II deposits. The dominant facies type, *type 5*, comprises coarse-grained (up to 0.5 m intermediate diameter) massive, cobble beds with little fabric or bedding. *Facies type 6* includes localized cross-bedded, coarse sand and fine gravel horizons. The two facies types are extensively cemented by calcite, but unlike the Fan I cements, those of Fan II appear to exhibit an opaque, clayey appearance with higher proportions of unaltered fine gravel particles in the matrix, and form a more friable and poorly indurated deposit.

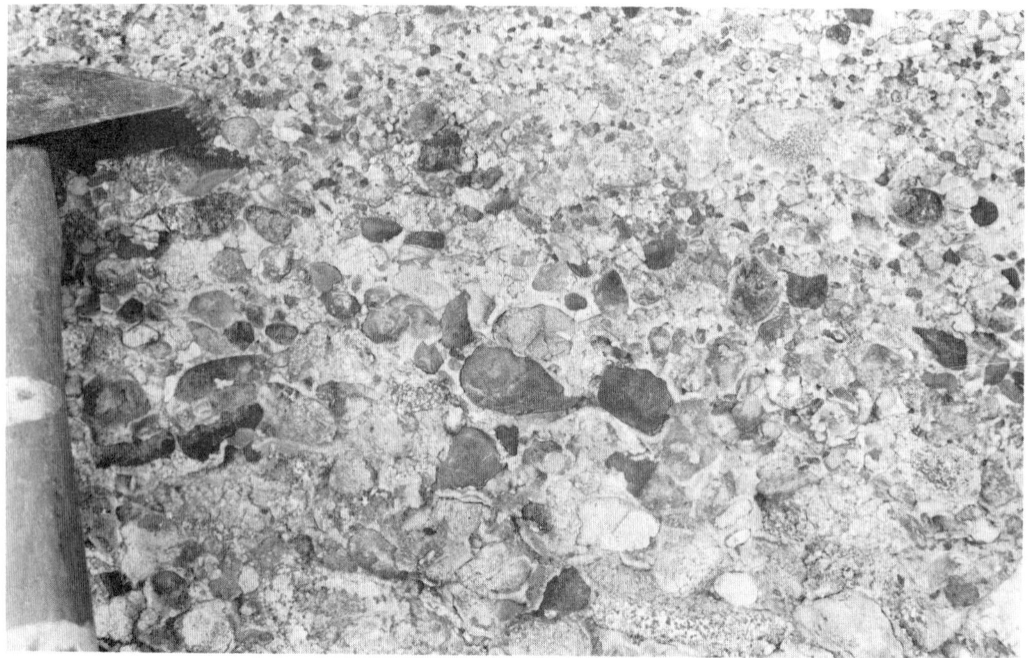

FIG. 9. Calcite crusts surrounding red-brown, chert-rich gravels of Fan I deposits, Site 18, Wadi Andam (see Fig. 6).

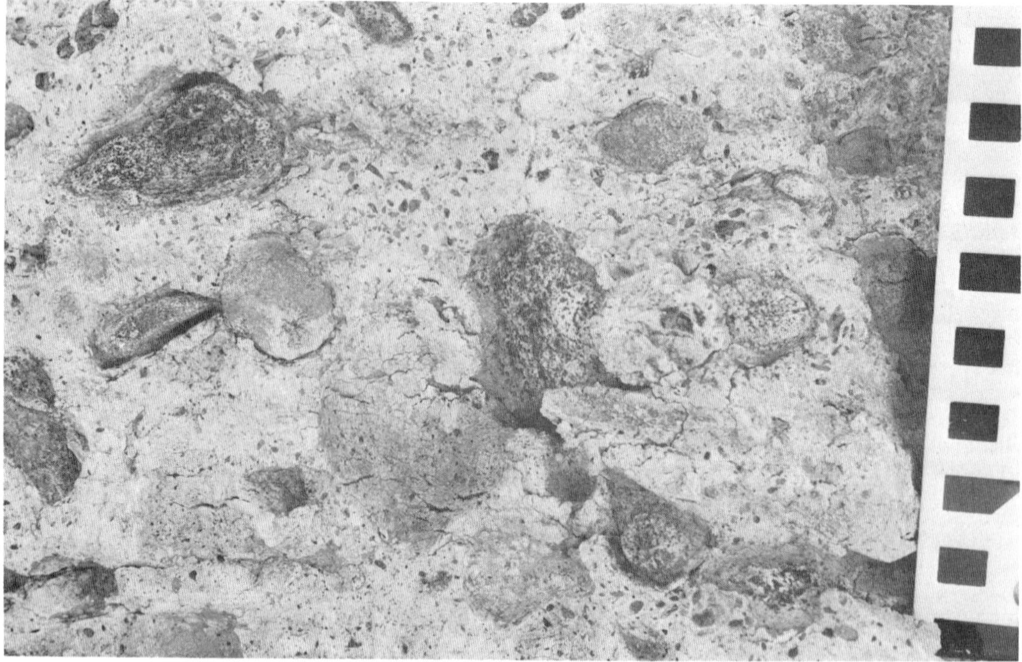

FIG. 10. Barzamanite, showing altered basic clasts and ghosts of pebbles, in fine-grained pink matrix with scattered gravel particles, Site 18, Wadi Andam.

X-ray diffraction indicates that the cement comprises a calcite–clay mineral mix (sample 79.4, Fig. 7).

(b) Sediment size characteristics

The Fan II gravels are poorly sorted. Maximum intermediate particle diameters decrease rapidly downstream, exceeding 40 cm in the proximal zone and reaching only 8 cm in the distal zone, representing an overall rate of size decrease of *c.* 5.5 mm km^{-1}.

(c) Lithological composition

The clast fraction of the Fan II gravels is dominated by well-rounded (*eg* see Fig. 6) gabbroic lithologies, with only minor percentages of cherts, limestones, serpentinite and other rock types.

(d) Surface weathering characteristics

The basic lithologies on the high channels and terrace deposits of Fan II exhibit evidence of extensive exfoliation, spalling and *in situ* splitting, processes that appear to be active at the present time. A major result of this rock disintegration is the production of a much finer grained surface deposit comprising increasingly small angular (see Fig. 6) fragments and flakes of cherts and ophiolites, with progressive weathering. Surface clasts exhibit desert varnish, faceting and surface solution effects depending on lithology.

General stratigraphic relations

Fan II deposits overlie and are cut into the lower Fan I deposits along the Wadi Andam drainage system. The Fan II deposits are found at three main topographic levels (3, 4*a* and 4*b*, Fig. 8*a*) along Wadi Andam, with the lower two levels cutting into both barzamanite and channel gravels of the Fan I system. The contact between the two fan deposits forms a sharp, erosional boundary, often marked by a series of irregular, longitudinal flutes scoured into the underlying barzamanite (*eg* see Fig. 6). The boundary surface is generally coated with a thin microcrystalline layer of brown micrite (see Goudie 1983). Boulders of older cemented gravels and barzamanite are occasionally incorporated into the basal gravels of the younger Fan System II.

Palaeohydrology of the raised channels of Fan Systems I and II

(a) Channel pattern and sinuosity

Preliminary examination of the planform characteristics of successive generations of palaeochannels suggests that the older channel courses were significantly more sinuous than the younger. It thus appears that overall channel sinuosities have gradually decreased through time (see Table 1). Similarly, many of the older fluvial deposits appear to have been preserved as separate single-thread channels, while the more recent channels appear to have been associated with extensive sheet gravel formation, possibly associated with braided flows (see Fig. 4).

(b) Palaeoflow parameters

Sources of error. Preliminary estimates of selected flow parameters have been computed for some of the Fan I and Fan II palaeochannels. At this stage, however, there are still many uncertainties and sources of error involved in the application of palaeohydraulic models to the raised channel deposits. For example, estimates of former flow hydraulics including flow depths, resistance to flow, velocity and discharge for different palaeochannels have a number of input requirements including, in particular, estimates of the cross-sectional area of flow, former energy gradients, and maximum sizes of particles that the stream was competent to move. Accurate determination of these primary parameters may not prove possible for many of the palaeochannels. Overbank sediments have largely been removed, so that former flood flow widths can rarely be correctly determined. Former energy gradients can be estimated only very approximately from the published topographic maps. In addition, since many of the channels appear to have been considerably dissected, diagenetically altered and let down on to an underlying surface both by deflation to form a pebble lag and by compaction associated with diagenesis, the original channel gradients are unlikely to have been preserved. Maximum particle sizes, at least of the surface materials, appear to decrease through time by progressive granular disintegration, and hence measurements of maximum clast sizes may not always represent the largest materials that were originally transported and deposited in the channel. Details of some of the procedures adopted in the palaeohydraulic analysis of the raised channel deposits, and discussion of the many assumptions and sources of error involved in this approach are outlined elsewhere (Maizels 1983, 1986, 1987*a* and *b*). The procedures for estimating selected palaeoflow parameters are briefly summarized below.

Procedures for palaeoflow estimation. Two approaches have been adopted for determining former flow conditions.

The first approach is based on initial estimates of the critical shear stress at which bed particles begin to move. This computation uses Shields' (1936) function and allows subsequent calculation of former flow depth using the Du Boys equation (see details in Maizels 1983), and resistance to flow and flow velocity using Darcy–Weisbach and Manning equations. This approach requires the field determination of maximum particle sizes, in order to provide a measure of bed roughness, former channel gradients, and former flow widths. The main assumptions of this critical shear stress approach are the former existence of steady uniform flow; the validity of the shear stress model itself; a known effect of packing and imbrication on the Shields coefficient; the absence of local bedforms or coarse clast concentrations; and the availability of all particle sizes for transport by competent Newtonian fluid flows (*eg* see Costa 1984). Most of these assumptions remain questionable.

This critical shear stress approach allows the estimation of former mean critical flow depth, Y_c, from a form of the Du Boys equation, such that

$$Y_c = \tau_c / \gamma S \quad (1)$$

where τ_c is the critical tractive force (N m^{-3}) defined as

$$\tau_c = \phi(\gamma_s - \gamma)D \quad (2)$$

where ϕ is Shields coefficient (taken as 0.056), γ_s is the specific weight of the sediment, γ is the specific weight of water and S is the former hydraulic gradient.

The determination of palaeoflow velocities has depended on obtaining a measure of flow resistance using well-established friction coefficients for small-scale roughness in channels, namely, Manning's n and the Darcy–Weisbach f. Manning's n has been computed here from three different models.

Strickler (1923) derived a simple functional relationship between n and the representative particle diameter such that

$$n = 0.039 D_{50}{}^{0.167} \quad (3)$$

where D_{50} = 50th percentile of the sediment size distribution. However, Limerinos (1970) demonstrated that n was best determined by the use of a logarithmic function relating the resistance coefficient to a measure of relative roughness (defined as the ratio of Y to D):

$$n = \frac{0.113 Y^{0.167}}{1.16 + 2.0 \log (Y/D_{84})} \quad (4)$$

where D_{84} = 84th percentile of sediment size distribution. More recently, Jarrett (1984) has derived a predictive equation for n based on multiple regression:

$$n = 0.32 S^{0.38} Y^{-0.16} \quad (5)$$

for $0.002 < S < 0.034$, $0.14 < Y < 2.01$ m, $0.028 < n < 0.159$, $0.09 < D < 0.79$ m.

The Darcy–Weisbach friction factor (f) has, however, a sounder theoretical basis as it is related to the von Karman–Prandtl velocity 'law', and is also dimensionally correct. Values of f were determined using the approach proposed by Hey (1979):

$$\frac{1}{\sqrt{f}} = C \log\left(\frac{aR}{D_s}\right) \quad (6)$$

where $C = 2.3/(\kappa/\sqrt{8})$, κ = the von Karman constant, and D_s = the representative grain size (m). The value of 'a' is a function of channel cross-sectional shape and spacing of roughness elements (Silberman *et al.* 1963), and may be determined either graphically (Hey 1979) or from:

$$a = 11.1 (R/Y_{max})^{-0.314} \quad (7)$$

where Y_{max} = maximum flow depth (m) (Thorne & Zevenbergen 1985).

Once the resistance coefficients are computed, they can be substituted into equations for critical mean flow velocity u_c, based either on Manning's n, where

$$u_c = \frac{Y^{0.67} S^{0.5}}{n} \quad (8)$$

or on the Darcy–Weisbach f, where

$$u_c = \left(\frac{8 g Y S}{f}\right)^{0.5} \quad (9)$$

where g is the acceleration due to gravity.

Further estimates of mean flow velocity were determined from Costa's (1983) equation which is based on the mean of four theoretical and empirical functions:

$$u = 0.18 (1000 Y)^{0.487}. \quad (10)$$

The validity of the flow depth and palaeovelocity estimates was assessed by calculation of the Froude number, Fr, where

$$Fr = v/(g Y_c)^{0.5}. \quad (11)$$

Froude numbers during high flows in coarse-grained gravel bed rivers rarely exceed values of about 2.5, and usually lie within 0.25 of the critical value of 1.0 (*eg* Boothroyd & Ashley 1975; Jarrett 1984; Bathurst 1985; Costa 1985; Thorne & Zevenbergen 1985). Palaeoflow depths and palaeovelocity computations that produced

Froude numbers within the range 0.75 to 1.25 were considered acceptable in this study.

Determination of palaeodischarge, Q, for the raised channel deposits was based on the flow continuity equation, where

$$Q = wY_c u_c. \quad (12)$$

The procedures outlined above for determining palaeovelocities therefore provided five different estimates of palaeodischarges, but each with error sources relating not only to field measurements but also to the validity of application of the various models themselves.

The second approach to the palaeohydrologic analysis of the raised channel deposits is based on the application of empirical discharge-form relations for meandering channels. These relations require determination of meander wavelength, L_M, radius of meander curvature, R_c, and bankfull channel width, w. Hence, major sources of error arise where former channel morphology is only partially or poorly preserved (*eg* see Ethridge & Schumm 1978; Rotnicki 1983). The equations adopted in this study are based on four empirical functions relating meander planform and channel morphological measures to discharge parameters.

Carlston (1965) (modified by Williams 1984) established an empirical relationship between meander wavelength and floods with a recurrence interval of 1.5 years ($Q_{1.5}$) for streams in the central USA, such that

$$Q_{1.5} = 0.011 L_M{}^{1.54}. \quad (13)$$

Schumm (1972) related channel width and maximum flow depth to bankfull discharge, defined as $Q_{2.33}$, for semi-arid and sub-humid channels in the USA and Australia, such that

$$Q_{2.33} = 2.66 w^{0.9} Y_{max}{}^{0.68}. \quad (14)$$

Estimates of flow depth calculated from equation (1) were substituted into Schumm's equation, while measurements of channel width and meander wavelength were taken from aerial photographs. Williams (1984) has provided two functions that relate channel width and meander radius of curvature to the maximum instantaneous discharge (Q), which more closely matches the types of discharges predicted by the critical shear stress approach. Williams found for meandering rivers in Sweden that

$$Q = 1.0 w^{1.3} \quad (15)$$

and that

$$Q = 0.28 R_c{}^{1.38}. \quad (16)$$

These relations provided four additional estimates of palaeodischarge for the meander channels, while only equations (14) and (15) were used for low-sinuosity channels. The two approaches to palaeodischarge prediction have allowed an envelope of discharge predictions based on nine different models to be produced for each of the raised channel systems.

Palaeoflow parameters associated with the raised channel systems. Palaeoflow depth, resistance, velocity, Froude number and discharge have been estimated for a number of raised channel deposits, while more detailed estimates have been made for successive generations of older palaeochannels in one area of Fan I (see Fig. 5), and for one area of Fan II channels bounding the present Wadi Andam (see Fig. 4 for locations). The results are summarized in Table 1.

Palaeoflow parameters of Fan I channel systems. Estimates of peak flow depths suggest that they averaged between about 1.2 and 4.5 m, although maximum depths may have exceeded 6 or 7 m in some channels during major flood events. Peak flow velocities averaged between about 2 and 7 m s^{-1} within the palaeochannel systems as a whole, with discharges ranging between about 160 and 1400 m^3 s^{-1}.

Palaeoflow estimates for successive generations of channels in 'Area 5' of Fan system I (Fig. 5, Table 1) are based on only approximate estimates of former energy gradients, derived from 1:100 000 topographic maps. This parameter probably represents the greatest source of error in the palaeoflow computations. Errors also arise in determination of former clast sizes, prior to weathering and rock breakdown, and measurement of channel width and planform characteristics from preserved channel fragments only.

The two earlier palaeochannel generations (I and II) exhibit significantly different channel form, sediment, and flow conditions from the youngest channel systems (V) in this area (Fig. 5). The earlier channels are at least twice as wide as the later channels, significantly more sinuous (system II), and comprise significantly finer clasts (representing long-term rock breakdown and/or finer sediments in transport). Predicted flow depths, and velocities were 30% to 85% greater in the younger, narrower, coarser-grained channels than in the broad sinuous, finer-grained older channels. However, because of similarities in cross-sectional areas of flow—the older channels exhibiting only marginally higher values—the peak discharges predicted for the different channels by the various methods are of a similar order of magnitude. The older channels were associated with palaeodischarges reaching about 1150 m^3 s^{-1} (with Froude numbers of about 0.9), while the younger channels were associated with those of about 1330 $m^3 s^{-1}$. Palaeodischarges predicted by the two approaches, at least for channel generation II, are also very similar in magnitude using equations (13) and (16).

Hence, significant changes in palaeoflow conditions appear to have occurred during fan aggradation. Palaeochannels became increasingly narrow, less sinuous, and probably carried coarser sediments; peak flow depths and velocities during flow events increased; while maximum discharges continued to be around a similar order of magnitude.

Palaeoflow parameters of Fan II channel systems. Fan II channel systems located at a similar distance from the mountain front as those examined on Fan I, exhibit significant contrasts with those of Fan I. In particular, the great widths of the preserved Fan II palaeochannel deposits, together with the nature of the constituent sedimentary structures and facies types (*eg* see Fig. 6), suggest that the deposits characterize broad braided channel systems. Estimates of peak flow depths of about 3.8 m suggest that width–depth ratios ranged between 200 and 400 (Table 1), almost an order of magnitude higher than in the Fan I channel systems. Channel gradients, particle sizes, flow resistance and velocity estimates are all similar in magnitude to those of the Fan I channels, but Froude numbers are much lower, suggesting that velocity estimates may be too low and/or depth estimates too high. The high palaeodischarges estimated by the different methods largely reflect, therefore, the broad lateral extent of the former channel systems. Peak palaeodischarges averaged between about 13 000 and 31 000 $m^3 s^{-1}$, according to most of the methods (particularly those with Froude numbers >0.7). The estimated peak flows therefore exceeded those of the Fan I channels by about one order of magnitude.

No detailed error analysis has yet been completed on these palaeohydraulic computations, but the palaeodischarge values were found to vary significantly according to the variability of the input parameters. For example, by varying each parameter by up to one standard deviation on either side of the mean in the case of S, w, and D and by varying ϕ values between 0.04 and 0.07 for the Fan II channels using equations (8), (9) and (10) resulted in palaeodischarge differences of up to $\pm 60\%$, $\pm 48\%$, $\pm 40\%$, and $\pm 67\%$, respectively. When these errors were combined, palaeodischarge values were found to range by up to 380% of the estimated discharges, although in this case the associated depth and velocity values appear excessively high (*ie* $Y_c > 7$ m, $u_c >$ 8 $m s^{-1}$).

Although there are few records that extend over more than a 5-year period, the available flood discharge data for recent floods in the Sharqiya suggest that peak flow conditions in the Fan I palaeochannels as a whole may have been fairly similar to those of the present day. Estimates of width, depth and velocity during recent wadi flood events in Oman (Curtis 1985) ranged between 5–405 m, 0.28–6.19 m, and 0.006–9.72 $m s^{-1}$, respectively; while peak discharges reached *c.* 4650 $m^3 s^{-1}$. These discharges are significantly lower than those estimated for the Fan II channel systems.

Chronological framework

The relative ages of the different palaeochannel and alluvial fan systems have been established on the basis of morphological and stratigraphical relationships, together with preliminary observations of the different degrees of cementation and weathering within each system. Since the PAWR borehole data (Jones 1986) suggest that the older fan gravels overlie the continental alluvial sediments of the Mio–Pliocene Upper Fars Group in the N, and Miocene limestones in the SW (Aubel 1983), the earliest possible date for commencement of gravel accumulation appears to be early Pliocene.

The gravels are also found in association with the aeolianite that underlies much of the present-day Sharqiya sand sea. The older gravels of Fan System I extend southeastwards beneath the western edge of the aeolianite outcrop. Hence part of this older gravel sequence clearly predates the formation of the aeolianite. Patches of younger gravel, and a single raised channel ridge, have also been located on a series of 'terrace' levels across the aeolianite outcrop and along the edge of the escarpment (Fig. 8*b* and Warren *et al.* 1985). These deposits indicate that flows occurred at levels of up to 50 m above the present wadi floor. These gravels are likely to be related to the final stages of formation of Fan System I, followed by successive stages of incision and deflation to produce the stepped terrace sequence along the escarpment. This period of land-surface lowering may have been contemporaneous with the formation of the terrace sequence within the Wadi Andam Fan System II (Fig. 8*a*).

There are no directly datable materials available at present from the raised channels and alluvial fan sequences. However, examination of the nature and occurrence of chert artefacts on (but not within) the raised channel surfaces was carried out by Edens (1986). Intense concentrations of worked cherts occur on the older, lower channel ridges of the Fan System I near Barzaman

(Fig. 4). According to Edens (1986) the artefacts are likely to date from some time between 4000 and 7000 BP; these dates are in turn likely to represent the latest period during which palaeochannel exhumation could have taken place.

Palaeoenvironmental and landscape change

Landscape evolution in the area of the raised channels appears to have occurred in five main phases, possibly commencing during the early Pliocene.

Phase 1 is represented by the initial development of a large piedmont alluvial plain extending southwards from the mountain edge. The earliest deposits on this surface were probably associated with the final stages of uplift of the Eastern Oman Mountains, and are now recorded in deep boreholes, sometimes directly overlying Miocene bedrock, at depths of 250–300 m and at the base of Fan Sequence I (Fig. 8).

Phase 2. The initial development of a piedmont alluvial fan during Phase 1 was followed by a prolonged period of overall fan aggradation to form the Fan I deposit. This aggradational phase resulted in the infilling of the piedmont basin to depths of $\geqslant$200 m with chert, ophiolite and limestone gravels from the Eastern Mountains. The lateral extent of this deposit was limited in the E by the presence of former sands, now forming the older aeolianites, although the eastern limit may have fluctuated through time.

Overall fan aggradation was associated with rapid cementation of the channel gravels through precipitation of calcite cements (Stalder 1975). The existence of a rising water table is indicated by the progressive long-term *in situ* chemical alteration of the gravel deposits, presumably lying near a rising water table. Gradually, the carbonate and ophiolite (especially serpentinite and peridotite) constituents of the gravels, together with the calcite cements and fine overbank and aeolian sediments, became altered or partially altered, to form the calcrete-like barzamanite deposits.

Progressive fan aggradation was accomplished by periodically rapid rates of sediment accumulation and by the successive development of new channels on the fan. Initiation of new channels probably occurred by blocking of channels with coarse sediments and aeolian sands, and by channel piracy and switching during high flow events.

It seems likely that substantial periods of time separated the development of successive channels during fan growth, since significant differences in the degree of weathering often occur in channels of different generations and at different elevations (Fig. 5). However, since some channels of different generations (*ie* with significantly different weathering characteristics) occur at the same elevations (*ie* with one channel ridge truncating a second), it seems likely that long-term fan aggradation was punctuated by periods of local channel downcutting into older channel deposits, allowing a new deposit to accumulate across its path.

Phase 3. The third phase of landscape evolution was characterized by a renewed period of fan sedimentation. These sediments represent the accumulation of thin (generally up to 3 m) ophiolite gravels of the Wadi Andam Fan System II. These younger gravels were deposited directly over the altered and partially altered older gravel of Fan System I. Distinctive unconformities and erosion surfaces mark the contact between the two deposits (Figs 6 and 8). The alluvial streams were very limited in extent compared with those of the older fan, disappearing southeastwards into the Sands.

Phase 4. The accumulation of the Wadi Andam gravels was followed rapidly by a phase of sudden and dramatic degradation of the fans (Fig. 8). Successive periods of large-scale incision of the Wadi Andam Fan System alternated with periods of thin gravel accumulation. Away from the wadi system and across the interfluves of the older fan systems, extensive lowering of the land surface took place. Along the Wadi Andam System, periodic incision was dominated by fluvial activity, resulting in the formation of two main terrace levels above the present wadi floor (Fig. 8). Up to 3 m of gravels were deposited on the terrace surfaces, unconformably overlying the older weathered gravel deposits and barzamanite of Fan System I (Fig. 6), and in more proximal zones, directly over bedrock. The weathering characteristics of the gravels on these terraces do not appear to be significantly different from those of the higher level gravels of Fan System II deposited during Phase 3 (above), suggesting that incision was accomplished relatively rapidly. Incision was also associated with relatively rapid lowering of the water table. Further S along the aeolianite escarpment, the lowering of the water table appears to have resulted in extensive deflation down to successive water table levels, forming a series of terrace surfaces (Gardner 1986). This deflation was accompanied by the occasional inundation by flows from the former wadi systems of Andam and Matam, initially extending for some 30 km to the SE. At times, more marked stream flow occurred closer to the wadi–aeolianite boundary, depositing channel sediments which subsequently became cemented and preserved as a raised channel ridge. In the interfluve areas of the fan deposits, land-surface lowering was accomplished largely by deflation, and only on a minor scale by water erosion. Extensive deflation of the barzamanite, the weathered cements, and the fine-grained matrix and overbank sediments resulted in the gradual exposure of formerly buried and diagenetically altered palaeochannel courses. Where successive palaeochannels had traversed the same areas of terrain, the younger channel deposits became exposed in a superimposed position over the older palaeochannels.

Phase 5. The final phase of landscape development is represented by the modern wadi system (Fig. 8). The modern Wadi Andam has cut into the Fan II deposits and continues to flow southwards towards the aeolianite escarpment, but now rarely, if ever, reaches the sea. There is little evidence of large-scale wadi channel modification or instability in recent times. In a number of wadi reaches, low terraces only 1 or 2 m high are found, and numerous bedrock outcrops occur in the floor of the wadi. These features suggest that the amount of incision and local sedimentation has been only minimal during periods of more recent wadi activity.

Palaeohydrological and palaeoclimatic implications of landscape evolution

Aggradation of Fan System I appears to have been associated with more sinuous channels, possibly related to less ephemeral flows than at present. Periods of flow extending for 2 or 3 months per year would probably have been sufficient to allow the development of an equilibrium relationship between the meandering channel planform and the associated channel capacity. However, peak discharges of some of the palaeochannels may have been of a similar order of magnitude to those of the present. Cementation of the channel gravels probably occurred soon after streamflow, and during periods of seepage and evaporation. The large-scale chemical alteration of the gravels would have required a high water table over prolonged periods of time, and hence is likely to have been associated with more humid conditions.

The formation of Fan System II and its subsequent sudden and deep entrenchment appear to have been associated with a more episodic, flashy flow regime. These flows may have occurred as intermittent periods of high flows acting to cut down through the fan deposits, followed by more stable flow conditions of sedimentary infilling or reworking. In addition, the apparent disappearance of the gravels into the Sands in the E may imply that semi-arid or arid conditions prevailed at that time. This period of fan downcutting was associated with a relatively rapid fall in the water table, possibly reflecting a combination of decreased sediment availability, decreasing climatic humidity, tectonic uplift and/or a glacial fall in sea level (Glennie pers. comm.). The extensive contemporaneous lowering of the interfluve fan areas is likely to have occurred during a period of increasing aridity, when sediments were exposed to wind erosion.

Chronological implications of landscape evolution

No dates are yet available for the older fan system, but the degree of widespread chemical alteration suggests that these deposits are likely to date from the Pliocene–early Pleistocene. Such a date would conform to the events proposed for this period by Hötzl *et al.* (1978*a*, *b*) and Anton (1984) for Saudi Arabia. In addition, aeolian sands and wadi gravels on the coast of Sabka Mati, Abu Dhabi, have been dated as early Pliocene (Glennie pers. comm.), suggesting that the period of alluvial fan aggradation in the Sharqiya could also have begun at this time. The hypothesis of progressively rising water tables during fan aggradation is further supported by the global rise of sea level during the Pliocene.

The overlying Fan II gravels are much younger since they are only minimally altered. Their relatively weak cementation may reflect either their relatively recent formation, or rapid downcutting, which cut off the supply of groundwater (Glennie pers. comm.). Their formation, and the period of fan degradation and land-surface lowering, probably date from arid and semi-arid phases of the mid- and late Pleistocene. The latest period of landsurface lowering and palaeochannel exhumation is likely to have been during the last glaciation (25 000 to 10 000 BP) when sea level is known to have been lower and desert winds are postulated to have been stronger (Glennie pers. comm.). According to the evidence provided by lithic components and artefacts, palaeochannel exhumation had virtually ceased by the early to mid-Holocene.

Conclusion

The model of palaeoenvironmental change proposed here invokes long-term fan aggradation during more humid phases (possibly during the

Pliocene and early Pleistocene), accompanied by cementation of the channel gravels and diagenesis of the ophiolite-rich and finer-grained sediments. Fluviatile entrenchment and landscape lowering by deflation of poorly cemented, fine-grained interfluve sediments appear to have occurred during subsequent alternating semi-arid and arid periods (possibly during the mid- and late Pleistocene). This model conforms with those proposed for Saudi Arabia by Hötzl *et al.* (1978*a*, *b*) and also for the Negev by Amit and Gerson (pers. comm.).

Many uncertainties remain in the model. In particular, the lithological, mineralogical and geochemical properties of palaeochannel sediments of different generations need to be determined in order to identify the temporal patterns of fan growth and channel evolution. Stratigraphical analysis of palaeochannel relationships also needs to be extended. Finally, the palaeohydrologic characteristics of successive channel systems need to be established more precisely, and possibly dated in order to be linked to likely patterns of palaeoclimatic change.

ACKNOWLEDGMENTS: The author would like to express particular appreciation to Dr K. Glennie for many invaluable discussions. Thanks are also extended to P. Considine, J. R. Jones, Dr G. Stanger and H. Weier, all of the Public Authority for Water Resources, Sultanate of Oman, and to B. Duff and Dr M. Hughes Clarke of Petroleum Development Oman for their helpful and stimulating discussions. Thanks are also due to C. McBean for valuable field assistance and to J. Cutler and J. McIntosh for help with the ground surveys. The logistic support of the Sultan of Oman's Armed Forces and the Coastal Security Force is also gratefully acknowledged. The author would like to thank all members of the Royal Geographical Society Eastern Sands Project for their enthusiasm and support during the progress of this research. Finally, the author would like to acknowledge assistance with laboratory analyses from Prof. I. Parsons, Dr G. Walkden and M. Lamb, and the thorough and helpful comments on the draft of this paper made by V. Baker, I. Gass, K. Glennie, C. C. Reeves, I. Reid and G. Stanger.

References

ANTON, D. 1984. Aspects of geomorphological evolution: paleosols and dunes in Saudi Arabia. *In:* JADO, A. R. & ZÖTL, J. G. (eds) *Quaternary Period in Saudi Arabia.* Vol. 2, Springer-Verlag, 275–296.

AUBEL, J. W. 1983. Bajada hydrological area. *In: The Hydrology of the Sultanate of Oman. A Preliminary Assessment.* Public Authorities Water Resource Report PAWR 83-1, 66–74.

BATHURST, J. C. 1985. Flow resistance estimation in mountain rivers. *Journal of Hydraulic Engineering, American Association of Civil Engineers* **111**, 625–643.

BEYDOUN, Z. R. 1980. Some Holocene geomorphological and sedimentological observations from Oman and their palaeogeological implications. *Journal of Petroleum Geology* **2**, 427–437.

BOOTHROYD, J. C. & ASHLEY, G. M. 1975. Processes, bar morphology, and sedimentary structures on braided outwash fans, northeastern Gulf of Alaska. *In:* JOPLING, A. V. & MCDONALD, B. C. (eds) *Glaciofluvial and Glaciolacustrine Sedimentation.* Society of Economic Palaeontologists and Mineralogists Special Publication **23**, 193–222.

BUTZER, K. W. & HANSEN, C. L. 1968. *Desert and River in Nubia.* Univ. of Wisconsin Press, 497 pp.

CARLSTON, C. W. 1965. The relation of free meander geometry to stream discharge and its geomorphic implications. *American Journal of Science* **263**, 864–885.

COSTA, J. E. 1983. Paleohydraulic reconstruction of flash-flood peaks from boulder deposits in the Colorado Front Range. *Geological Society of America Bulletin* **94**, 986–1004.

—— 1984. Fluvial paleoflood hydrology (abs.): *EOS* **65**, 890.

—— 1985. Interpretations of the largest rainfall-runoff floods measured by indirect methods on small drainage basins in the conterminous United States. *US-PRC Bilateral Symposium on the Analysis of Extraordinary Flood Events, Nanjing, China.* 42 pp + figs and tables.

CURTIS, W. F. 1985. *Fluvial sediments in northern Oman. Public Authorities Water Resource Report* PAWR 85-15, 76 pp.

EDENS, C. 1986. Prehistory. *In:* DUTTON, R. (ed.) *Rapid Assessment Document, Oman Wahiba Sands project 1985/86. Royal Geographical Society.* Chap. 3.3, 14 pp.

ETHRIDGE, R. G. & SCHUMM, S. A. 1978. Reconstructing paleochannel morphologic and flow characteristics: Methodology, limitations, and assessment. *In:* MIALL, A. D. (ed.) *Fluvial Sedimentology. Memoir of the Canadian Society of Petroleum Geologists, Calgary* **5**, 703–721.

GARDNER, R. 1986. Aeolianite. *In:* DUTTON R. (ed.) *Rapid Assessment Document, Oman Wahiba Sands Project 1985/86. Royal Geographical Society.* Chap. 3.1, 13 pp.

GLENNIE, K. W. 1970. *Desert Sedimentary Environments.* Elsevier, 222 pp.

——, BOEUF, M. G. A., HUGHES CLARKE, M. W., MOODY-STUART, M., PILAAR, W. F. H. & REINHARDT, B. M. 1974. The Geology of the Oman Mountains. *Transactions of the Royal Dutch Geological and Mining Society* **31**, 423 pp.

GOUDIE, A. S. 1983. Calcrete. *In:* GOUDIE, A. S. & PYE, K. (eds) *Chemical Sediments and Geomorphology.* Academic Press, 93–131.

HEY, R. D. 1979. Flow resistance in gravel-bed rivers. *Proceedings of the American Society of Civil Engineers, Journal of the Hydraulics Division* **105**, HY4, 365–379.

HOLM, D. A. 1960. Desert geomorphology in the Arabian peninsula. *Science* **132**, 1369–1379.

HOPSON, C. A., COLEMAN, R. G., GREBORY, R. T., PALLISTER, J. S. & BAILEY, E. H. 1981. Geologic section through the Samail ophiolite and associated rocks along a Muscat–Ibra transect, southeastern Oman Mountains. *Journal of Geophysical Research* **86**, 2527–2544.

HÖTZL, H., MAURIN, V. & ZÖTL, J. G. 1978*a*. Geologic history of the Al Hasa area since the Pliocene. *In:* AL-SAYARI, S. S. & ZÖTL, J. G. (eds) *Quaternary of Saudi Arabia*, Vol. 1. Springer-Verlag, 58–77.

——, KRAMER, F. & MAURIN, V. 1978*b*. Quaternary sediments. *In:* AL-SAYARI, S. S. & ZÖTL, J. G. (eds) *Quaternary of Saudi Arabia*, Vol. 1. Springer-Verlag, 264–301.

JARRETT, R. D. 1984. Hydraulics of high-gradient streams. *Journal of Hydraulic Engineering, American Society of Civil Engineers* **110**, 1519–1539.

JONES, J. R. 1986. *Results of test drilling for water in northwestern Sharqiyah area, Sultanate of Oman, 1983–84. Public Authorities Water Resources Report* PAWR 85-21, 20 pp.

KASSLER, P. 1973. The structural and geomorphic evolution of the Persian Gulf. *In:* PURSER, R. H. (ed.) *The Persian Gulf.* Springer-Verlag, 11–32.

KING, L. C. 1942. *South African Scenery.* Oliver & Boyd.

KNETSCH, G. 1954. Allgemein-geologische Beobachtungen aus Ägypten (1950–53). *Neues Jahrbuch Geologische Paläontologie Abh* **99**, 287–297.

LIMERINOS, J. T. 1970. Determination of the Manning coefficient from measured bed roughness in natural channels. *United States Geological Survey Water Supply Paper.* 1898-B, 47 pp.

MCCLURE, H. A. 1976. Radiocarbon chronology of late Quaternary lakes in the Arabian desert. *Nature* **263**, 755–756.

—— 1978. Ar Rub' Al Khali. *In:* AL-SAYARI S. S. & ZÖTL, J. G. (eds) *Quaternary of Saudi Arabia*, Vol. 1. Springer-Verlag, 252–263.

MAIZELS, J. K. 1983. Palaeovelocity and Palaeodischarge determination for coarse gravel deposits. *In:* GREGORY, K. J. (ed.) *Background to Palaeohydrology.* John Wiley and Sons, Chichester, 101–139.

—— 1986. Palaeochannels. *In:* DUTTON, R. (ed.) *Rapid Assessment Document, Oman Wahiba Sands Project 1985/86. Royal Geographical Society.* Chap. 3.2, 28 pp.

—— 1987*a*. Large-scale flood deposits associated with the formation of coarse-grained, braided terrace sequences. *In:* ETHRIDGE, F. G., FLORES, R. M. & HARVEY, M. D. (eds) *Society of Economic Palaeontologists & Mineralogists Special Publication* **39** (in press).

—— 1987*b*. Modelling of palaeohydrologic change during deglaciation. *Géographie Physique et Quaternaire* **XL**, 263–277.

MILLER, R. P. 1937. Drainage lines in bas-relief. *Journal of Geology* **45**, 432–438.

REEVES, C. C. JR 1983. Pliocene channel calcrete and suspenparallel drainage in West Texas and New Mexico. *In:* WILSON, R. C. L. (ed.) *Residual deposits: surface related weathering processes and materials.* Geological Society of London, Special Publication **11**, 179–183.

ROTNICKI, K. 1983. Modelling past discharges of meandering rivers. *In:* GREGORY, K. J. (ed.) *Background to Palaeohydrology.* John Wiley & Sons, Chichester, 321–354.

SCHUMM, S. A. 1972. Fluvial palaeochannels. *In:* RIGBY, J. K. & HAMBLIN, W. K. (eds) *Recognition of Ancient Sedimentary Environments.* Society of Economic Palaeontologists and Mineralogists, Special Publication **16**, 98–107.

SHIELDS, A. 1936. Änwendung der Ähnlichkeitsmechanik und der Turbulenzforschung auf die Geschiebebewegung. (Application of similarity principles and turbulence research to bedload movement.) (Transl. OTT, W. P. & VAN UCHELEN, J. C.) *United States Department of Agriculture, Soil Conservation Service*, Cooperative Laboratory, California Institute of Technology, 70 pp.

SILBERMAN, E., CARTER, R. W., EINSTEIN, H. A., HINDS, J. & POWELL, R. W. 1963. Friction factors in open channels: Progress Report Task Force on friction factors in open channels of the Committee on Hydromechanics of the Hydraulics Division. *Proceedings of the American Society of Civil Engineers, Journal of the Hydraulics Division* **69**, 97–143.

STALDER, P. J. 1975. Cementation of Pliocene–Quaternary fluviatile clastic sediments in and along the Oman Mountains. *Geologie en Mijnbouw* **54**, 148–156.

STRICKLER, A. 1923. Beiträge zur Frage der Geschwindigkeitsformel und der Rauhigkeitsamen für Ströme, Kanäle und geschlossene Leitungen. *Mitteilungen Amtes Wasserwerkes Bern* **16**, 77 pp.

THORNE, C. R. & ZEVENBERGEN, L. W. 1985. Estimating mean velocity in mountain rivers. *Journal of Hydraulic Engineering, American Society of Civil Engineers* **111**, 612–624.

WARREN, A., JONES, D., GOUDIE, A. & COOKE, R. 1985. Preliminary report of the geomorphological team. *In: Royal Geographical Society Mapping Phase Report, Oman Wahiba Sands Project 1985/86.* 41–124.

WATSON, A. 1983. Gypsum crusts. *In:* GOUDIE, A. S. & PYE, K. (eds) *Chemical Sediments and Geomorphology.* Academic Press, 133–162.

WATTS, N. L. 1980. Quaternary pedogenic calcretes from the Kalahari (southern Africa): mineralogy, genesis and diagenesis. *Sedimentology* **27**, 661–686.

WILLIAMS, G. P. 1984. Palaeohydrologic equations for rivers. *In:* COSTA, J. E. & FLEISHER, P. J. (eds) *Developments and Applications of Geomorphology.* Springer-Verlag, 343–367.

J. K. MAIZELS, Department of Geography, University of Aberdeen, St Mary's, High Street, Old Aberdeen AB9 2UF, Scotland, UK.

Tectonism, climatic change and fluviatile desert sediments

Tectonic control of desert sediments in rift basins ancient and modern

L. E. Frostick & I. Reid

SUMMARY: Desert sediments are frequently associated with continental rifts or aulacogens in the stratigraphic record. In this circumstance, rift structure and the timing and magnitude of tectonic events exert an overriding control on the distribution and character of alluvial sediments. Recent advances in the understanding of rift structure make a reassessment of ideas on the relationship between faulting, drainage patterns and alluvial facies distribution timely. Existing models of symmetrical graben with large alluvial fans shed from the footwalls of planar, normal boundary faults do not accord with observations in recent rifts. For example, half-graben dominate the structure throughout the African rifts. Boundary faults are mostly listric and the associated back-tilting of fault-blocks, combined with the doming that frequently accompanies rifting, causes back-shedding of rivers away from the basins. Alluvial fan development at fault scarps is, therefore, limited. The maximum observed fan radius in Lake Turkana is, for example, only 8 km, or 10% of the half-graben width at that point. The supply of sediment to rift basins can be dominated by axial rivers (*eg* Sugata R. flowing into L. Logipi in Kenya, the Awash R. of the Afar in Ethiopia, and the Rhine R. flowing through the Rhine rift), but there are a surprising number of basins that do not have major axial inputs (*eg* L. Tanganyika and L. Malawi in East Africa and L. Baikal in Siberia). In these cases, the importance of smaller, but more numerous rivers draining across the roll-over becomes obvious. These rivers are subjected to rapid changes in base level and longitudinal gradient: movement of a boundary fault alters the geometry of the lake basin and base-level falls as the lake adjusts. This is accompanied by the aggradation of coarse alluvial deposits across the roll-over. Clast size gradients in these deposits are well defined for the region between the basin margin and the main depocentre, with typical measured values of -3 mm km^{-1}. During the long, intervening periods of tectonic quiescence these and other deposits are reworked and redistributed downslope and offshore. Deltas build out into the lake, but microstratigraphy reveals that they are not of the classic Gilbert type. Short- and long-term fluctuations in lake level cause large-scale migration of the shoreline so that lake muds and alluvial sands are interdigitated over many kilometres in sequences which vary from a few metres to tens of metres in thickness. A synthesis of detailed work carried out on the sedimentology and geomorphology of a modern rift of the common half-graben type provides a new model of sedimentation that will prove useful in interpreting ancient deposits with hydrocarbon potential.

Continental rifts represent an important early phase of passive margin development (Le Pichon & Sibuet 1981). Economically, they are often the location of significant reserves of oil and gas (*eg* in the North Sea). They also represent an early end-member in the stages of crustal evolution that culminate in continental fragmentation (Burke 1977; Milanovsky 1981). Consequently, the formation of continental rifts is almost invariably associated, either directly or indirectly, with the initial stages of ocean development. For example, the Benue Trough of W and Central Africa is a Cretaceous rift associated with the opening of the S Atlantic (Olade 1975; Wright 1981; Nwachukwa 1985; the Viking Graben of the North Sea is an older rift dating back at least to the Permian and a product of early development of the N Atlantic suture (Ziegler 1981; Dewey 1982; Badley *et al.* 1984).

Continental rifts provide topographical sinks that both control the deposition, and ensure the preservation of non-marine sediments world-wide. In the absence of the active basin subsidence that is associated with these highly localized features of present-day and ancient peneplaned cratons, aeolian and fluvial deposits would be reworked constantly, leaving little record in the stratigraphical column. It is not surprising, therefore, that desert and semi-desert sequences are often best and most completely preserved throughout geological time in continental rifts, *eg* the Old Red Sandstone of Scotland, the Permo–Triassic rocks of the northern North Sea, and the Mio–Plio–Pleistocene sediments of the East African Rift. In fact, prior to the evolution of the higher land plants, all continental sediments were probably developed in environments where surface processes, if not necessarily the climate, were more akin to present-day drylands, devoid as they would have been of the protection

From FROSTICK, L. & REID, I. (eds), 1987, *Desert Sediments: Ancient and Modern*, Geological Society Special Publication No. 35, pp. 53–68.

afforded by vegetation (*eg* Torridonian deposits of Scotland). For this reason, there are great benefits to be gained from a detailed comparison of modern rift basins in desert settings and their ancient counterparts.

The comparison is made even more rewarding because the geomorphic setting that dictates patterns of sedimentation—controlled as it is by structure— is remarkably similar from one rift system to another. Even at a gross continental scale there are common characteristics. For example, a juxtaposition of the fault patterns of the North Sea rifts (Permian to Jurassic) and the East African system (Tertiary to Recent) (Fig. 1) reveals a common pattern of bifurcation and widening northward from a triple point (Quads 16 & 22 in the North Sea; the north Kenya dome in East Africa). In both cases, the boundary faults are discontinuous and often curved (a pattern that is more obvious in the North Sea structure, but only due to the differing scaling in Fig. 1). Besides this, major faulting occurs generally on only *one* rift margin. As a result, the basins that make up the rift are typically asymmetrical, *ie* half-graben.

This similarity in structural style must be reflected in the character of the sedimentary fills. In East Africa, the influence of faulting on topography and, therefore, on drainage pattern and direction, on river character and sediment supply, and on the location of depocentres has left strong marks on the Cenozoic sedimentary record. There are lessons here for the interpretation of ancient sequences, especially since the relationships between tectonic adjustment and sedimentation are often blurred by the consequences of later events and are therefore rarely clear in these older rift deposits. Recent seismic data from both Africa (Rosendahl & Livingstone 1983; Rosendahl *et al.* 1986) and the North Sea (Gibbs 1984) have clarified ideas about rift structure. This, together with extensive research on rift basin sediments and sedimentation in East Africa (Frostick & Reid 1979, 1980; Cohen *et al.* 1986; Reid & Frostick 1986; Tiercelin 1986; Williams *et al.* 1986; Williamson & Savage 1986) has inspired this reassessment of arid-zone continental rift sedimentation. It provides a new model that gives a satisfactory explanation for lithotype distributions that have remained anomalous in existing sedimentary models.

Rift structure and geomorphology

The topographic expression of continental rifts—deep and steep-sided valleys often developed in areas of otherwise subdued relief (*eg* the Rhine Rift, Cloos 1955; Sittler 1969; and the Suez Rift, Robson 1971)—has provoked considerable geomorphological and geological interest. These topographic lows are long-lived because successive periods of crustal extension re-utilize existing

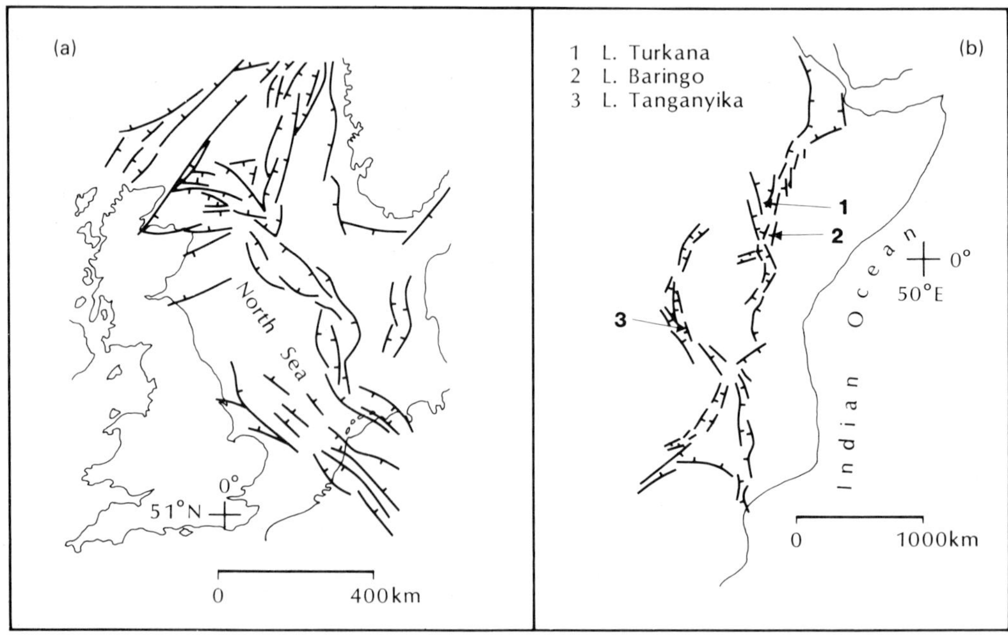

FIG. 1. A comparison of the major faults of the North Sea (*a*) and East Africa (*b*). Note the different scales.

fault-lines. As a result they become sites for the accumulation of thick and varied sequences of sediment. The cross-sectional morphology of those rift valleys that were recognized as such by early geologists led to the development of a classic structural model. In this, the steep valley sides are projected subsurface as two normal planar faults (Fig. 2) and the floor of the valley is depicted as a down-faulted keystone. This is the full graben model that has become embedded in the literature (*eg* Cloos 1930; Cogne *et al.* 1966; Illies 1981). The facies models that have accompanied it have focused attention on valley symmetry and especially on the importance of the flanking alluvial fans that issue from opposing fault scarps (Fig. 2*b*).

However, not everyone has subscribed faithfully to these ideas. Carey (1958), for example, not only noted the cross-sectional asymmetry of some rift basins but also suggested that there was a curious curvature to the boundary faults. With more recently available seismic data, it has been possible to reassess the universality of the classical graben model and its implications for patterns of sedimentation.

In plan, the fault patterns in rifts are always complex (Fig. 1) and at first sight appear to be only vaguely ordered. Closer examination reveals not only that the faults are frequently curved (shown schematically in Fig. 3*b*), but also that the rift is composed of segments. In any single segment, *one* boundary fault is dominant. Subsurface data confirm that the major structure is a half-graben, rather than a full graben (Gibbs 1984; Rosendahl *et al.* 1986; Fig. 4). Furthermore, seismic sections show that the boundary faults are generally listric in form and that they sole in low angle, deep crustal detachments. The roll-over created by the half-graben is disrupted by smaller antithetic and synthetic faults, and these lead to the localized development of small-scale horst and graben. The back-tilting associated with listric boundary faulting and accompanying footwall collapse, combines with the regional doming that frequently accompanies rifting (Fairhead 1986) to produce a surface gradient *away* from the rift basin, diverting drainage and starving it of sediment on the faulted margin (Fig. 3*b*). In addition, 'saddle' basins may develop perched between a series of back-tilted blocks. Again, these act as important drainage diversions.

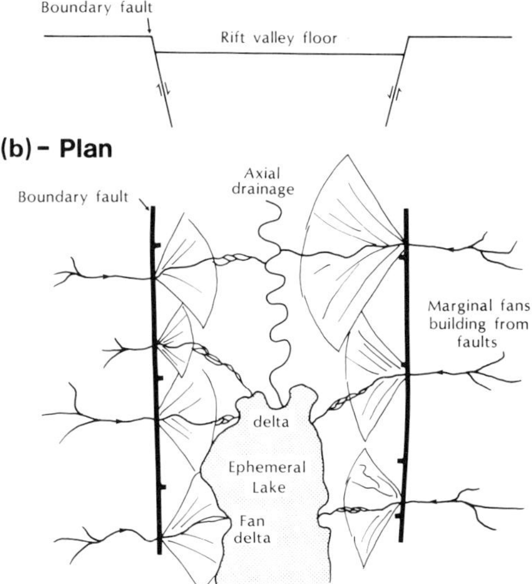

FIG. 2. Schematic diagram of the classic full-graben rift valley model: (*a*) structure and (*b*) sedimentation.

The faults that bound the half-graben are discontinuous. Both Gibbs (1984) and Rosendahl *et al.* (1986) have shown that they alternate from one side of the rift to the other. This pattern is shown schematically in Fig. 3*b*. Each half-graben segment is rarely more than 100 km long (Fig. 4) and is separated from its neighbours by features running transverse to the main rift trend. The structure of these features remains enigmatic; they have been variously described as transfer faults (Gibbs 1984) and as accommodation zones (Rosendahl *et al.* 1986). Often, the many faults which must occur in these zones are too small to be detected in seismic sections. Their structural function is to facilitate the alternating polarity of the boundary faults. Their geomorphic function is to provide a corridor through which diverted 'saddle' drainage can finally gain access to the rift. Because of this, they are often the sites of large deltas (*eg* the Turkwel delta of L. Turkana, Kenya) and of fans (*eg* the large Jurassic submarine fans in the West Shetland Basin of the European continental shelf). The sedimentary function of transfer zones is to partition the rift into separate sub-basins, in each of which a different, yet contemporary set of sediments can be deposited.

The linked half-graben structure of rifts is well illustrated by two African lake basins—Turkana and Tanganyika (Fig. 4). The shape of the lakes is strikingly similar. The alternating polarity of the listric boundary faults and the cross-sectional asymmetry of the sub-basins both show very clearly on the seismic interpretations (Tanganyika) and sketch cross-sections (Turkana) of Fig. 4. Lake Turkana consists of three structurally distinct sub-basins; Tanganyika, a much larger lake, contains at least six.

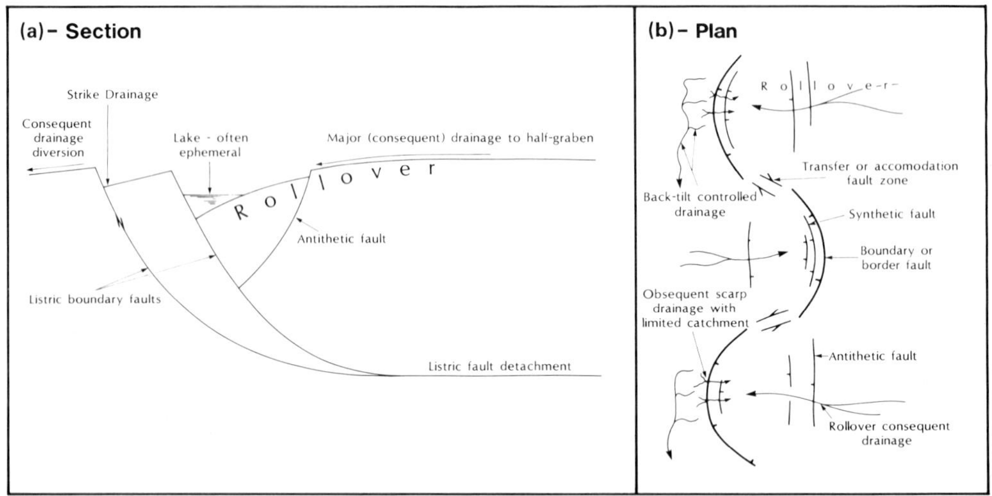

FIG. 3. Schematic diagram of the listric half-graben model of rift structure: (*a*) cross-section and (*b*) plan (after Gibbs 1984 and Rosendahl *et al.* 1986). The effect of faulting on patterns of drainage are shown in (*b*).

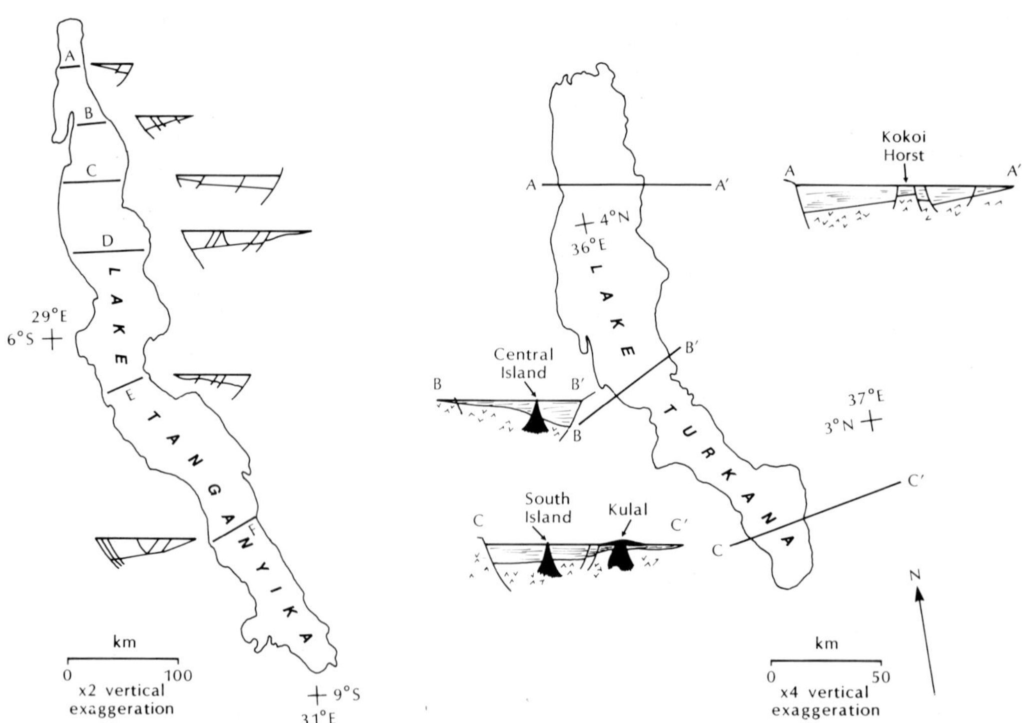

FIG. 4. Sketch sections through the sedimentary basins of L. Turkana (right) and L. Tanganyika (left). Those of Tanganyika are derived from the seismic sections in Rosendahl *et al.* (1986).

Rifting and regional drainage

In arid-zone continental rift basins, rivers are the main suppliers of sediment. There may be long-travelled exotic perennial rivers; these will depend for their flow upon a different rainfall regime upstream and beyond the desert margin, *eg* the Jordan draining towards the Dead Sea and the Omo draining the Ethiopian highlands and entering L. Turkana at its northern end. However, ephemeral streams dominate the landscape (Reid & Frostick 1986). Despite the sporadic nature of their flood regime, these streams transport large quantities of sediment. For example, the measured concentrations of suspended sediment of flash floods in the L. Turkana catchment exceed 15 g l^{-1} (Frostick *et al.* 1983; Reid & Frostick 1987). Sediment yields are characteristically high, lying between 850 and 1600 t km^{-2} a^{-1} (Reid & Frostick 1986). Because these streams are ubiquitous, and because their sediment loads are so high, they are responsible for depositing a larger fraction of rift sediments than would be imagined from their comparatively short length and from the fact that they might flow on only a few occasions in any one year.

Rift structure exerts a dominant control on both local and regional drainage and, therefore, on erosion and deposition. This is seen clearly in recent rifts. Figure 5 shows the present-day catchment of L. Tanganyika; major boundary faults, as determined by Rosendahl *et al.* (1986), are also shown. Although L. Tanganyika is set in an environment that is far from desertic, it provides important lessons about the interplay of structure and drainage. This part of the E African rift system is narrow, and the contributing water catchment is confined by the proximity of the Congo watershed to the W and the L. Victoria basin and L. Rukwa rift to the E. Focusing at the scale of the individual river basins, it is clear that each boundary fault acts as a barrier to drainage and, hence, to sediment transfer across the basin margin. Alluvial fans associated with the rift scarps are few and small. Instead, major river inputs come from the roll-over zones (*eg* the Malagarasi R.). There is little axial drainage.

In other sections of the East African Rift, both axial and diverted 'saddle' drainage are more important. Figure 6 depicts the main structural elements and drainage of the northern Gregory Rift between L. Bogoria and L. Turkana. Here, the rift is broad and diffuse. At latitude 0°40′N (L. Baringo), the main boundary fault lies on the western margin. This is the Kamasia Fault with a throw of over 2000 m (Chapman *et al.* 1978). N and S of the Kamasia Fault, half-graben polarity switches and the major faults lie on the eastern margin. The drainage is more complex than that around L. Tanganyika. For instance, L. Baringo does not sit against the main local boundary fault. Instead, it lies in a shallow basin between antithetic and synthetic faults on the Kamasia roll-over (Fig. 6). The Kerio R. is a strike stream draining a saddle on the back-tilted blocks that lie between the Kamasia and its accompanying

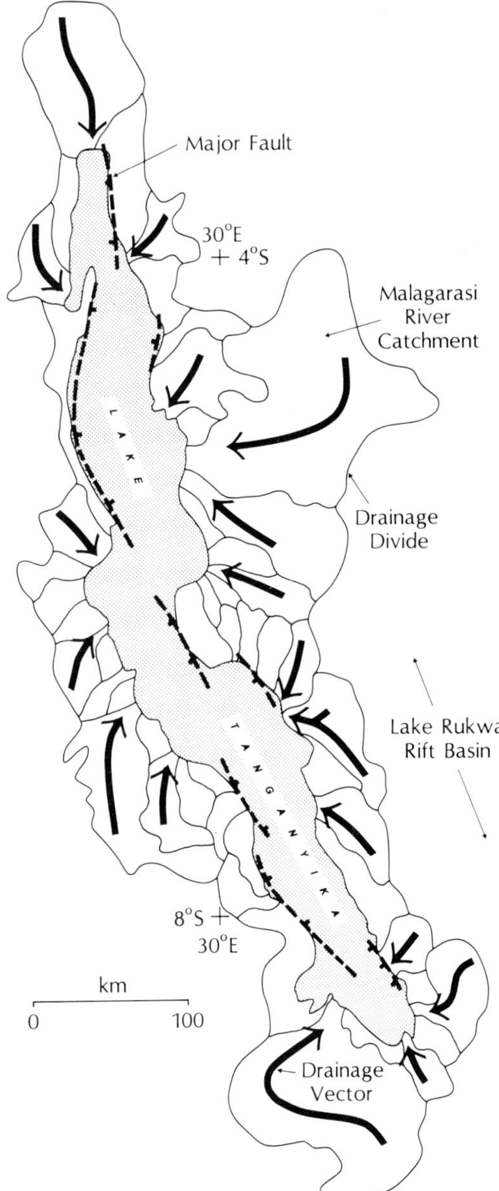

FIG. 5. Map of L. Tanganyika showing the relationships between the main faults, the main river basins feeding the lake, and the main directions of river drainage (drainage vectors).

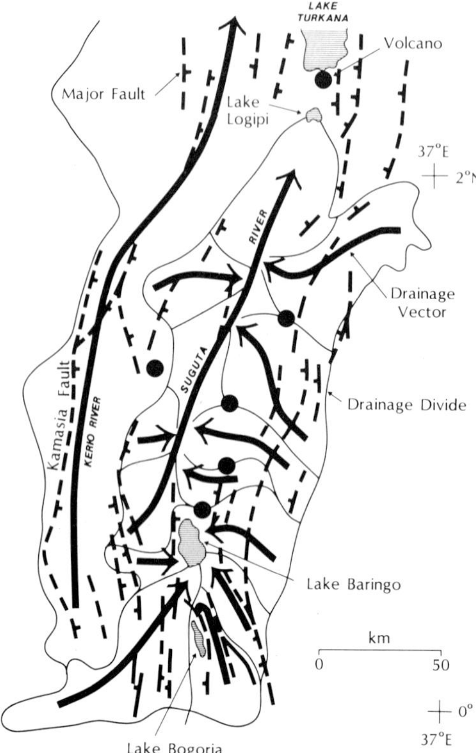

FIG. 6. Map of the Gregory Rift between Lakes Bogoria and Turkana showing the relationships between the main faults, the main river catchments, and main directions of river drainage (drainage vectors).

synthetic faults. Back-tilting ensures that the Kerio receives few left bank tributaries as it flows northwards, and it eventually gains access to the rift in the L. Turkana basin after a course of 250 km. By the time it reaches L. Turkana, transmission losses and evaporation have significantly reduced its discharge. As a result, its sediment input to the rift is much smaller than would be suggested by its length (Cohen *et al.* 1986). The Suguta R. is axial in this section of the East African Rift. It has its headwaters close to L. Baringo, but it is diverted away from this lake by faulting and volcanism (Fig. 6). Because of this, it drains northwards towards L. Turkana, only stopping short because of a volcanic barrier formed approximately 10 000 years ago (Truckle 1976). However, in reality, the extreme aridity of the Lower Suguta Valley combines with the uncertainty of rainfall in its semi-arid headwaters to ensure that it rarely runs to L. Logipi. Axial drainage is, therefore, less significant as a determinant of sedimentation here than would appear merely from an inspection of a map of drainage. It is interesting to speculate that the same will have been true of ancient continental rift systems in arid zones.

These two examples from the East African Rift highlight many of the important regional affects of tectonic activity on drainage and, therefore, on sedimentation. Boundary faults are *not* the major source of large volumes of detritus dumped directly into the basins, at least during the early phases of rifting. In fact, rather than representing the edge of a topographical sink that attracts regional drainage, they often become barriers to drainage. As a result, the faulted margin may become sediment starved. Instead of being the site of extensive wedges of coarse sediment they are characterized by small, steep alluvial fans that are built by short scarp-face streams. The main supply of sediment to the rift basin comes from two other directions: there may be significant axial drainage which contributes fine material brought as suspended load in low-gradient, far-travelled rivers (*eg* the Omo, Butzer 1971); on the other hand, there are undoubtedly the numerous shorter but steeper streams which cross the half-graben roll-over and carry with them heavy loads of coarser material.

The importance of roll-over rivers is clear on satellite imagery and air photographs of the L. Turkana basin shown in Figs 7 and 8. One of the most extensive sediment accumulations in the basin is that of Koobi Fora on the NE margin of the lake (Figs 7*b* and 8). It is situated on a half-graben roll-over, opposite the main boundary fault that defines the E-facing scarp of the Laburr Range (Figs 7*a* and 9). It is across this roll-over that the Sagan R., at one time in the past, carried Ethiopian water through the Bakate Gap to L. Turkana (Frostick & Reid 1980; Harvey & Grove 1982). Further S, basin polarity changes, with the boundary fault now on the eastern margin (Fig. 7). Here, the Turkwel crosses a roll-over, this time from the W, and has contributed in part to the extensive Miocene–Recent sediments of the Lothagam embayment.

Half-graben sedimentation: a case study

Patterns of sedimentation in the listric half-graben of arid-zone continental rifts are complex. Data on which facies models might be based are sparse. The Koobi Fora sedimentary basin is a useful exception. It has been studied intensively because of a well-preserved fossil vertebrate fauna that include numerous hominids (Leakey & Leakey 1978). After many years of detailed research by ourselves and others (see *eg* Bishop

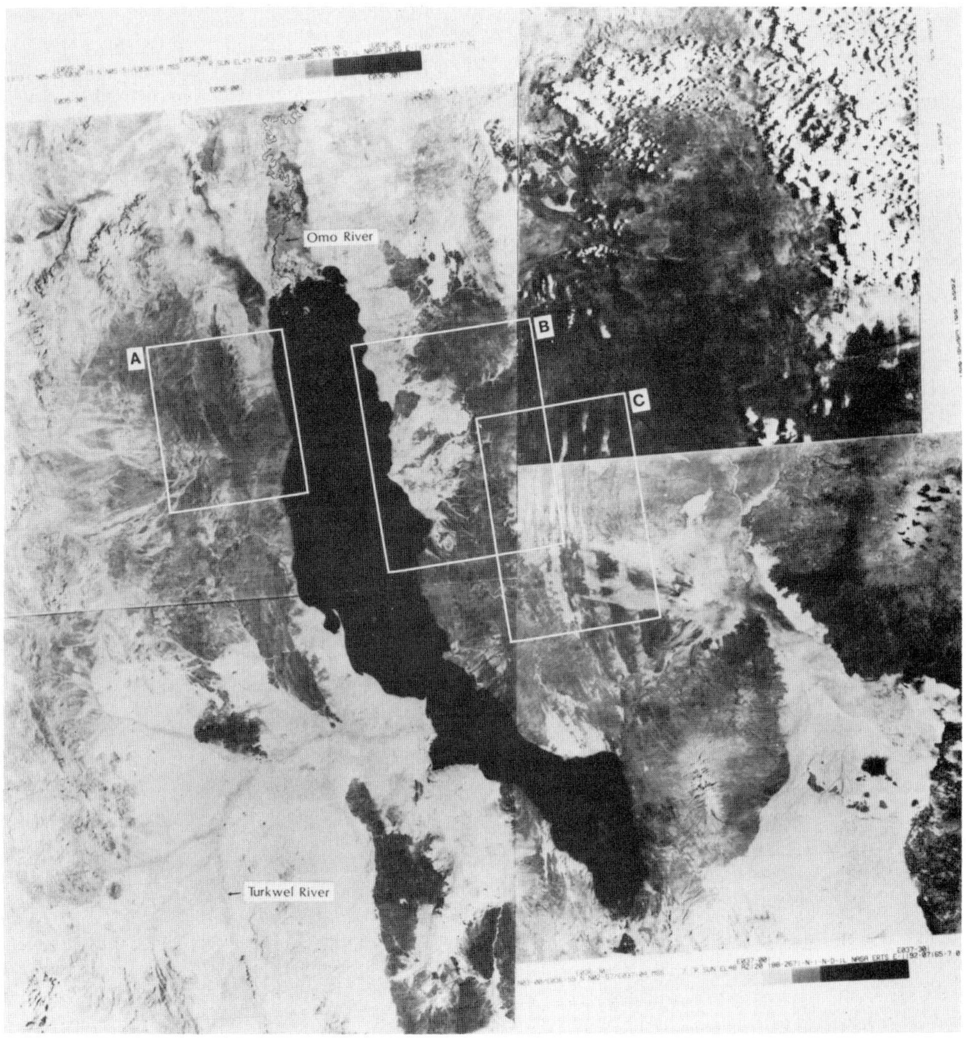

FIG. 7. *Landsat* mosaic of the L. Turkana extension of the Gregory Rift. Areas labelled A and B are shown in more detail in Figs 9 and 8, respectively. Area C encloses the complex *en echelon* faulting of Kino Sogo.

1978; Fostick *et al.* 1986), it is now possible to present a synthesis of sedimentation in an arid-zone continental rift that might have general value.

Satellite imagery (Fig. 7) shows that the Koobi Fora basin is situated at a point where the East African Rift broadens northwards. The most active tectonic zone appears to have shifted eastwards since the Miocene (Cerling & Powers 1977) and now sits largely E of L. Turkana, running from the southern end of the lake through Kino Sogo (Fig. 7, labelled C) up to Chew Bahir (formerly L. Stefanie). The breadth of the L. Turkana basin—over 80 km at its widest part—has helped to clarify patterns of sedimentation. This is because contributions from the eastern and western margins are largely kept separate, and only interbed with deposits brought by the axial Omo. The lake itself occupies the deepest part of the basin, its present western shore is within 12 km of the Laburr fault. The Omo R. enters from the N and supplies large volumes of fine sediment (Yuretich 1979) but it contributes little to the thick sequences of silts, sands and gravels which are widespread in the basin. This coarser material is supplied in the main by steeper streams which drain, or have drained in the past, across the roll-over, debouching into the lake to

FIG. 8, Aerial photo-mosaic of the Koobi Fora sedimentary basin, NE of L. Turkana. The area depicted is that of Box B in Fig. 7. The Bakate corridor, now a wind gap but at one time cut by the Sagan R. draining from Ethiopia, can be seen crossing the basalt plateau of Suregei-Gombe in the upper right quadrant.

form a series of deltas. Sedimentation in these deltas, and along the rest of the shore, is complex (Frostick & Reid 1986). This is due to various factors which include the sporadic supply of sediment from ephemeral flows, the influence of periodic tectonism, and the inherent instability of base-level in closed basins where the vagaries of climatic change can lead, together with tectonism and volcanism (Watkins 1986) to substantial changes in lake level (Butzer 1971; Reid & Frostick 1985, 1986). The significance of the last factor cannot be overstressed. Changes in lake level range from an annual rise and fall of *c*.1 m to the dramatic changes in excess of 100 m that have occurred over only a few thousands of years during the Pleistocene. These fluctuations

Fig. 9. Aerial photo-mosaic of the NW shore of Lake Turkana showing the fault scarp (Laburr Range) and small alluvial fans. Note the saddle drainage of the Kachoda River behind the main fault scarp. The area depicted is that of Box A in Fig. 7.

have left sedimentary signatures in the stratigraphical record which have been documented (Frostick & Reid 1986; Owen & Renaut 1986).

The faulted margin

Turning first to the faulted basin margin (Fig. 9), it is possible to see that most of the drainage is diverted away from the lake to run N–S in strike saddle basins. Only small fan-deltas of less than 8 km radius build out from the scarp. Their extent is less than 10% of the total basin width. They encroach on fine lake sediments which have been supplied to a large extent by the axial Omo R. As a result, coarse and fine sedimentary facies pass abruptly into each other both laterally and vertically.

The Roll-over

During periods of active rifting, the power of the rivers that drain across the roll-over suddenly increases as their gradients are increased and as the lake regresses to occupy a deepened basin. This encourages erosion and produce a 'wave' of coarse detritus which progrades into the rift (Phase I, Fig. 10). These tectonically-induced periodic influxes of coarse sediment are readily identified in the Plio–Pleistocene record of the Koobi Fora basin. During the past 3 Ma, there have been two such waves of sediment (*c.*2.4 and 0.8 Ma, Fig. 11). Both have produced similar flat-topped basin-wide spreads of pebble conglomerate (Fig. 11). Detailed petrography shows that these conglomerates were deposited from braided streams, close to base-level and heavily loaded with locally-derived detritus (Frostick & Reid 1980). The particles are mainly of basalt torn from the Miocene to Pleistocene flows which ring the basin. They are characterized by a rapid, linear, downstream decrease in clast size (Fig. 11).

Interjacent with these coarse spreads of conglomerate are more complex sequences of finer sediments. They are interbedded gravels, sands and silts with numerous erosional surfaces. They represent reworking of the coarser episodic sediment inputs during the long periods of quiescence that separate the short sharp shocks of tectonism. During these long intervening periods, sandy alluvial, deltaic and shoreline sediments encroach upon the lake (Fig. 12). This is Phase II of Fig. 10. These sequences become progressively more quartz-rich through repeated cycles of reworking, even though the climate is arid and weathering rates are low. The labile lithic fragments and feldspars that are an original component are gradually broken down by attrition (Frostick & Reid 1980; Cohen *et al.* 1986). This is consistent with the patterns found in ancient semi-arid and arid zone continental rift sediments (*eg* the Torridonian sandstones of NW

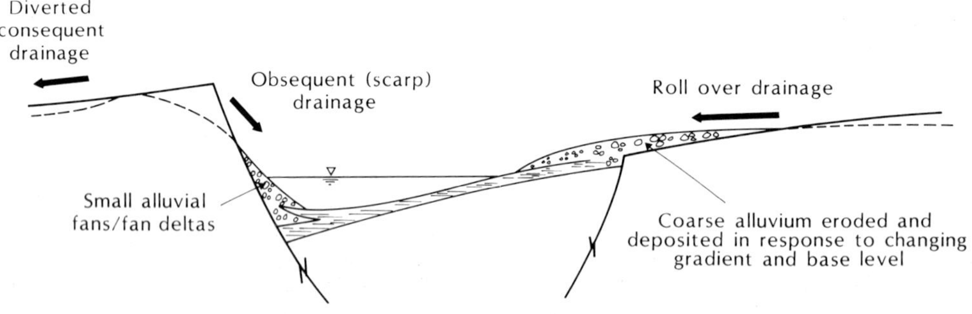

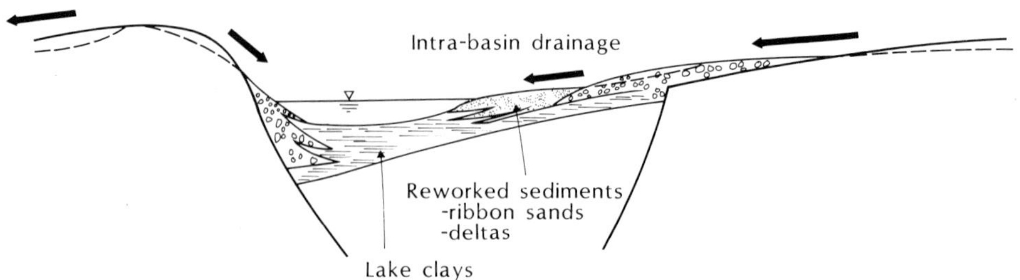

FIG. 10. Schematic diagram of the phases of sedimentation in an active rift setting. Phase I immediately postdates fault movement; Phase II occurs in the long periods between successive spasms of fault activity.

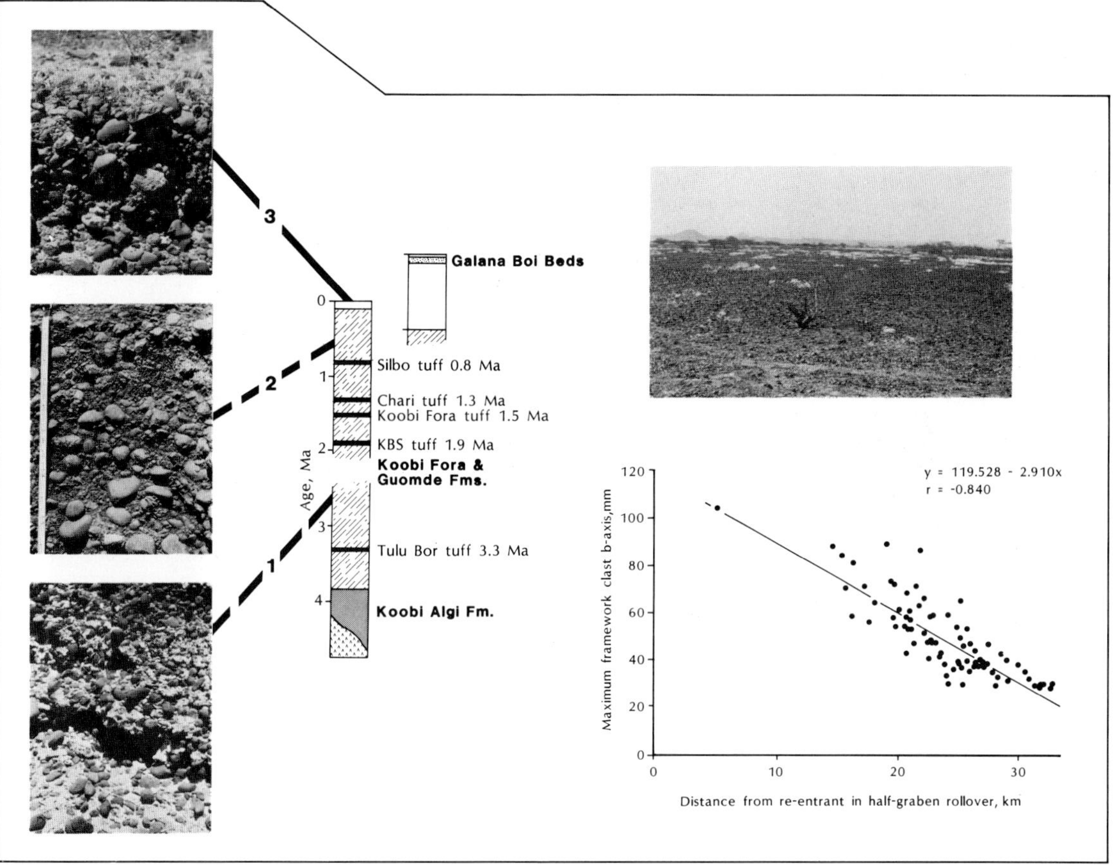

FIG. 11. Plio–Pleistocene stratigraphy of the Koobi Fora sedimentary basin (centre) after Cerling & Brown (1982); sections (left) through basin-wide gravel sheets laid down during the early (1) and late (2) Pleistocene compared with the gravel train of a modern braided stream (3); view of the 0.8 Ma gravel sheet (top right); grain size of the 0.8 Ma gravel sheet showing the linear and rapid downstream fining that is characteristic of a braided stream deposit.

Scotland). However, it is of interest that the rapid induration that characterizes desert environments on the one hand (Frostick & Reid 1979; Walker *et al.* 1978), and the violent action of flash floods on the other, can mean that the sediments are intraclast-rich. This is common, *eg* in the desert sediments of the Trias of the Moray Firth of NE Scotland. The intraclasts are short-travelled, moving no more than tens of metres from one reworked position to another within Phase II of Fig. 10.

The complexity of the reworked deposits increases downstream towards the lake shore (Fig. 13). In proximal locations, interbeds are simply coarse and fine river channel and overbank deposits. More distally, the alluvial sediments are interbedded with deltaic and beach sands and with lagoon and lake silts and clays. Bioturbation is common here, and organic material (rare in most desert deposits) increases in importance (Fig. 13).

The close-set interbedding is caused by periodic fluctuations in lake level. Because gradients on the delta plain are generally low, the shoreline migrates over considerable distance in response to very small changes in water stage. As a consequence, delta sequences carry the marks of frequent transgressions and regressions (Frostick & Reid 1986).

Conclusions and general model

Recent advances in our understanding of rift structure have led us to reappraise available models of arid-zone rift sedimentation. The

FIG. 12. View from the eastern shore of L. Turkana across the Tulu Bor delta towards the fault scarp of the Laburr range on the western shore.

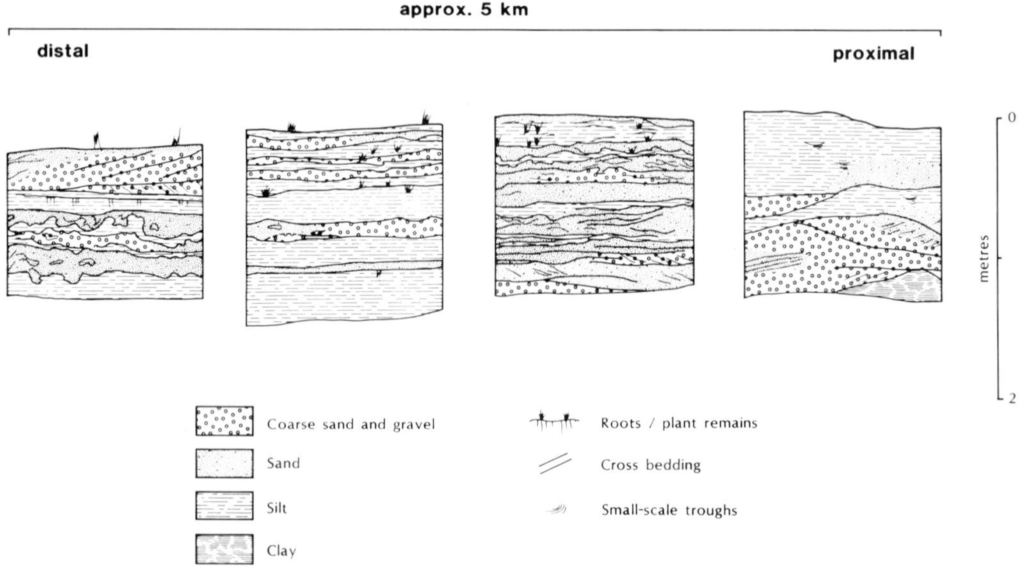

FIG. 13. Representative measured sections through the sediments of the Tulu Bor delta, showing downstream changes in sediment type and in sequence complexity as the lake shore is approached (towards left). Note the bioturbation of the lake shore deposits (far left) and the relatively simple sequence of the alluvial deposits (far right).

fundamental structural unit is a half-graben produced largely by movement along a listric boundary fault. As a result, the basin-fill is asymmetrical and the facies that develop on the two margins of the rift are markedly different. Besides this, the rift 'valley' is made up of a series of *linked* half-graben whose polarity alternates. In this structural setting, axial drainage is not always a dominant component of the sediment delivery system as is implied by existing models (*eg* that of Bridge & Leeder 1979). Individual half-graben may, in fact, form localized and separate depocentres. The integration of axial drainage comes later, only if sedimentation rates exceed rates of subsidence and the half-graben fill with sediment.

A new model of arid-zone continental rift sedimentation is presented in Fig. 14. A notable difference between this and its precursors is that the alluvial fans building out from the boundary fault are limited in extent. The back-tilting of the rift margin in response to both complex listric faulting and the doming that accompanies early rifting tends to divert rivers away from this margin. As a result, it is comparatively starved of detritus. Fans and fan-deltas are built by short, scarp-face streams which head close to the boundary fault. They prograde onto fine lake sediments with which they are interbedded, forming a relatively simple alternating coarse–fine sequence (Fig. 14, left). The fines generally represent periods of tectonic quiescence, while

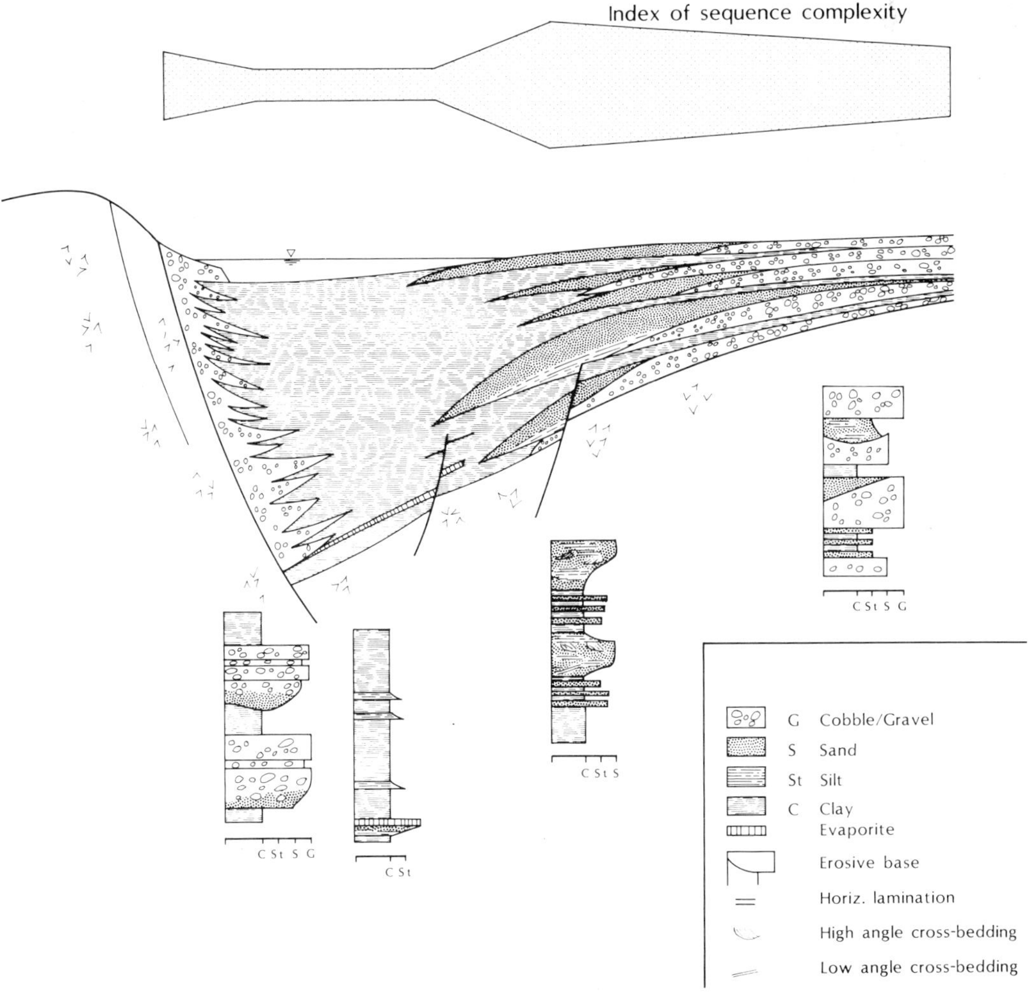

FIG. 14. Model of sedimentation in a half-graben containing a lake. Characteristic but schematic sedimentary sequences are given for four locations in the rift cross-section. The index of sequence complexity is based on the number of facies expected in a sequence typical of each location.

the encroachment of coarse debris in the fans/fan-deltas reflects an increase in potential energy in the scarp drainage following episodes of faulting. Further from the margin, the lake deposits are largely laminated silts and clays. There may also be evaporite sequences developed during periods of increased aridity and low lake levels (Fig. 14, centre left).

Much of the coarser sediment in the basin is supplied by ephemeral streams, or possibly by far-travelled seasonal and even perennial rivers, draining across the roll-over. Periods of active rifting are marked by the progradation of very coarse alluvial sediments (Fig. 14, right). Intervening periods of tectonic quiescence are marked by the progressive reworking of these coarse sediments into sandy alluvial, shoreline and deltaic deposits which become interleaved with lake fines (Fig. 14, centre right).

During much later phases, as rifting gives way to passive margin development, the patterns of sedimentation will inevitably alter. The infilling of saddle basins on the faulted margin may combine with general subsidence of the rift valley shoulders to allow larger quantities of coarse sediment across the faulted margin (see *eg* the Jurassic of the Brae field in the Viking Graben, Harmes *et al.* 1981). These later phases may also be marked by transition from continental to marine conditions.

The sedimentary sequences which develop at different distances from an active rift fault during the early phases of activity are very characteristic of their location. Where the basin contains an ephemeral or permanent lake, they are most complex close to the lake margin but on the side *away* from the boundary fault. Even in the absence of a lake, streams crossing the roll-over in this same general part of the basin cross-section must respond to fault movements; the resulting deposits will show evidence of complex cut-and-fill at both small and large scale. Away from the roll-over lake margin in both directions the sedimentary sequence becomes simpler. This has been shown schematically in Fig. 14 (top) where broad categories of facies type making up a recognizable and repeated sequence are summed for each location to given an index of 'sequence complexity'. This index largely reflects our observations of sediments and sedimentation in a recent rift. However, it offers the possibility of identifying approximate position relative to a controlling boundary fault in ancient arid-zone rifts where data are incomplete and scanty. It should prove useful, for example, where information is limited to a number of isolated offshore wells.

References

BADLEY M. E., EGEBERT, T. & NIPEN, O. 1984. Development of rift basins illustrated by the structural evolution of the Oseberg feature, Block 30/6, offshore Norway. *Journal of the Geological Society of London* **141**, 639–649.

BISHOP, W. W. (ed.) 1978. *Geological Background to Fossil Man.* Scottish Academic Press, Edinburgh.

BRIDGE, J. S. & LEEDER, M. R. 1979. A simulation model of alluvial stratigraphy. *Sedimentology* **26**, 617–644.

BURKE, K. 1977. Aulacogens and continental breakup. *Annual Review Earth and Planetary Sciences* **8**, 371–396.

BUTZER, K. W. 1971. *Recent history of an Ethiopian delta. University of Chicago Research Paper of the Department of Geography* **136**, 184 pp.

CAREY, S. W. 1958. The tectonic approach to continental drift. *In:* CAREY, S. W. (ed.) *Continental drift—a symposium.* Hobart, Australia, University of Tasmania, 173–363.

CERLING, T. E. & BROWN, F. H. 1982. Tuffaceous marker horizons in the Koobi Fora region and the Lower Omo valley. *Nature* **299**, 216–221.

—— & POWERS, D. W. 1977. Palaeorifting between the Gregory and Ethiopian rifts. *Geology* **5**, 441–444.

CHAPMAN, G. R., LIPPARD, S. J. & MARTYN, J. E. 1978. The stratigraphy and structure of the Kamasia Range, Kenya Rift Valley. *Journal of the Geological Society of London* **135**, 265–281.

CLOOS H. 1930. Zur experimentellen Tektonik. *Natwwissenschaften* **18**, 741–747.

—— 1955. Geologische struckturkarte der Mittelgebirge. *Geologische Rundschau* **44**, 480.

COGNE, J., GEZE, B., GOGUEL, J., GROLIER, J., LETOURNEUR, J., PELLET, J., ROTHE, J. & SITTLER, C. 1966. Les "rifts" et les failles de decrochement en France. *Revue Geographie et Physical Geologie Dynamique sere* **28**, 123–131.

COHEN, A. S., FERGUSON, D. S., GRAM, P. H., HUBLER, S. L. & SIMS, K. W. 1986. The distribution of coarse-grained sediments in modern Lake Turkana, Kenya: implications for clastic sedimentation models of rift lakes. *In:* FROSTICK, L. E., RENAUT, R. W., REID, I. & TIERCELIN, J. J. (eds) *Sedimentation in the African Rifts. Special Publication of the Geological Society* **25**, 127–139.

DEWEY, J. F. 1982. Plate tectonics and the evolution of the British Isles. *Journal of the Geological Society of London* **139**, 317–412.

FAIRHEAD, J. D. 1986. Geophysical controls on sedimentation within the African rift systems. *In:* FROSTICK, L. E., RENAUT, R. W., REID, I. & TIERCELIN, J. J. (eds) *Sedimentation in the African Rifts*. Special Publication of the Geological Society **25**, 19–27.

FROSTICK, L. E. & REID, I. 1979. Drainage net control of sedimentary parameters in sand-bed ephemeral streams. *In:* PITTY, A. F. (ed.) *Geographical Approaches to Fluvial Processes*. Geobooks, Norwich, 173–201.

—— & —— 1980. Sorting mechanisms in coarse-grained alluvial sediments: fresh evidence from a basalt plateau gravel, Kenya. *Journal of the Geological Society of London* **137**, 431–441.

—— & —— 1986. Evolution and sedimentary character of lake deltas fed by ephemeral rivers in the Turkana basin, northern Kenya. *In:* FROSTICK, L. E., RENAUT, R. W., REID, I. & TIERCELIN, J. J. (eds) *Sedimentation in the African Rifts*. Special Publication of the Geological Society **25**, 113–125.

——, —— & LAYMAN, J. T. 1983. Changing size distribution of suspended sediment in arid-zone flash floods. *Special Publication of the International Association of Sedimentologists* **6**, 97–106.

——, RENAUT, R. W., REID, I. & TIERCELIN, J. J. (eds) 1986. *Sedimentation in the African Rifts*. Special Publication of the Geological Society **25**.

GIBBS, A. D. 1984. Structural evolution of extensional basin margins. *Journal of the Geological Society of London* **141**, 609–620.

HARMES, J. C., TACKENBERG, P., PICKLES, E. & POLLOCK, R. E. 1981. The Brae oil field. *In: Petroleum Geology of the Continental Shelf of North-West Europe*. Institute of Petroleum, London, 352–357.

HARVEY, R. & GROVES, A. T. 1982. A prehistoric source of the Nile. *Geographical Journal* **148**, 327–336.

HAY, J. T. C. 1978. Structural development in the north North Sea. *Journal of Petroleum Geology* **1**, 65–77.

ILLIES, J. H. 1981. Mechanism of graben formation. *Tectonophysics* **73**, 249–266.

LE PICHON, X. & SIBUET, J. C. 1981. Passive margins: a model of formation. *Journal of Geophysical Research* **86**, 3708–3720.

LEAKEY, M. G. & LEAKEY, R. E. (eds) 1978. *Koobi Fora Research Project 1. The Fossil Hominids and an Introduction to their context—1968–74*. Clarendon Press, Oxford, 191 pp.

MILANOVSKY, E. E. 1981. Aulacogens of ancient platforms; problems of their origin and tectonic development. *Tectonophysics* **73**, 213–248.

NWACHUKWA, J. I. 1975. Petroleum prospects of the Benue Trough, Nigeria. *American Association of Petroleum Geologists Bulletin* **69**, 601–609.

OLADE, M. A. 1975. Evolution of Nigeria's Benue Trough (aulacogen): a tectonic model. *Geological Magazine* **112**, 575–583.

OWEN, R. B. & RENAUT, R. W. 1986. Sedimentology, stratigraphy and palaeoenvironments of the Holocene Galana Boi Formation, NE Lake Turkana, Kenya. *In:* FROSTICK, L. E., RENAUT, R. W., REID, I. & TIERCELIN, J. J. (eds) *Sedimentation in the African Rifts*. Special Publication of the Geological Society **25**, 311–332.

REID, I. & FROSTICK, L. E. 1985. Beach orientation, bar morphology and the concentration of metalliferous placer deposits: a case study, Lake Turkana, N. Kenya. *Journal of the Geological Society* **142**, 837–848.

—— & —— 1986. Slope processes, sediment derivation and landform evolution in a rift valley basin, northern Kenya. *In:* FROSTICK, L. E., RENAUT, R. W., REID, I. & TIERCELIN, J. J. (eds) *Sedimentation in the African Rifts*. Special Publication of the Geological Society **25**, 99–111.

—— & —— 1987. Flow dynamics and suspended sediment properties in arid zone flash floods. *Hydrological Processes* **1**, 239–253.

ROBSON, D. A. 1971. The structure of the Gulf of Suez (Clysmic) rift, with special reference to the eastern side. *Journal of the Geological Society* **127**, 247–276.

ROSENDAHL, B. R. & LIVINGSTONE, D. A. 1983. Rift Lakes of East Africa: new seismic data and implications for future research. *Episodes* **83**, 14–19.

——, REYNOLDS, D. J., LORBER, P. H., BURGESS, C. F., MCGILL, J., SCOTT, D., LAMBIASE, J. J. & DERKSEN, S. J. 1986. Structural expressions of rifting; lessons from Lake Tanganyika, Africa. *In:* FROSTICK, L. E., RENAUT, R. W., REID, I. & TIERCELIN, J. J. (eds) *Sedimentation in the African Rifts*. Special Publication of the Geological Society **25**, 29–43.

SITTLER, C. 1969. The sedimentary trough of the Rhine graben. *Tectonophysics* **8**, 543–560.

TIERCELIN, J. J. 1986. The Pliocene Hadar Formation, Afar depression of Ethiopia. *In:* FROSTICK, L. E., RENAUT, R. W., REID, I. & TIERCELIN, J. J. (eds) *Sedimentation in the African Rifts*. Special Publication of the Geological Society **25**, 21–40.

TRUCKLE, P. H. 1976. Geology and late Cainozoic lake sediments of the Suguta trough, Kenya. *Nature* **263**, 380–383.

WALKER, T. R., WAUGH, B. & CRONE, A. J. 1978. Diagenesis in first cycle desert alluvium of Cenozoic age, southwestern United States and northwestern Mexico. *Bulletin of the geological Society of America* **89**, 19–32.

WATKINS, R. T. 1986. Volcano-tectonic control of sedimentation in the Koobi Fora sedimentary basin, Lake Turkana. *In:* FROSTICK, L. E., RENAUT, R. W., REID, I. & TIERCELIN, J. J. (eds) *Sedimentation in the African Rifts*. Special Publication of the Geological Society **25**, 85–95.

WILLIAMS, M. A. J., ASSEFA GETANEH & ADAMSON, D. A. 1986. Depositional context of Plio-Pleistocene hominid-bearing formations in the Middle Awash valley, southern Afar Rift, Ethiopia. *In:* FROSTICK, L. E., RENAUT, R. W., REID, I. & TIERCELIN, J. J. (eds) *Sedimentation in the African Rifts*. Special Publication of the Geological Society **25**, 241–251.

WILLIAMSON, P. G. & SAVAGE, R. J. G. 1986. Early rift sedimentation in the Turkana basin, northern Kenya. *In:* FROSTICK, L. E., RENAUT, R. W., REID, I. & TIERCELIN, J. J. (eds) *Sedimentation in the*

African Rifts. Special Publication of the Geological Society **25**, 267–283.

WRIGHT, J. B. 1981. Review of the origin and evolution of the Benue Trough in Nigeria. *Earth Evolution Sciences* **2**, 98–103.

YURETICH, R. F. 1979. Modern sediments and sedimentary processes in Lake Rudolf (Lake Turkana) Eastern Rift Valley, Kenya. *Sedimentology* **26**, 313–331.

ZIEGLER, P. A. 1981. Evolution of sedimentary basins in North-West Europe. *In:* ILLING, L. V. & HOBSON, G. D. (eds) *The Petroleum Geology of the Continental Shelf of North-West Europe*. Heyden, London, 3–39.

L. E. FROSTICK, Royal Holloway and Bedford New College, University of London, Egham Hill, Egham, Surrey TW20 0EX, UK.

I. REID, Department of Geography, Birkbeck College, University of London, Malet Street, London WC1E 7HX, UK.

Ancient ephemeral stream deposits: a local terminal fan model from the Bunter Sandstone Formation (L. Triassic) in the Tønder-3, -4 and -5 wells, Denmark

H. Olsen

SUMMARY: Facies studies of a *c*. 30 m thick fluvial sand body in the Tønder area, Denmark indicate that the sand body was formed by an ephemeral stream system comparable to modern terminal fans. The sediments are grouped into three genetic facies associations (A, B, C) reflecting two styles of channel flow (A and B) and overbank sheet flow (C). The sediments are dominated by parallel laminations (28%) and large-scale cross-bedding (27%) in association with small-scale cross-bedding (17%), massive bedding (14%) and intraformational conglomerates (5%). Mudstones and siltstones form only 9% of the deposits. Anhydrite (formerly gypsum) is associated with the fine-grained deposits which are mainly of overbank origin. It is commonly observed as intraformational clasts in the channel deposits. The inferred ephemeral stream complex probably formed a link between a more permanent braided stream system towards the N (the Skagerrak Formation) and a sabkha system towards the S (the Middle Buntsandstein in northern Germany).

For many decades sedimentologists have been attracted by ephemeral streams and their deposits (*eg* McGee 1897; Davis 1938; McKee *et al*. 1967; Williams 1971; Karcz 1972; Picard & High 1973; Frostick & Reid 1977; Parkash *et al*. 1983; Sneh 1983; Stear 1985). In spite of this the facies models for ephemeral stream systems are only poorly developed compared to perennial stream systems (*eg* Miall 1977).

The aim of this paper is to describe the facies and sequences of a supposed ancient sandy ephemeral stream complex (terminal fan) which may serve as a reference for further studies of ephemeral stream deposits. This paper also documents an example in which plane-bed lamination is a less important constituent compared with many other supposed ancient sandy ephemeral stream deposits (*eg* Stear 1983; Tunbridge 1984) and serves as an alternative to the terminal fan model erected by Tunbridge (1984) on the basis of the Devonian Trentishoe Formation.

Geological setting

The Bunter Sandstone Formation is a part of the L. Triassic Bacton Group (Rhys 1974; Bertelsen 1980). It is *c*. 210 m thick in the Tønder area, southern Denmark (Fig. 1) and is divisible into a lower Bunter sandstone sequence and an upper Bunter sandstone sequence each, 20–40 m thick and separated by a *c*. 100 m thick mudstone sequence (Fig. 1). The sediments are penetrated by three closely spaced wells, Tønder-3, -4 and -5. The general description below is based on sedimentological core analyses by Clemmensen (1985) who interpreted the sediments as desert sand-plain, sabkha and inland basin deposits reflecting a marginal position in the Permo–Triassic North German Basin (Fig. 1). The lower sandstone sequence is dominated by aeolian sand sheet and ephemeral stream deposits. The upper sandstone sequence consists of two major parts. The lower part is entirely ephemeral stream deposits, whereas the upper part is made up of sabkha and shoreline deposits. The Bunter Sandstone Formation is equivalent to part of the sandy and conglomeratic alluvial Skagerrak Formation to the N (Bertelsen 1980) and grades into sabkha and inland basin deposits towards the S (Clemmensen 1979). The fluvial transport direction was southwestwards (Fig. 2). Only parts of the Ringkøbing–Fyn High appear to have been exposed in the Early Triassic and the main source region is believed to be the Fennoscandian Shield.

This paper focuses on the ephemeral stream deposits forming the lower part of the upper sandstone sequence (Fig. 1). The fluvial deposits rather abruptly overlie the sabkha and inland basin deposits of Tønder-5 whereas a more gradual transition from ponded water mudflat to sheetwash mudflat deposits (*cf* Hardie *et al*. 1978) is observed in Tønder-3 and -4. This transition could be related to the major Hardegsen unconformity (Baartman pers. comm.) but its nature in these wells suggests that it is mainly depositional in this, marginal, part of the basin. The sequence reflects the progradation of a stream complex

From Frostick, L. & Reid, I. (eds), 1987, *Desert Sediments: Ancient and Modern*, Geological Society Special Publication No. 35, pp. 69–86.

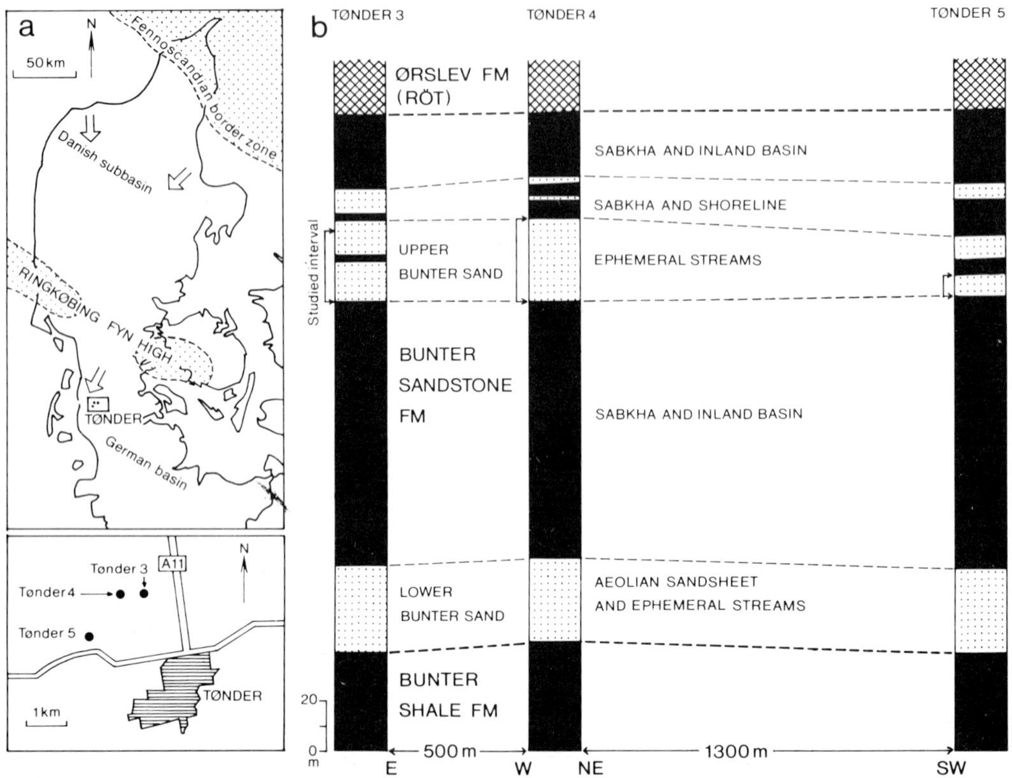

FIG. 1. (*a*) Location map. The inferred fluvial transport directions in the Danish Subbasin are based on Pedersen & Andersen (1980). (*b*) Stratigraphy, well correlation and general geological interpretation of the Bunter Sandstone Formation, after Clemmensen (1985).

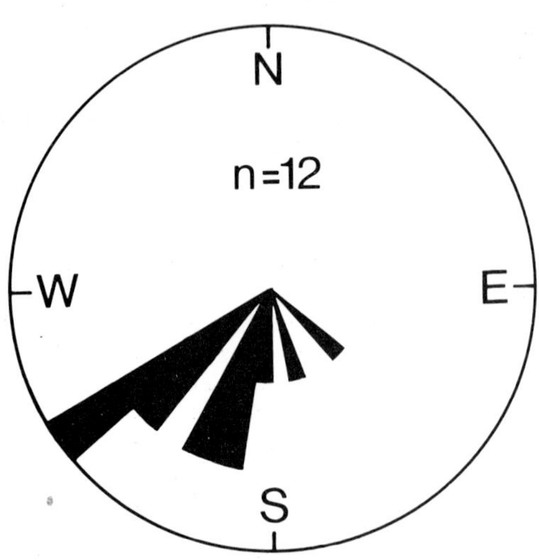

FIG. 2. Inferred fluvial transport directions in the upper Bunter sandstone sequence. The palaeocurrents are based on dipmeter readings from large-scale cross-bedded sandstones (facies St and Sx).

which decays downslope into a major sabkha and an associated shallow lake or inland sea (Fig. 3; Clemmensen 1985).

Sedimentary facies

The sediments in the cores investigated are grouped into eight facies types.

Small-scale cross-bedded sandstone (Sr)

Small-scale cross-bedded sandstone, facies Sr (Fig. 4*b*), forms 17% of the deposits. Facies Sr comprises very fine and fine sand and occurs as 2–105 cm thick cosets. Individual cross-sets are 0.5–1.5 cm thick and planar or planar wedge-to-trough shaped. The dip direction of cross-laminae is always unidirectional within the cosets. Form-sets are asymmetric, ripple symmetry indices being of the order of 4. Climbing ripple lamination commonly occurs with a climb-angle of up to 15°. Intraformational mudclasts are rare or absent.

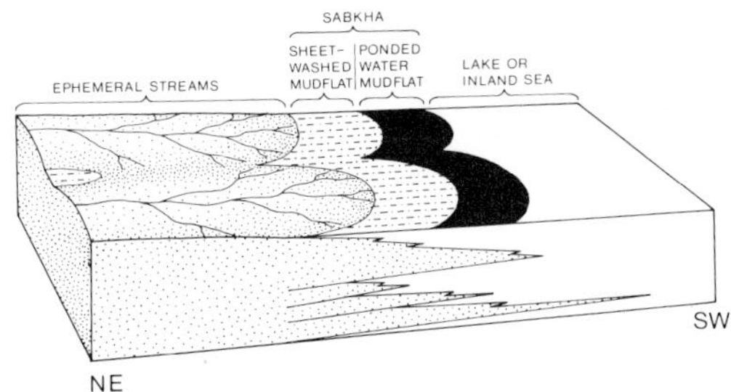

FIG. 3. Tentative depositional model for the upper Bunter sandstone sequence (modified after Clemmensen 1985).

Facies Sr is interpreted as deposited from small migrating current ripples under lower flow regime conditions. Facies Sr may form opposing cross-laminations in the lower part of large cross-sets (Fig. 6*a* and *b*). In such cases the small-scale cross-bedding is interpreted as back-flow ripple lamination (*cf* Boersma *et al.* 1968).

Horizontally laminated sandstone (Sh)

Facies Sh (Fig. 4*c*) forms 15% of the sections and is composed of very fine sand. Horizontal lamination (inclination less than 5°) occurs in 2–125 cm thick beds. The lamination is formed of 0.5–2 mm thick continuous laminae. Parting lineations occur locally. Upward and downward transitions to small-scale cross-bedding is common. Intraformational mudclasts are rare.

The association with parting lineation and upward transitions from small-scale cross-bedding without draping of ripple forms suggest that facies Sh was deposited under higher flow velocities than facies with small-scale cross-bedding, *ie* upper plane-bed phase. According to *eg* Harms *et al.* (1982) the upper plane-bed phase for very fine sand occurs at supercritical flow conditions. Reid & Frostick (1987) have, however, shown that horizontal lamination in ephemeral streams commonly are laid down under subcritical flow conditions. The flow conditions of the plane bed phase and associated horizontal lamination remain unsolved.

Low-angle cross-bedded sandstone (Sl)

Facies Sl (Fig. 4*d*) constitutes 13% of the sections. It comprises 2–125 cm thick beds of very fine and fine sandstones. The laminations are inclined at 5–15°. Individual laminae are usually fractions of a millimetre to 5 mm thick although exceptionally thick laminae (>1 cm) were observed. The laminations can be either continuous or discontinuous. Low-angle laminations intimately associated with low-angle inclined small-scale cross-bedding is sporadically observed (Fig. 4*e*). Intraformational mudclasts are rare.

Low-angle cross-bedding records the deposition of planebeds on sloping surfaces (point and longitudinal bars) at or near upper flow regime conditions (Picard & High 1973). The intimate association of facies Sl and Sr suggests flow conditions at the boundary between the upper flow regime (Sl) and lower flow regime (Sr). Palaeocurrents of the small-scale cross-bedding indicate directions obliquely up or across the palaeoslope (as determined from the dips of facies Sl).

Large-scale trough cross-bedded sandstone (St)

Large-scale trough cross-bedding comprises 11% of the rock sequence (Fig. 4*f*). The range of grain sizes is from very fine to medium sand. Facies St occurs in 10–70 cm thick cosets of 3.5–14 cm thick cross-sets (within the core diameter). The lower bounding surfaces are always concave and erosive. The cross-laminae are 0.5–3 cm thick, inclined at angles of up to 25° and concave or tangential to the lower set boundaries. Small-scale backflow cross-sets are rather common in the lower part of trough sets (observed in seven trough sets). Sub-rounded to well-rounded intraformational mudclasts occur frequently as small flat pebbles oriented parallel to the cross-lamination.

The concave set boundaries, the thickness of individual sets, the high dips of cross-laminae and the backflow cross-bedding suggest that facies St was formed by sinuous crested dunes with frontal vortices under the upper part of the lower flow regime (*cf* Harms *et al.* 1982).

Large-scale cross-bedded sandstone (Sx)

Facies Sx constitutes 16% of the deposits (Fig. 4*g*). Grain size varies from very fine to medium sand with a dominance of fine-grained sets. This facies is similar in general appearance to facies St except for the planar bases of individual sets. Furthermore foresets are either angular or tangential, the latter usually observed as an increase of foreset dip from base to top, with a maximum dip of 30°. Facies Sx frequently occurs as up to 60 cm thick single sets.

Facies Sx sediments are interpreted as megaripple and transverse bar deposits. It is impossible to assess whether the set boundaries are truely planar or gently curved due to the small diameter of the cores (10 cm). Single sets probably repre-

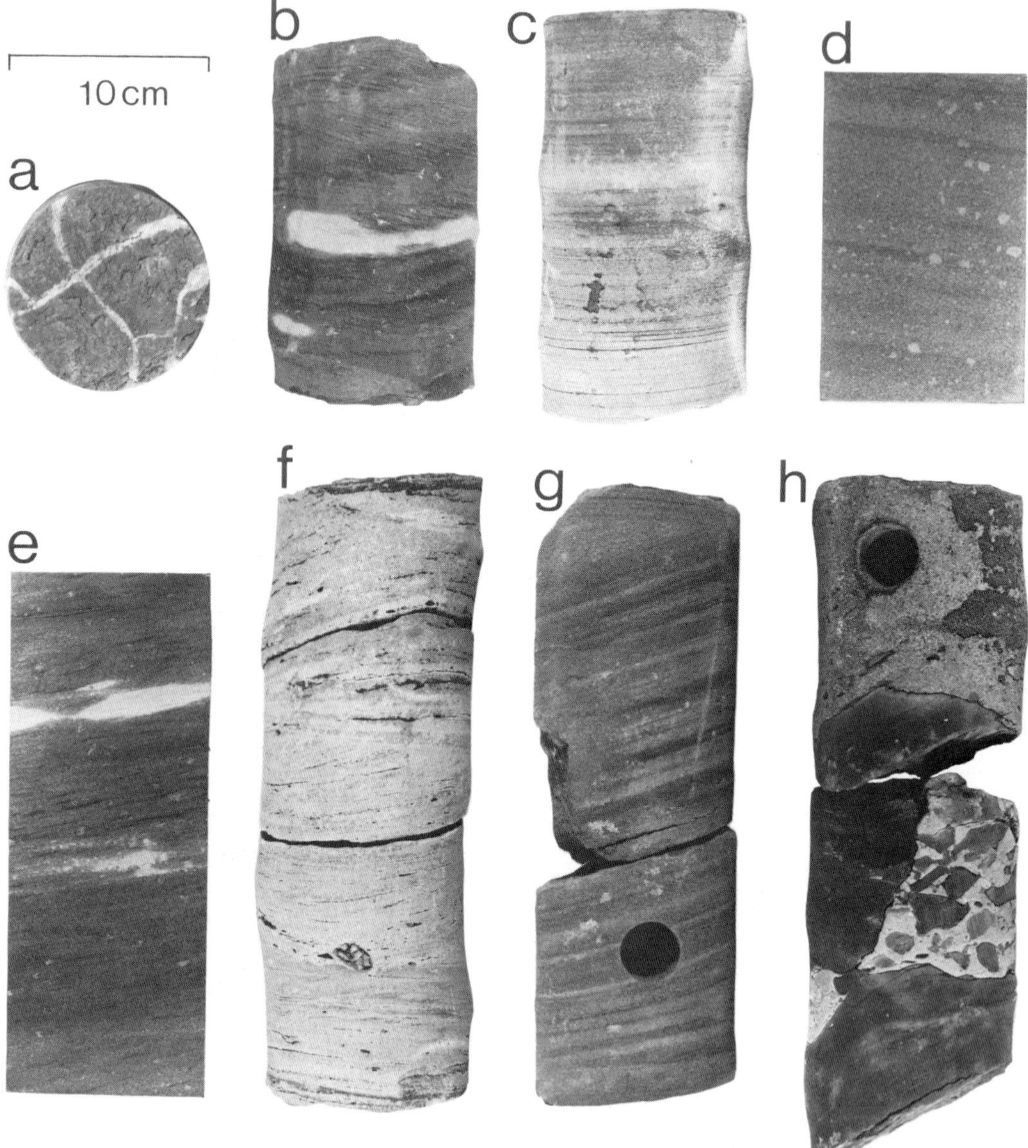

FIG. 4. Examples of facies types. (*a*) Mudstone (facies M) with polygonal desiccation cracks. (*b*) Small-scale cross-bedded sandstone (facies Sr), a few anhydrite crystals are seen. (*c*) Horizontally laminated sandstone (facies Sh). (*d*) Low-angle cross-bedded sandstone (facies Sl). (*e*) Low-angle cross-bedded sandstone (facies Sl) closely associated with small-scale cross-bedding (facies Sr) indicating palaeoflow across the local slope (as determined from the dip of facies Sl). (*f*) Large-scale trough cross-bedded sandstone (facies St). (*g*) Large-scale cross-bedded sandstone of undifferentiated type (facies Sx). (*h*) Intraformational conglomerate (facies Ci) and massive sandstone (facies Sm, upper part of photo). In examples (*d*) and (*e*) the cores are cut lengthwise.

sent transverse bars whereas cosets of rather thin sets more closely resemble facies St cosets and are interpreted accordingly as straight or slightly sinuous crested megaripple deposits of the intermediate part of the lower flow regime (*cf* Harms *et al.* 1982).

Massive sandstone (Sm)

Massive fine to medium sandstones average 14% of the cored interval (Fig. 4*h*). The bed thickness varies from 3 cm to 166 cm. Intraformational mudclasts up to 7.5 cm are very common constit-

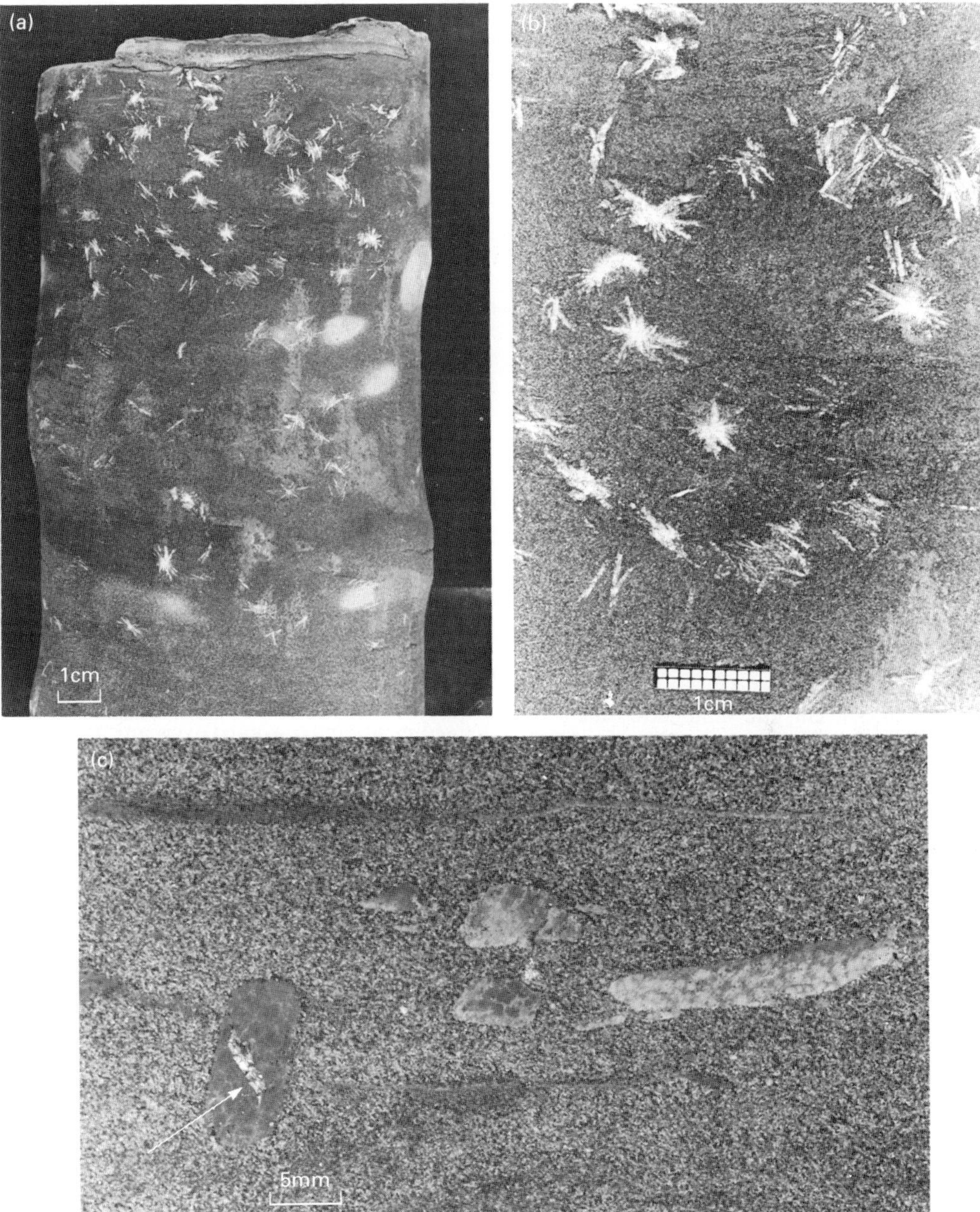

FIG. 5. Anhydrite. (*a*) Overbank deposits (facies Sr) with displacively grown gypsum now replaced by anhydrite. (*b*) Close-up of (*a*), in which several clusters of anhydrite are seen. (*c*) Intraformational mudclasts (reworked overbank mud). One of the clasts contains an anhydrite crystal (arrowed).

uents of facies Sm and gradual transitions to intraformational conglomerates (facies Ci) are frequently observed. The mudclasts are flat and generally very angular and may either float or form a framework. The clasts are commonly concentrated as either basal lags or layers 5–55 cm above the bases of Sm beds. The lower bed boundaries of facies Sm are always erosional. Weakly developed small-scale cross-laminae are observed in some of the beds.

The massive sandstones are interpreted as deposited from hyperconcentrated flows in which the rapid deposition hindered the development of bedforms. In hyperconcentrated flows pebble- and cobble-sized clasts will be supported by the combined effect of fluid turbulence, hindered settling due to the high concentration of grains, matrix buoyant lift and dispersive pressure resulting from grain collision (see *eg* Lowe 1982). Intraformational clasts could therefore be kept moving until flow deceleration had resulted in a decrease in particle concentration below a certain critical value. Hereafter the large intraformational clasts would be deposited and the resulting sedimentation unit would comprise a clast concentration several centimetres above the erosive base. Massive beds with occasional cross-laminae were probably deposited from hyperconcentrated flows under lower flow regime conditions. Actually most deposition in ephemeral streams seems to occur when Froude numbers are below 0.8 (Frostick pers. comm.).

Mudstone and siltstone (M)

Mudstones and siltstones (Fig. 4*a*) form 10% of the successions. Facies M consists of grain sizes ranging from clay to silt/very fine sand, usually as a poorly sorted mixture. Better sorted siltstones are, however, also common. Beds range from less than 1 mm to 90 cm, the millimetre-thin beds occurring as flasers in small-scale cross-beds. Facies M occurs as massive and, more commonly, thinly laminated beds. The lamination is usually horizontal but inclined lamination (up to 20°) and small-scale cross-bedding is also seen. Desiccation cracks, sometimes in polygonal patterns, are associated with several beds (Fig. 4*a*).

The mudstones and siltstones are interpreted in terms of deposition from suspension in slowly moving or ponded water. However, the small-scale cross-bedded siltstones record some bedload transport and migration of silt ripples. The inclined lamination is interpreted as the draping of concave channel bases. This special variation of facies M is dealt with in more detail later (facies association B).

Intraformational conglomerate (Ci)

Intraformational conglomerates comprise 5% of the cored sediments (Fig. 4*h*). The conglomerates consist of sub-rounded to, more commonly, very angular clasts ranging from a few millimetres to 16 cm in diameter. The clasts vary from flat to almost equant. Facies Ci beds are composed of between 25% and almost 100% mudclasts. The conglomerates usually rest on scoured surfaces but also occur several centimetres above the base of the bed. Conglomerate beds range up to 46 cm in thickness.

The intraformational conglomerates are generally interpreted as channel lag deposits. The dominance of very angular clasts suggests that transport was of short duration. Some of the conglomerate beds seem to be almost *in situ* mud layers (facies M) only slightly broken up. Intraformational conglomerates several centimetres above the base of massive sandstone beds were probably deposited from hyperconcentrated stream-flows.

Anhydrite

Throughout the fluvial deposits scattered crystals of anhydrite are found. The anhydrite occurs as elongate prismatic crystals up to several millimetres long, often forming radiating clusters (Fig. 5). The anhydrite crystals are found either in primary growth positions in facies M and Sr in the upper part of sedimentation units of association C and distal A (Fig. 4*a* and *b*), or as part of intraformational mudclasts in facies associations A and B (Figs 5*c* and 7*b*).

Morphological and textural relationships indicate that the anhydrite replaced gypsum crystals (Fine pers. comm.). The presence of anhydrite exclusively in shallow overbank and distal channel sediments and reworked overbank sediments (intraformational clasts) indicates that crystal growth was a primary process near the surface in the overbank environment and in the distal and abandoned channel environments. Similar displacive gypsum needles have been reported from modern sabkha and associated mudflat environments (Hardie *et al.* 1978; Handford 1982) and modern ephemeral stream environments (Picard & High 1973).

Genetic facies associations

The eight sedimentary facies have been grouped into three genetic facies associations: A—Stratified sandy channel deposits; B—Massive sandy and stratified silty channel deposits; C—Overbank deposits. Facies associations A and C are almost equally important, forming 40% and 37% of the deposits, respectively. Association B forms 23%.

Stratified sandy channel deposits (association A)

Facies association A (Figs 6, 7) is dominated by large-scale cross-bedded and low-angle cross-bedded sandstones (facies St, Sx and Sl). Small-scale cross-bedded (Sr) and horizontally laminated sands (Sh) are also important constituents

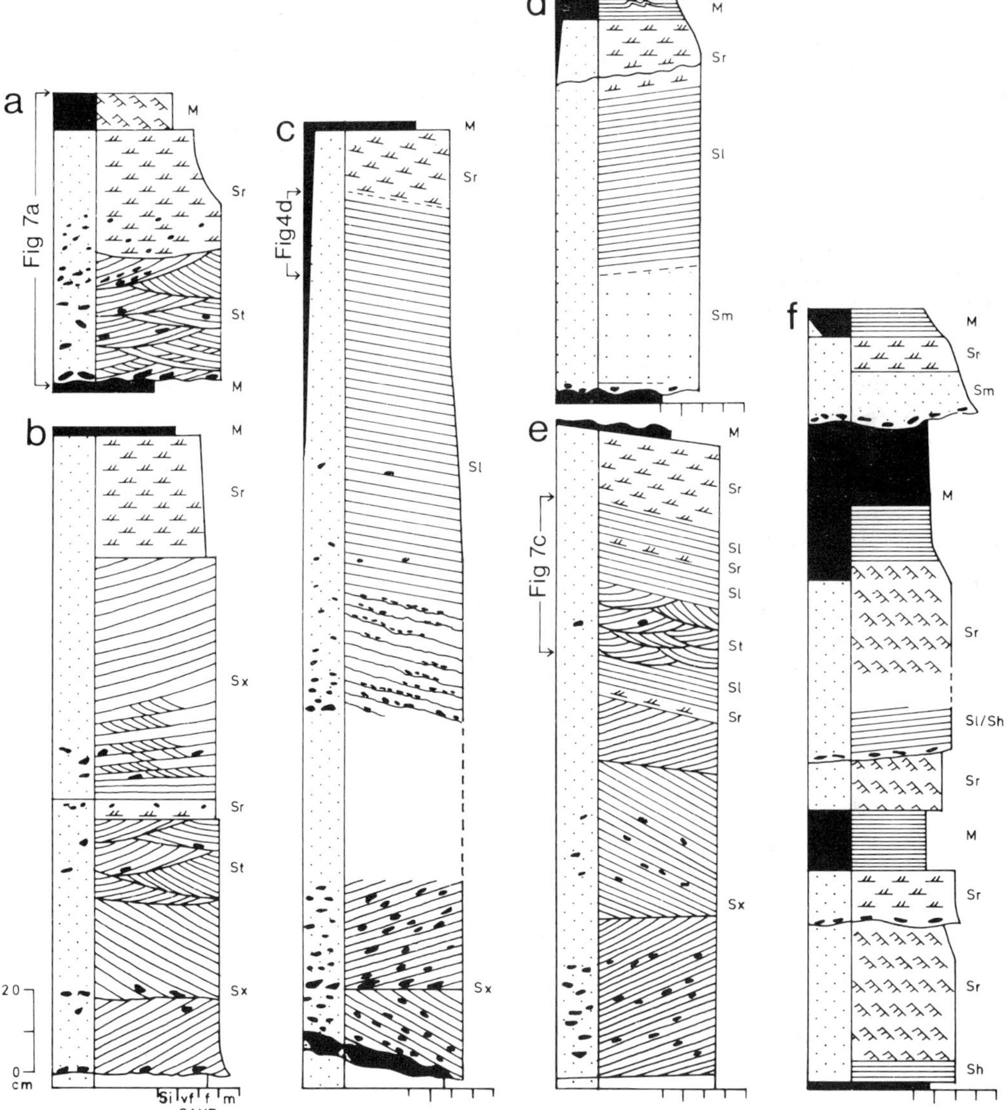

FIG. 6. Examples of facies association A. (*a*) Fining upward sequence indicating vertical accretion. (*b*) Composite fining upward sequence composed of two amalgamated sequences, both indicating vertical accretion. (*c, d, e*) Fining upward sequences indicating vertical to lateral accretion. (*f*) Distal/marginal examples of facies association A, dominated by facies Sr and facies M. Except for backflow cross-lamination in examples (*a*) and (*b*) and the cross-lamination and horizontal lamination of siltstones in examples (*a*), (*d*) and (*f*) all symbols are explained in Fig. 12.

of this association. The sediments are organized in fining upward sequences 20 cm to 450 cm thick (mean: 110 cm). The fining upward trend is accompanied by a change in stratification, usually from larger scale cross-bedding to smaller scale cross-bedding. An 'ideal' sequence comprises a scoured surface overlain in turn by facies Sx, St, Sr and M.

This sequence reflects initial channel scouring followed by fluvial bar or megaripple migration (Sx). The trough cross-bedded sands (St) may indicate dune migration on bar tops or, more likely, a general change in flow conditions (? shallowing). Small-scale cross-bedding, usually associated with decreasing grain size, indicates falling stage conditions followed by flow cessation

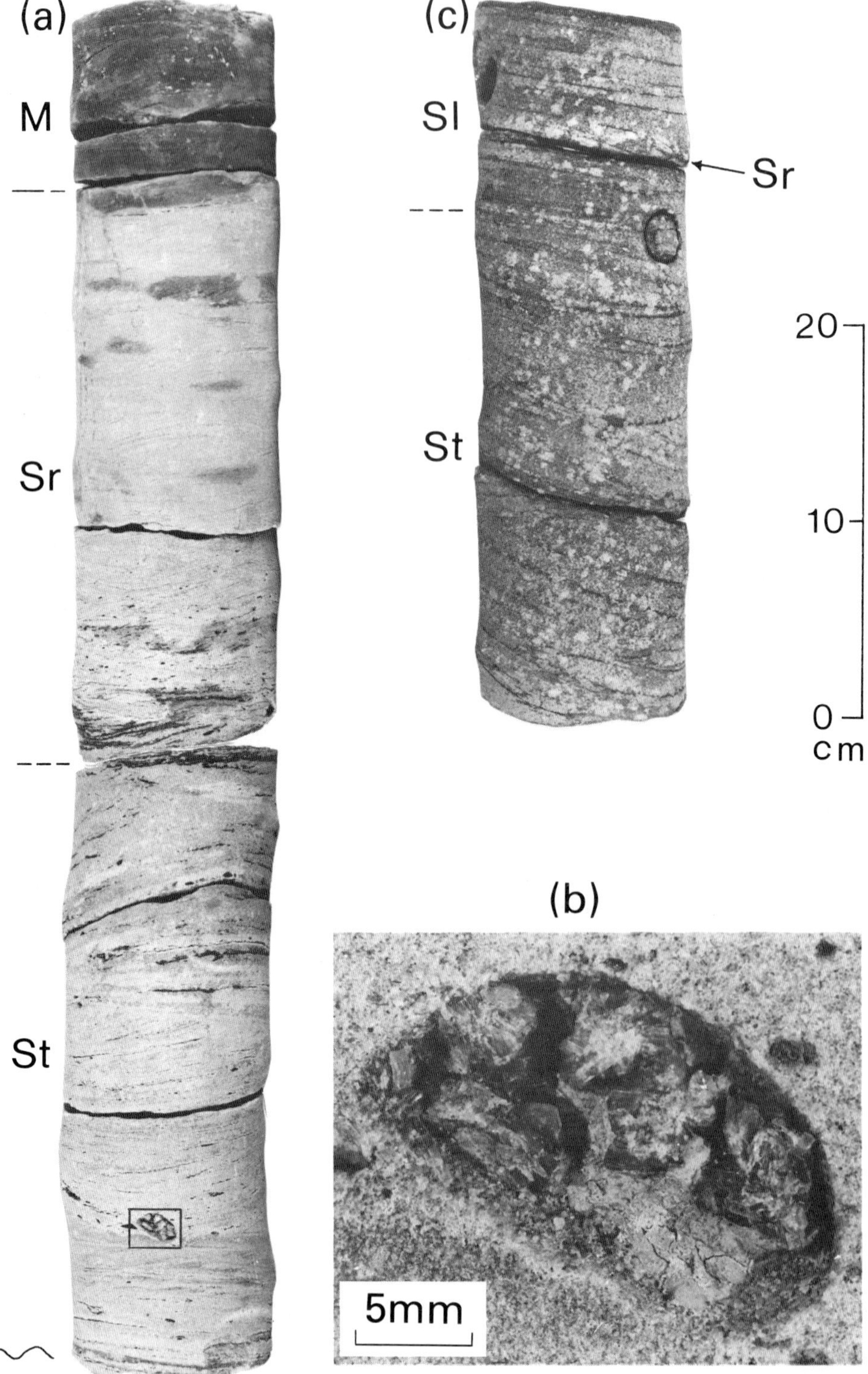

FIG. 7. Examples of facies association A. (*a*) Fining upward sequence indicating vertical accretion (see also Fig. 5*a*). (*b*) Close-up of intraformational mudclast in (*a*) containing a cluster of anhydrite crystals. (*c*) Upper part of fining upward sequence (see Fig. 6*e*). Note that the set boundaries of facies Sr and St are inclined parallel to the dip of facies Sl and indicate palaeoflow along the dip of facies Sl (lateral accretion).

and fall-out of suspended mud (M). Most fining upward sequences are, however, incomplete variants of this 'ideal' sequence (Figs 6*a*, 7*a*).

Sometimes fining upward sequences comprise two or more bedsets each defined by an upward change in stratification type (large-scale to small-scale) and separated by weak, almost mudclast-free, eroded surfaces (Fig. 6*b*). Such composite sequences were probably formed by several successive flood episodes within the same channel.

Another variant of facies associated A is characterized by the low-angle cross-bedding of facies Sl (Figs 6*c*, *d*, *e* and 7*c*). Low-angle cross-bedding tends to occupy the middle and upper parts of the fining upward sequences and may dominate them. Fining upward sequences in which facies Sl occurs usually comprise a large-scale cross-bedded lower part and facies Sl almost always exhibits an upward transition to facies Sr. The set boundaries of the small-scale cross-bedding are always inclined almost parallel to the low-angle cross-bedding and the foreset dips indicate palaeoflows obliquely up, or across, the local slope. Occasionally facies Sr and St are interbedded with facies Sl and show similar relations of set boundaries and foreset dips to those mentioned above (Figs 6*e*, 7*c*).

Low-angle cross-bedding is well-known from recent ephemeral streams (*cf* Picard & High 1973). The thickness (up to 1.25 m) of the present examples is, however, larger than those observed by Picard & High (1973). Ancient examples of moderately dipping (average 18°), up to 2.5 m thick, cross-sets have been described from ephemeral stream deposits by Williams (1966). He also noticed a common upward transition to small-scale cross-bedding and occasional interbedding of small- and large-scale cross-bedding with set boundaries parallel to the low-angle cross-laminae. Furthermore palaeocurrents inferred from the overlying and interbedded cross-strata were parallel to the strike of the low-angle cross-laminae (Williams 1966). Both Picard & High (1973) and Williams (1966) relate low-angle cross-bedding predominantly to point bar migration. The common upward transition from large-scale cross-sets (facies Sx or St) to low-angle cross-bedding suggests a shift from vertical to lateral accretion. This pattern bears little resemblance to the deposits of perennially flowing meandering streams in which thin channel lags are overlain by thick, laterally accreted, point bar deposits (*cf* for example Collinson 1978). A better analogy is found in ephemeral streams. Ephemeral streams are often straight or braided during periods of high discharge but tend to meander during waning flood stages (Fig. 8). The observed vertical-to-lateral accretion trends are thus interpreted as reflecting peak and falling stages of floods in ephemeral streams. In continuation of this idea the more commonly occurring, vertically accreted, fining upward sequences (Fig. 6*a*, *b*) may have been formed in a broad part of the stream course which was only covered by water during periods of high discharge (Fig. 8).

Thin fining upward sequences dominated by small-scale (often climbing ripple) cross-bedding and interbedded with relatively thick mudstones (Fig. 6*f*) are interpreted as distal and marginal channel fill deposits. Anhydrite crystals in primary growth position only occur in this subtype of facies association A. This feature probably reflects prolonged periods in which the distal and marginal parts of the fluvial system were not inundated, allowing evaporation and the growth of gypsum crystals close to the surface. This subtype is transitional to facies association C.

Massive sandy and stratified silty channel deposits (association B)

Facies association B occurs in two variants. The most common variant is characterized by massive sandstones, usually with a high intraformational mudclast content (Figs 9*a*, 10*a*). The silty variant of facies association B is dominated by low-angle laminated siltstones and is found closely associated with the sandy variant (Figs 9*b*, 10*b*).

The sandy beds occur in fining upward sequences 6–166 cm thick (mean: 50 cm). The beds are dominated by massive bedding (facies Sm) and intraformational mudclasts are very common. The massive facies commonly passes upwards into facies Sr, St and Sl, a transition accompanied by a decrease in grain size.

A detailed description and interpretation of the dominant sandy facies in association B, facies Sm, has already been given. In short, the massive sandstones are interpreted as deposited rapidly from hyperconcentrated flows on the evidence of floating mudclasts. The upward transitions to stratified finer grained sediments probably reflect a change to normal traction sedimentation in the more dilute residual flows. Hyperconcentration of the flood flows is interpreted as due to an initial rapid infiltration of water into the bed, a process commonly observed in ephemeral streams (*eg* Knighton 1984).

The silty beds in association B are composed of low- to medium-angle (<20°) cross-stratified siltstones. Grain size variations from lamina to lamina are sometimes very marked (coarse silt/fine silt). Desiccation cracks were observed in one bed. Individual beds are 10–30 cm thick with erosional bases. Stratification is generally parallel

FIG. 8. Ephemeral stream in North Greenland during falling flood stage (photo: J. Lautrup, with permission from the Geological Survey of Greenland). The stream is meandering and deposition is occurring on point bars. The width of the channel is 20–30 m. The stream was straight and braided during peak discharge (flood). The deposits of the last major flood are slightly lighter coloured and indicate a width of the stream at flood peak of the order of 300 m. The margins of this flood-stream are indicated. Hypothetical positions of the fining upward sequences comparable with Fig. 6 are indicated.

to the erosive bases of beds but may flatten towards the top. Individual beds are variously non-graded, normally graded or inversely graded (Fig. 9*b*).

The silty, stratified beds resemble the plug-deposits of abandoned river channels, chute channels and crevasse splay channels described by Singh (1972) and inferred by Plint (1983) and Tunbridge (1984). Normal grading of beds suggests deposition from waning currents, whereas inverse grading suggests either accelerated flow during plugging, or a change in the grain size of sediment supplied. The overall deposit geometry and stratification suggests the draping of sediment over concave channel bases. The presence of desiccation cracks and grain-size changes between laminae suggest multiple floods during the process of plugging. Abandonment of the channels may have occurred due to avulsion (river-channel plugging, *cf* Plint 1983, Tunbridge 1984), blocking upstream (chute channel plugging, *cf* Plint 1983, McGowen & Garner 1970) or local blocking by the deposition of fines from suspension on the recession limb of the flood hydrograph (*cf.* Singh 1972). The associated debris flow-like deposits (facies Sm) indicate that a process similar to the chute channel plugging may have occurred due to deposition from hyperconcentrated flows blocking the upstream parts of the fluvial channels. The downstream parts of the channels were thus abandoned before bedload deposition could occur to any extent. The channels were later filled by overbank silts during a series of flood events.

Facies association B reflects in general very swift channelized ephemeral flow and channel blocking.

Overbank deposits (association C)

Facies association C (Fig. 11) is dominated by small-scale cross-bedded and horizontally laminated very fine to fine-grained sandstones (facies Sr and Sh). Minor constituents are large-scale cross-bedded sandstones (Sl, St, Sx) and mudstones (M). The association differs from association A and B in the paucity of scour surfaces and

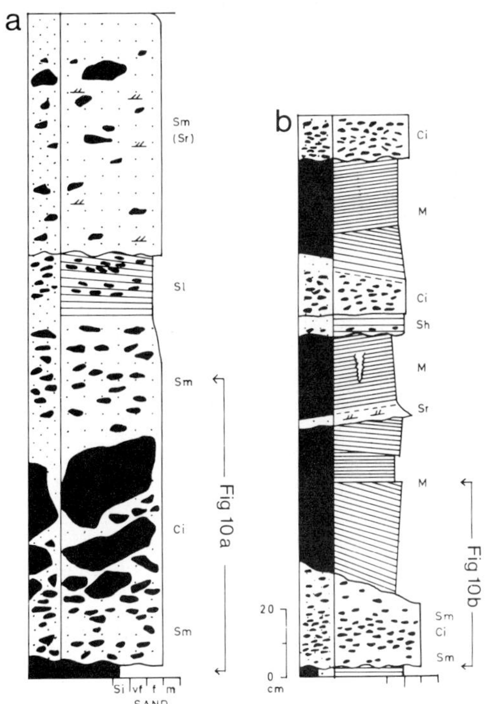

FIG. 9. Examples of facies association B. (*a*) Fining upward sequences dominated by massive sandstones (Sm) and intraformational conglomerates (Ci). Note the inverse grading of intraformational clasts at the base of the lower sequence probably reflecting hyperconcentrated stream flow. (*b*) Association B dominated by stratified siltstone filling up supposed small channel forms. A large internal desiccation crack is shown in one of the stratified siltstone beds indicating a multi-episodic channel infilling. See Fig. 12 for explanation of the other symbols.

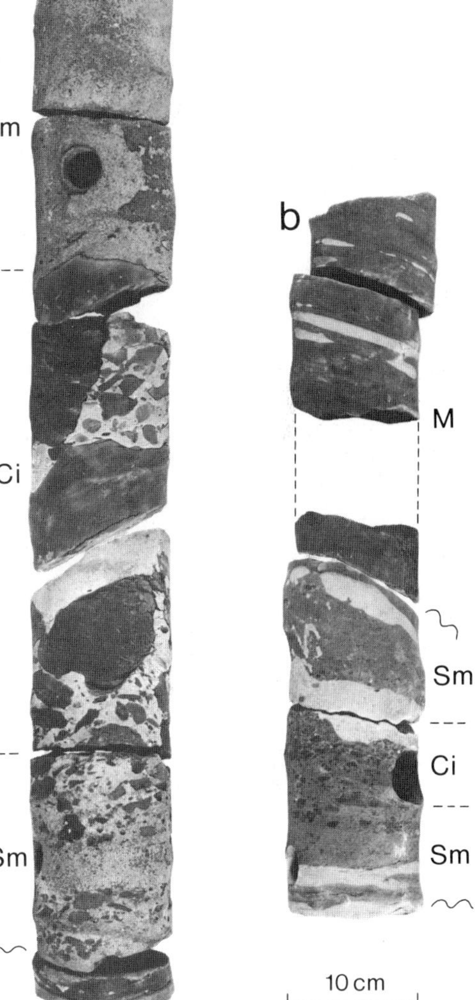

FIG. 10. Details of facies association B from Fig. 8. (*a*) Facies Sm and Ci, note inverse grading above scour surface. (*b*) Stratified siltstone (M) inclined parallel to supposed channel margin. Below the siltstone there is a massive sandstone bed with a high concentration of intraformational clasts in the middle part (hyperconcentrated flow deposit).

fining upward sequences. Typically association C occurs as metre-thick successions of facies Sr and/or Sh (Fig. 10*a*, *b*). Bedset contacts are usually obscure. Occasionally thinly interbedded very fine grained sands (facies Sr) and muds are observed (Fig. 11*c*).

The dominance of horizontal lamination and small-scale cross-bedding, the lack of channel scours and associated lags, the rare fining upward sequences and the fine grain size suggest deposition in an overbank environment resembling that of the sandy ephemeral Bijou Creek (McKee *et al*. 1967) and the overbank environments during the Laingsburg flood (Stear 1985). The dominantly horizontally laminated units are interpreted as proximal overbank sediments deposited under upper flow regime conditions. Occasional scour surfaces overlain by units with low-angle cross-bedding probably indicate local scour-and-fills. Dominantly small-scale cross-bedded units are interpreted as intermediate overbank deposits whereas interbedded sand and mud deposits are interpreted as distal overbank sediments. These interpretations are in close accord with the observations of Smith (1980) and Stear (1983) made in deposits interpreted as overbank from the Permian Lower Beaufort. This is also thought to be associated with an ephemeral stream system (Stear 1983).

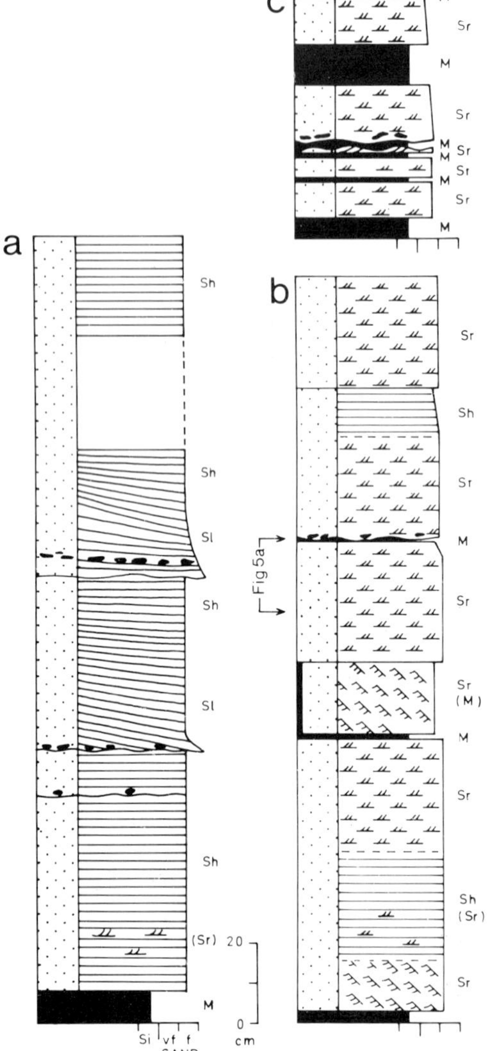

FIG. 11. Example of facies association C. (*a*) Proximal overbank deposits dominated by facies Sh. (*b*) Intermediate overbank deposits dominated by facies Sr. (*c*) Distal overbank deposits composed of thinly interbedded mud and sand. See Fig. 12 for explanation of symbols.

Sediment-body composition

The fluvial sediment-body at Tønder is subdivided into four units, all of which are recognized in the three wells (Fig. 12). Unit 1 is characterized by an upward succession of facies associations A, C and A in Tønder-3 and -4. This sequence suggests the dominance of channel flow (association type A) followed by a period of mainly overbank flow (association C) and finally a return to channel flow (association A). In Tønder-5 unit 1 comprises a single fining upward sequence of association A. It may be correlated with the upper association A deposits of unit 1 in the other two wells and reflects deep incision of a main channel.

Unit 2 is characterized by a succession of association A followed by association C in Tønder-4. The unit is bounded by a mudclast-strewn erosional surface and exhibits a stepwise increase in grain size from unit 1. In Tønder-5, unit 2 shows a similar succession of association A overlain by association C. The mudstone component is, however, more common than in Tønder-4 and this may reflect the distal or more marginal position of Tønder-5 during the deposition of unit 2. In Tønder-3 unit 2 is entirely composed of association C indicating a dominance of overbank sheet flooding and thus a position marginal to the main channels.

Unit 3 is entirely built up of facies association A in Tønder-3 and Tønder-4. It is not cored in Tønder-5 but from the gamma ray log (Fig. 12*b*) it can be seen that unit 3 is composed entirely of mudstones. These finer-grained deposits also reflect a marginal position for Tønder-5 during deposition.

Unit 4 comprises more than 50% of the sand body. It is only cored in Tønder-3 and Tønder-4 (Fig. 12*a*). Here the unit is composed of an alternation of the three facies associations apparently without a systematic upward trend. In Tønder-3 unit 4 is different from Tønder-4 in that there is a 3 m thick mudstone/siltstone interval in the lower part. These finer-grained deposits suggest a distal or marginal position for Tønder-3 during the initial deposition of unit 4. In Tønder-5 the unit is seen to be composed of 40–50% mudstones and siltstones (petrophysical log interpretation). This high fines content is probably responsible for the compaction of unit 4 in Tønder-5 (a 33% thinning from Tønder-4 to Tønder-5 is observed).

Taking the sediment body as a whole Tønder-5 differs significantly from Tønder-3 and Tønder-4 in the generally finer-grained nature of the units. These fine-grained deposits may, as already mentioned, be interpreted as either distal or marginal deposits. Dipmeter readings (Fig. 2) suggest southwesterly palaeocurrents, *ie* from Tønder-4 to Tønder-5. In the light of this the fine-grained sediments of units 2, 3 and 4 in Tønder-5 are interpreted as distal deposits rather than lateral–marginal. Therefore, in a reconstruction of the *sandstone* body at Tønder (Fig. 13) the sandy components of the units and the sandstone body as a whole are interpreted as thinning towards the SW.

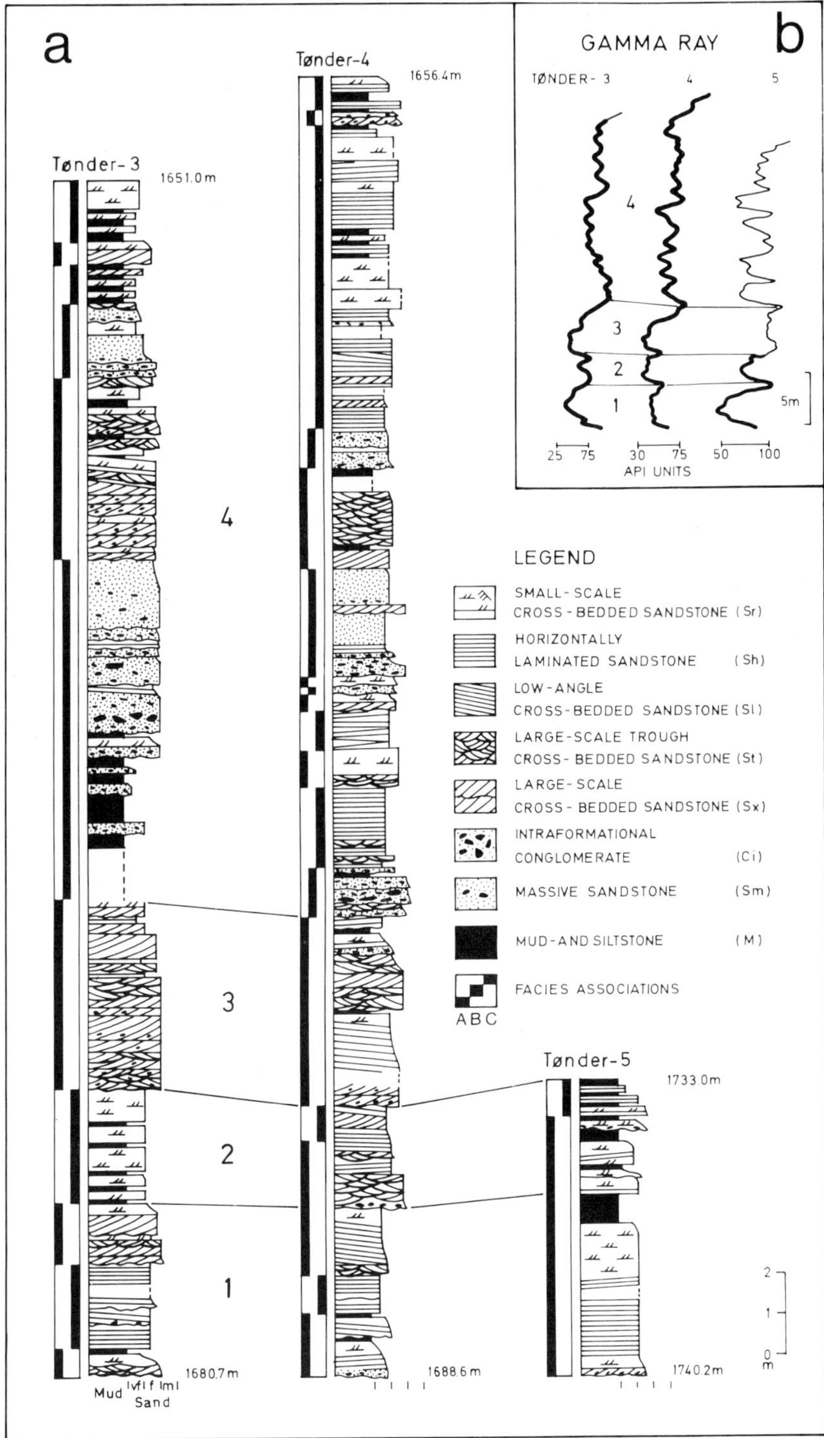

FIG. 12. Sedimentological facies logs (*a*) from the cored intervals of the fluvial part of upper Bunter sandstone sequence in Tønder-3, -4 and -5. Correlation of sedimentary units (1, 2, 3, 4) is shown. The units are defined by lithology and facies associations. Genetic facies associations (A, B, C) are indicated by bars. (*b*) Gamma log motifs from the entire fluvial part of upper Bunter Sandstone in the three wells. Cored intervals are indicated by thick lines. For the cored intervals sedimentological criteria are used to define the units (1, 2, 3, 4). The units correspond to segments easily recognized in the gamma logs.

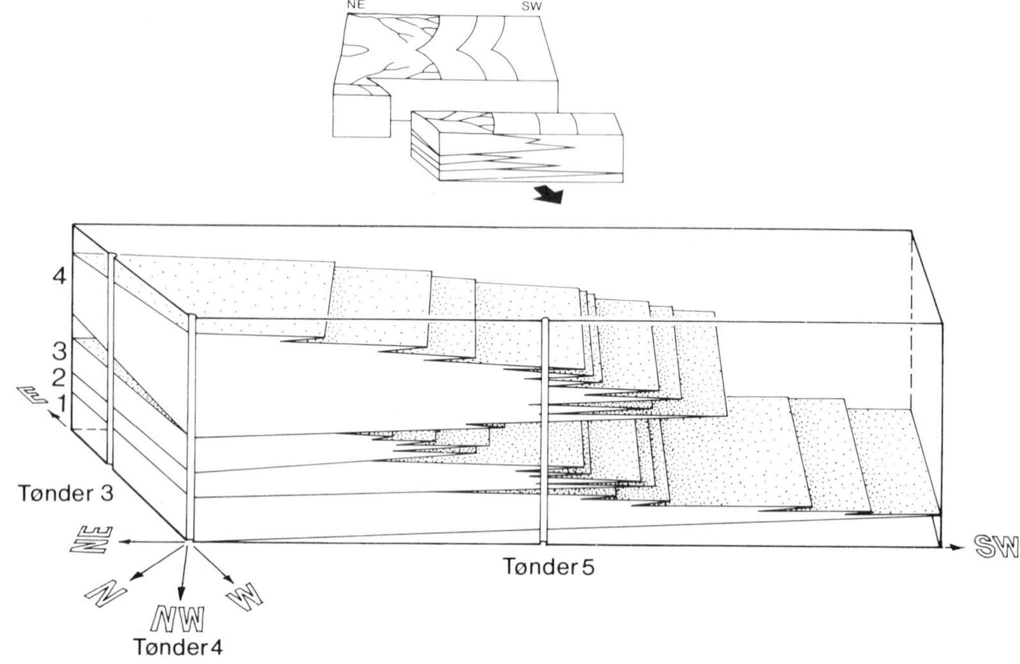

FIG. 13. Reconstruction of the sand-body geometry at Tønder based on cores and gamma ray logs. The positions of the wells (Tønder-3, -4 and -5) are indicated. The sand body thins rapidly towards the SW. Compare inset with Fig. 3.

Depositional system

Downstream changes in the character of the sediment body over very short distances (*c.* 1300 m) suggests a rapid downstream decay of flood flow. Rapid downstream decay of stream flow within a limited area occurs in three different types of depositional systems: deltas, alluvial fans and terminal fans.

A deltaic origin is ruled out in view of a lack of a permanent water body in which a delta could form. The inland basin to the south of Tønder (Fig. 3) was subject to extreme water level fluctuations and frequent sub-aerial exposure (Clemmensen 1979).

The area immediately N of the Tønder area is cut by a number of WNW–ESE trending faults which generally throw down to the S (Michelsen *et al.* 1981). Activity along these faults during deposition of the fluvial sands might favour the development of alluvial fans. This idea is, however, at odds with the well-sorted nature of the sands and with their distribution on both sides of the fault zone (Nielsen 1983).

A terminal fan origin might help explain the downstream change in depositional style. Terminal fans have a lobate form and are not more than a few kilometres in width and downstream length. They have a radiating channel pattern in which the channel width and depth decrease downstream suggesting falling discharges (Mukerji 1976; Friend 1978; Parkash *et al.* 1983). Terminal fans develop due to a number of factors including highly variable discharge, abundant availability of loose detritus and high sediment loads in the streams during rainy seasons, loss of water by infiltration into permeable alluvium and by evaporation, and lower precipitation in the depositional basin than in the hinterland. These factors result in channel choking and repeated bifurcation of the channels into distributaries that ultimately disappear downstream (Friend 1978; Parkash *et al.* 1983).

The ephemeral stream deposits at Tønder show a downward transition from mudflats deposited from overbank flows to those which accumulated in ponded water (*cf* Hardie *et al.* 1978), suggesting that the ephemeral channels decayed into the sabkha before reaching a permanent body of water. This, combined with the inferred fixed position of the fluvial depositional system and rapid downstream change of depositional style suggests that a terminal fan origin for the ephemeral stream deposits at Tønder is most appropriate.

Mukerji (1976) and Parkash *et al.* (1983) have made the only detailed study of a modern terminal fan at the termination of the Markanda river. The Markanda terminal fan is composed of several fanlobes. The four-part sediment body in the Tønder area could be interpreted as resulting from repeated shifting of the depositional area from one fanlobe to another due to avulsion. The correlation of the four units in the three wells suggests that they all penetrate the same fanlobes. Alternatively the four depositional units may simply reflect four progradational events of a single lobe.

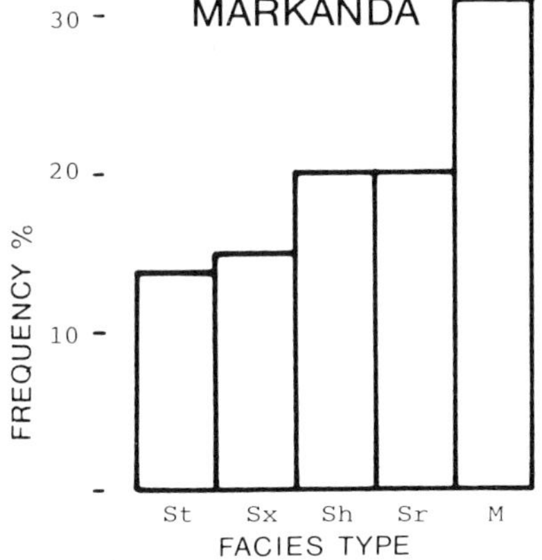

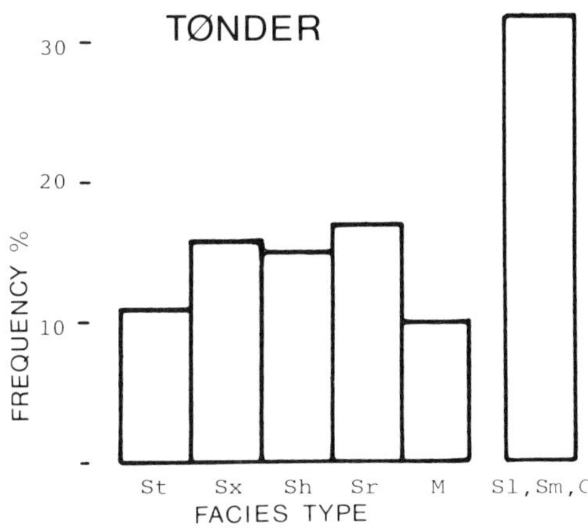

FIG. 14. Comparison between the facies proportions of the Markanda terminal fan (from Parkash *et al.* 1983, their fig. 3) and the terminal fan deposits of the Bunter Sandstone Formation at Tønder. The facies codes used for the Markanda terminal fan cover the following sedimentary structures of Parkash *et al.* (1983):
St = trough cross-bedding, Sx = planar cross-bedding, Sh = horizontal bedding, Sr = climbing ripple lamination + trough cross-lamination, M = horizontal laminated silt + massive mud.

A direct comparison of the facies proportions of the Markanda terminal fan deposits (Parkash *et al.* 1983, their fig. 3) and the sediments from Tønder-3, -4 and -5 (Fig. 14) reveals a comparable picture. Similar proportions of facies St, Sx, Sh and Sr are seen whereas facies M forms a much larger part of the deposits of the Markanda terminal fan. Mudstones and siltstones would however, constitute a larger proportion of the deposits if the whole of the sediment body had been cored in Tønder-5, judging from petrophysical logs. Facies Sl, Sm and Ci were not quantified for the Markanda terminal fan. Low-angle crossbeds (Sl) were rarely observed probably due to the absence of point bars and the rare presence of side bars in the channels as observed by Parkash *et al.* (1983). 'Poorly bedded sands with mud pebbles' (? facies Sm and Ci) were encountered only on the upper fan (Parkash *et al.* 1983, p. 340). These sediments may be equivalent to the hyperconcentrated flow deposits (Sm and Ci in association B) of Tønder. Finally the thickness of fining upward sequences measured in the Markanda fan (15 to 200 cm) are comparable with those in the deposits at Tønder (association A, mean: 110 cm, association B, mean: 50 cm).

The facies and sequences found in the fluvial Bunter Sandstone deposits of Tønder thus form the basis for a local model of terminal fan sedimentation comparable to a modern terminal fan system (Fig. 15). Facies association A dominates the channels. Fining upward sequences decrease in thickness from the upper terminal fan to the lower terminal fan, associated with an increase in the proportion of mudstones and siltstones. Facies association B is formed by deposition from hyperconcentrated flows in channels on the upper terminal fan. Choking of channels by these sediments may cause the channels to bifurcate and separate fanlobes may thus be created. The rapidly abandoned channels are then filled by mudstones and siltstones (association B). Facies association C is formed in the overbank areas with increasing mud content away from the channels.

Discussion

Fluvial processes in semi-arid regions are very variable. The stream may be largely channelized (*eg* Karcz 1972; Picard & High 1973), purely non-channelized (*eg* McGee 1897; Davis 1938) or show a combination of channel flow and overbank flow (*eg* McKee *et al.* 1967; Stear 1985). Ephemeral streams commonly occur on alluvial fans (*eg* Bull 1972). They may be single-thread (*eg* Karcz

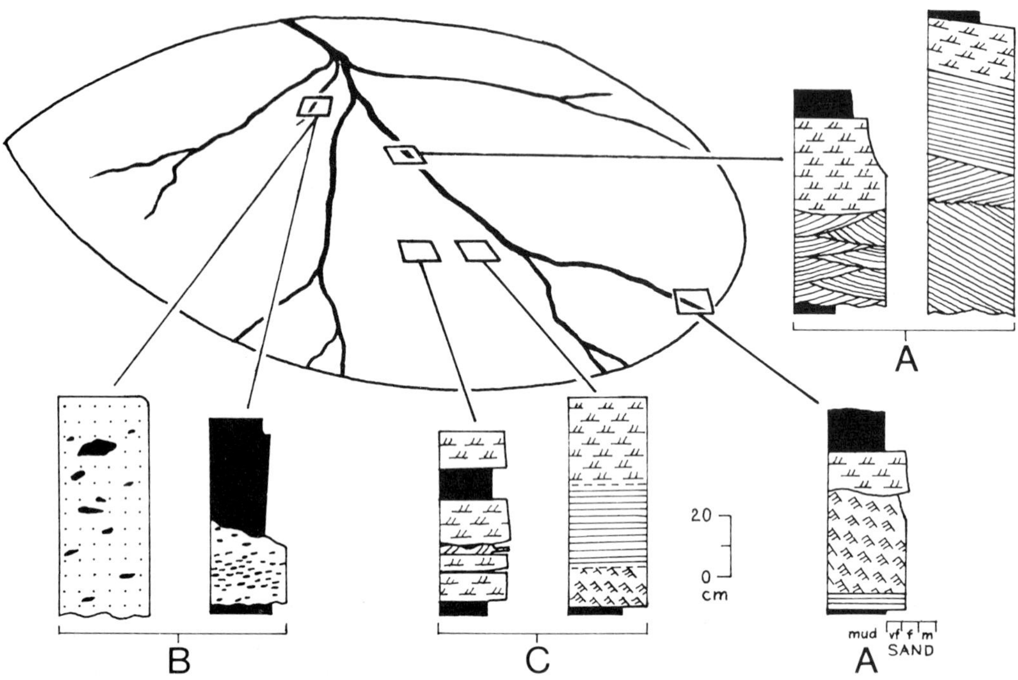

FIG. 15. Facies model for terminal fan deposits of the Bunter Sandstone Formation at Tønder. In the majority of channels facies association A is formed, with the vertically accreted subtype dominant. In the lower part of the fan the distal/marginal subtype of association A is formed. Facies association B is formed in rapidly abandoned channels in the upper part of the fan. The fills of these channels are dominated by massive sand in their upstream part and siltstone in their downstream part. Facies association C is formed in the overbank areas with increasing mud content away from the channels. See Fig. 12 for explanation of symbols.

1972) or may split up into a radiating channel pattern from a single large river channel and form a terminal fan (*eg* Mukerji 1976; Friend 1978; Parkash *et al.* 1983).

The sediments of the fluvial sequence described here were probably laid down on a terminal fan. The terminal fan is believed to be linked to braided systems emanating from the Fennoscandian Border Zone and represented by the deposits of the Skagerrak Formation and Bunter Sandstone Formation (Pedersen & Andersen 1980) in the Danish Subbasin. Towards the S the sediments grade into sabkha and associated shallow lake or inland sea deposits (Clemmensen 1979, 1985). The Skagerrak and Bunter Sandstone Formations are thus an example of a stream system which decays downslope into a major sabkha–inland basin system and are closely analogous to the depositional environment inferred for the Devonian Trentishoe Formation (Tunbridge 1984). The sediments in the Tønder area differ, however, in several ways. Firstly plane-bed lamination (facies Sh and Sl) is much less important. Secondly channel flow deposits (association A and B) are more dominant than overbank flow deposits (association C). This does not necessarily indicate a dominance of channelized flow but rather a higher preservation potential for the channel sands. The fluvial style is considered to be consistent with deposition from a combination of channel flow and overbank flow (*cf* McKee *et al.* 1967; Stear 1985). Thirdly the presence of evaporitic minerals in mudstones and very fine sandstones suggests a close genetic relation between the overbank and abandoned channel environments and the sabkha environment. The differences between these two models is not surprising. Tunbridge (1984) predicted a great variability of facies models for terminal fans. Unfortunately the source of information on modern terminal fans is restricted to observations in only one location, the Markanda terminal fan (Mukerji 1976; Parkash *et al.* 1983). The sedimentary features of the supposed ancient terminal fan deposits described here seem to resemble the Markanda terminal fan environment more closely than the inferred terminal fan deposits described by Tunbridge (1984).

Conclusions

The depositional environment of the fluvial sediment-body in the Bunter Sandstone Formation of the Tønder area is thought to be an ephemeral stream complex comparable to modern terminal fans.

The fluvial sediment body is composed of four units. Each unit reflects either the progradation of a fanlobe or of the entire fan system.

Three flow types occurred: channelized flow with a tendency to meander during falling flood stages (association A), very swift hyperconcentrated channelized flow associated with channel choking (association B), and overbank flow (association C).

Precipitation of gypsum took place near the surface in the overbank areas, distal channels and abandoned channels.

ACKNOWLEDGMENTS: I thank Dansk Olie- og Gasproduktion A/S and Mærsk Olie & Gas A/S for placing core and petrophysical logs at my disposal. Lars Clemmensen, Lynne Frostick and three unknown referees made many suggestions for improving the manuscript. Technical assistance was provided by Pernille Andersen, Ole B. Berthelsen and Rene Madsen. The study was funded by a grant from the Danish Ministry of Energy as part of a PhD thesis supervised by L. Clemmensen.

References

BERTELSEN, F. 1980. Lithostratigraphy and deposition history of the Danish Triassic. *Danmarks Geologiske Undersøgelse, Serie B*, **4**, Copenhagen.

BOERSMA, J. R., VAN DE MEENE, E. A. & TJALSMA, R. C. 1968. Intricate cross-stratification due to interaction of a mega ripple and its lee-side system of backflow ripples (upper-pointbar deposits, Lower Rhine). *Sedimentology* **11**, 147–162.

BULL, W. B. 1972. Recognition of alluvial-fan deposits in the stratigraphic record. *In*: RIGBY, J. K. & HAMBLIN, W. K. (eds) *Recognition of Ancient Sedimentary Environments*. Special Publications of the Society of Economic Palaeontology and Mineralogy, Tulsa, **16**, 63–83.

CLEMMENSEN, L. B. 1979. Triassic lacustrine red-beds and palaeoclimate: The "Buntsandstein" of Helgoland and the Malmros Klint Member of East Greenland. *Geologische Rundschau* **68**, 748–774.

—— 1985. Desert sand plain and sabkha deposits from the Bunter Sandstone Formation (L. Triassic) at the northern margin of the German Basin. *Geologische Rundschau* **74**, 519–536.

COLLINSON, J. D. 1978. Alluvial sediments. *In*: READING, H. G. (ed.) *Sedimentary Environments and Facies*. Blackwell Scientific Publications, Oxford, London, Edinburgh, Boston, Melbourne, 15–60.

DAVIS, W. M 1938. Sheetflood and streamfloods. *Bulletin of the Geological Society of America* **49**, 1337–1416.

FRIEND, P. F. 1978. Distinctive features of some ancient river systems. *In*: MIALL, A. D. (ed.) *Fluvial Sedimentology*. Canadian Society of Petroleum Geologists, Memoir **5**, 531–542.

FROSTICK, L. E. & REID, I. 1977. The origin of horizontal laminae in ephemeral stream channel-fill. *Sedimentology* **24**, 1–9.

HANDFORD, C. 1982. Sedimentology and evaporite genesis in a Holocene continental-sabkha playa basin—Bristol Dry Lake, California. *Sedimentology* **29**, 239–254.

HARDIE, L. A., SMOOT, J. R. & EUGSTER, H. P. 1978. Saline lakes and their deposits: a sedimentological approach. *In*: MATTER, A. & TUCKER, M. E. (eds) *Modern and Ancient Lake Sediments*. Special Publication of the International Association of Sedimentologists **2**, Blackwell Scientific Publications, Oxford, 7–42.

HARMS, J. C., SOUTHARD, J. D. & WALKER, R. G. 1982. *Structures and Sequences in Clastic Rocks*, SEPM Short Course No. 9, Society of Economic Palaeontology and Mineralogy, Calgary.

KARCZ, I. 1972. Sedimentary structures formed by flash floods in southern Israel. *Sedimentary Geology* **7**, 161–182.

KNIGHTON, D. 1984. *Fluvial Forms and Processes*. Edward Arnold, London.

LOWE, D. R. 1982. Sediment gravity flows: II. Depositional models with special reference to the deposits of high-density turbidity currents. *Journal of sedimentary Petrology* **52**, 279–297.

MCGEE, W. J. 1897. Sheetflood erosion. *Bulletin of the Geological Society of America* **8**, 87–112.

MCGOWEN, J. H. & GARNER, L. E. 1970. Physiographic features and stratification types of coarse-grained point bars: Modern and ancient examples. *Sedimentology* **14**, 77–111.

MCKEE, E. D., CROSBY, E. J. & BERRYHILL, H. C. 1967. Flood deposits. Bijou Creek. Colorado, June 1965. *Journal of sedimentary Petrology* **37**, 829–851.

MIALL, A. D. 1977. A review of the braided-river depositional environment. *Earth Science Reviews* **13**, 1–62.

MICHELSEN, O., SAXOV, S., LETH, J. A., ANDERSEN, C., BALLING, N., BREINER, N., HOLM, L., JENSEN, K., KRISTIANSEN, J. I., LAIER, T., NYGAARD, E., OLSEN, J. C., POULSEN, K. D., PRIISHOLM, S., RAADE, T. B., SØRENSEN, T. R. & WÜRTZ, J. 1981. *Kortlægning af potentielle geotermiske reservoirer i Danmark. Danmarks Geologiske Undersøgelse, Serie B*, **5**, Copenhagen.

MUKERJI, A. A. 1976. Terminal fans of inland streams in Sutlej–Yamuna Plain, India. *Zeitschrift für Geomorphologie N. F.* **20**, 190–204.

NIELSEN, B. L. 1983 *Tønder-4 and Tønder-5. Bunter Sandstone Formation, Core Analysis and Sedimentology*. Dansk Olie & Naturgas A/S, Internal Report.

PARKASH, B., AWASTHI, A. K. & GOHAIN, K. 1983. Lithofacies of the Markanda terminal fan, Kurukshetra district, Haryana, India. *In*: COLLINSON, J. D. & LEWIN, J. (eds) *Modern and Ancient Fluvial Systems*. Special Publication of the International Association of Sedimentologists **6**, 337–344.

PEDERSEN, G. K. & ANDERSEN, P. R. 1980. Depositional environments, diagenetic history and source areas of some Bunter Sandstones in northern Jutland. *Danmarks Geologiske Undersøgelse, Årbog 1979*, 69–93.

PICARD, M. D. & HIGH, L. R. 1973. *Sedimentary Structures of Ephemeral Streams*. Developments in Sedimentology 17. Elsevier, Amsterdam.

PLINT, A. G. 1983. Sandy fluvial point-bar sediments from the Middle Eocene of Dorset, England. *In*: COLLINSON, J. D. & LEWIN, J. (eds) *Modern and Ancient Fluvial Systems*. Special Publication of the International Association of Sedimentologists **6**, 355–368.

REID, I. & FROSTICK, L. E. 1987. Flow dynamics and suspended sediment properties in arid zone flash floods. *Hydrological processes*, **1**, 239–253.

RHYS, G. H. 1974. *A proposed standard lithostratigraphic nomenclature for the southern North Sea and an outline structural nomenclature for the whole of the (UK) North Sea*. Report of the Institute of Geological Science, London, 17.

SINGH, I. B. 1972. On the bedding in the natural-levee and point-bar deposits of the Gomti River, Uttar Pradesh, India. *Sedimentary Geology* **7**, 309–317.

SMITH, R. M. H. 1980. The lithology, sedimentology and taphonomy of flood-plain deposits of the Lower Beaufort (Adelaide Subgroup) strata near Beaufort West. *Transactions of the Geological Society of South Africa* **83**, 399–413.

SNEH, A. 1983. Desert stream sequences in the Sinai Peninsula. *Journal of sedimentary Petrology* **53**, 1271–1279.

STEAR, W. M. 1983. Morphological characteristics of ephemeral stream channel and overbank splay sandstone bodies in the Permian Lower Beau-Group, Karoo Basin, South Africa. *In*: COLLINSON, J. D. & LEWIN, J. (eds) *Modern and Ancient Fluvial Systems*. Special Publication of the International Association of Sedimentologists **6**, 405–420.

—— 1985. Comparison of the bedform distribution and dynamics of modern and ancient sandy ephemeral flood deposits in the southern Karoo Region, South Africa. *Sedimentary Geology* **45**, 209–230.

TUNBRIDGE, I. P. 1984. Facies model for a sandy ephemeral stream and clay playa complex; the Middle Devonian Trentishoe Formation of North Devon, UK. *Sedimentology* **31**, 697–715.

WILLIAMS, G. E. 1966. Planar cross-stratification formed by the lateral migration of shallow streams. *Journal of sedimentary Petrology* **36**, 742–746.

—— 1971. Flood deposits of the sand-bed ephemeral streams of central Australia. *Sedimentology* **17**, 1–40.

H. OLSEN, Institute of General Geology, Øster Voldgade 10, DK-1350 Copenhagen K, Denmark. *Present address:* Geological Survey of Greenland, Øster Voldgade 10, DK-1350 Copenhagen K, Denmark.

Alluvial fan dissection: relationships between morphology and sedimentation

A. M. Harvey

SUMMARY: Aggradation and dissection on alluvial fans influence the coupling between sediment source areas and main channels. Previous alluvial fan models emphasize the influence of drainage area and sediment character on overall fan morphometry and proximal–distal sedimentary variations. A model is proposed for the morphology and sedimentology of dry-region alluvial fans, based on crusted and non-crusted Quaternary fans in SE Spain, differentiating between aggradational and dissectional tendencies. Many fans show fan-head trenching, with the channel emerging on the fan surface at a mid-fan intersection point. Distally either aggradation or headcut development may occur, with headcut development and trenching leading ultimately to the total dissection of the fan surface. Three sets of factors determine intersection-point behaviour: fan surface crusting, discrepancy between fan and channel slopes, and channel morphology. Multiple regression models are developed to account for variations in fan and channel slopes and channel morphology, taking into account source area and sediment variations. Intersection point trenching is most common on steep crusted fans, where the discrepancy between fan slope and channel slope is greatest, and where channel widths are limited. Arid region fans from the American Southwest are contrasted with the Spanish fans. Neither the non-crusted nor the crusted American fans examined show any tendency towards intersection-point trenching. Their fan slope/channel slope discrepancies are less and their channels are wider, the result of both differing Quaternary sequences and contemporary environments between Spain and the American Southwest. The implications are important for the continuity of dry-region fluvial systems and for the stratigraphy of alluvial fan sequences. Simple fans with distal area aggradation show a simple stacked stratigraphy, but fans with a propensity for mid-fan and distal trenching may show a complex inset stratigraphy.

Alluvial fans are important zones in the geomorphology and sedimentology of dry-region fluvial systems. Whether they aggrade or are dissected determines the strength of coupling between mountain sediment course areas and either distal axial drainage or the sites of basinal deposition. Aggrading fans may act as buffers within the system separating source areas from axial drainage, and although their deposits may contain a complete sedimentary record relating to the supply of sediment from the source area, there may be only poor correlation with axial or basinal alluvial sequences. On the other hand, dissected fans offer no buffer and there will be continuity from source area to axial drainage; as a result, there may be better correlation of alluvial sequences throughout the system.

Fan-head trenches are common on many fans. In the apex area, the channel has a lesser slope than the fan surface, and is entrenched, emerging onto the fan surface at a mid-fan intersection point (Hooke 1967; Bull 1977). Behaviour at the intersection point is important in determining whether the distal part of the fan dissects or aggrades. Most fans described in the literature are characterized by intersection point and distal aggradation (Bull 1977; Wasson 1974), but in Spain intersection-point headcut development and distal fan trenching have been observed (Harvey 1984*a*, *b*), which, when it occurs throughout the distal zone, creates channel and sedimentary continuity through the fan system.

Alluvial fan models, based largely on fans in the American Southwest, deal primarily with fan sedimentology or fan morphology. Tectonics are often seen as a primary control of fan deposition (Bull 1964), with climate influencing the depositional regime (Williams 1973; Wasson 1979). Depositional processes include debris-flow, sheetflood and stream-flood (Blissenbach 1954; Hooke 1967) and sediments may show proximal to distal facies and particle size variations (Rust 1979). Vertical sequences are generally of the 'Trollheim' or 'Scott' types (Miall 1978) and may show upward coarsening (Steel *et al.* 1977; Rust 1979). Morphological studies of fans relate morphometric properties to source-area characteristics. In particular, fan area has been related to drainage area (*eg* Kesel 1985; Denny 1965), and, more importantly, fan slope has been related to drainage area, to rock type and to sediment calibre (Blissenbach 1952; Bull 1962; Denny 1965; Hooke 1968). Some studies deal with fan-head trenching and surface segmentation, attrib-

From FROSTICK, L. & REID, I. (eds), 1987, *Desert Sediments: Ancient and Modern,* Geological Society Special Publication No. 35, pp. 87–103.

uting trenching to external changes such as tectonic activity or climatic change, to ageing of the system and the progressive and consequential diminution of sediment yields (Eckis 1928), or to intrinsic thresholds related to critical fan-head slopes and alternating depositional mechanisms (Schumm 1977).

There have been relatively few studies of the relationships between fan sedimentology and geomorphology especially in the context of the progressive development of alluvial fans (Harvey 1984*a,b*). This paper addresses the question of intersection-point trenching thresholds, and attempts to identify the factors distinguishing *distally trenching* from *distally aggrading* fans. It is based on evidence from Quaternary alluvial fans in semi-arid SE Spain, contrasting their morphometry and development with fans in the arid American Southwest.

Based on Spanish and American fans, a simple classification of intersection point and distal fan behaviour can be proposed that relates fan surface and channel longitudinal profiles (Fig. 1). The profile segments are depicted in Fig. 1 as straight segments, fan surface slopes and channel slopes reflecting the influence of appropriate controlling factors. If these should vary down-fan, for example through change in process, a diminution in sediment calibre, or a drainage increment brought about by confluence, the segments might each be concave but the general principles of the model would still apply. Aggradational or dissectional behaviour at the intersection point would be determined by the relationship between sediment supply and stream power, in the context of the threshold of critical stream power (Bull 1979). Under conditions of excessive sediment supply, profile A might apply, where the channel emerging onto the fan surface at the intersection point aggrades the distal portion of the fan, and by progradation buries the original fan surface. Ultimately trenched channel and distal channel profiles might be similar. This relationship is often assumed to apply to fans of the American Southwest and has been ascribed to tectonically-induced fan-head trenching (Bull 1964). In areas that are less active tectonically, or with a lesser sediment supply, profiles B and C might apply. Here the trenched channel emerges onto the fan surface at the intersection point and inherits a slope from the fan surface steeper than that of the channel within the fan-head trench. However a rapid increase in width on the unconfined distal parts of the fan (Harvey 1984*c*), may cause a loss of unit steam power and therefore distal deposition at a slope intermediate between proximal fan and channel slopes (profile B). If there is no erosion and little deposition on the distal fan, profile C results. On some of the Spanish fans intersection-point headcuts develop (profile D). In these circumstances the increase in slope has caused channel incision. These headcuts penetrate calcrete-crusted fan surfaces and, once formed, are persistent features protected by the indurated, calcrete 'caprock', unlike ephemeral discontinuous gullies in unconsolidated alluvium. The trenched channels formed by the incision prevent development of excessive channel width, thereby maintaining unit stream power. Down-fan, a secondary intersection point will form at which further headcutting may occur (profile D). Continuous development of intersection point headcuts will lead to the total trenching of the fan surface (profile E), linking the source-area channel system with the basin axial drainage. Development of profiles D and E might be expected where sediment supply is low, where fan surfaces are crusted and where a major discrepancy exists between fan and channel slopes at the intersection point.

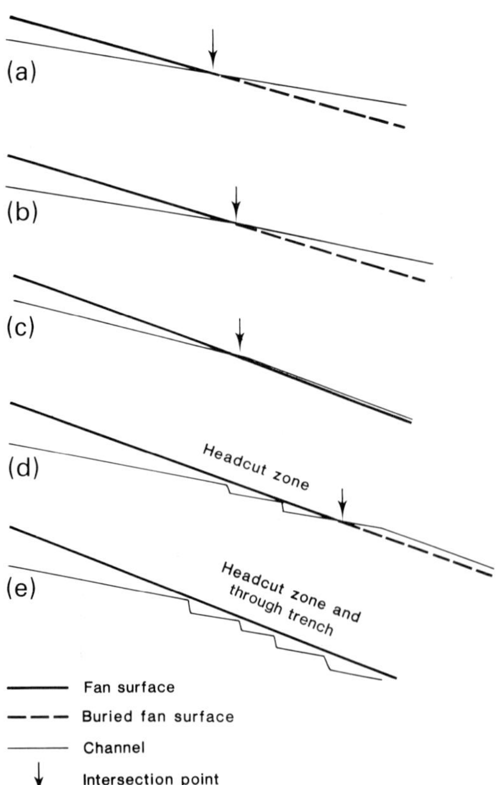

FIG. 1. Schematic models of fan surface–channel profile relationships, for explanation see text.

The Quaternary alluvial fans of SE Spain

Quaternary alluvial fans occur throughout SE Spain (Fig. 2) within an area that, today, has a semi-arid climate. Mean annual precipitation ranges from 330 mm in Alicante to 170 mm in parts of Almeria province (Geiger 1970). During the Quaternary the area experienced a cool, aridic climate (Amor & Florschutz 1964; Sabelberg 1977). The fans occur at the margins of mountain ranges of the Betic Cordillera *sensu lato*, in diverse tectonic settings, some at faulted Neogene basin margins, others mantling early Quaternary pediment surfaces (Harvey 1978, 1984*b*). Few show evidence of tectonic disturbance. The tectonic setting influences fan location rather than fan morphology or sedimentology.

The fans studied can be categorized broadly into four groups, on the basis of source-area geology (Fig. 2):

(1) North of Alicante, fans issue from the limestones of the Pre-Betic ranges.

(2) In central Murcia, fans issue from the low-grade metamorphic rocks and Triassic sedimentary rocks of the Sierra de Carrøscoy.

(3) In S Murcia and Almeria provinces, fans issue from Neogene sedimentary rocks and Triassic sedimentary and low-grade metamorphic rocks, groups occur near Lorca, near Vera, and near Tabernas (N of Almeria).

(4) In southern Murcia and Almeria provinces, fans issue from the high-grade metamorphic rocks of the Sierra de Almenara, Sierra de Filabres and the Sierra de Alhamilla, and from the volcanic rocks of the Cabo be Gata ranges, E of Almerøa.

Throughout the area, the fans show some similarity of development, with earlier phases predominantly of aggradation followed by later phases dominantly of dissection (Harvey 1984*b*). The aggradation phases appear to be predomi-

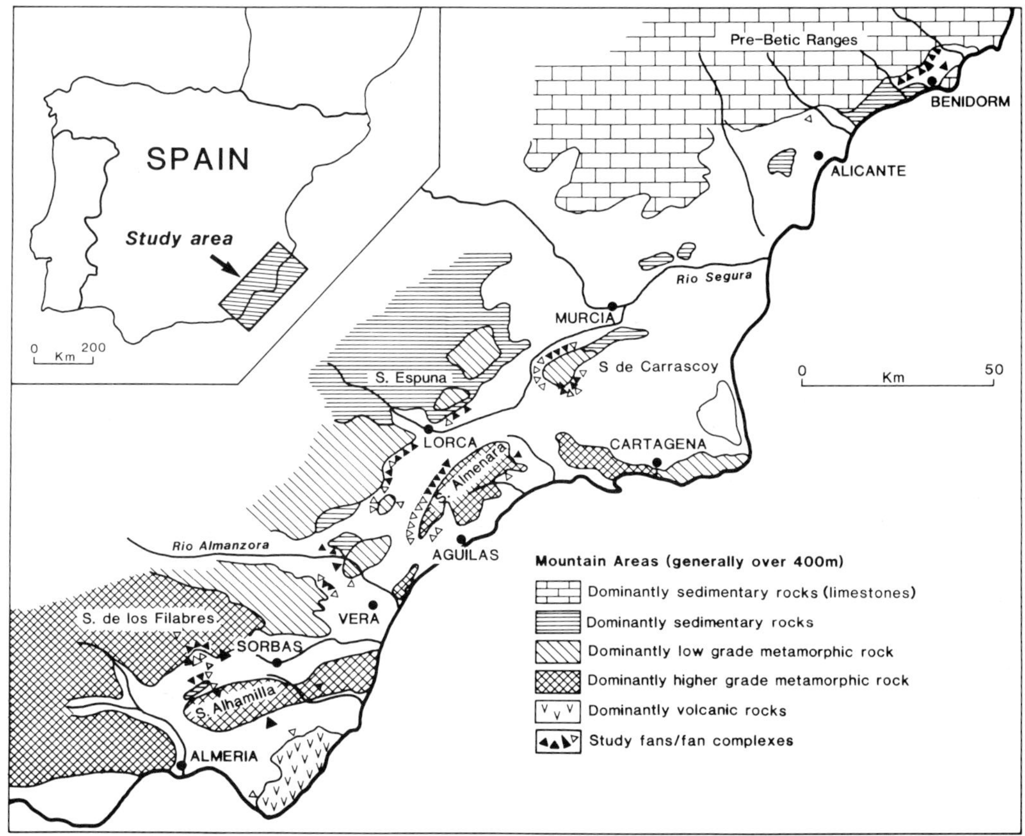

FIG. 2. Location of study fans in SE Spain. Closed triangles indicate fans for which both fan surface and fan channel data are available; open triangles indicate that fan slope data only are available.

nantly pre-Würm in age but locally may be younger. They include a range of deposits from debris flows, especially in small fans from limestone and other sedimentary rock sources, to braided stream deposits, especially in larger fans and in those fed by the high-grade metamorphic areas (Harvey 1984*a*). Within the overall aggradational sequences, there is evidence for major periods of dissection, especially in the Pre-Betic region and Carrascoy fan groups. Within these groups, and those southern fans fed by sedimentary areas, there is a progressive vertical decrease in the importance of debris flow and other fines-rich deposition, similar to that described by Wasson (1979) from Australian fans. The aggradation sequences are interpreted as the result of long-term, Quaternary erosion from the source areas, onto which climatically induced fluctuations in sediment production are superimposed (Harvey 1984*b*).

On many fans, both capping the aggradational sequences and occurring buried within the sequences, are calcrete crusts, with strongly indurated crusts on pre-Würm surfaces and less mature, weaker crusts capping the Würm deposits (Dumas 1969). They are best developed on the Pre-Betic region and Carrascoy fan groups and least well developed on some fans fed from high-grade metamorphic areas (Harvey 1984*b*). The induration of the calcrete crusts on many of the Spanish fans creates contrasts with the fans of the American Southwest, where although a number do have crusted surfaces (Lattman 1973), many are much less strongly cemented.

The later phases of development on the Spanish fans are characterized by dissection, by major fan-head trenching and in some cases by trenching throughout the fan from fan-head to distal locations. During overall dissection, major periods of aggradation have occurred in both the Würm and the Holocene (Harvey 1984*b*), forming terraces within the fan-head trenches. The dissection sequence is most complex in the Pre-Betic region fans, where up to four cut-and-fill stages have been recognized. Elsewhere two stages are common, but to the S, and especially in the high-grade metamorphic areas, the complexities within the dissection sequence are represented by only one terrace deposit, or simply by an erosional bench. This regional variation may reflect a later onset of dissection on some southern fans or a lower sensitivity to trenching thresholds (Harvey 1984*b*). The long-term trend from aggradation to dissection reflects long-term tectonic, climatic and developmental trends (Eckis 1928). The style of dissection, whether simply by fan-head, or by both fan-head and distal trenching is controlled by intersection-point behaviour.

At the present time, channel behaviour is wholly dissectional, at least within the fan-head trenches. For 38 fans on which the main channels have been examined, four channel classes can be identified (Table 1). Headcut channels (types D and E, Fig. 1) are most common in the Pre-Betic region and Carrascoy fan groups whereas simple channels (type B), with fan-head trenches and distal aggradation are most common in the southern fan groups on both crusted and non-crusted fans.

Fan morphometry

Four factors might be expected to lead to the development of intersection-point trenching: low sediment yields; crusted fan surfaces; large discrepancies between fan and channel slopes; and confined narrow channels. Of these, the first two relate to the climatic and geological setting. The second two relate to fan morphometry. Studies of American fans demonstrate, within major types of source-area geology, statistical correlations between drainage area and fan morphometric properties, including fan slope (Denny 1965; Hooke 1968; Bull 1962; Blissenbach 1952). Similarly for both fan and non-fan channels in climatically and geologically similar arid and semi-arid regions, the channel slope and

TABLE 1. *Occurrence of channel types, number by fan group*

	Crusted fans			Non-crusted fans
Fan group	Headcut channel types D/E*	Scoured channel type C	Simple channel type B	Simple channel type B
Pre-Betic region fans	5	1	1	0
Carrascoy region fans	3	2	3	0
Southern fans, sedimentary source	0	3	3	3
Southern fans, metamorphic source	1	0	9	7

*For fan type (A, B, *etc.*), see Fig. 1.

width correlate with drainage area (Wells 1977; Leopold & Bull 1979; Denny 1965).

Fan slopes on proximal fan surfaces have been measured on 77 Spanish fans (Fig. 2). Channel slope and width within the fan-head trench have also been measured on 36 of these. On 10 channels where complete channel surveys were made, a mean value for the slope and width of non-incising, non-aggrading reaches within the fan-head trench is used. Elsewhere, similar sites were selected in mid-trench. In addition to the 36 fan channels, 44 non-fan channels were measured as a basis for comparison, giving a total of 80 channels.

The logarithmic regression relationships may be expressed in the general form:

$$G=aA^b$$
$$S=fA^g$$
$$W=hA^k$$

where G and S are fan and channel slopes respectively, W is channel width (m) and A is drainage area (km^2).

For fan slopes (Fig. 3), there is a moderate overall correlation with drainage area for the log-normalized data ($r=-0.638$, Table 2) when all data points are included, but the scatter is considerable. When the fan groups and sub-groups are considered separately, the correlations generally improve, with the exception of those in the Pre-Betic region and the Carrascoy groups. On its own the Carrascoy relationship is not statistically significant, perhaps due to localized tectonic influences on this fan group, but when considered with the Pre-Betic region group (all northern fans, Table 2) the correlation improves. These two fan groups are rich in debris flows (Harvey 1984*a*) and generally have steeper slopes per drainage area than the fluvially-dominated fans further S. As a whole, the values of the constant a and the exponent b accord generally with the range of values derived for N American fans ($a=0.027$ to 0.14; $b=-0.16$ to -0.32).

For the relationship between channel slope and drainage area (Fig. 3, Table 2) there is again considerable scatter about the overall best-fit regression line. There is little difference between channels on crusted and non-crusted fans, nor between fan and non-fan channels, although non-fan channels show a much stronger correlation. Again, channel slopes on the northern fans tend to plot higher than the rest; but the difference is less than it is in the fan slope plot.

For the relationship between channel width and drainage area (Fig. 3, Table 2) there is again considerable scatter and little difference between fan and non-fan channels, nor between the separate groups. However, as has been observed in the American Southwest (Van Arsdale 1982), channels on crusted fans are markedly narrower than those on non-crusted fans. Residuals from the 'all fan channels', width *vs* drainage area regression (Table 2), for channels on the eight non-crusted fans are significantly different (at the 1% level by Students 't' test), from those for channels on the crusted fans. As a whole the Spanish channels are much narrower than those on many American arid-area fans (Denny 1965).

Fan and channel slopes correlate inversely with drainage area, reflecting the greater propensity for debris-flow generation, and the coarser clasts delivered from smaller source areas (Harvey 1984*a*). Channel width correlates directly with drainage area reflecting the greater runoff generated by large catchments. These tendencies are emphasized when multiple, rather than bivariate, correlations for selected groups are considered (Table 3). In addition to drainage area, drainage basin relief and slope are considered as independent variables determining fan slope; a channel sediment index is added to the channel slope and width correlations; while channel slope is seen to be an influence on channel width.

The multiple correlation coefficients are a significant improvement on the bivariate correlations, in all cases statistically significant to at least the 5% level. In most cases most of the variance is accounted for by taking into account drainage area, drainage basin relief and drainage basin slope. Only for the width prediction on the northern fan channels do other channel variables improve the explanation that is based on drainage-basin variables alone. The best regression relationships are quoted in Table 3, which, although they allow reasonable prediction, can only be used to differentiate between distally trenching and aggrading tendencies in the broadest sense. More important, perhaps, are how the variables relate to one another within individual fans.

Controls of distal fan trenching

For 10 of the fans, complete channel profiles have been surveyed, four of which are illustrated (Fig. 4) together with the variations in width down-channel (Fig. 5). Corachos, in the Carrascoy group, has multiple headcuts and distal trenching, and approximates to a type E profile shown in Fig. 1. Ceporro, in the southern group from near Tabernas, fed from an area of sedimentary rocks, has intersection point scour and only limited distal aggradation (Type C, Fig. 1). Villaescusa, in the southern group, fed from a metamorphic area in the Sierra de Almenera, is a simple distally-aggrading fan (Type B, Fig. 1). On

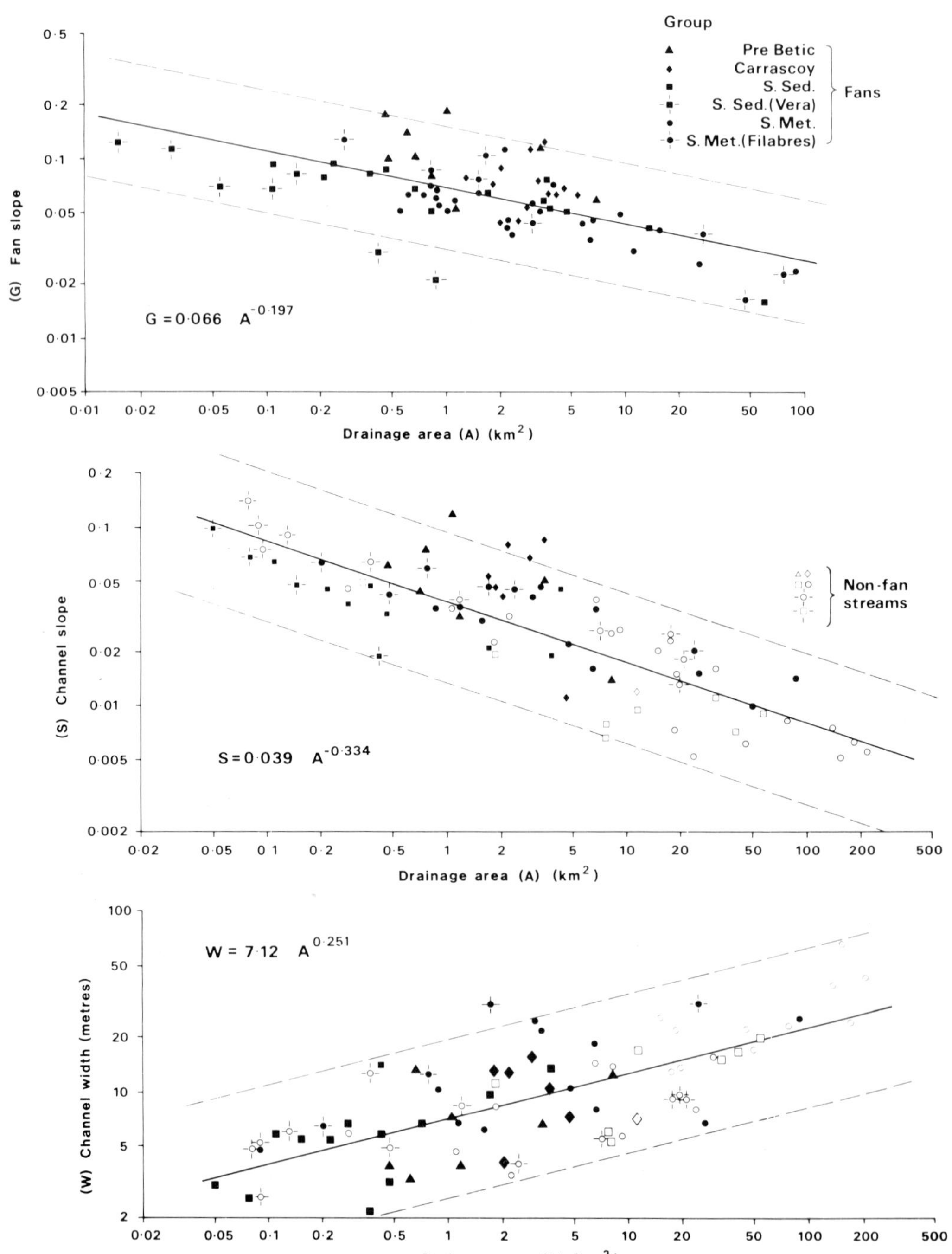

FIG. 3. Plots of fan slope, channel slope and channel width against drainage area for fans in SE Spain; fan or channel groups are identified by different symbols. Regression lines shown are for the overall relationships (see Table 2); dashed lines are the 95% confidence limits.

TABLE 2*a*. *Fan morphometry, bivariate regression analyses of fan slope* vs *drainage area (km²)*, $G = aA^b$

	n	Correlation coefficient (r)	Const. a	Exp. b	Standard error of the estimate (log units)
All fans	77	−0.638	0.066	−0.197	0.169
Pre-Betic region fans	9	−0.447†	0.102	−0.225	0.177
Carrascoy region fans	15	−0.373†	0.092	−0.286	0.153
All northern fans	24	−0.533	0.095	−0.291	0.162
Southern fans (sedimentary areas)	19	−0.708	0.054	−0.178	0.171
Vera sub-group	6	−0.943	0.023	−0.436	0.106
Southern fans (metamorphic areas)	34	−0.839	0.063	−0.248	0.106
Filabres sub-group	8	−0.937	0.084	−0.038	0.121

TABLE 2*b*. *Fan morphometry, bivariate regression analyses of channel slope* vs *drainage area (km²)*, $S = fA^g$

	n	r	f	g	$s.e.$
All channels	80	−0.831	0.039	−0.334	0.202
All fan channels	36	−0.550	0.042	−0.220	0.208
Non-fan channels	44	−0.910	0.034	−0.346	0.168
Northern fan channels	13	−0.511*	0.058	−0.410	0.265
Southern sedimentary areas (all channels)	21	−0.895	0.027	−0.327	0.169
Southern metamorphic areas (all channels)	45	−0.883	0.043	−0.342	0.173
Southern sedimentary areas (fan channels)	8	−0.522†	0.032	−0.180	0.191
Southern metamorphic areas (fan channels)	15	−0.805	0.043	−0.244	0.128
Crusted fans, channels	28	−0.440	0.041	−0.210	0.229
Non-crusted fans, channels	8	−0.860	0.047	−0.253	0.135

TABLE 2*c*. *Fan morphometry, bivariate regression analyses of channel width (m)* vs *drainage area (km²)*, $W = hA^k$

	n	r	h	k	$s.e.$
All channels	80	0.721	7.12	0.251	0.218
All fan channels	36	0.552	7.69	0.258	0.243
Non-fan channels	44	0.816	6.57	0.260	0.196
Northern fan channels	13	0.504*	6.31	0.336	0.221
Southern sedimentary areas (all channels)	21	0.838	6.83	0.273	0.184
Southern metamorphic areas (all channels)	45	0.676	7.72	0.230	0.235
Southern sedimentary areas (fan channels)	8	0.722	7.58	0.313	0.195
Southern metamorphic areas (fan channels)	15	0.398*	9.77	0.176	0.287
Crusted fans, channels	28	0.443	6.97	0.221	0.239
Non-crusted fans, channels	8	0.716	12.47	0.198	0.172

† Not significant.
* Significant only at the 10% level. All other correlations significant to at least the 5% level and above.

Corachos, the intersection point coincides with a local steepening in the fan surface profile, accentuating the distal fan-trenching tendency. Width is fairly constant throughout, but with a local increase where aggradation is taking place upstream of the upper headcut. There is a local decrease in width in the steeper reach at the intersection point, accentuating scour towards the lower headcut. On Ceporro, both fan and channel slopes are slightly concave but with some discrepancy between the two near the intersection point. Width shows a slight decrease towards the zone of intersection-point scour, then a rapid increase on the distal fan surface. On Villaescusa, the difference between fan and channel slopes is small. Width shows a decrease in mid-fan but then a rapid increase through the intersection point into the zone of distal-fan aggradation. Also shown on Figs 4 and 5 is Lisbona fan, an immature scouring fan, where dissection of the fan surface is only partial. Dissected zones of lower slope and sieve-lobe deposition (Bull 1977) contrast with the steeper-scouring zones. Width shows considerable variation, partly reflecting localization of aggradational and scouring tendencies.

From the evidence of the detailed profiles, intersection-point scour and trenching appear to be closely related to large discrepancies between fan and channel slopes, occuring on crusted fans with relatively narrow channels. Evidence of the importance of profile steepening is indicated by Table 4. On 21 fans, data are available for distal fan slopes below the intersection points. Of the eight headcut and scoured channels, all but one show a greater than 20% slope increase at the intersection point, whereas the majority of the simple channels show a slope increase of less than 20%. The discrepancy between fan and channel slopes, indicative of the potential steepening at the intersection point, may be expressed by residuals from a regression equation expressing the relationship between the two variables. Similarly, in the absence of hydraulic information, an indication of the relative stream power may be obtained by examining the residuals from the channel width to drainage-area regression (Fig. 3, Table 2). Using these overall relationships, residuals for the headcut channels can be contrasted with those for the other channels. For the 36 fans for which fan slope, channel slope and width data are available, the regression relationship between the channel and fan slopes has been calculated (Fig. 6). The headcut channels tend to plot with high negative residuals, indicative of large fan–channel slope discrepancies. They are significantly different at the 5% level (Students 't' test) from those of simple and scoured channels. Similarly the same fan type tends to plot with high negative residuals from the overall width to drainage area regression ($W=7.12\,A^{0.25}$, Fig. 3, Table 2), thus suggesting the maintenance of higher unit stream power downfan. The residuals for headcut and scoured channels are significantly different from those for the simple channels at the 1% level.

TABLE 3. *Multiple correlation and regression for selected fan groups*

	Correlation coefficients*						Best regression equations
Fan groups	n	$R_{G\cdot A}$	$R_{G\cdot AR}$	$R_{G\cdot ARB}$	$R_{G\cdot AP}$	$R_{G\cdot APS}$	
All fans	77	−0.638	0.788	$\underline{0.801}$			$G=0.0145\ A^{-0.276}\ R^{0.352}\ B^{0.282}$
Northern fans	24	−0.533	0.852	$\underline{0.878}$			$G=0.0047\ A^{-0.464}\ R^{0.635}\ B^{0.547}$
Southern fans,							
sedimentary areas	19	−0.708	$\underline{0.744}$	0.744			$G=0.0044\ A^{-0.375}\ R^{0.464}$
metamorphic areas	34	−0.839	$\underline{0.894}$	0.894			$G=0.0066\ A^{-0.361}\ R^{0.403}$
	n	$R_{S\cdot A}$	$R_{S\cdot AR}$	$R_{S\cdot ARB}$	$R_{S\cdot AP}$		
All channels	80	−0.831	$\underline{0.884}$	0.885	0.832		$S=0.0012\ A^{-0.487}\ R^{0.609}$
Northern fan channels	13	−0.511	$\underline{0.776}$	0.777	0.615		$S=0.0003\ A^{-0.679}\ R^{0.866}$
Southern channels,							
sedimentary areas	21	−0.895	0.897	$\underline{0.902}$	0.896		$S=0.0082\ A^{-0.415}\ R^{0.099}\ B^{-0.328}$
metamorphic areas	45	−0.883	0.885	$\underline{0.908}$	$\underline{0.905}$		$S=0.185\ A^{-0.179}\ R^{-0.100}\ B^{0.526}$
						or	$S=0.054\ A^{-0.351}\ P^{-0.058}$
	n	$R_{W\cdot A}$	$R_{W\cdot AR}$	$R_{W\cdot ARB}$	$R_{W\cdot AP}$	$R_{W\cdot APS}$	
All channels	80	0.721	0.723	$\underline{0.740}$	0.722	0.734	$W=0.111\ A^{-0.028}\ R^{0.553}\ B^{-0.576}$
Northern fan channels	13	0.504	0.512	$\underline{0.519}$	0.519	$\underline{0.655}$	$W=0.072\ A^{-0.030}\ R^{0.668}\ B^{-0.331}$
Southern channels,						*or*	$W=32.78\ \mathrm{A}^{0.530}\ P^{-0.111}\ S^{0.421}$
sedimentary areas	21	0.835	0.839	$\underline{0.906}$	0.838	0.863	$W=0.017\ A^{-0.134}\ R^{0.756}\ B^{-0.974}$
metamorphic areas	45	0.676	0.679	$\underline{0.681}$	0.677	0.678	$W=0.204\ A^{-0.025}\ R^{0.477}\ B^{-0.531}$

G=fan slope, S=channel slope, W= width (m), A=drainage area (km^2), R=drainage basin relief (m), B=drainage basin mean slope.
P=course sediment index (cm), 5th largest clast in 0.25 m^2 quadrat on bar surface.
*Correlation coefficients underlined are those for the best regression equations (see right), those with the lowest standard errors of the estimate.

FIG. 4. Examples of profiles for three Spanish fans: Corachos is type E, Ceporro of type C and Villaescusa is type B (see Fig. 1). Lisbona illustrates an immature dissecting fan.

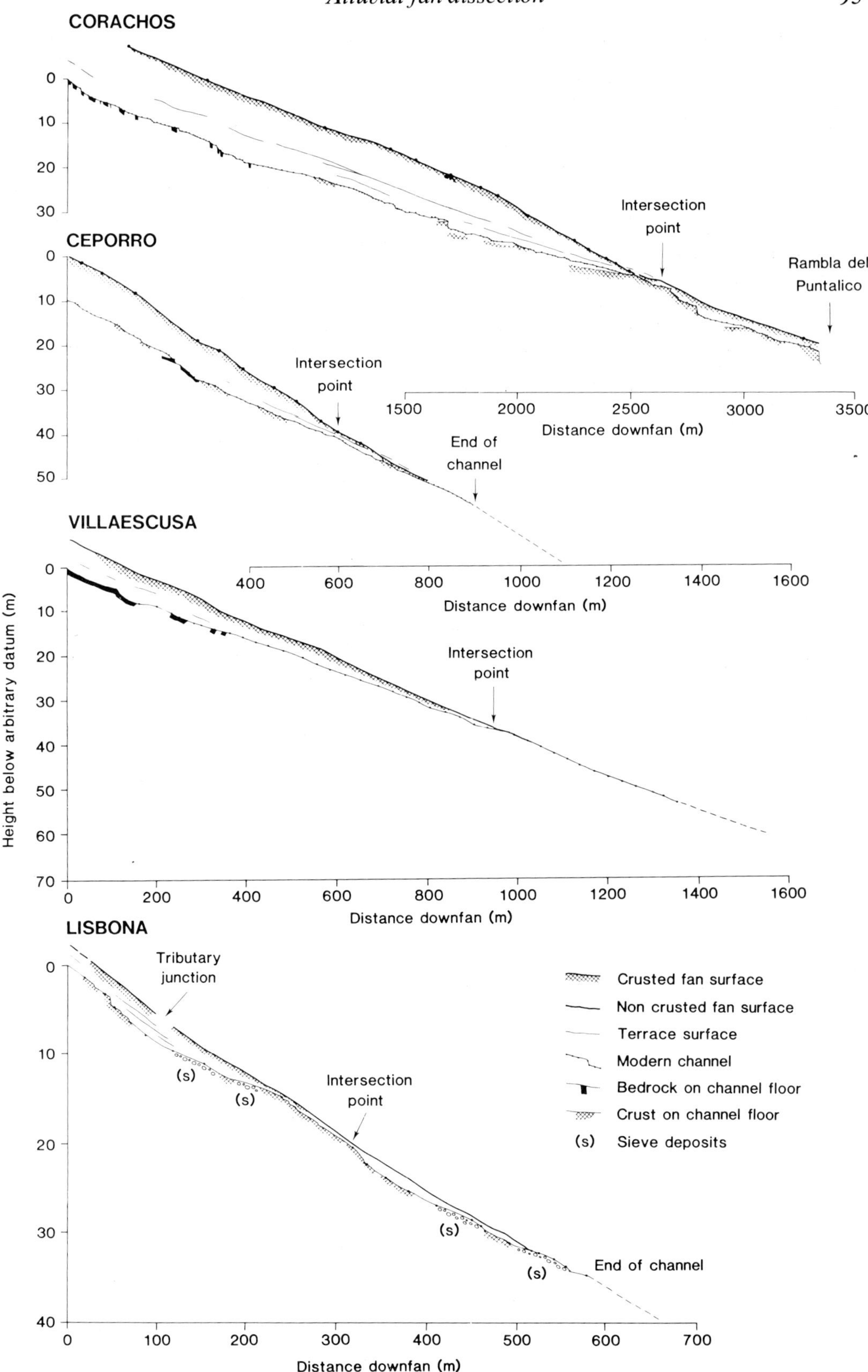
CORACHOS
Intersection point
Rambla del Puntalico
1500 2000 2500 3000 3500
Distance downfan (m)
CEPORRO
Intersection point
End of channel
400 600 800 1000 1200 1400 1600
Distance downfan (m)
VILLAESCUSA
Height below arbitrary datum (m)
Intersection point
0 200 400 600 800 1000 1200 1400 1600
Distance downfan (m)
LISBONA
Tributary junction
(s)
Intersection point
End of channel
0 100 200 300 400 500 600 700
Distance downfan (m)
Crusted fan surface
Non crusted fan surface
Terrace surface
Modern channel
Bedrock on channel floor
Crust on channel floor
(s) Sieve deposits

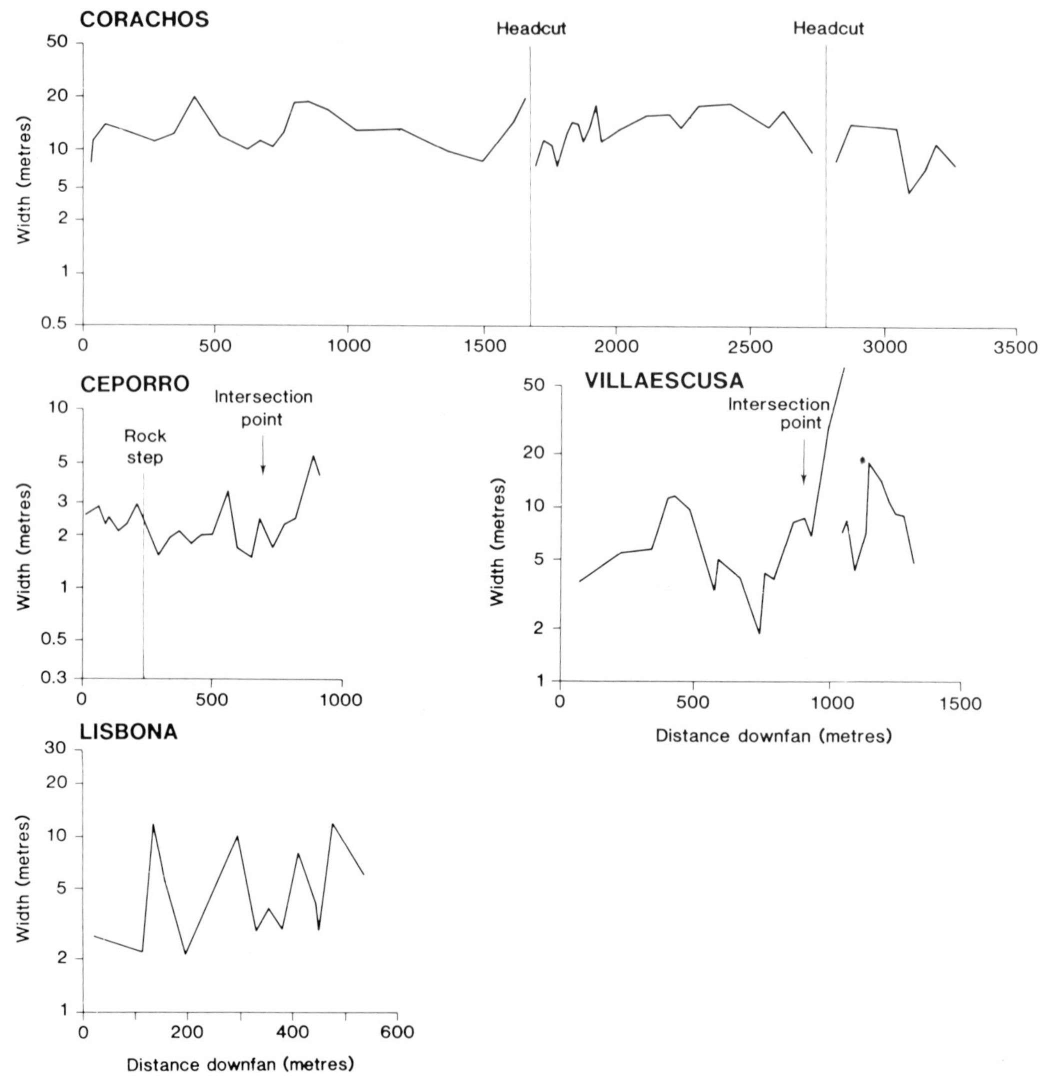

FIG. 5. Channel width characteristics on Corachos, Ceporro, Villaescusa and Lisbona fans.

TABLE 4. *Downfan channel slope percentage change at intersection points, number of fans by channel type*

	Increase		Decrease
	>20%	0–20%	
Headcut channels	2	1	0
Scoured channels	5	0	0
Simple channels			
(crusted fans)	2	4	2
(non-crusted fans)	1	3	1

When the two sets of residuals are plotted (Fig. 7), a clear distinction is apparent. The diagonal shown on Fig. 7 separates high fan–channel slope discrepancies and/or channels that are comparatively narrow per unit drainage area from low fan–channel slope discrepances and/or comparatively wide channels. All but one of the headcut channels plot to the lower left of the diagonal and the majority of the others plot to the upper right, clearly indicating the importance of large slope discrepancies and small widths in headcut channels. Chi-squared tests have been applied to these distributions, using the four cells defined by abscissa and ordinate zero, and using two and three parallel diagonal cells, the latter being defined at values equivalent to 0.2 orthogonal to, and on either side of, the central diagonal. In all cases the distributions of headcut channels are significantly different from the other groups at the 5% level at least.

Also shown on Fig. 7 and in Table 5 are data for seven fans from the American Southwest. There are three crusted and four non-crusted fans, but all have simple channels. All but one crusted fan plot well to the upper right in Fig. 7, further confirming the discriminant role of the fan–channel slope difference and width drainage area relationship in intersection-point behaviour. Interestingly, although all the American fans appear to be aggrading in their distal portions, channel slopes increase at the intersection points. Characteristic morphology is illustrated by the Grotto Canyon fan (Fig. 8). The constituent deposits of this fan, exposed in the fan-head trench, are dominantly channel gravels. The fan has backfilled into the mountain catchment, but subsequently has been dissected by a fan-head trench. Successive stages of the dissection have left erosional terraces which grade into the fan surface at intersection points progressively further down. The discrepancy between fan and channel slopes is small when compared with many Spanish fans, but even here the channel slope increases by *c*.17% downfan from the intersection point. The major contrast with the Spanish fans is the channel width, which is large and increases progressively away from the fan apex. Mean width in the fan-head is an order of magnitude greater than on Spanish fans issuing from a similar drainage area.

The morphometric contrasts between the American and Spanish fans throw some light on

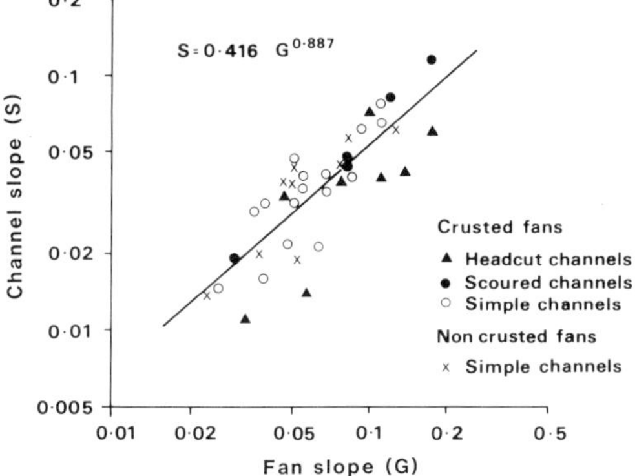

FIG. 6. Channel slope *vs* fan slope for selected Spanish fans according to channel type. ($r=0.831$, standard error of the estimate = 0.132 log units).

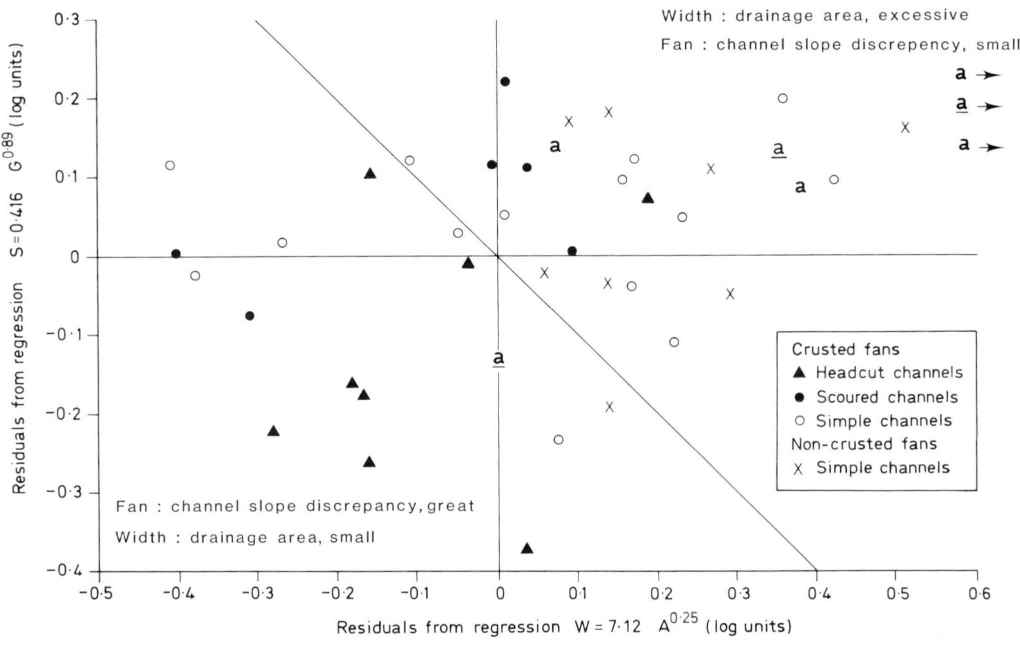

FIG. 7. Plot of residuals from channel slope (*S*) *vs* fan slope (*G*) and channel width (*W*) *vs* drainage area (*A*) regressions. Spanish fans are differentiated according to channel type; **a** indicates plotting positions for selected American fans; **a̲** indicates crusted American fans. Arrows indicate plotting position off the diagram to the right.

the controls over the distal aggradation *vs* trenching behaviour. Fan slopes as a function of unit drainage area for the American fans plot near the regression line for Spanish examples established in Fig. 3 (Fig. 9). Channel slopes plot generally above the equivalent line, and channel widths markedly above (Fig. 9). Comparable American and Spanish fans have surface slopes that are broadly similar; but the American fans have steeper channel slopes, and therefore less fan–channel slope difference; they also have much wider channel cross-sections. If these seven American fans are representive of both crusted and non-crusted arid-area fans in the Mohave and Sonoran deserts, intersection-point trenching would seem unlikely as a regional characteristic. This may reflect the greater aridity of the American desert environment, the higher sediment mobility, the greater effectiveness of extreme storm conditions (Baker 1977), and the large width of desert stream channels.

Discussion

The distinction between distally aggrading and distally trenching fans has two important, and related implications: (1) in the context of the coupling of sediment source and sink areas and sediment transport continuity within dry-region fluvial systems; and (2), in the context of the development of alluvial fan sedimentary sequences. An aggrading fan will trap sediment derived from the mountain source area, especially the coarse fraction, and hence break the continuity of the fluvial system that would otherwise exist between sediment source and the axial drainage of the basin. Therefore the fan response to sediment supply variations may be recorded *within* the fan but not, in the short term, within the behaviour and deposits of the axial drainage. Likewise, the upstream transmission of changes in axial drainage such as a tectonically induced change in base level, may be inhibited, again in

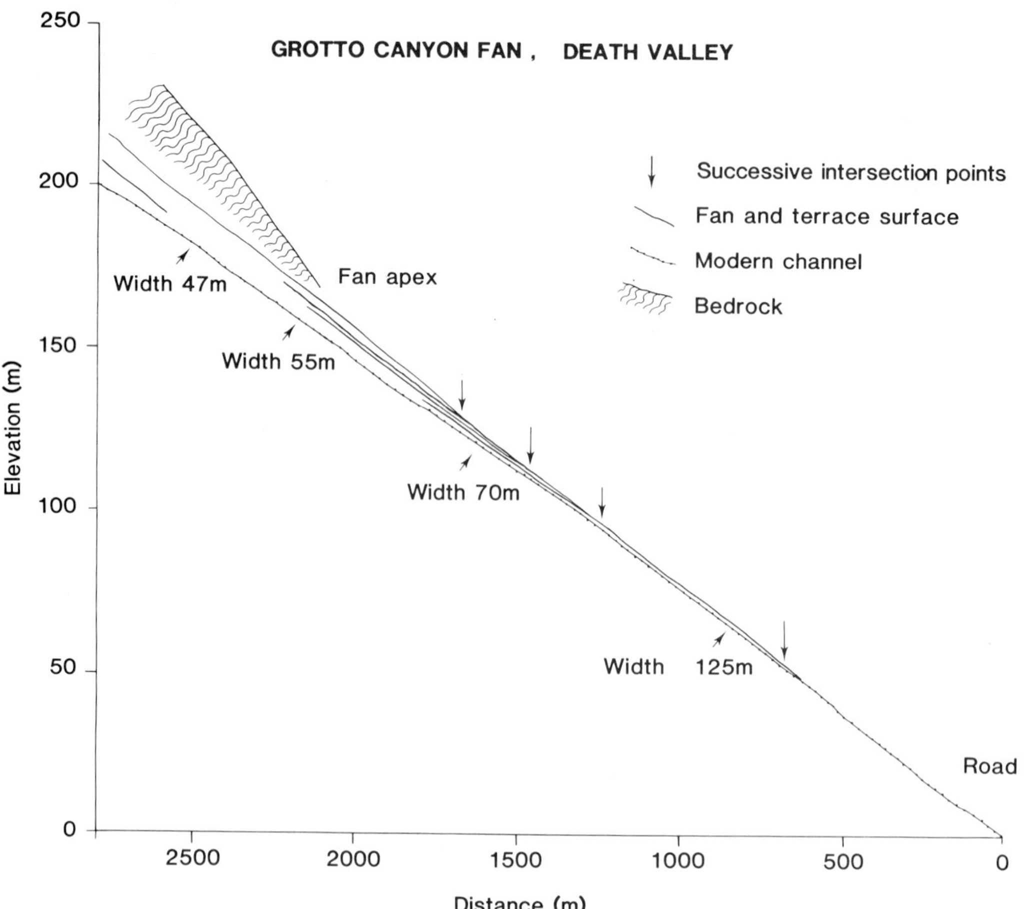

FIG. 8. Fan surface and channel profiles for Grotto Canyon fan, Death Valley California; note progressive increase in channel widths downfan.

the short term, by the presence of an aggrading fan. Over longer periods, tectonically-induced distal dissection may restore continuity of the fluvial system but only after a considerable time-lag. The correlation between sedimentary sequences in fan and axial drainage may therefore be poor. On the other hand, there will be continuity from mountain source to axial drainage in distally trenched fans. The axial channels of the basin are more likely to respond sensitively to changes in sediment supply from mountain source areas and correlation between basin marginal and axial sedimentary sequences are likely to be better.

The mountain front zone within which alluvial fans occur is particularly sensitive to local variations in critical stream power (Bull 1979). The slopes of alluvial fan surfaces and of the fan channels are a response to prevailing sediment supply and transport mechanisms. They are susceptible to modification in conditions where sediment supply changes. The sedimentary sequences of alluvial fans will reflect variations in sediment supply, but their stratigraphy will differ according to their aggradational or dissectional style. Non-trenched fans will build up stacked sedimentary sequences, with vertical variations that are a response to sediment supply conditions. Unconformities will be shallow or planar, and will reflect primarily local switching of channels or depositional zones over the fan surface. If such fans are dissected only by fan-head trenches, there will be contrasts between proximal and distal sites, with distal sites showing stacked sedimentary sequences, and proximal sites either simply an erosional sequence after early aggradation, or inset cut-and-fill terraces (Fig. 10). On fans whose morphometry is conducive to distal trenching, sedimentary sequences may be more complex. Distal sites may show inset and overlapping stratigraphy with high amplitude and steep disconformities (Fig. 10).

This contrast may be illustrated by the sedimentary sequences from the aggradation phases of two Spanish fans (Fig. 11). The Honda fan, supplied by the high-grade metamorphic terrain of the Sierra de los Filabres (Fig. 2), shows a sequence above the modern intersection point that has no sign of trenching prior to the modern dissection (Fig. 11, above). There is a simple

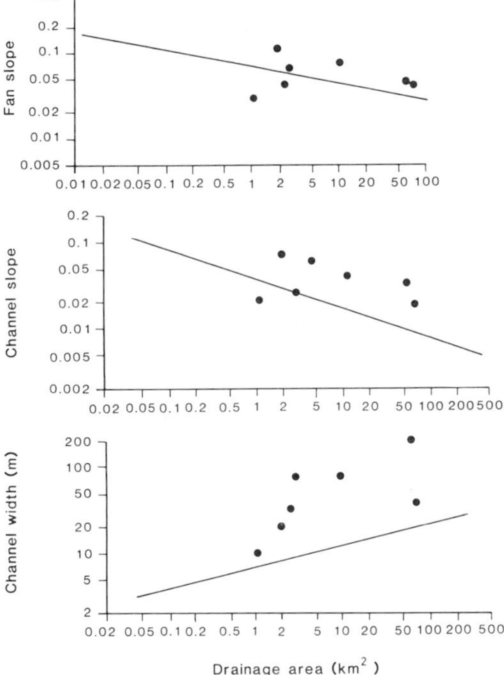

FIG. 9. Comparison of American fan parameters with least-squares regression lines for the parameters of the Spanish fans (regression lines from Fig. 3).

TABLE 5. *Data from American fans*

Fan	Location		Drainage area (km^2)	Fan* slope	Channel† slope	Distal‡ slope	Channel§ width
Crusted fans.							
Searles Pass (1)	Panamint	Calif.	2.0	0.135	0.093	0.102	19
Searles Pass (2)	Panamint	Calif.	80.0	0.052	0.047	0.045	*c.* 200
Kyle Canyon	Las Vegas	Nev.	90.0	0.046	0.020	0.025	22
Non-crusted fans							
Grotto Canyon	Death Valley	Calif.	10.6	0.085	0.064	0.075	*c.* 70
Badwater	Death Valley	Calif.	3.5	0.075	0.072	0.075	*c.* 75
River Mountain	Henderson	Nev.	2.5	0.051	0.036	0.045	23
Osborne Wash	Parker	Ariz.	1.3	0.031	0.021	0.028	6

* Fan slope (mean slope of the fan surface above the intersection point).
† Channel slope (within the fanhead trench).
‡ Distal slope (downfan from the intersection point).
§ Mean channel width, m, (within the fanhead trench).

stacked sequence of braided stream deposits. In contrast, the Tapia fan in the Pre-Betic range (Fig. 2), shows a distal sequence (Fig. 11, below) characterized by inset and overlapping stratigraphy. A basal unit of silt with occasional sheets of gravels is deeply trenched. The trench is filled with channel gravels, after which further silt layers were deposited and on which a calcrete crust developed. This crusted sequence was dissected later, after which further channel gravels and silt were deposited, to be followed by the formation of a second calcrete crust. Today, Tapia is a dissecting fan of type D (Fig. 1) whereas Honda is a distally aggrading fan of type B (Fig. 1).

Factors influencing the distal-fan regime operating at present have also worked in the past, so stamping their mark on the sedimentary sequences. The fan groups currently most prone to distal trenching also show inset, as opposed to stacked, stratigraphies in deposits of the aggradation phases. All six fans in the Pre-Betic group and all seven in the Carrascoy group which have been examined for stratigraphic sequences (Harvey 1984*b*), show evidence of deep dissection within aggradational sequences. These two groups are most prone to distal dissection (Table 1). Only one out of seven southern fans fed from sedimentary rocks, and one out of eleven southern fans fed from metamorphic rocks show evidence of deep trenching within exposures. Only one of the American fans examined showed evidence of deep trenching.

The morphometry of fans is a product of identifiable sequences of geomorphic processes; but these processes have to be placed in the context of Quaternary climatic change. Fan surface slopes produced by debris flow and sheet-flood (gravel-sheet) deposition, are steeper than those produced by channel processes, hence the greater discrepancy between fan and channel slopes on debris-flow and sheet-gravel fans than on fluvial fans (Harvey 1984*a*). This is one possible reason for the contrast between the Spanish and American fans studied here: American fans are dominated by channel gravels and there is a relatively small discrepancy between fan and channel slopes. However, there is a progressive change in sedimentary style identifiable in many of the Spanish fans from debris flow and fines-rich deposition in early phases to gravel-rich deposition later and finally to a phase of dissection. This suggests a progressive diminution in the availability of fines and a progressive increase in the water: sediment ratios yielded by the mountain source areas. Such trends have not

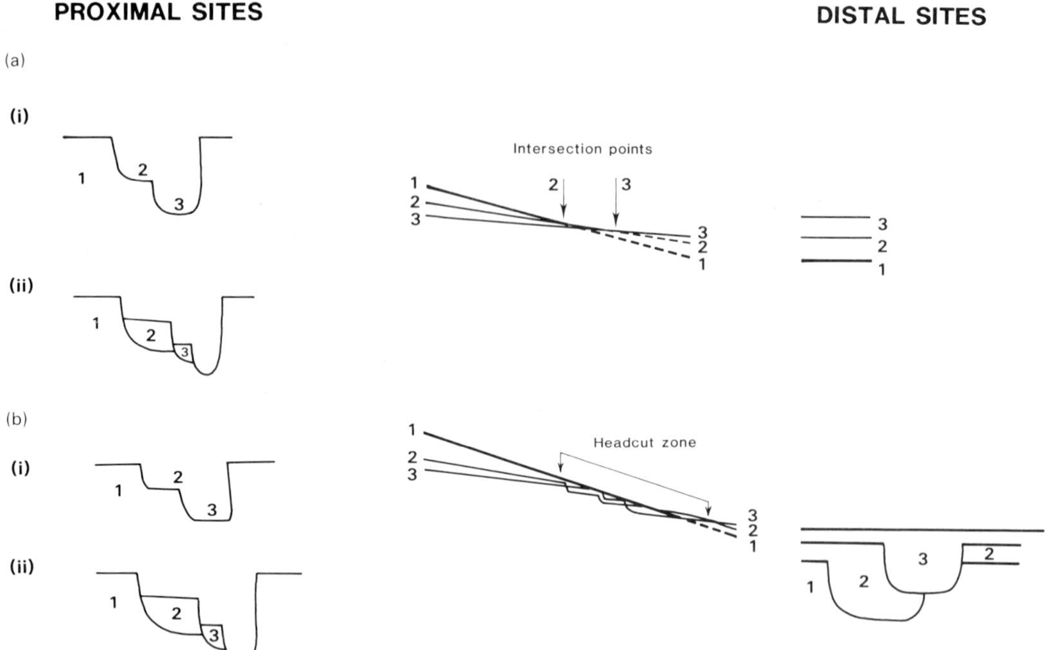

FIG. 10. Schematic model for alluvial fan stratigraphy. (*a*) simple fans, (*b*) distally trenched fans. In each case: (i) without proximal cut-and-fill sequences; (ii) with proximal cut-and-fill sequences.

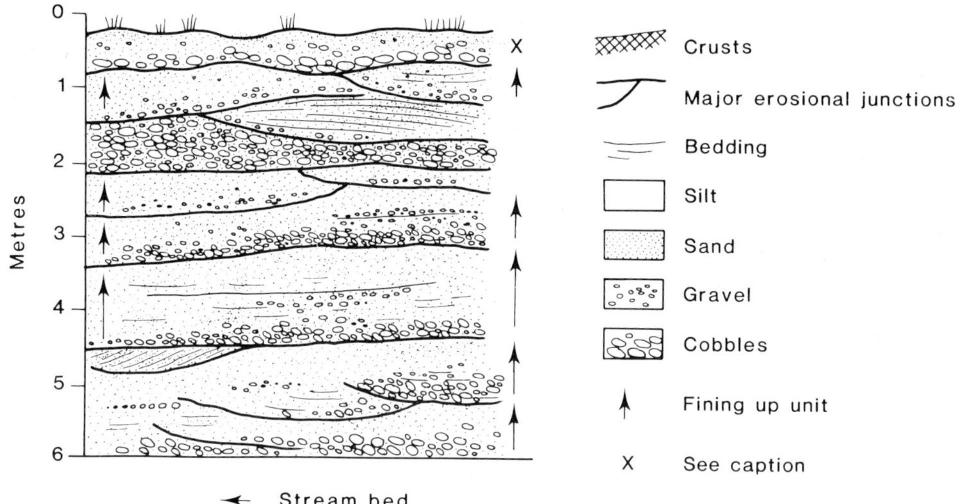

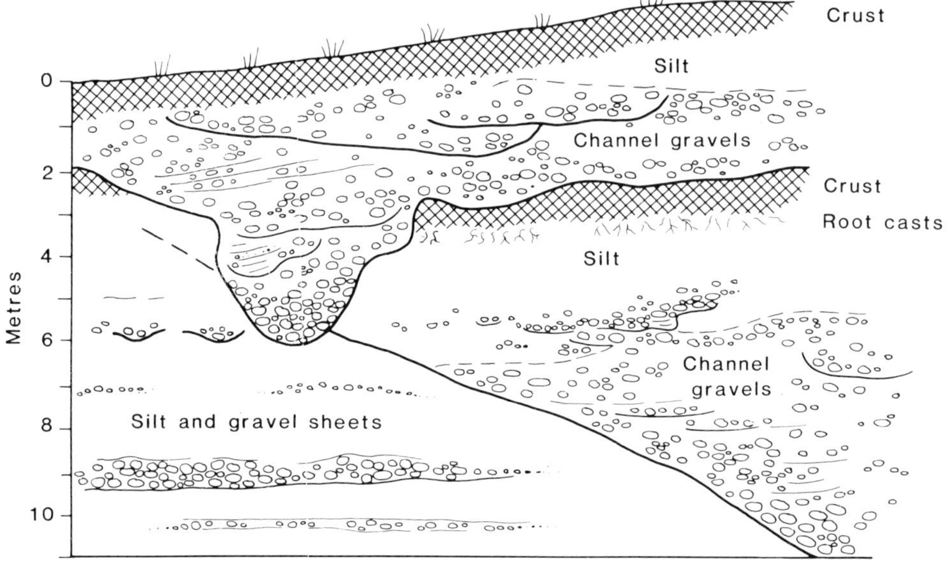

FIG. 11. Examples of sedimentary sequences. Above: Honda fan, Tabernas, Almeria, a simple stacked fluvial sequence; × marks a younger erosional terrace veneer. Below: Tapia fan, Benidorm, Alicante, an inset distal fan sequence.

been identified on the American fans. Superimposed on the long-term trends on the Spanish fans are major fluctuations in sediment yield that reflect Quaternary climatic change. The major periods of aggradation are thought to relate to cool and dry Quaternary 'glacials' (Harvey 1984*b*). These alternate with periods of dissection, during which the surface crusts formed. Again, there appear to be contrasts between the Spanish and American fans, with the more arid American environment apparently less prone to such large climatically induced fluctuations and characterized, generally, by less well indurated calcrete crusts. The crusts appear to be important for channel confinement, preventing the development of excessive width, and accentuating any tendency for intersection-point scour in the Spanish fans.

The present morphology of Quaternary alluvial fans can be parameterized and contrasts between Spanish and American examples can be seen to reflect climatic differences. However, the effects of these differences can only be understood in the context of the sequential development of the fans. In turn, the sedimentary sequence of alluvial fans must be seen against a back-cloth of sediment and source-area characteristics and morphometry controlled surface processes.

ACKNOWLEDGMENTS: The author is grateful to the research fund of the University of Liverpool for financial assistance towards the costs of the fieldwork; also to the staffs of the drawing office, particularly Sandra Mather, and of the photographic section of the Department of Geography, University of Liverpool, for producing the diagrams.

References

AMOR, J. M. & FLORSCHUTZ, F. 1964. Results of the preliminary palynological investigation of samples from a 50 m boring in southern Spain. *Boletin de la Real Sociedad de Historia Natural (Geologia)* **62**, 251–255.

BAKER, V. R. 1977. Stream channel response to floods, with examples from central Texas. *Geological Society of American Bulletin* **88**, 1057–1071.

BLISSENBACH, E. 1952. Relation of surface angle distribution to particle size distribution on alluvial fans. *Journal of Sedimentary Petrology* **22**, 25–28.

—— 1954. Geology of alluvial fans in semi arid regions. *Geological Society of America Bulletin* **65**, 175–190.

BULL, W. B. 1962. Relations of alluvial fan size and slope to drainage basin size and lithology in western Fresno County, California. *United States Geological Survey Professional Paper* **450B**, 51–53.

—— 1964. Geomorphology of segmented alluvial fans in western Fresno County, California. *United States Geological Survey Professional Paper* **352E**, 89–129.

—— 1977. The alluvial fan environment. *Progress in Physical Geography* **1**, 222–270.

—— 1979. Threshold of critical stream power. *Geological Society of America Bulletin* **90**, 453–464.

DENNY, C. C. 1965. Alluvial fans in Death Valley Region, California and Nevada. *United States Geological Survey Professional Paper* **466**, 59 pp.

DUMAS, M. B. 1969. Glacis et croutes calcaires dans le levant espanol. *Association de Geographie Francais Bulletin* **375**, 553–561.

ECKIS, R. 1928. Alluvial fans of the Cucamunga district, southern California. *Journal of Geology* **36**, 225–247.

GEIGER, F. 1970. Dei ariditat in sudostspanien. *Stuttgarter Geographische Studien* **77**, 173 pp.

HARVEY, A. M. 1978. Dissected alluvial fans in southeast Spain. *Catena* **5**, 177–211.

—— 1984*a*. Debris flows and fluvial deposits in Spanish Quaternary alluvial fans: implications for fan morphology. *In*: KOSTER, E. H. & STEEL, R. J. (eds) *Sedimentology of gravels and conglomerates*. Canadian Society of Petroleum Geologists Memoir **10**, 123–132.

——1984*b*. Aggradation and dissection sequences on Spanish alluvial fans: influence on morphological development. *Catena* **11**, 289–304.

—— 1984*c*. Geomorphological response to an extreme flood: a case from southeast Spain. *Earth Surface Processes and Landforms* **9**, 267–279.

HOOKE, R. LE B. 1967. Processes on arid-region alluvial fans. *Journal of Geology* **75**, 438–460.

—— 1968. Steady-state relationships on arid region alluvial fans in closed basins. *American Journal of Science* **266**, 609–629.

KESEL, R. H. 1985. Alluvial fan systems in a wet-tropical environment, Costa Rica. *National Geographic Research* **1**, 450–469.

LATTMAN, L. H. 1973. Calcium carbonate cementation of alluvial fans in southern Nevada. *Geological Society of America Bulletin* **84**, 3013–28.

LEOPOLD, L. B. & BULL, W. B. 1979. Base level, aggradation, and grade. *Proceedings of the American Philosophical Society* **123**, 168–202.

MIALL, A. D. 1978. Lithofacies types and vertical profile models in braided river deposits: a summary. *In*: MIALL, A. D. (ed.) *Fluvial Sedimentology*. Canadian Society of Petroleum Geologists Memoir **5**, 597–604.

RUST, B. R. 1979. Facies models 2: Coarse alluvial deposits. *In*: WALKER, R. G. (ed.) *Facies Models*. Geoscience Canada, Reprint Series 1, Kitchener, Ontario, 9–21.

SABELBERG, U. 1977. The stratigraphic record of late Quaternary accumulated series in southwest Morocco and its consequences concerning the pluvial hypotheses. *Catena* **4**, 204–215.

SCHUMM, S. A. 1977. *The Fluvial System*. J. Wiley & Sons, New York, 338 pp.

STEEL, R. J., MOEHLE, S., NILSEN, H., RØE, S. L. & SPINNANGR, A. 1977. Coarsening-upwards cycles in the alluvium of Hornelen Basin (Devonian), Norway: Sedimentary response to tectonic events. *Geological Society of America Bulletin* **88**, 1124–1134.

VAN ARSDALE, R. 1982. Influence of calcrete on the geometry of arroyos near Buckeye, Arizona. *Geological Society of America Bulletin* **93**, 20–26.

WASSON, R. J. 1974. Intersection point deposition on alluvial fans: an Australian example. *Geografiska Annaler* **56A**, 83–92.

—— 1979. Sedimentation history of the Mundi Mundi alluvial fans, western New South Wales. *Sedimentary Geology* **22**, 21–51.

WELLS, S. G. 1977. Geomorphic controls of alluvial fan deposition in the Sonoran Desert, southwestern Arizona. *In*: DEOHRING, D. O. (ed.) *Geomorphology in Arid Regions*. Allen & Unwin, London, 51–78.

WILLIAMS, G. E. 1973. Late Quaternary piedmont sedimentation, soil formation and palaeoclimates in arid South Australia. *Zeitschrift für Geomorphologie* **17**, 102–125.

A. M. HARVEY, Department of Geography, University of Liverpool, PO Box 147, Liverpool L69 3BX, UK.

Quaternary landforms, sediments, depositional environments and gastropod isotope ratios at Adrar Bous, Tenere Desert of Niger, south-central Sahara

M. A. J. Williams, P. I. Abell & B. W. Sparks

SUMMARY: Sediments deposited in a variety of environments including fluviatile sands and gravels, aeolian sands, lacustrine clays and diatomites, and swamp clays crop out in and around Adrar Bous, an isolated ring-complex in the northern Tenere Desert of Niger. These sediments range from Recent to at least Middle Pleistocene in age, and contain both reworked and primary-context stone artefact assemblages testifying to recurrent prehistoric occupation of the plains around the massif from at least Upper Acheulian times onwards. During the early Holocene a small lake, stable at 710 m a.s.l. supported a community of Epi–Palaeolithic hunter-gatherers who also ate aquatic foods until the lake dried out at around 7500–7000 BP. Occupation of the piedmont plains resumed during 6000–4000 BP when Neolithic pastoralists grazed their cattle around a smaller lake stable at 700 m a.s.l. The Holocene mollusc assemblage and the stable isotope ratios obtained for different species of Holocene gastropods are entirely consistent with, and help to refine, the palaeoenvironmental reconstruction based upon stratigraphic and sedimentary analyses supplemented by the use of prehistoric stone tool assemblages as 'zonal fossils'.

The existence of prehistoric sites and associated lake or river deposits in the heart of the Sahara, and of now vegetated dunes over 500 km beyond the present southern limit of active dunes has prompted considerable speculation as to the magnitude and frequency of Quaternary climatic fluctuations in the Sahara (Williams 1984) as well as in other great deserts of the world. An immediate difficulty confronting anyone seeking to infer former climate from the often fragmentary depositional legacy of rivers, lakes and dunes, which may or may not contain plant and animal fossils, lies in distinguishing the degree to which the sedimentary record may reflect climatic, tectonic, and volcanic events. In dealing with the remains of plant and animal fossils, immediate taphonomic questions spring to mind, including the possibility of reworking older fossils, differential destruction and preservation, and the selective concentration of vertebrate remains by prehistoric humans and other predators. One paradoxical solution to this dilemma is deliberately to seek out localities known to have been occupied by Palaeolithic hunter-gatherers or Neolithic herders, so that the investigator can benefit from as diverse an array of field evidence as possible, including stone tool concentrations, middens, occupation sites, and rock art. These can be used to reinforce evidence obtained by more conventional methods such as stratigraphic mapping, fossil analysis, and radiometric dating. One locality in which this type of analysis is possible is Adrar Bous ring-complex in the south-central Sahara. The aim of this paper is to examine the late Quaternary depositional history of Adrar Bous for evidence of former climatic fluctuations. In so doing we also consider the palaeoecological significance of the Holocene mollusca, including some of the ways in which the stable isotopic composition of the mollusc shells may reflect former hydrological events.

Location

Adrar Bous is an early Silurian ring complex located 65 km ENE of Mont Greboun, a 2000 m basalt-capped plateau which forms the highest part of the northern Aïr Mountains (Fig. 1). It rises over 300 m above the adjacent plains, which themselves range in elevation from 700 to 730 m. Adrar Bous consists of a central massif of granite flanked by a series of broad valleys and shallow piedmont depressions, with an outer, discontinuous rim of rocky hills (Fig. 3). The main massif is aligned SW to NE and forms an ellipse 12 km wide and 16 km long (Figs 2 and 3).

Situated at 8°57′E and 20°18′N and surrounded by the drifting sands of the Tenere desert, it is hot and dry for much of the year, although from December to February the mornings are cold, with dawn temperatures ranging from 0°C to 5°C, but rising to over 30°C after midday. Spasmodic summer storms may reach Adrar Bous with the northward arrival of the July–September monsoonal rains, but even Iferouane in the heart of the Aïr has only 50 mm annual rainfall (Lefèvre 1960). Highly localized storms result in the flash

From FROSTICK, L. & REID, I. (eds), 1987, *Desert Sediments: Ancient and Modern,* Geological Society Special Publication No. 35, pp. 105–125.

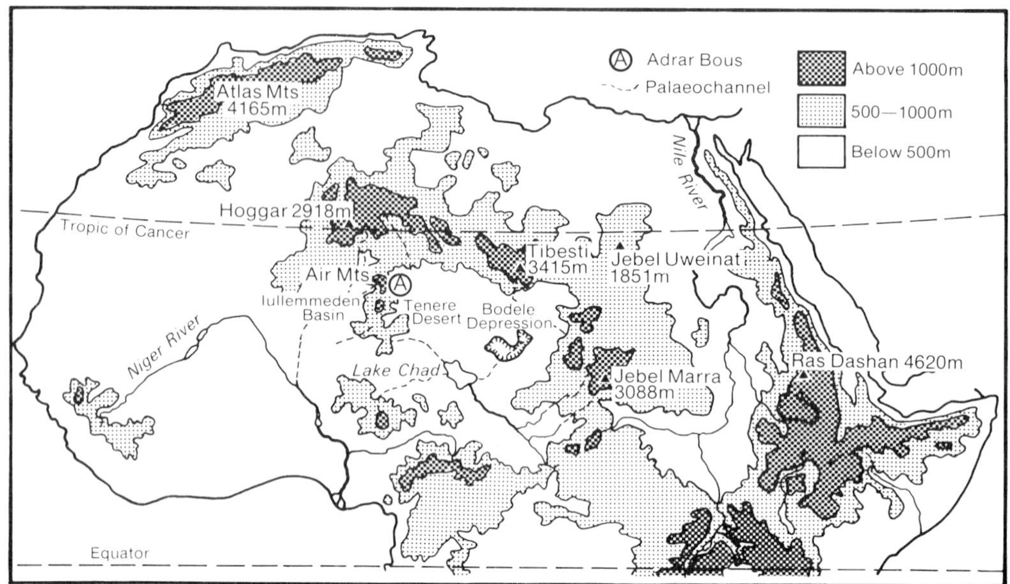

FIG. 1. Relief map of northern Africa showing location of major Saharan uplands. Adrar Bous ring-complex lies immediately E of the northern Aïr Mountains, in the Tenere Desert of Niger.

floods responsible for transporting the coarse quartz sands and occasional boulders which presently line the wadis in and around the mountain.

Dominant surface processes are deflation (particularly from March to May), episodic slopewash and sporadic fluvial erosion and the deposition of sandy alluvia. That erosion is limited is shown by the presence of unabraded proto-historic tumuli in certain hillside gullies and more particularly by the persistence of a mantle of unconsolidated, sub-horizontal, Upper Quaternary deposits of fluviatile, lacustrine and aeolian origin around the margins of the main massif (Williams 1971). These sediments are described in this paper. They appear to reflect regional controls of erosion and deposition rather than purely local events.

Geological history of Adrar Bous

The oldest known rocks in this region are those of the Lower Precambrian Suggarian Series which consists of strongly folded metasediments, migmatites and granites (Greigert & Pougnet 1967). A major unconformity separates the Suggarian formation from the weakly metamorphosed, sub-horizontal basal conglomerates and arkoses of the overlying Middle to Upper Precambrian Pharusian Series, which are widespread to the E of the Aïr and underlie the sand plains around Adrar Bous. The western flank of Adrar Bous appears to be partly comprised of steeply-tilted and displaced portions of the Precambrian basement. The schists and intercalated amphibolites N of the 1970 camp (see Fig. 3) belong to the Suggarian Series and the nearby conglomerates to the younger Pharusian Series (Raulais 1959).

Erosion of the Precambrian basement was followed by deposition of Palaeozoic continental, and occasional marine, sandstones, shales and tillites which crop out as low plateaux N of Adrar Bous. Whether the sandstone mesa 20 km NW of Adrar Bous is Palaeozoic or younger is unclear, and Cornet (1962) even considered the isolated quartzite butte 35 km SE of Adrar Bous as Pharusian, although it may be far younger.

Over three quarters of the Aïr consist of Younger Granite ring-structures intruded through the Precambrian basement which attain elevations of up to 1900 m. There are over 25 such ring-complexes in the Aïr, Adrar Bous being the most northerly, and they range in diameter from 2 to 30 km, with a mean width of 15 km. The Younger Granite province (Black & Girod 1970) comprises 60 separate ring-structured massifs concentrated in a narrow belt running from northern Nigeria to northern Aïr (from latitude 5°N to 21°N (1800 km) and from longitude 8°E to 10°E (200 km)). The Nigerian massifs are early to mid-Jurassic in age, but the Aïr ring complexes developed during the Silurian and Devonian periods. Rubidium–strontium whole rock and

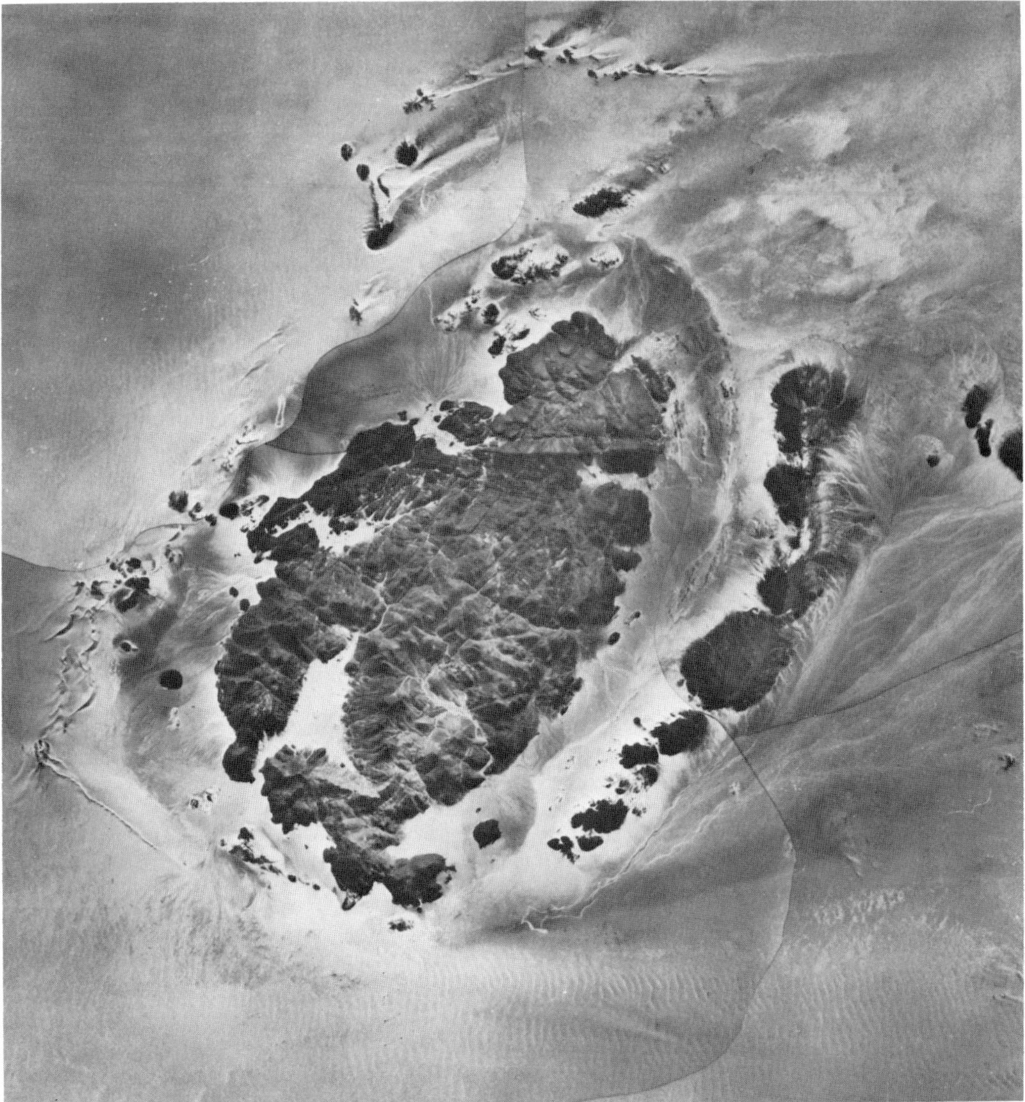

FIG. 2. Uncontrolled photo-mosaic of Adrar Bous ring-complex. See Fig. 3 for geomorphic interpretation and for location of stratigraphic sections.

feldspar age determinations give an age of 429 ± 4 Ma for Adrar Bous (Bowden *et al.* 1976). The southward younging of ring-complexes from Northern Aïr (early Silurian) to northern Nigeria (early to mid-Jurassic) may reflect northward drift of the African plate over one or more hot spots.

Emplacement of the ring-complexes was generally initiated by outpourings of rhyolites and tuffs, followed by anorthosites, syenites and then by intrusions of hyperalkaline aegirine– and riebeckite–granites, culminating in the intrusion of alkaline biotite– and hornblende–biotite–granites as ring-dykes or as crescent-shaped masses (Black 1963).

The core of Adrar Bous consists of alkaline granites and microgranites rich in micropegmatite and microperthite. The flanks are made up of rhyolites, dolerites and quartziferous microgabbros to the SW and SE, of massive quartzites and basalts to the E, and of steeply-tilted Precambrian mica–schists and amphibolites to the W (Raulais 1959; Cornet 1962; Black 1963; Williams 1976). Intrusion of the Silurian granites would have caused localized updoming of the older Palaeozoic cover, the resulting erosion leading to the

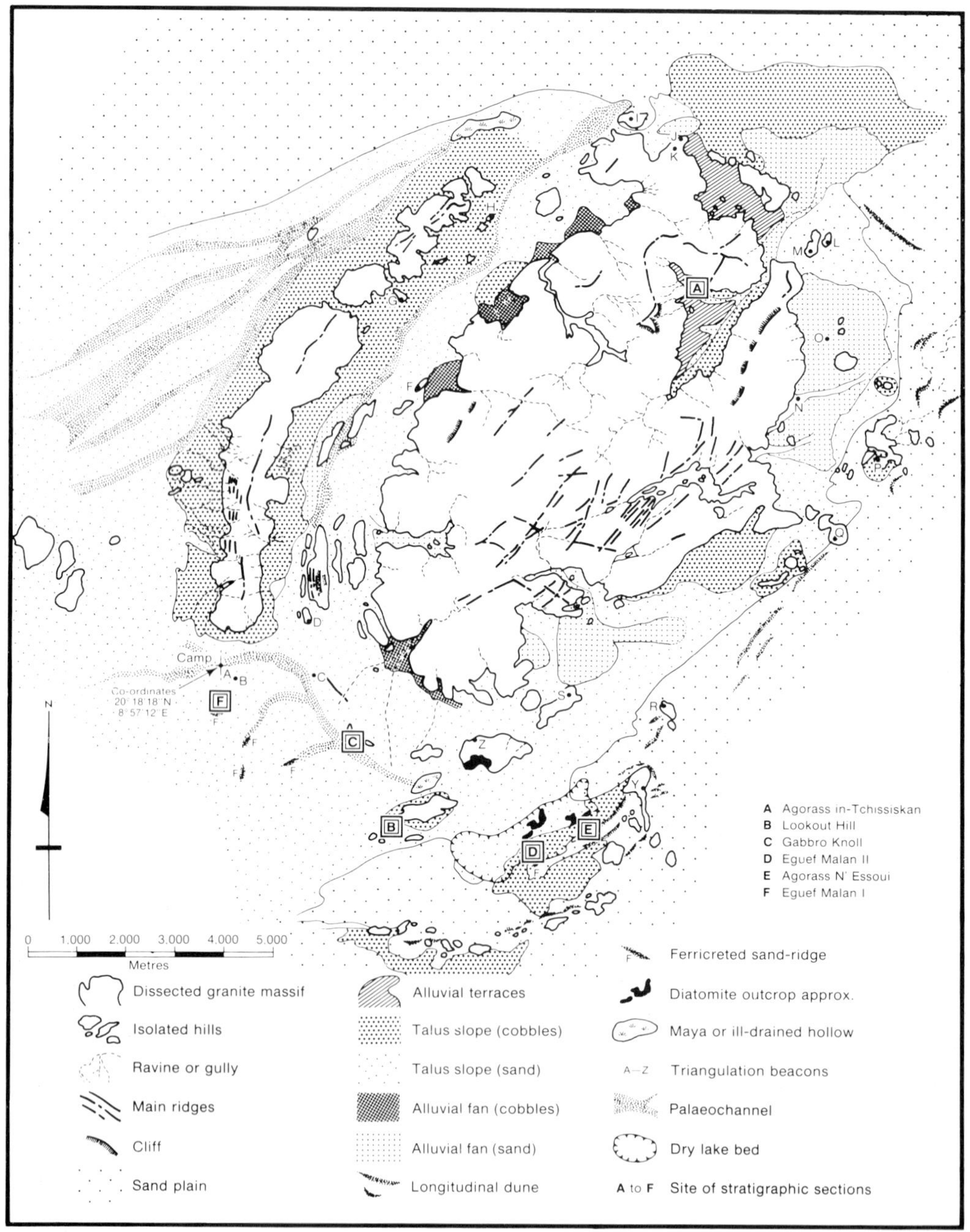

FIG. 3. Geomorphic map of Adrar Bous showing location of the six key Quaternary stratigraphic sites discussed in the text.

eventual exposure of the Younger Granites. Away from the ring-complex the Palaeozoic sediments would have remained undisturbed, resting directly upon the Precambrian basement.

The unresistant sandstones N of Adrar Bous may belong to the poorly dated *'Continental intercalaire'* formation (Kilian 1931). They are composed of continental clays and sandstones which pre-date the earliest Cretaceous marine transgression (Albian) and post-date the youngest Palaeozoic rocks. They include Permo–Triassic arkoses with frequent silicified wood fragments,

Jurassic sandstones with dinosaur fossils and Lower Cretaceous red clays and fluvial sandstones. They crop out N of Adrar Siret some 65 km SW of Adrar Bous, and are widespread S and W of Agades and the Aïr (Dresch 1959; Ginsburg *et al.* 1966; Greigert & Pougnet 1967, pp. 106–112).

The first major advance of the Cretaceous sea was during the Albian, and after a slight recession in Lower Cenomanian times the sea reached its highest level during the Upper Cenomanian and Lower Turonian transgression when it reached the margins of the Hoggar (Raulais 1951) and penetrated through the narrow gulf between Tibesti and the Hoggar into the Ténéré du Tafassasset depression N of the Aïr, and as far W as the Adrar des Iforas (Furon 1963, p. 50; Faure 1962). Adrar Bous may well have become submerged at this time, for Vogt & Black (1963) have suggested that the sedimentary remnants at 1400 m on the Younger Granite Tamgak plateau 25 km ENE of Iferouane and high on the Goundai ring-complex are Cretaceous. If so, then considerable post-Cretaceous uplift must have occurred in and around the Aïr. Faure (1959*a*) suggests that gentle upwarping of the Aïr is probably still continuing. Significantly, both Bordet (1952) and Rognon (1967) have stressed the comparatively recent updoming of the Hoggar. Whether Adrar Bous underwent similar Cainozoic uplift remains unknown. Certainly it seems possible, for the plains slope steadily away from Adrar Bous, descending from an elevation of 700–710 m at *Agorass N'Essoui* (Fig. 3) to 495–500 m at Locality 4, 50 km to the SE.

Late Tertiary volcanism and associated faulting resulted in reactivation of Palaeozoic and older structural lineaments and in massive flows of the rhyolites and basalts which constitute the older plateau basalts of the Aïr (*eg* Greboun), the Hoggar (Rognon 1967; Conrad 1969) and Tibesti (Vincent 1963). Downfaulting of the fossé du Tefidet (Faure 1959*b*), in which are preserved over 1000 m of Lower Cretaceous sandstones, and of two more extensive buried graben which run from NW to SE across the Tenere (Louis & Rechenman 1966), probably also occurred at about this time.

Preceding and accompanying the Tertiary volcanism there was a prolonged period of deep-weathering which is reflected in the kaolinitic and/or bauxite weathering profiles up to 45 m thick developed on Eocene to Precambrian formations in central Niger. With post-Eocene uplift the dominantly chemical and biogenic sedimentation ceased (Faure 1962; Greigert & Pougnet 1967, p. 157) and during the ensuing erosion major rivers from the Aïr, the Hoggar and Tibesti laid down the extensive fluviatile gravels, sands and clays of the *Continental terminal* formation around the central Saharan uplands. The origin of the Sahara as a continental desert, and of Adrar Bous as a desert inselberg, may be said to stem from the Miocene Alpine orogeny and the subsequent stripping of the Eocene deep weathering profile. Evolution of the xeric flora endemic to the Sahara probably began in Middle Miocene times (Maley 1980) and not, as often assumed, in Pleistocene times alone, for xerophytic fossil pollen is recorded in Early Villafranchian Saharan deposits (Rossignol & Maley 1969) implying that plants adapted to aridity were already in existence at the start of the Pleistocene. Any Tertiary formations which may exist around Adrar Bous are masked by the Upper Quaternary deposits described below.

Quaternary sediments of Adrar Bous

The Quaternary deposits of Adrar Bous vary widely in lithology and provenance, and include fluviatile sands and gravels, deltaic sands, lacustrine clays, marls and diatomites, aeolian sands, colluvial sands and silts, gravelly fanglomerates, and swamp clays.

The massif rises steeply from the plains at about 700 m a.s.l. to over 1000 m, and is flanked by a now dry peripheral depression which contained swamps, small lakes and streams up to approximately 4000 years ago (Fig. 3). Earlier depositional environments reflect the varied physical environment of the massif, which ranges from steep boulder-mantled ridges (Fig. 4) separated by broad sandy valleys, to smaller, isolated hills (Lookout Hill and Gabbro Knoll) fringed by broad pediments and locally flanked by fluvio-lacustrine deposits (*Agorass N'Essoui*). Alluvial fans made up of alternating boulders and gravelly sands debouch from the mouths of narrow mountain channels or *kori* in which ephemeral streams rarely flow. Occasional dunes, both fixed and active, separate the plains and closed depressions adjacent to the main massif from the moving sands of the Tenere (*Eguef-Malan I and II*). Within the massif proper there is a wide, flat, structurally-controlled valley (*Agorass in-Tchississkan*) in which several generations of alluvium have been eroded into a series of low terraces. The deposits characteristic of these different localities will now be considered, beginning with the most recent. The identification of individual layers is based upon field observations of sediment or soil texture, colour, structure, macro-fabric and lithology, upon the vertical and lateral

FIG. 4. Boulder-mantled hillslope of joint-controlled granite corestones, Adrar Bous. In the foreground two partly exhumed corestones are emerging from the *in situ* weathered mantle and its overlying colluvium.

relationships between horizons, laboratory grain-size determinations and the recognition of Early, Middle and Late Stone Age artefacts resting *in situ* within the relevant horizons (Clark 1971; Clark *et al.* 1973; Clark pers. comm.).

Alluvial terrace deposits at *Agorass in-Tchissiskan*

Three non-paired terraces have been eroded into five sedimentary horizons (Fig. 5*a*) which mantle the northwestern corner of *Agorass in-Tchissiskan*. The terrace heights above the modern channel-floor are 3.3, 1.2 and 0.4 m for the high, middle and low terraces respectively, and each terrace step is overlain by a layer of alternating coarse gritty sands and silts which forms the uppermost horizon throughout Adrar Bous. These sands were 120, 20 and 5 cm thick on the top, middle and low terraces, and two Neolithic burials (Fig. 6) were located on the glacis which is graded to the top terrace. Beneath the stratified coarse sands and silts (a) there were, from the top down;

(b) a layer of brown, gritty, sandy loam up to 130 cm thick;
(c) a layer of black, weakly vertisolic loam or grey silt 25–40 cm thick;
(d) a layer of grey and brown, mottled and iron-stained loamy sand 30 cm thick;

(e) a well-developed black cracking clay loam or vertisol up to 80 cm thick;
(f) a second grey and brown, mottled, gritty, sandy loam, either massive or columnar in structure, and weakly iron-cemented, up to 40 cm thick; and
(g) a basal cobble layer of rounded granite boulders set in a matrix of coarse sand.

Mixed Upper Acheulian and Levallois-type artefacts were found in the basal cobble layer, and Levallois-type flakes occurred in a discontinuous lens of angular gravel between the lower vertisol and lower mottled sandy loam. The clay content of the lower mottled layer was 10%–15.5% and that of the upper layer was 11%–18%. Cumulative grain-size distribution was almost identical in both layers. The upper vertisolic loam had 60%–61% sand, and the lower vertisol was slightly heavier in texture, with 52.5%–60% sand. The coarsest horizons were the brown sandy loam (9% clay, 84.5% sand) and the stratified coarse sands and silts (8% clay, 81%–84.5% sand).

Colluvial and aeolian sands in the western valley

West and south of *Agorass in-Tchississkan* there is a clay-lined depression set in a grove of *Acacia seyal* trees. A dry well 6.5 m deep, and now sand-filled, extends through over 5 m of grey-brown vertisolic clay loam, showing that the area around the well has long been a run-on site. Further S, along the E of the valley, channel-bank sections reveal a sequence comparable to that described above, with a basal cobble layer, two vertisol layers each over a mottled sandy loam, a brown sandy loam and the modern stony laminated coarse sands and silts. Along the *kori* floors and on the higher ground 0.5 to 1 km NE of the 1970 camp (Fig. 3) the brown sandy loam was darker, deeper, siltier and more compacted, with a coarse surface network of vertical cracks. The clay content was 8%–11% in the top 20 cm, decreasing to 4% at 50 cm. There was twice as much silt as clay in this layer, some of it stemming from the ash of the extensive Neolithic middens in this locality (Hugot 1962*a*, *b*).

Beneath the Neolithic loam is a white to very pale brown well-sorted aeolian sand with over half of its well-rounded quartz grains ranging in size between 0.5 and 1.0 mm. Charcoal and burnt bone from Neolithic hearths at Adrar Bous have yielded ^{14}C ages of 5050 ± 150 BP and 4910 ± 135 BP (Table 1; Delibrias & Hugot 1962). The aeolian sands are therefore at least 5000 years old.

Hillslope deposits flanking Gabbro Knoll and Lookout Hill

About 5 km SE of the 1970 camp is a prominent isolated hill. Along the lower slopes of both this hill (Lookout Hill) and a tor 2 km farther N (Gabbro Knoll) a massive mottled grey and brown sandy loam (Fig. 5*b*) crops out. It is covered locally by the eroded remains of a dark vertisolic loam (Fig. 5*c*). The Gabbro Knoll sequence is, from the top:

(a) 20 cm of stratified coarse and fine sands;
(b) up to 20 cm of friable brown sandy loam;
(c) up to 15 cm of grey-brown loam, with Neolithic artefacts above it;
(d) up to 15 cm of gravels;
(e) an olive-green calcareous loam, with 16% clay and 17.5% silt, at least 60 cm thick;
(f) a grey and brown mottled sandy loam over 15 cm thick; and
(g) a yellow-brown sandy clay loam at least 60 cm thick.

Fresh Upper Acheulian hand-axes were found above layer (f) and fresh Levallois–Mousteroid 'Middle Stone Age' artefacts were found in the top 5 cm of the green calcareous loam (layer e). With the possible exceptions of (e) and (g) all the above layers were present immediately NW of Lookout Hill. In addition, over a metre of yellow and grey sands containing fresh Aterian implements covered the middle footslopes and were underlain by a massive calcareous yellow-brown loam with 18% clay and 19% silt. The Lookout Hill sequence may be summarized as follows, beginning at the top:

(a) recent wind-blown sands over 15–100 cm of stratified coarse and fine sand;
(b) 5 cm of dark grey vertisolic loam, with Neolithic artefacts on the surface;
(c) yellow-brown to grey sands (with Aterian artefacts) at least 190 cm thick, merging laterally with
(d) yellow-brown sandy loam with occasional gravel lenses and Aterian tools, 10 cm thick;
(e) 45 cm of pale yellow-brown mottled sandy loam;
(f) 100 cm of calcareous yellow-brown loam, greyer and sandier with depth; and
(g) grey and brown mottled sandy loam at least 75 cm thick.

The calcareous loam (f) is probably coeval with that at Granite Knoll, and both layers were laid down after the departure of Upper Acheulian groups and before the arrival of Aterian communities in this area. Dated Upper Acheulian sequences elsewhere in Africa are at least 150 000 years old and Aterian cultural remains date back to more than 40 000 BP.

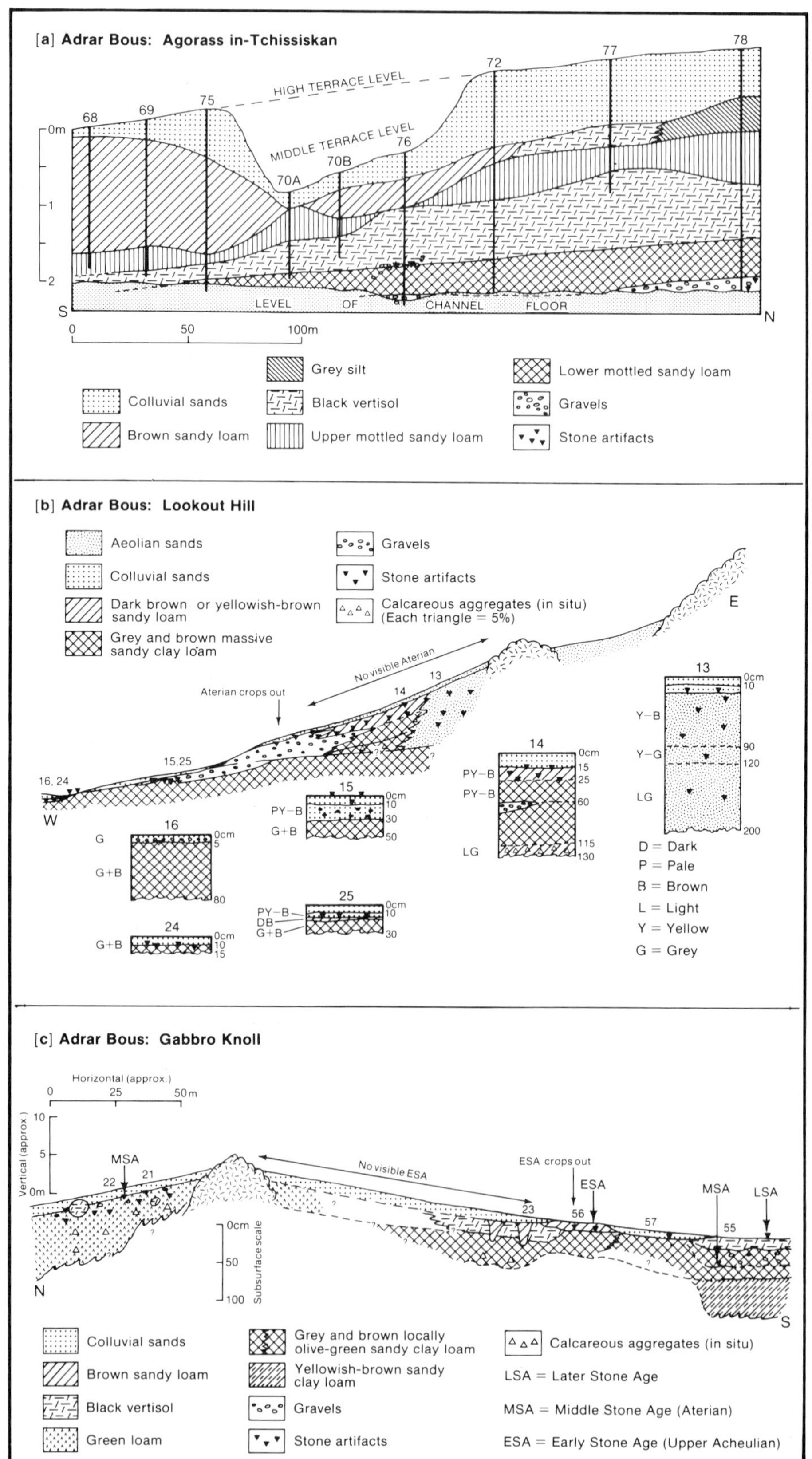

FIG. 5. Quaternary stratigraphy, Adrar Bous. (*a*) *Agorass in-Tchissiskan*; (*b*) Lookout Hill; (*c*) Gabbro Knoll; (*d*) *Eguef-Malan II*; (*e*) *Agorass N'Essoui*; (*f*) *Eguef-Malan I*.

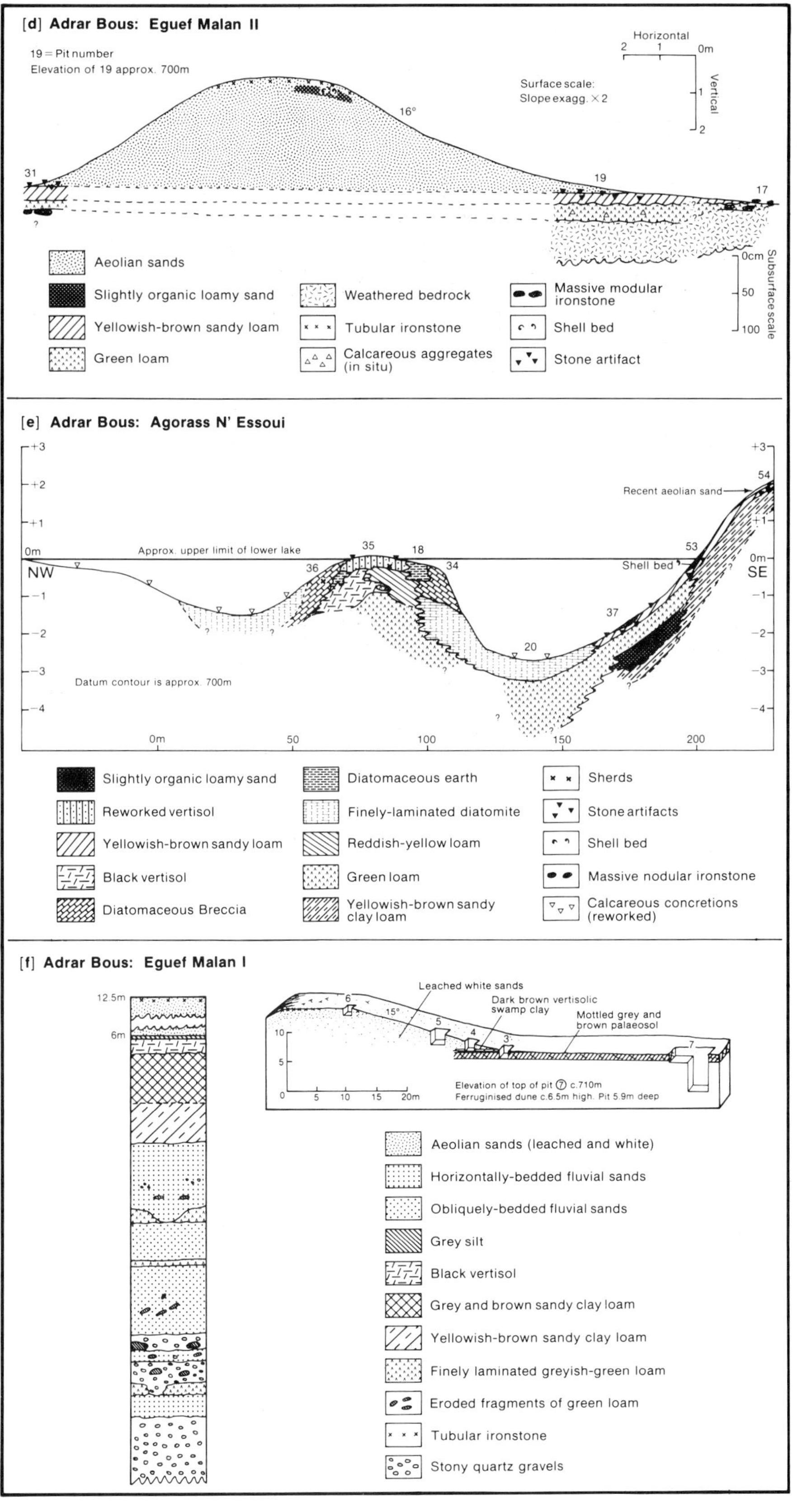
[d] Adrar Bous: Eguef Malan II
19 = Pit number
Elevation of 19 approx. 700m
Horizontal
2
1
0m
Vertical
1
2
Surface scale:
Slope exagg. × 2
16°
31
19
17
0cm
50
100
Subsurface scale
Aeolian sands
Slightly organic loamy sand
Yellowish-brown sandy loam
Green loam
Weathered bedrock
Tubular ironstone
Calcareous aggregates (in situ)
Massive modular ironstone
Shell bed
Stone artifact
[e] Adrar Bous: Agorass N' Essoui
Approx. upper limit of lower lake
Recent aeolian sand
Shell bed
NW
SE
Datum contour is approx. 700m
Slightly organic loamy sand
Reworked vertisol
Yellowish-brown sandy loam
Black vertisol
Diatomaceous Breccia
Diatomaceous earth
Finely-laminated diatomite
Reddish-yellow loam
Green loam
Yellowish-brown sandy clay loam
Sherds
Stone artifacts
Shell bed
Massive nodular ironstone
Calcareous concretions (reworked)
[f] Adrar Bous: Eguef Malan I
12.5m
6m
Leached white sands
Dark brown vertisolic swamp clay
Mottled grey and brown palaeosol
15°
Elevation of top of pit ⑦ c.710m
Ferruginised dune c.6.5m high. Pit 5.9m deep
Aeolian sands (leached and white)
Horizontally-bedded fluvial sands
Obliquely-bedded fluvial sands
Grey silt
Black vertisol
Grey and brown sandy clay loam
Yellowish-brown sandy clay loam
Finely laminated greyish-green loam
Eroded fragments of green loam
Tubular ironstone
Stony quartz gravels

FIG. 6. Neolithic grave on the lower talus slope, *Agorass in-Tchissiskan*, Adrar Bous. Charcoal within this grave yielded a radiocarbon age of 4440 ± 120 BP (Carter & Clark 1976).

Lacustrine clays and diatomites of *Agorass N'Essoui*

Immediately N of Lookout Hill is a low ill-drained col at about 705 m a.s.l. which collects the rare runoff from the hill. A patch of diatomaceous earth on the edge of the col contained 17.5% clay and 19% silt, and appears to be part of a discontinuous bed of diatomite which crops out S of hill 'Z' 1.5 km NE of Lookout Hill (Fig. 3). The maximum elevation of this diatomite ('upper diatomite') is about 708 ± 2 m a.s.l., and along the palaeo-shoreline of the lake in which it formed there are ferruginized fish vertebrae and the barbed bone points and stone microliths of an Epi–Palaeolithic microlithic culture dated elsewhere in this region at about 8000 BP. Calcareous concretions overlying the upper diatomite (Fig. 7) have a radiocarbon age of 7310 ± 120 BP (Faure *et al.* 1963).

About 2.5 km SE of hill 'Z' a low sand ridge (*Eguef Malan II*, Fig. 5*d*) marks a period during which the dwindling lake was stable, this time at about 700 m a.s.l. The dune surface is fixed by numerous ferruginous tubules which resemble fossil Phragmites reeds. Beneath the tubular ironstone, and often embedded in it, are freshwater *Melanoides tuberculata* shells comparable to those found at the same elevation 1.5 km further E (Fig. 5*e*). Beneath the leached white dune sands a yellow-brown loamy sand containing derived Aterian and Levallois-type artefacts lies above a calcareous olive-green or yellow-brown loam with 65% sand, 16%–19% silt and 9%–11% clay (Fig. 8). Beneath the green loam a zone of massive ferruginous nodules sometimes occurs, separating the loam from the underlying decayed granite bedrock. Neolithic artefacts and fish vertebrae occur on and behind the ferruginized dune, here named *Eguef-Malan II*.

It seems that as the lake receded slightly over 7000 years ago, small dunes migrated northwards across the exposed lake floor and became fixed during a subsequent lake stillstand roughly contemporaneous with the occupation of this area by Tenereen Neolithic man about 4000–6000 years ago.

About 1.5 km NE of *Eguef-Malan II*, at *Agorass N'Essoui*, there is a low terrace of black clay which rises nearly 3 m above the diatomite-floored hollow at the foot of the pediment and which extends southeastwards beneath a conspic-

FIG. 7. Rolled calcium carbonate concretions on the floor of the 710 m a.s.l. lake at *Agorass N'Essoui*, Adrar Bous. These nodules have yielded a radiocarbon age of 7310 ± 120 BP (Faure *et al.* 1963).

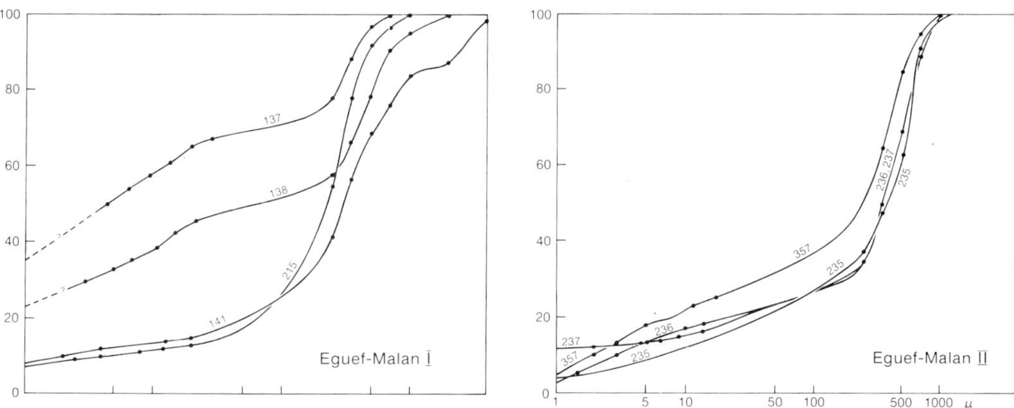

FIG. 8. Cumulative grain size curves for selected stratigraphic units, *Eguef-Malan I* and *II*.

uous knife-edged rock-cored dune. Diatomaceous breccia containing Neolithic potsherds is banked against the black clay bench (Fig. 5*E*), which is itself underlain by a calcareous green clay loam containing 16% clay, 12% silt and 42% very fine sand. The black vertisol, by contrast, contains 6.5%–10% clay, 23.5%–36.5% silt and 58%–67.5% sand.

A green loam underlies the diatomite and extends up the southern pediment slopes where it overlies a lens of sand and a massive yellow-brown loam, on the surface of which are fresh Upper Acheulian hand-axes. Above the green loam, which contains fresh 'Middle Stone Age' artefacts in its upper portion, is a yellow-brown sandy loam containing derived Aterian and

Levallois-type implements, and above this in turn is a grey-brown slightly organic loamy sand containing freshwater mollusc shells concentrated at a level of about 700 m a.s.l. (Fig. 5*e*). The overall sequence of this admittedly complex locality is as follows from the top:

(a) Modern wind-blown sands at the foot of the dune, up to 15 cm thick on the pediment;
(b) diatomaceous breccia with occasional Neolithic shards, at least 50 cm thick, *or* dark grey-brown loamy sand, locally shelly, 15 cm thick;
(c) very dark grey-brown vertisolic clay loam 105 cm thick *or* light grey finely-laminated diatomite 65 cm thick;
(d) yellow-brown sandy loam (15 cm) *or* gypsiferous sandy loam (20 cm);
(e) olive-green massive loam, locally calcareous, more than 110 cm thick;
(f) 40 cm of grey-brown or yellow-brown sand, sealed by 5 cm of massive nodular ironstone;
(g) massive yellow-brown loam, over 60 cm thick.

Layer (g) is similar in colour, texture and structure to layer (g) at Gabbro Knoll, and both underlie sediment containing unabraded Upper Acheulian hand-axes. Layer (e) is almost certainly the same as the green loam at *Eguef-Malan II* and layer (e) at Gabbro Knoll, all of which either contain fresh Middle Stone Age artefacts in their upper levels or underlie a sandier layer with derived Aterian and Levallois-type implements.

Sequence on the plains: *Eguef-Malan I*

Eight kilometres NW of the dark clay bench at *Agorass N'Essoui* and 1 km S of the 1970 camp is *Eguef-Malan I*, a ferruginized sand-ridge (Fig. 5*f*). The total section depth is 12.5 m of which the top 6.5 m are leached aeolian sands which unconformably overlie 20 cm of dark grey vertisolic loam (27%–41% clay, 24.5%–28% silt). The vertisol is preserved beneath the dune, but had been eroded elsewhere to reveal a massive mottled grey and brown sandy loam 55 cm thick over a yellow-brown fine sandy loam. No prehistoric artefacts were found, and the lower part of the section appears to reflect two main erosion–aggradational events, with progressive decrease in grain size throughout the cycle. There is a corresponding upward change in the degree of sorting, the σ_1 values (Folk 1965, p. 46) ranging from 3.10 ϕ (very poorly sorted), through 1.50 ϕ (poorly sorted) to 0.92 ϕ (moderately sorted) (Fig. 8).

Correlation of deposits around the mountain (Fig. 9)

At *Agorass in-Tchissiskan* there are two vertisols separated by a mottled sandy loam. The brown sandy loam which overlies the upper vertisol is granulometrically similar to the Neolithic horizon in the western valley. In one terrace section the upper vertisol is replaced laterally by a grey diatomaceous silt comparable to that found near the high lake shoreline just NW of Lookout Hill. The eroded low-lying dark loams flanking Gabbro Knoll and Lookout Hill appear to be co-extensive with the vertisol protected by the ferruginized dune at *Eguef-Malan I*, which must be roughly 5000–7000 years old, since the similar feature at *Eguef-Malan II* formed after the recession of the high lake and before the low lake stillstand. Were the vertisol much older than the dune, which was stabilized during Neolithic times, it would doubtless have been eroded, so that the vertisolic loam W of Adrar Bous seems to have formed during, or shortly after, the high lake stillstand about 8000 BP.

The green loam of Gabbro Knoll, *Eguef-Malan I* and *Agorass N'Essoui* is considerably older than the vertisol. It is clearly younger than the mottled grey and brown sandy loam or the yellow-brown loam upon which fresh Upper Acheulian hand-axes occur, and older than the Aterian cultures of Lookout Hill and *Eguef-Malan I*, and like the lower or main vertisol at *Agorass in-Tchissiskan* contains 'Middle Stone Age' and fresh Levallois-type assemblages in its top few decimetres. A generalized scheme for the unconsolidated sediments at Adrar Bous would therefore be:

Holocene

(1) Modern stratified coarse and fine sands; aeolian sands.
(2) As above. Proto-historic graves.
(3) Brown sand/silt loam. Low lake diatomite. Neolithic 4000–5000 BP.
(4) Dunes and aeolian sands. (Recession of high lake).
(5) Younger vertisol. High lake diatomite. (?) Epi-Palaeolithic *c.* 8000 BP.

Pleistocene

(6) Upper mottled sandy loam or yellow-brown sandy loam. Aterian $\geq$40 000 to $\leq$150 000 BP.
(7) Lower vertisol or green loam. Levalloisø–Mousteroid.
(8) Lower mottled sandy loam. Levallois/Upper Acheulian.
(9) Yellow-brown loam or basal cobbles. Upper Acheulian >150 000 BP.
(10) Sands (fluvial).

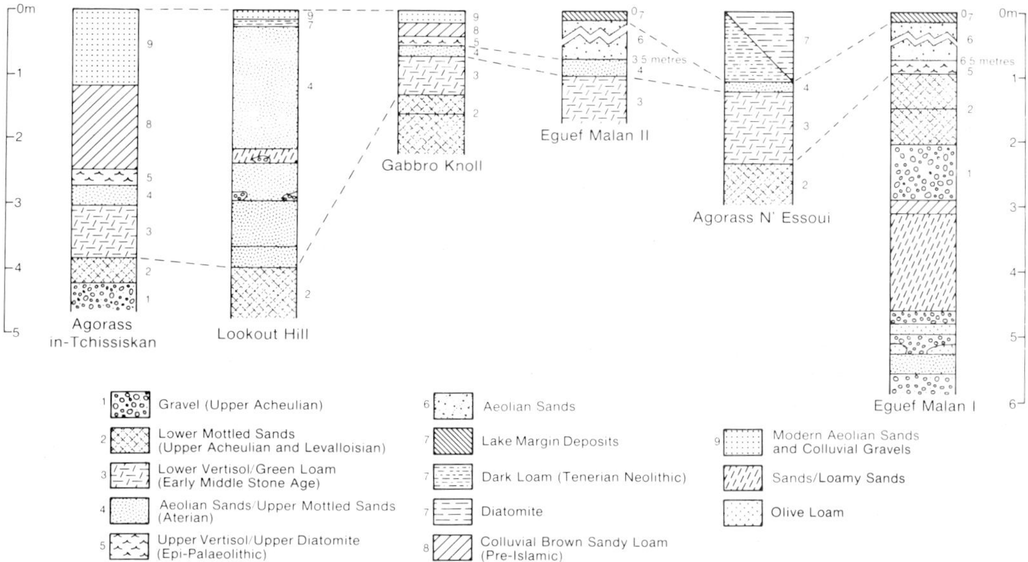

FIG. 9. Provisional litho-stratigraphical correlations between the six Quaternary sites shown on Fig. 3, supported by prehistoric stone tool assemblages used as zonal fossils.

(11) Loam, sands, gravels.
(12) Loam, sands, gravels.

Palaeoclimatic implications

The oldest sediments seen were the quartz gravels, fluvial sands and intercalated olive-grey loams in the lower 3–4 m of the *Eguef-Malan I* sequence, in which at least two erosional–depositional episodes may be recognized. The poor sorting and high content of fine sub-angular quartz gravels in the basal layers suggests rapid erosion of a granite weathering profile, and the fine laminations, better sorting and finer grain-size of the upper (or sub-gravel) layers suggests slow erosion of vegetated soil mantles and/or deposition under backslope or lacustrine swamp conditions. The evidence does not negate Chavaillon's view (1964) that Saharan valley aggradation occurs during the transition from wet to dry conditions and vertical erosion during the change from dry to moister conditions (see also Alimen 1969, pp. 203–205). The sequence is probably in age Middle Pleistocene.

There then ensued a period of prolonged and continuous deposition of fine sandy loams on the plains, culminating in a period of marked stability and pedogenesis, with the development of a gleyed and iron-stained coarse prismatic-structured 'pseudo-gley' soil—the grey and brown mottled sandy loam underlying the horizon containing Upper Acheulian hand-axes. The gleying and mottling is indicative of a seasonally-fluctuating water table, with alternating reducing and oxidizing conditions, as in the seasonally-wet tropics today.

In the mountain valleys a basal cobble layer of local provenance underlies a grittier version of the 'pseudo-gley', which is widespread around the immediate periphery of the massif. A long swampy phase ensued and dark clays accumulated in local swamps at about the time that green clays and loams accumulated in an extensive lake system S of the massif and across the col between Granite Knoll and Lookout Hill. The green colour suggests reduction of trivalent (ferric) iron to the divalent ferrous form under perennial reducing conditions in acidic lake waters. The main vertisol probably developed in seasonally-flooded swamps fringing this early lake, in much the same way that vertisols formed in the Gezira plains of central Sudan (Tothill 1946).

The 'Middle Stone Age' or Levallois–Mousteroid industrial assemblage which occurs within the top 20 cm of this green loam in sites near to higher ground, was probably contemporary with the drying-out of the lake. With the onset of drier conditions erosion of the weathered rock was rapid, and an admixture of colluvial clayey sands and of aeolian sands was laid down over large areas. Apart from a weak translocation of iron, pedogenesis was minimal.

With cooler and wetter conditions the former lake-basin filled once more, to the level of the high lake. Diatoms flourished in the lake waters, and mountain streams built deltas out into the lake. Epi–Palaeolithic hunters and fishers subsisted on Nile perch, mud fish, crocodile, hippo and tortoise from the lake (Clark 1971; Smith 1976). The upper vertisolic loams probably formed then, and swamps doubtless persisted as the high lake dwindled, some 7000 years ago, causing the formation of large $CaCO_3$ concretions on the exposed lake floor. Small barchans advanced from the S and occupied the now dry surface of the former lake and swamps. Rivers draining the western valley became increasingly overloaded with sand and eventually ceased to flow. Extensive aeolian sand sheets formed.

A renewal of slightly moister/cooler conditions saw the stabilization of the dunes, the presence of a lower lake, and the arrival of Tenereen Neolithic herders over 5000 years ago (Table 1). The deposition of up to 120 cm of brown silty loam by slopewash at *Agorass in-Tchissiskan* in no more than 4000 years gives a mean minimum rate of accumulation of 0.3 mm a^{-1}, or roughly ten times the expected rate of sediment yield from gentle granite slopes in the seasonally-wet tropics, where sediment yields are in any case very high (Douglas 1967). This suggests that the clearing and grazing activities of Neolithic man had already accentuated the effects of the progressively deteriorating climate. A 30% reduction in grass cover is known to be able to cause a 20-fold increase in sediment yield (Williams 1969) when the grass is initially of the tall open savanna type.

Weak pedogenesis took place at this time (Fig. 10), the former dunes became fixed, and wave-action reshaped *Eguef-Malan II*, throwing up freshwater shells onto the flanks of what was now a sandy beach-ridge.

Drier conditions resumed, and have continued with only minor interruptions to the present-day. Iron, no longer mobilized by humic acids from the shore reed vegetation, was precipitated around the roots and stems of plants growing on the dunes, and at *Eguef-Malan II* littoral shells are incorporated into the ferruginous matrix which now caps the summit of the sand-ridge.

Minor vertical erosion preceded and followed the arrival of the Proto-historic tumulus-builders in pre-Islamic times, and the lower two terraces at *Agorass in-Tchissiskan* were probably cut during the past 1500 years. Elsewhere, gully erosion seems ineffectual, and morphogenesis is limited to rare, highly localized, stream erosion, minor and sporadic slopewash, and, in exposed sites, deflation. See Table 2 for a very tentative summary of late Quaternary climatic fluctuations at Adrar Bous.

Holocene lakes and associated sub-fossil mollusca

The foregoing attempt to reconstruct the Quaternary environmental fluctuations at Adrar Bous was based upon inferences drawn very largely from field mapping of the litho–stratigraphical sequences around the mountain, aided greatly by the use of prehistoric artefact assemblages as both 'zone fossils' and cultural markers. In the case of the Holocene lower lake, which was reasonably stable at 700 m a.s.l. during at least part of the time that Neolithic herders roamed Adrar Bous from 6000 to 4000 BP, it is possible to supplement the purely stratigraphic data with palaeoecological inferences drawn from the sub-fossil mollusca collected from the 700 m strandline at *Agorass N'Essoui* (Loc. 3) and *Eguef-Malan II* (Loc. 2). In addition, mollusc bulk samples were collected from two closed depressions well away from the mountain, one (Loc. 1) 35 km to the SE, the other (Loc. 4) 50 km to the SE. Table 3 shows the numbers and percentages of individuals from different species identified.

The species present in these faunas are typical common sub-fossil Saharan ones (Sparks & Grove 1961). There is a restricted number of species but often large numbers of individuals from each of those species. All the species are freshwater ones with the exception of *Succinea* which is a marsh genus. The number of freshwater snails seems to

TABLE 1. *Radiocarbon age determinations from Adrar Bous*

Age (years BP)	Lab. no.	Material	Event dated	Reference
7310±120	T-361	$CaCO_3$	Desiccation of 710 m lake	(1)
5760±500	UCLA-1658	Bone collagen	Neolithic *Bos*	(2)
5740±580	n.a.	Bone collagen	Neolithic *Bos*	(3)
5050±150	n.a.	Burnt bone	Neolithic midden	(3)
4910±135	N-870	Charcoal	Neolithic midden	(4)
4440±120	n.a.	Charcoal	Neolithic burial	(5)

References: (1) Faure *et al.* 1963, p. 52. (2) Carter & Clark 1976, p. 488. (3) Clark *et al.* 1973, p. 278. (4) Carter & Clark 1976, p. 487. (5) Smith, 1980, p. 455.

FIG. 10. Neolithic pots within the mid-Holocene slightly organic loamy sand unit, Adrar Bous. The Tenerian Neolithic is dated between *c.* 6000 and *c.* 4000 BP (Clark *et al.* 1973).

indicate some continuity in the shallow lakes or ponds, but more land snails would normally have been expected unless every sample represented conditions some way from the contemporary shore. Perhaps the rainfall was too low for a sufficiently richly vegetated environment which would have encouraged snails.

Of interest is the genus *Hydrobia*, a snail characteristic of brackish conditions. It seems indistinguishable from British *Hydrobia ventrosa*, a species having a fairly wide range in southern European coastal waters. It is certainly the same as the sub-fossil species recorded from Kharga as *H. stagnalis* (Gardner 1935) and probably the same as *H. peraudieri*, found in a number of localities in the northern half of the Sahara.

TABLE 2. *Late Quaternary climatic fluctuations at Adrar Bous*

Years BP		Inferred climate
0–1500	(1) Modern wind action, minor slopewash and gullying	Arid and hot
1500–2500	(2) Minor vertical erosion; slopewash	Semi-arid
4000–6000+	(3) Weak pedogenesis, then iron segregation. Lower lake.	Sub-humid
c. 7500	(4) Dune migration, deflation, carbonate precipitation.	Arid
≥8000	(5) Upper lake. Swamp soils formed.	Humid/cool
>12000	(6) Prolonged deflation, minor slopewash, weak iron segregation, carbonate precipitation.	Arid
≥40000 to ≤150000	(7) Extensive lake(s). Major period of swamp soil formation.	Humid
	(8) Deflation, slopewash, pedogenesis ('pseudo-gley').	Sub-humid to semi-arid
>150000	(9) (10) Erosion in the massif, deposition on the plains	
	(11) Erosion–aggradation episode.	Moist → dry?
	(12) Erosion–aggradation episode.	Moist → dry?

Too little is known of the ecology of most African snails to infer much from the changing percentages of the species, although it would be reasonably safe to conclude that the higher percentages of *Hydrobia* represent more brackish conditions. On the whole, however, the faunas of the various samples appear to be reasonably uniform.

Stable isotope ratios in the Holocene gastropods of Adrar Bous

The shells at Adrar Bous afford an opportunity of examining the climatic conditions in that locality in the mid-Holocene. Approximately 80 analyses were made by PIA for oxygen- and carbon-stable isotope ratios on the aragonitic shells of five species of gastropods collected from four localities at or near Adrar Bous. The isotope ratios were obtained by sequential sampling along the growth spiral of the shells whenever the shell was large enough to permit multiple samples from a single shell (Abell 1985). With very small shells (many were 1 to 2 mm in diameter) the requisite sample size for the mass spectrometry necessitated using the whole shell for a single analysis. The carbon dioxide was liberated on a vacuum line using 100% phosphoric acid, and the isotope ratios measured in a V.G. Micromass 602-D mass spectrometer. The raw isotope ratios were corrected to the PDB standard. The oxygen isotope ratios have been corrected to aragonite (Grossman 1982), the unaltered aragonitic character of the shells having been established by XRD analyses of selected specimens. The isotope ratios are recorded in Table 4. When the size of the shell permitted sequential sampling as many samples as possible were taken along the growth spiral (Table 3). Of the four localities where shells were collected, two were within the Adrar Bous ring-complex (Localities 2 and 3 at *Eguef-Malan II* and *Agorass N'Essoui* respectively), and come from the strandline of the 700 m a.s.l. palaeolake. This lake has been provisionally assigned an age of 6000–4000 BP (Williams 1976), and is associated with a Neolithic culture. Locations 1 and 4 are 35 and 60 km SE of Adrar Bous respectively, on the fringes of the Tenere desert. The elevation of Locality 4 is 200 m below the lake levels of Localities 2 and 3. Both Localities 1 and 4 occur in diatomite-lined depressions.

Shells from the same climatic regime generally show a weak correlation between oxygen and carbon isotopic ratios (Abell unpublished data). All of the analyses recorded in Table 3 are plotted in Figs 11 and 12. In Fig. 11 isotope data are plotted differentiating among the species. Little can be concluded from this plot except that *Melanoides tuberculata* occurs exclusively in Locality 4, and that it and the *Lymnaea* specimens from that locality fall well outside the main body of data points. In Fig. 12 where locality is the significant variable, it is clear that Locality 1 shows a preponderance of data points with more

TABLE 3. *Holocene fossil non-marine mollusca identified from Adrar Bous and the adjacent Tenere desert*

Sample Number			Loc. 1			Loc. 3	Loc. 2	Loc. 4
			A193	A194	A195	A369	A355	T480
Melanoides tuberculata	(Müller)		86	147	94	161	12	4
		%	(20)	(24)	(31)	(35)	(4)	
Lymnaea natalensis	(Krauss)		9	18	9	4	3	2
		%	(2)	(3)	(3)	(1)	(1)	
Biomphalaria pfeifferi	(Krauss)		144	216	91	37	60	8
		%	(34)	(35)	(30)	(8)	(21)	
Gyraulus costulatus	(Krauss)		1	1	—	1	1	—
Anisus chudeaui	(Germain)		6	4	—	—	—	—
Anisus dallonii	(Germain)		22	23	12	1	1	—
		%	(5)	(4)	(4)			
Segmentina angusta	(Jickeli)		7	6	—	—	—	—
		%	(2)	(1)				
Bulinus truncatus	(Audouin)		23	31	16	71	73	2
		%	(5)	(5)	(5)	(16)	(25)	
Hydrobia ventrosa	(Montagu)		133	166	85	180	136	—
		%	(31)	(27)	(28)	(40)	(48)	
Succinea sp.			—	1	—	—	—	—
	Totals		431	613	307	455	286	16

negative oxygen and carbon isotope ratios, while Localities 2 and 3 tend toward more positive values of both isotopes. The shells from Locality 4 are split between two groups; the *Biomphalaria* and *Bulinus* specimens form part of the main trend while the *Lymnaea* and *Melanoides* are in a quite separate grouping.

The slightly different groupings by locality along the main trend of Fig. 12 probably reflect somewhat different patterns of water accumulation for the lakes within the same climatic regime. The dichotomy for the Locality 4 shells, however, undoubtedly reflects two distinctly different climatic episodes. The main trend points to an episode probably concurrent with the 6000–4000 BP humid phase which enabled Neolithic herders to graze their cattle at least seasonally around the mountain. The much more negative oxygen isotope ratios at Locality 4 are almost certainly representative of a different, far wetter and/or cooler, regime and may well be representatives of the 7500–7000 BP episode of wet, cool weather in the Sahara identified by Pachur & Braun (1980). However, a second possibility exists. In the oases of the Sahara today, gastropods are found in springs and wells which show oxygen isotope ratios near −8% to −10% (PDB), and which are certainly sampling 'fossil' water from the Saharan aquifers (Sonntag *et al.* 1980). The pattern which Sonntag and co-workers observe is a progressive depletion of both hydrogen and oxygen isotopes along the meteoric water line sampling progressively across the Sahara from W to E. Sonntag *et al.* consider that there has been no significant recharge of the deeper aquifers of the central Sahara in recent times, so that the isotope ratios should be remarkably constant from springs or wells where evaporation is not important. Could these aberrant oxygen isotope ratios from Locality 4 indicate a spring in or near the mid-Holocene lake? We discount this possibility, having analysed sequentially a number of *Melanoides tuberculata* shells from large springs (Buffalo Springs and Mzima Springs) in Kenya, and found little variation in oxygen isotope ratios (about 0.3‰ to 0.5‰) over their lifetimes, in contrast to the 1.3‰ to 1.5‰ variation we find in the Locality 4 shells.

TABLE 4. *Oxygen and carbon isotope ratios, Adrar Bous gastropods*

Sample site	Species	$\delta^{18}O$ (PDB)	$\delta^{13}C$ (PDB)
A-193	*Bulinus truncatus*	−4.09, −4.56, −5.29, −2.64	−5.59, −5.02, −9.09, −6.26
Loc. 1	*Lymnaea natalensis*	−2.64, −3.19, −3.49	−9.70, −9.98, −11.36
		{−3.13, −3.34, −3.52, −3.15}	{−2.33, −3.86, −3.99, −4.41}
	Hydrobia ventrosa	−2.91, −3.27, −4.45, −3.26	−5.47, −5.28, −6.20, −4.90
		−2.95, −2.55, −3.28	−5.07, −4.52, −5.09
A-194	*Lymnaea natalensis*	−2.36, −2.87, −4.36	−7.05, −8.71, −8.34
Loc. 1	*Hydrobia ventrosa*	−3.84. −2.27, −3.47, −2.19	−6.13, −5.42, −5.10, −5.23
		−3.49, −2.44	−5.13, −4.99
A-195	*Biomphalaria pfeifferi*	−4.47, −3.66, −4.31, −5.18	−8.66, −7.33, −6.79, −6.89
Loc. 1	*Lymnaea natalensis*	−4.44, −3.36, −5.50, −5.08	−9.86, −4.75, −7.48, −5.44
	Hydrobia ventrosa	−3.37, −3.51, −3.13, −2.55	−5.55, −5.09, −5.18, −5.52
		−3.13, −4.26, −2.29	−5.42, −6.37, −4.78
A-355	*Lymnaea natalensis*	−0.09, −1.51, −2.15	−5.10, −4.51, −4.20
Loc. 2			
A-369	*Bulinus truncatus*	−4.06, +1.25, −1.67, −0.99	−8.75, −3.28, −7.80, −4.47
Loc. 3	*Lymnaea natalensis*	−1.00, {−3.06, −2.87, −3.01}	−1.99, {−7.58, −6.66, −7.12}
	Hydrobia ventrosa	−1.30, −1.10, −2.70, −2.79	−2.33, −1.30, −2.76, −0.22
		+1.27, −3.51, +0.15	−2.31, −3.46, −2.13
T-480	*Biomphalaria pfeifferi*	−1.68, −3.03	+1.64, −0.79
Loc. 4		{−2.86, −2.61, −3.69, −3.37, −3.06}	{−1.98, −2.05, −4.98, −4.48, −5.52}
	Bulinus truncatus	−3.39, −3.01, {−0.15, +0.35}	−3.23, −2.12, {+1.44, +0.13}
	Lymnaea natalensis	−8.65, {−7.83, −7.14, −6.53}	−0.90, {−2.65, −3.16, −2.71}
	Melanoides tuberculata	{−9.23, −9.04, −9.55, −9.35, −9.10	{−3.00, −3.34, −3.08, −3.37, −3.43
		−8.85, −7.71, −7.40, −7.37}	−2.61, −2.70, −2.47, −2.73}

Bracketed series are sequential analyses along the growth spiral of an individual gastropod shell, starting at apex. Unbracketed values are whole shell analyses.

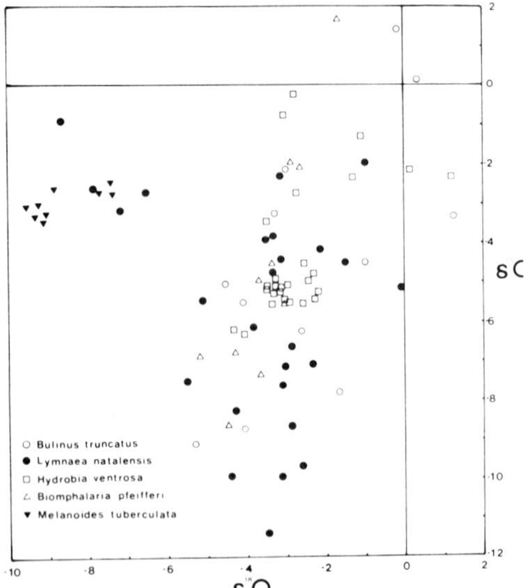

FIG. 11. Oxygen and carbon isotope ratios plotted according to gastropod species, Adrar Bous.

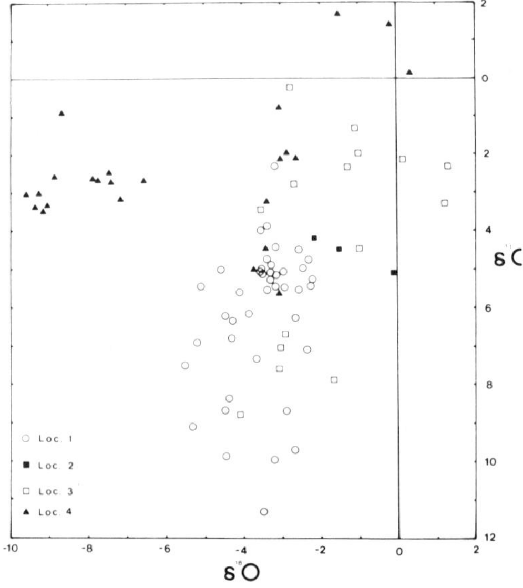

FIG. 12. Oxygen and carbon isotope ratios plotted according to locality of origin, Adrar Bous.

The relationship between oxygen and carbon isotope ratios observable in Figs 11 and 12 is worth comment. During the existence of a particular climatic weather pattern, such as that at Adrar Bous about 6000 to 4000 years ago, the rainfall for a given area will show a relatively narrow range of oxygen isotope values (Dansgaard 1964; Yurtsever 1975), which will be largely a function of distance from the oceanic source of the water and the amount of precipitation en route. After precipitation, there will be various chances for evaporative fractionation of the water, depending on surface *vs* groundwater transport, depth of the water body, humidity, wind speed, *etc.*, and a continuous movement of the oxygen isotope values toward more positive values. At the same time, gastropods will be growing in these bodies of water, and flourishing according to nutrient supply, temperature, pH, *etc.* As evaporation proceeds and the oxygen isotope ratio rises, indicating a diminished supply of freshwater, the supply of nutrients may be expected to decrease, and a greater reliance for shell growth be placed on atmospheric carbon dioxide rather than the re-use of carbon dioxide produced by recycling of the biomass now highly depleted in ^{14}C. Thus, a rough parallelism is often observed in the oxygen and carbon isotope ratios of freshwater gastropods, but with oxygen isotope variation being significantly larger than carbon isotope variation.

Finally, the climatic range recorded by the sequential isotope measurements along the growth spiral of the shells is of interest in indicating the span of climatic conditions. Sequences in specimens from Locality 1 have a range of only 0.4‰. This is quite limited compared with equatorial African gastropods, and probably reflects a limited temperature range (about 2°C) over the life of the shell. (The temperature coefficient for change in oxygen isotope ratio is about −0.2‰ per °C, Friedman & O'Neil 1977). Similarly in Localities 2 and 3 the one sequence of only 0.2‰ is also probably associated with temperature change. Such small changes suggest a fairly equable climatic regime, although the rather small size of the shells may mean that they represent only a small and restricted portion of the annual climatic cycle. Certainly the wider range among the individual shells, where only whole shell analyses were possible, indicates that the climatic conditions were considerably more variable, but not necessarily within a single year. In Locality 4, the two sequential analyses for shells in the main group show fairly large ranges of oxygen isotope ratios—up to 1.1‰—which could be ascribed to temperature control, although not with any certainty. The two sequences obtained for shells outside the main trend suggest that evaporation was likely to be in control at the time these shells grew, as temperature differences sufficient to produce ranges up to 2.2‰ are unlikely. Furthermore, the shift toward positive values at the aperture is often indicative of increasingly evaporative conditions which can decimate gastropod populations. However, in the absence of other evidence, this is speculative. Certainly cool weather necessary to produce a shift of 2.2‰ (about 11° or 12°C) could be lethal, but is more likely to have simply brought about a halt in shell growth, and be exhibited in our isotope ratio *vs* growth diagrams only in the last few increments of shell growth rather than in several whorls.

Conclusions

Although there is good evidence of late Cainozoic volcanism and continuing epeirogenic uplift in the Aïr Mountains region of the south-central Sahara, there is no compelling evidence that the pattern of Quaternary sedimentation around Adrar Bous ring-complex is in any way a result of tectonic uplift. Rather, a climate fluctuating from arid to semi-arid seems adequate to account for the vertical and lateral facies changes from piedmont alluvial fan deposits to lacustrine muds or swamp loams during more humid times, and from colluvial–alluvial slopewash granule sands through ephemeral channel gravel-sands to well-sorted dune and sand-plain medium sands during more arid times. Both the mollusc assemblages and the stable isotope ratios obtained on sub-fossil gastropods from the Holocene lakes indicate that there may have been significant annual climatic variation within the life of the lakes in question but that such variation is not usually evident within the lifetime of any one individual gastropod. The use of shell isotope ratios is also useful in discriminating between palaeo-hydrologically distinct populations of gastropods, although to an untrained observer such shell concentrations may appear relatively homogeneous. Finally, in this part of Saharan Africa at least the use of primary-context, or little disturbed, prehistoric stone tool assemblages proved to be an invaluable adjunct to conventional stratigraphic mapping of Middle and Upper Quaternary formations.

ACKNOWLEDGMENTS: M. A. J. Williams extends grateful thanks to David Hall, expedition leader; to Professor Desmond Clark for enlarging his cultural horizons; to the government of Niger for the privilege of working at Adrar Bous; to Tony Piggott, John Rogers, Mike Saunders and Geoff Parkes for careful surveying of mountain and palaeo-lake; to Macquarie University for their enlightened encouragement of field research; to Gary Swinton and Tony Miller (Monash, Geography) for excellent maps and plates; and to Annette Rogers for her care with the manuscript.

References

ABEBE, B., CHAVAILLON, J. & SUTTON, J. E. G. (eds) 1976. *Proceedings of the 7th Panafrican Congress of Prehistory and Quaternary Studies*, Addis Ababa 1971. Ministry of Culture, Antiquities Administration, Addis Ababa.

ABELL, P. I. 1985. Oxygen isotope ratios in modern African gastropod shells: a data base for paleoclimatology. *Chemical Geology (Isotope Geoscience Section)* **58**, 183–193.

ALIMEN, H. 1969. Les études francaises sur le Quaternaire d'Afrique. *In:* CNRS, *Études Françaises sur le Quaternaire*. INQUA, Paris, 201–214.

BLACK, R. 1963. Note sur les complexes annulaires de Tchouni-Zarniski et de Gouré (Niger). *Bulletin, Bureau de recherches géologiques et minières* **1**, 31–45.

—— 1965. Sur la signification pétrogénétique de la découverte d'anorthosites associées aux complexes annulaires subvolcaniques du Niger. *Compte rendu de l'Académie des sciences. Paris* **260D**, 5829–5832.

—— *Esquisse géologique de l'Adrar Bous*. (1 cm rep. 1 km.)

—— & GIROD, M. 1970. Late Palaeozoic to Recent igneous activity in West Africa and its relationship to basement structure. *In:* CLIFFORD, T. N. & GASS, I. G. (eds) *African Magmatism and Tectonics*. Oliver & Boyd, Edinburgh, 185–210.

BORDET, P. 1952. *Les appareils volcaniques récents de l'Ahaggar*. Mon. Rég. Algérie 1: No. 11, 19ᵉ Congr. Géol. Int., Alger.

BOWDEN, P., VAN BREEMEN, O., HUTCHISON, J. & TURNER, D. C. 1976. Palaeozoic and Mesozoic age trends for some ring complexes in Niger and Nigeria. *Nature* **259**, 297–299.

CARTER, P. L. & CLARK, J. D. 1976. Adrar Bous and African cattle. *In:* ABEBE, B. *et al.* (eds) *Proceedings of the 7th Panafrican Congress of Prehistory and Quaternary Studies*. Ministry of Culture, Antiquities Administration, Addis Ababa, 487–493.

CHAVAILLON, J. 1964. *Étude stratigraphique des formations quaternaires du Sahara nord-occidental*. CNRS, Centre Recherches Zones Arides **5**, 1–394.

CLARK, J. D. 1971. An archaeological survey of Northern Aïr and Tenere. *Geographical Journal* **137**, 455–457.

—— 1976. Epi-Palaeolithic aggregates from Greboun Wadi, Aïr, and Adrar Bous, north-western Tenere, Republic of Niger. *In:* ABEBE, B. *et al.* (eds) *Proceedings of the 7th Panafrican Congress of Prehistory and Quaternary Studies*. Ministry of Culture, Antiquities Administration, Addis Ababa, 67–78.

—— & BRANDT, S. A. (eds) 1984. *From Hunters to Farmers. The Causes and Consequences of Food Production in Africa*. University of California Press, Berkeley & Los Angeles.

——, WILLIAMS, M. A. J. & SMITH, A. B. 1973. The geomorphology and archaeology of Adrar Bous, central Sahara: a preliminary report. *Quaternaria* **17**, 245–297.

CONRAD, G. 1969. *L'évolution continentale post-hercynienne du Sahara algérien.* CNRS, Centre Recherches Zones Arides **10**, 1–527.

CORNET, A. 1962. Rapport géologique. *In:* HUGOT, H. J. (ed.) *Missions Berliet Ténéré-Tchad.* Arts et Métiers Graphiques, Paris, 33–70.

DANSGAARD, W. 1964. Stable isotopes in precipitation. *Tellus* **16**, 436–468.

DELIBRIAS, G. & HUGOT, H. J. 1962. Datation par le méthode dite 'du C 14' du Néolithique de l'Adrar Bous Ténéréen. *In:* HUGOT, H. J. (ed.) *Missions Berliet Ténéré-Tchad.* Arts et Métiers Graphiques, Paris, 71–72.

DOUGLAS, I. 1967. Man, vegetation and the sediment yield of rivers. *Nature* **215**, 925–928.

DRESCH, J. 1959. Notes sur la géomorphologie de l'Air. *Bulletin de l'Association de Géographies Français* **280–281**, 2–20.

FAURE, H. 1959*a*. Géologie des formations sédimentaires à l'E de l'Aïr (Niger). *Bulletin de la Société géologique de France 7e Ser.* **1**, 143–149.

—— 1959*b*. Une hypothèse sur la structure de Ténéré (Niger). *Compte rendu de l'Académie des sciences. Paris* **249D**, 2591–2593.

—— 1962. Esquisse paléogéographique du Niger oriental depuis le Crétacé. *Compte rendu de l'Académie des sciences. Paris* **254D**, 4485–4486.

——, MANGUIN, E. & NYDAL, R. 1963. Formations lacustres du Quaternaire supérieur du Niger oriental: Diatomites et âges absolus. *Bulletin Bureau de Recherches géologiques et minières (Dakar)* **3**, 41–63.

FOLK, R. L. 1965. *Petrology of Sedimentary Rocks.* University of Texas, Austin, Texas.

FRIEDMAN, I. & O'NEIL, J. R. 1977. Compilation of stable isotope fractionation factors of geochemical interest. *Geological Society Professional Paper 440-KK.* US Government Printing Office, Washington, D.C.

FURON, R. 1963. *The Geology of Africa.* Trans. HALLAM, A. & STEVENS, L. A. Oliver & Boyd, London.

GARDNER, E. W. 1935. The Pleistocene fauna and flora of Kharga Oasis, Egypt. *Quarterly Journal of the Geological Society of London* **91**, 479–518.

GINSBURG, L., DE LAPPARENT, A. F., LORNET, B. & TAQUET, P. 1966. Empreintes de pas de vertébrés tétrapodes dans les séries continentales à l'ouest d'Agadès (République du Niger). *Compte rendus de l'Académie des sciences, Paris* **263D**, 28–31.

GREIGERT, J. & POUGNET, R. 1967. Essai de description des formations géologiques de la République du Niger. *Mémoires Bureau de recherches géologiques et minières (Dakar)* **48**, 1–236.

GROSSMAN, E. 1982. *Stable isotopes in live benthic foraminifera from the Southern California Borderland.* PhD thesis, University of Southern California, Los Angeles, California.

HUGOT, H. J. (ed.) 1962*a*. *Missions Berliet Ténéré-Tchad.* Arts et Métiers Graphiques, Paris.

—— 1962*b*. Premier apercu sur la préhistoire du Ténéré du Tefassasset. *In:* HUGOT, H. J. (ed.) *Missions Berliet Ténéré-Tchad.* Arts et Métiers Graphiques, Paris, 49–178.

KILIAN, C. 1931. Des principaux complexes continentaux du Sahara. *Compte rendu de la Société géologique de France 1928–31*, May 4, 109–111.

LEFÈVRE, R. 1960. *Etude d'écoulement dans le Massif de l'Aïr.* ORSTOM, Paris.

LOUIS, P. & RECHENMAN, J. 1966. Interprétation géologique de certaines anomalies gravimétriques du Ténéré (République du Niger). *Compte rendu de l'Académie des sciences, Paris* **263D**, 476–479.

MALEY, J. 1980. Les changements climatiques de la fin du Tertiaire en Afrique: leur conséquence sur l'apparition du Sahara et de sa végétation. *In:* WILLIAMS, M. A. J. & FAURE, H. (eds) *The Sahara and the Nile. Quaternary environments and prehistoric occupation in Northern Africa.* Balkema, Rotterdam, 63–86.

PACHUR, H. J. & BRAUN, G. 1980. The palaeoclimate of the Central Sahara, Libya and the Libyan Desert. *Palaeoecology of Africa* **12**, 351–363.

RAULAIS, M. 1951. Du Crétacé probable sur les hauts reliefs sahariens. *Compte rendu sommaire des séances de la Société géologique de France, 1951, January 22*, 22–23.

—— 1959. Esquisse géologique sur le massif cristallin de l'Aïr (Niger). *Bulletin de la Société géologique de France 7e Ser.* **1**, 207–223.

ROGNON, P. 1967. *Le massif de l'Atakor et ses bordures (Sahara central). Étude géomorphologique.* CNRS, Centre Recherches Zones Arides **9**, 1–560.

ROSSIGNOL, M. & MALEY, J. 1969. L'activité hors de France des palynologues et paléobotanistes français du Quaternaire. *In:* CNRS, *Études françaises sur le Quaternaire.* INQUA, Paris, 265–274.

SMITH, A. B. 1976. A microlithic industry from Adrar Bous, Tenere desert, Niger. *In:* ABEBE, B. *et al. Proceedings of the 7th Panafrican Congress of Prehistory and Quaternary Studies.* Ministry of Culture, Antiquities Administration, Addis Ababa, 181–196.

—— 1980. The Neolithic tradition in the Sahara. *In:* WILLIAMS, M. A. J. & FAURE, H. (eds) *The Sahara and the Nile. Quaternary environments and prehistoric occupation in Northern Africa.* Balkema, Rotterdam, 451–465.

SONNTAG, C., THORWEIHE, U., RUDOLPH, J., LÖHNERT, E. P., JUNGHAUS, C., MÜNNICH, K. O., KLITSCH, E., SHAZLY, E. M. & SWAILEM, F. M. 1980. Isotopic identification of Saharian groundwaters, groundwater formation in the past. *Palaeoecology of Africa* **12**, 159–171.

SPARKS, B. W. & GROVE, A. T. 1961. Some Quaternary fossil non-marine mollusca from the central Sahara. *Journal of the Linnaean Society London, Zoology* **44** (298), 355–364.

THORP, M. B. 1969. Some aspects of the geomorphology of the Aïr Mountains, southern Sahara. *Transactions of the Institute of British Geographers* **47**, 25–46.

TOTHILL, J. D. 1946. The origin of the Sudan Gezira clay plain. *Sudan Notes and Records* **27**, 153–183.

VINCENT, P. M. 1963. Les volcans tertiaires et quaternaires du Tibesti occidental et central (Sahara du Tchad). *Mémoires. Bureau de recherches géologiques et minières* **23**, 1–307.

VOGT, J. & BLACK, R. 1963. Remarques sur la géomorphologie de l'Aïr. *Bulletin, Bureau de recherches géologiques et minières (Dakar)* **1**, 1–29.

YURTSEVER, Y. 1975. *Worldwide survey of stable isotopes in precipitation.* IAEA Report of the Section for Isotope Hydrology, 53 pp.

WILLIAMS, M. A. J. 1969. Prediction of rainsplash erosion in the seasonally wet tropics. *Nature* **222**, 763–765.

—— 1971. Geomorphology and Quaternary geology of Adrar Bous. *Geographical Journal* **137**, 449–455.

—— 1976. Upper Quaternary stratigraphy of Adrar Bous (Republic of Niger, south-central Sahara). *In:* ABEBE, B. *et al.* (eds) *Proceedings of the 7th Panafrican Congress of Prehistory and Quaternary Studies.* Ministry of Culture, Antiquities Administration, Addis Ababa, 435–441.

—— 1984. Late Quaternary prehistoric environments in the Sahara. *In:* CLARK, J. D. & BRANDT, S. A. (eds) *From Hunters to Farmers. The Causes and Consequences of Food Production in Africa.* University of California Press, Berkeley & Los Angeles, 74–93.

—— & FAURE, H. (eds) 1980. *The Sahara and the Nile. Quaternary environments and prehistoric occupation in Northern Africa.* Balkema, Rotterdam.

M. A. J. WILLIAMS, Dept of Geography and Cenozoic Research Unit, Monash University, Clayton, Victoria 3168, Australia.

P. I. ABELL, Chemistry Dept, University of Rhode Island, Kingston, R.I. 02881, USA.

B. W. SPARKS, Jesus College, University of Cambridge, UK.

AEOLIAN SEDIMENTS
Dust dynamics and deposits

Aeolian abrasion of quartz particles and the production of silt-size fragments: preliminary results

W. B. Whalley, B. J. Smith, J. J. McAlister & A. J. Edwards

SUMMARY: Experimental abrasion of quartz sand in an air stream shows that particles with initially high angularity are progressively rounded by the chipping of edges and corners. Grains closely resemble dune sand grains from deserts with respect to rounding and microscopic surface textures after 48 hours of abrasion. Results show an initially high weight loss from the original grains (primarily by edge chipping), compared with subsequent abrasion. High percentages of coarse and medium silt are produced but there is a gradual increase in the percentage of fine silt as rounding increases and the quantity of fines generated declines. Silica extraction from the fines using water shows concentrations of several hundred ppm. Implications for loess production and silcrete formation are suggested.

It has long been suggested (although not universally agreed, *eg* Folk (1978)) that desert sand grains, both ancient and modern, are predominantly rounded in outline and that this property can be used as a diagnostic tool for their recognition. Consequently, much effort has been expended trying to identify and understand the mechanisms by which rounding is produced. Early experimental work on the effects of aeolian attrition on residual grain shape was reported, for example, by Kuenen (1960) and Kuenen & Perdok (1962). Such experimental work has, however, taken on a new impetus with the advent and use of the scanning electron microscope (SEM), primarily for the identification of surface textures on sand-sized material (*eg* Kaldi *et al.* 1978; Krinsley *et al.* 1979; Wellendorf & Krinsley 1980; Lindé & Mycielska-Dowgiałło 1980; Whalley *et al.* 1982). This has allowed the repetition and extension of earlier work and examination of new questions concerning the production of rounded grains (Folk 1978; Goudie & Watson 1981). In addition, it allows detailed examination of the small fragments produced during attrition (especially those which fall in the size range characteristic of loess) (*eg* Smalley & Krinsley 1978; Whalley & Smith 1981).

In view of these developments, it would seem that there are now three broad areas that need to be addressed by future studies of attrition:

(1) Aeolian attrition of original quartz grains.
  (i) Can we better understand the mechanisms of attrition?
  (ii) How effective is aeolian attrition in reducing original grains (grus) to a rounded outline?
(2) Production of fine sediment by aeolian action.
  (i) Is coarse (loess-size) silt produced in deserts? if so, then:
  (ii) What are the mechanisms of silt production in deserts?
  (iii) What happens to the silt produced?
(3) Can silica dissolution be enhanced by aeolian action on either sand grains or silt?

These are not necessarily exclusive questions, nor, in the widest context of deserts, are they solely concerned with direct aeolian (grain-on-grain) action. For example, some silt production can be accounted for by salt and water action in deserts (Goudie *et al.* 1979; Pye 1983; Smith *et al.* 1987).

The main problems in answering these questions are the investigation of mechanisms of inter-grain activity, catching the products of this action and the examination of both products and original grains for size and shape variation as the process proceeds. This is not easy and would seem to present almost insuperable difficulties in the field. In the laboratory, the construction of a wind tunnel like that used by Kuenen (1960) is expensive and is likely to lead to the loss of attrition products.

In this paper an experiment is described designed to investigate the attrition questions outlined above. This uses equipment much simpler than that of Kuenen, but designed specifically to retain all products of attrition as well as allow examination of the original, 'core' sand grains.

Experimental procedure

The apparatus used is shown in Fig. 1 (a modified version of that described in Whalley *et al.* (1982)). A large 'test tube' reaction chamber was used as a holder for the grains under test. These were agitated by a controlled, constant stream of air

From FROSTICK, L. & REID, I. (eds), 1987, *Desert Sediments: Ancient and Modern*, Geological Society Special Publication No. 35, pp. 129–138.

entering through a jet in the glass tube and venting through an electrostatic precipitator. The precipitator is of simple cylindrical ($\phi = 25$ mm) type with a central electrode operating at 7 kV. Efficiency is estimated as better than 95%.

Runs were made for 1, 2, 4, 8, 16, 32 and 48 hours at several pressures—representing various levels of particle flux/contact. Results from the highest attrition rate experiment are presented in this paper.

About 12 g of crushed Brazil quartz was used in the runs, this was sieved to the range 350–500 μm with any adhering fines removed by gentle washing after swirling in an ultrasonic bath. Some of these 'core' grains were collected from the bottom of the chamber for SEM examination after each time period; all attrition fragments were removed from the precipitator, weighed and used for Coulter Counter size analysis, SEM examination and silica solubility determination. Standard Coulter Counter methods were used for the size analysis based on a 280 μm tube and using a 100 μm tube to check the distribution of the finest material where necessary down to a minimum size of 2 μm.

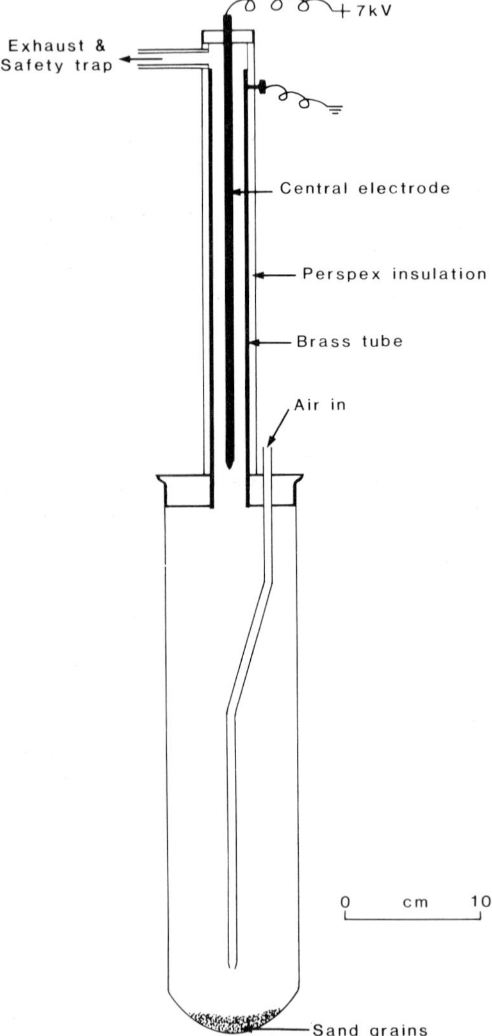

FIG. 1. The experimental apparatus.

For the SEM analyses, a number of grains were taken for a general assessment of outline change, six 'core' grains were examined in detail at magnification of about 100 and 500 times; for the fines a magnification of approximately 1000 was generally used. Silica was extracted from the attrition products by swirling 0.1 g in 10 ml of deionized water for 15 min. Analysis was achieved using standard colorimetric (molybdenum blue) methods.

No attempt has been made here to quantify the grain outline changes.

Experimental rationale

Most investigations of the dynamic aspects of aeolian action have considered the energy requirements of particle entrainment and transport. It is obviously important that such work be taken into account in any overall model of particle attrition, but the complexity of grain trajectories, landing characteristics and particle flux suggests that this is some way off. Instead, it seems more appropriate, as a first step, to study the effects of abrasion under simplified conditions. In particular, constant particle flux lends itself as a surrogate for a variety of field variables. In the experiment described here none of the dynamic factors is considered directly. The results are purely used to suggest some answers to the questions concerning the nature of the sand grains and fines found in nature.

The experiment is an idealization because of the difficulties in looking at grains moving freely over a two-dimensional surface. The practical problems of examining the fines under these conditions are considerable. Although it is not possible to relate the conditions of the experiment to any actual wind velocity, we consider that in any desert the grains will be subjected to a wide variety of wind conditions and that, in any case, the main control on the attrition of particles will be the grain contacts at a given flux. Furthermore, the interactions between saltating grains and those in suspension will be complex, being a function of free stream velocity, particle size, overall shape and surface roughness. Extra complication results from: (1) the length of time that any grain will be subjected to a variety of (time-integrated) conditions and (2) the rotation of grains, neither of which can be evaluated in the field. The crucial factor, therefore, is the number of grain–grain interactions in the air, or grain–surface contacts on the ground. This complexity explains why in these and previous experiments, (*eg* Wellendorf & Krinsley 1980; Lindé & Mycielska-Dowgiałło 1980), it has been impossible to evaluate fully these actual contact velocities. Future experiments are planned to address this question directly.

TABLE 1. *Characteristics of material produced and caught by the precipitator in the experiment.*

Time interval (hr)	Fines* produced (g)	Percentage of fines† by size class		Water soluble Si (ppm)
		10–50 μm	20–60 μm	
0–1	0.416	79	74	300
1–2	0.100	82	90	—
2–4	1.077	69	45	1000
4–8	0.697	80	67	1600
8–16	2.040	77	60	450
16–32	1.459	69	51	950
32–48	0.664	62	45	600

* Original weight was 11.99 g.
† Mean of two analyses.

Results

Particle outline changes

Figure 2 illustrates a selection of outline changes and edge detail for various samples taken at standard times after the start of the experiment. The results are similar to those reported previously (Lindé & Mycielska-Dowgiałło 1980; Whalley *et al.* 1982) and discussed further in Whalley & Marshall (1986). After four hours, abrasion is still confined to the edges of the grains with the basic shape remaining substantially unaltered. For the most part, the original particle shape largely determines the final shape. However, for some particularly long particles, with *a*-axis > 500 μm (Fig. 1*a*), it appears that total breakage can occur rather than just chipping of the edges. Hence, more new particles of sand size may enter the system. This is reflected in the quantity of debris produced in the first hour (Table 1). After 48 hours, there is sufficient edge modification to produce well-rounded grains. At this stage there is a decline in the quantity of fines produced (Table 1) as well as a shift to smaller size ranges (Fig. 3). Although the assessment of angularity–roundness change is entirely qualitative, easily discernable modification can take place over a relatively short time period.

Production of fines

Particle morphologies of some of the fine sediment produced over several run times are shown in the photomicrographs in Fig. 4.

The data from the experiments are best exemplified by reference to Table 1 and Figs 3 and 4. In Table 1, Coulter Counter results (280 μm tube) for the various run times are given together with the weights of attrition fragments produced in each interval. The table shows the percentages of the fines in the size ranges 10–50 μm and 20–60 μm, these being the two size ranges commonly used to define loess (Smalley & Smalley 1983). The high percentage of particles within these limits is clear; the percentage only starts to decline after about 16 hours as the grains become more rounded and fragments < 10 μm are produced more prolifically.

Figure 5 shows the cumulative percentage of fines (the original weight less the removed material) produced for each of the experimental time periods. There appears to be a decline in attrition towards the end of each run.

Silica dissolution

It has been suggested (Waugh 1970; Folk 1978; and see Whalley & Smith 1981 for discussion) that the products from early aeolian abrasion could be a source of silica for the formation of silcretes. Table 1 shows the total free silica in solution obtained by water extraction from the attrition fragments produced in these experiments. There is considerable variability and the results are complicated by the time-integrated nature of the samples removed from the precipitator, but the results approximate the quantities of dissolved silica which might be obtained purely as a result of the abrasion mechanism if, for example, water flows through the dust or the dust is scavenged from the atmosphere by rainfall.

Discussion and implications

Attrition of quartz grains

The quartz grains used in the experiment, although relatively defect-free, are sufficiently similar in shape and surface morphology to grus grains weathered from outcrops to provide an

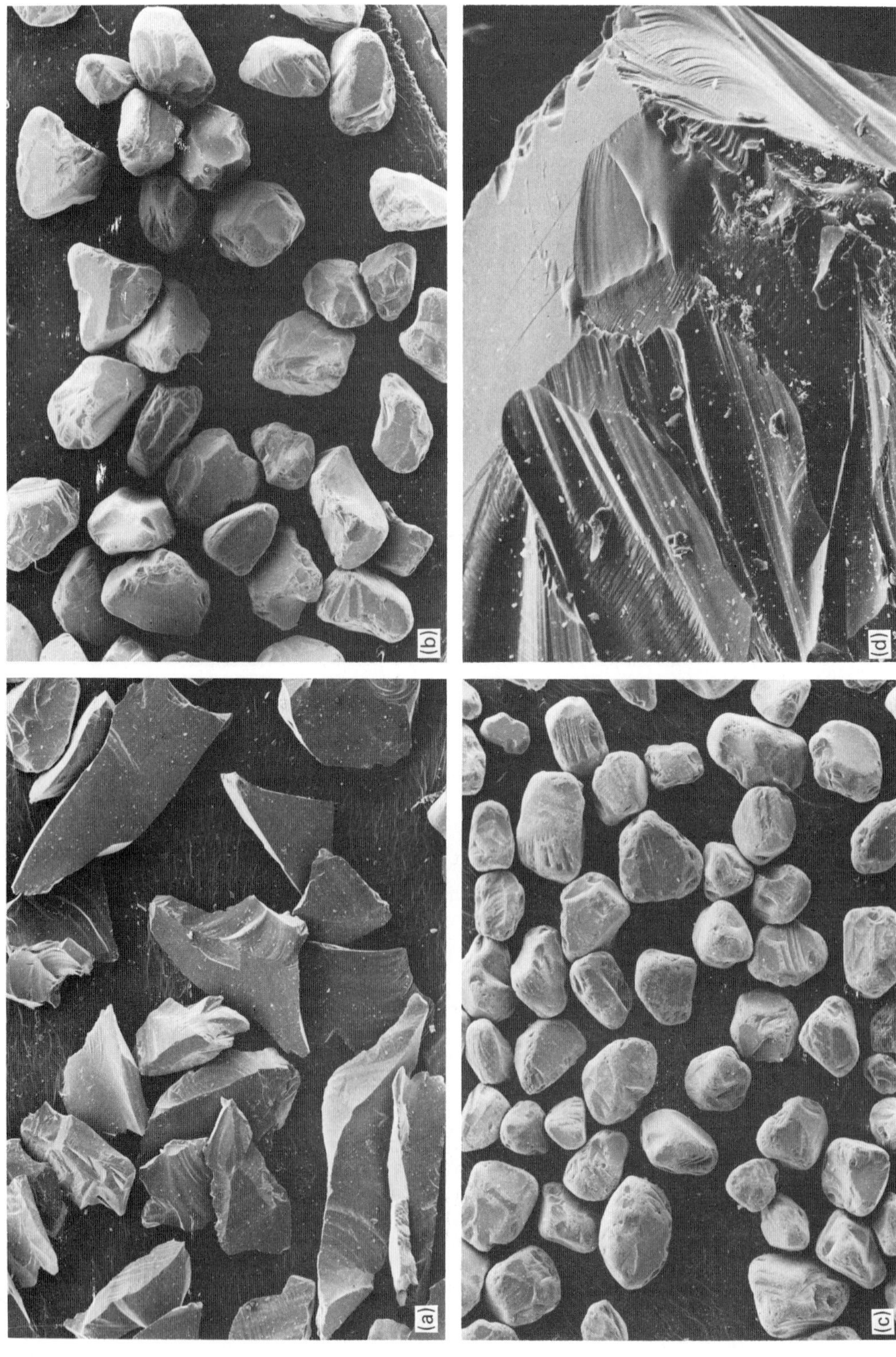
(b)
(d)
(a)
(c)

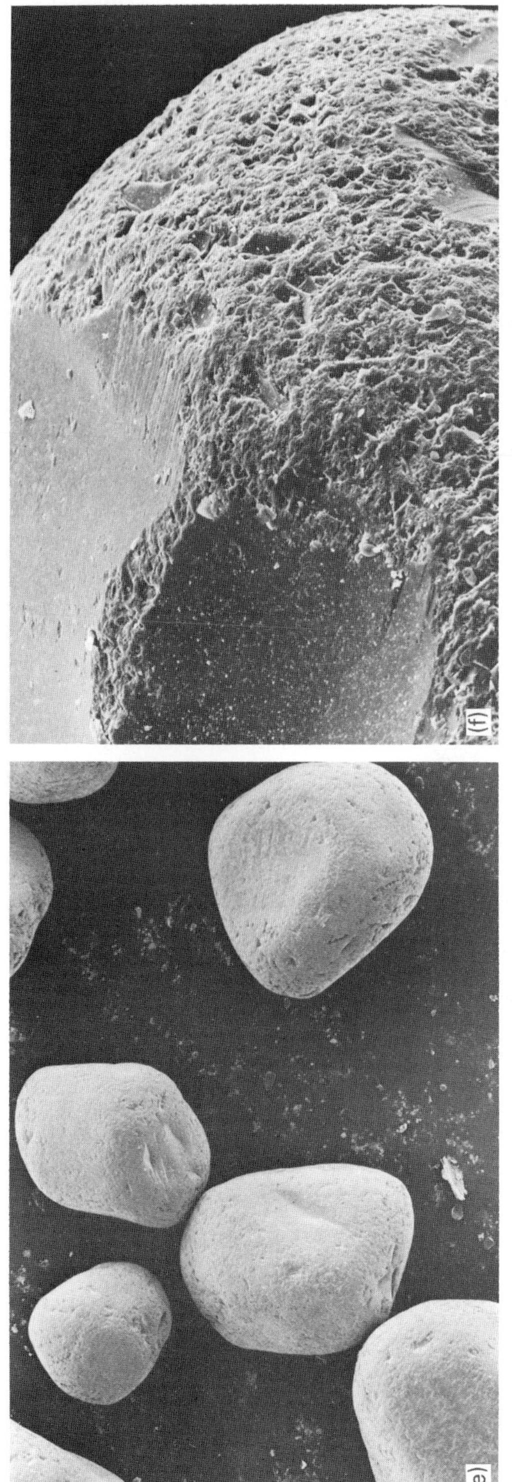

FIG. 2. (*a*) Original grains of crushed quartz used in the experiment. Picture width = 4.7 mm. (*b*) Grains after 2 hours of high energy attrition. Picture width = 4.7 mm. (*c*) Grains after eight hours attrition. Picture width = 4.7 mm. (*d*) Edge of original, crushed quartz, grain. Picture width = 200 μm. (*e*) Grains after 48 hours of abrasion. Picture width = 2 mm. (*f*) Detail of edge of grain after 4 hours of abrasion. Picture width = 200 μm.

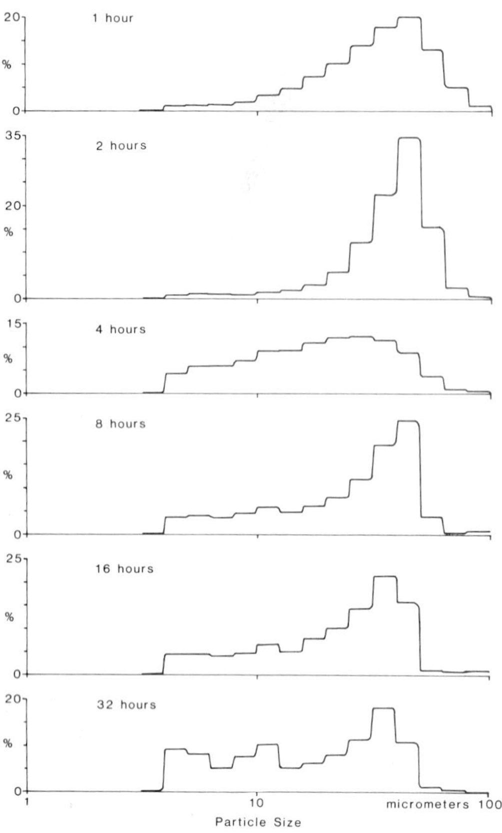

FIG. 3. Size distributions of grains during course of experimental abrasion (Coulter Counter analysis, 280 μm tube).

estimated minimum attrition and a convincing starting point for the experiment (Fig. 6*a*). If the mechanism of rounding can be modelled then it is possible to examine abrasion quantitatively as well as qualitatively.

It is clear from previous experiments (Chepil 1945; de Felice 1956; White & Schultz 1977) that sand-sized grains in an air-stream rotate rapidly. We have not yet been able to show conclusively that the grains in our experiment are spinning, although evidence from micrographs strongly suggests that they are. If so, rotation could provide an important and previously neglected additional mechanism for grain attrition in addition to the effects of direct impact, both between grains in transport and stationary ones and between mobile saltating grains. This is because even when grains touch with the same forward velocity, any protruberances will collide at their combined angular velocities (Fig. 7). Indeed, a rotational element in grain attrition would logically make a significant contribution to the observed rounding of grains in aeolian environments. The mechanism of the removal of fragments is discussed in greater detail in Whalley & Marshall (1986). However, the experiment shows very effectively the removal of edges and corners and the substantial shape modification which occurs. Abrasion runs at lower pressures show that edge rounding occurs even at much lower saltation fluxes than those reported above (Whalley *et al.* 1982) as seen, for example, in Fig.6*b*.

Rounding of quartz grains

Despite the assertions of, for example, Folk (1978), it is undeniable that the rounding of quartz grains does occur in deserts (Fig. 6*c*) and the results of these experiments suggest that even limited saltation can achieve some edge abrasion. The factors determining the extent of rounding will include the amount of energy imparted to grains in saltation or suspension and the frequency with which this is achieved. In addition, the modelling of collisions of grains attempted previously (*eg* Rumpel 1985) has not considered rotation and the effect this might have not only on energy dissipation when impacts occur but also on enhancing particle trajectory—this last making more time available for the collisions (the Magnus effect: White & Schulz 1977). Absence of grain rounding may reflect the lack of opportunities for rounding energy to be imparted to grains. For example, the vast majority of grains on a sand sea lie, at any one time, beneath the surface and are abraded only infrequently. In turn, the type of dune field (sand-passing or sand-trapping) will therefore be important in controlling how frequently grains appear at the surface and are subject to abrasion. Furthermore, mixed populations of angular and rounded grains may derive from a combination of well-rounded grains which have been subjected to abrasion for some time, with grus recently derived from bedrock and thus showing much less in the way of modification.

Production of fine sediment (attrition fragments)

The corollary of the above is that silt-sized fines can be produced by aeolian attrition of angular quartz sand. Table 1 shows that much of the fines removed from angular crushed Brazil quartz is of medium to coarse silt size (Fig. 5), and that for a given quantity of quartz grains of medium sand size, the amount produced could be considerable. The implications of these results are significant in the debate over the provenance of loess from deserts (Kuenen 1960; Smalley & Smalley 1983).

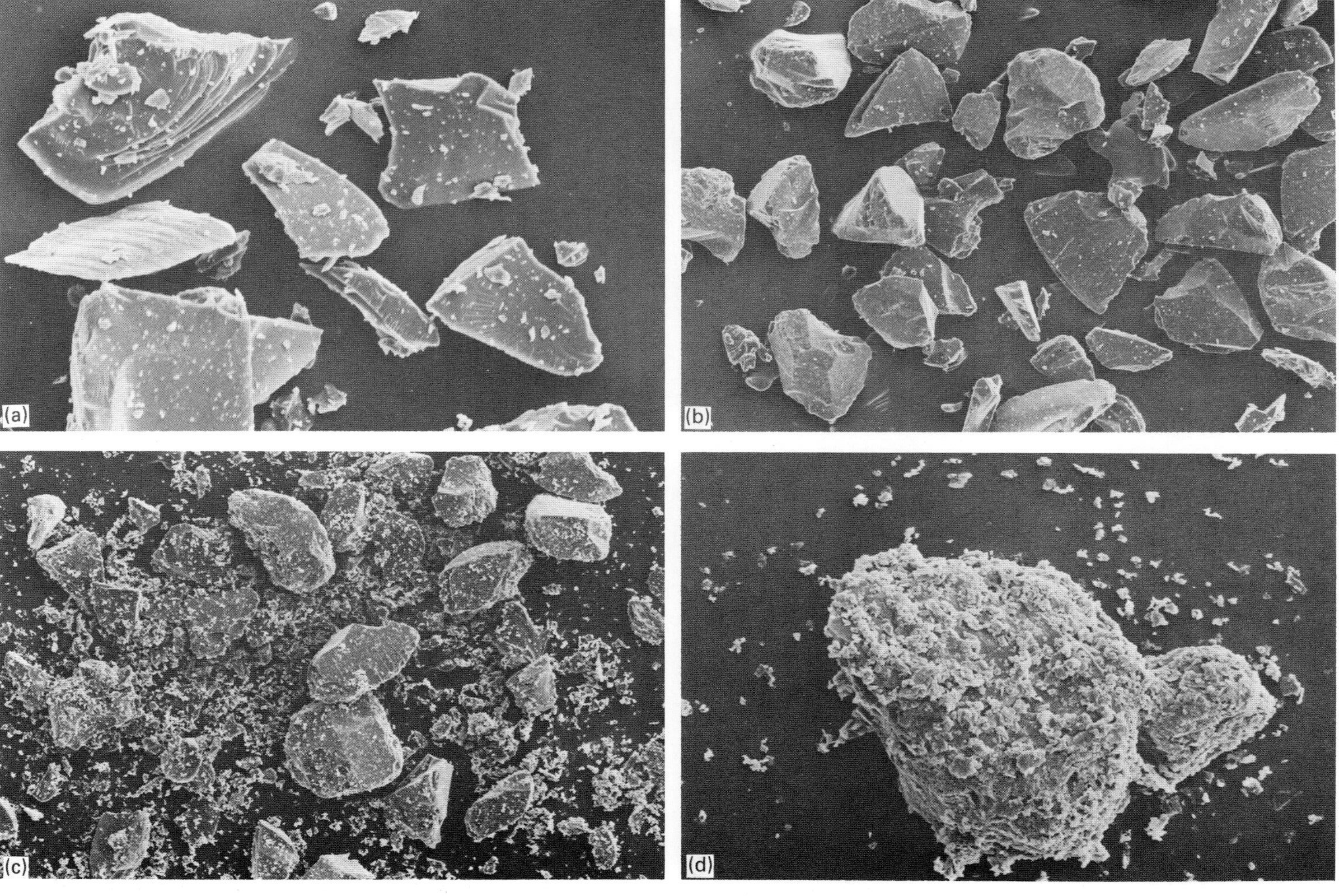

FIG. 4. (*a*) Fines produced after 1 hour of abrasion. Picture width = 100 µm. (*b*) Fines produced after 2 hours abrasion. Picture width = 300 µm. (*c*) Fines produced after 8 hours abrasion. Picture width = 300 µm. (*d*) Fines produced after 48 hours of abrasion. Picture width = 100 µm.

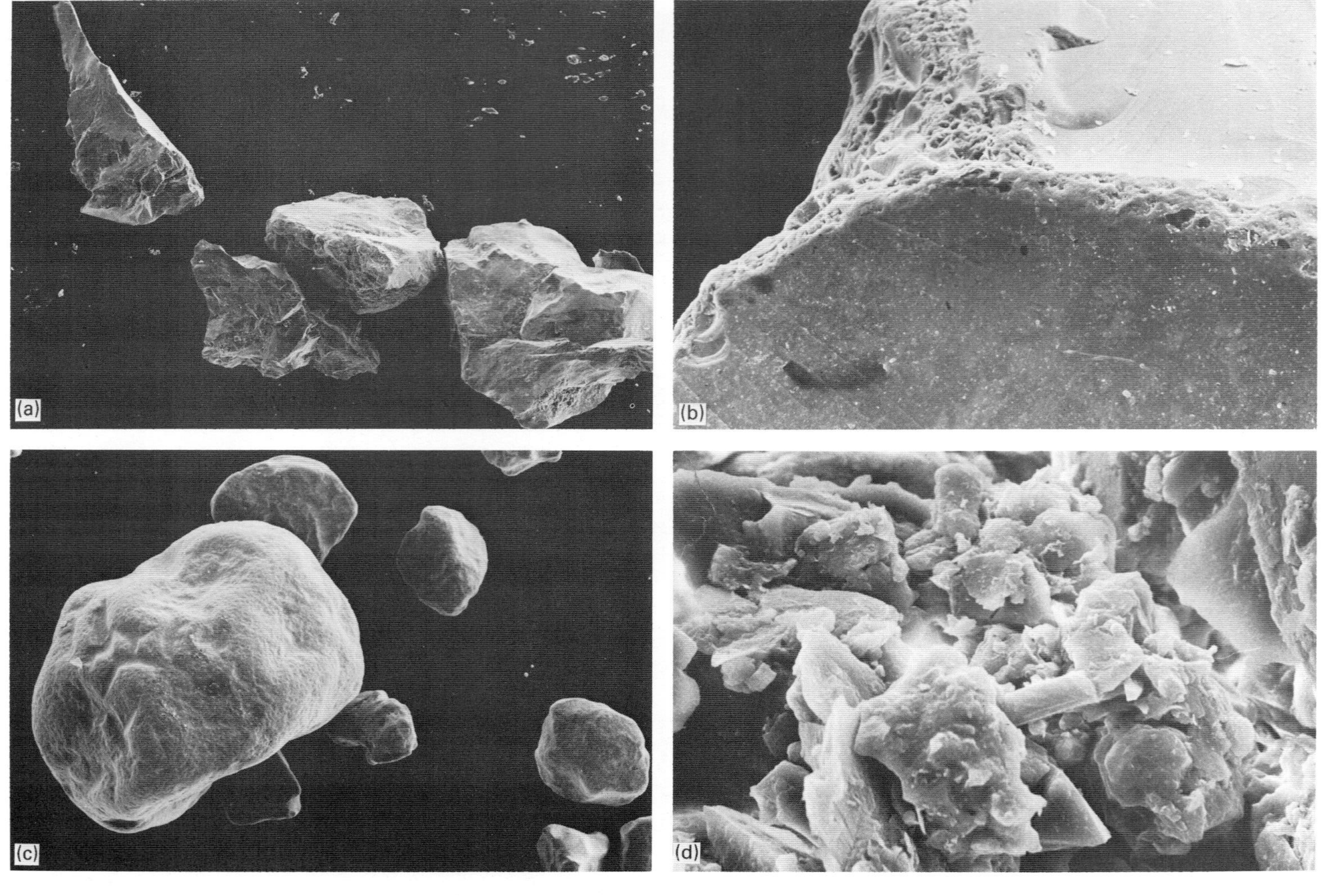

FIG. 6. (*a*) Typical quartz grus grains, Powell Springs, Arizona. Picture width = 1.6 mm. (*b*) Low-energy experiment, grain after 36 hours. Picture width = 170 μm. (*c*) Desert dune sand grain, El Golea, Tunisia. Picture width = 750 μm. (*d*) Loess deposit, Danube terrace. Picture width = 65 μm.

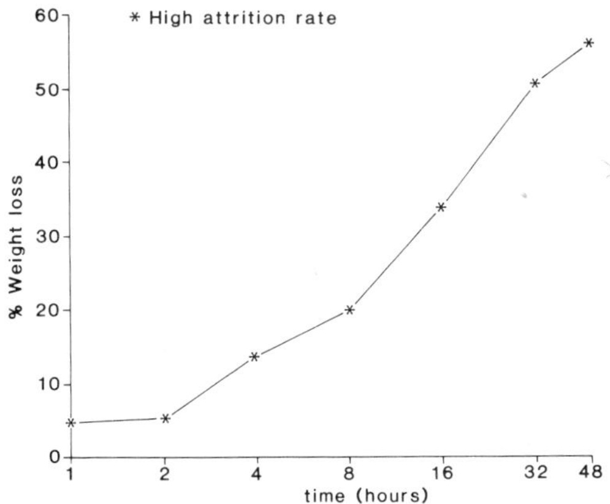

FIG. 5. Cumulative percentage weight loss of original grains (corrected for material removed) against log time of attrition. Note log time-scale.

Despite some findings in support of the existence of 'desert loess' (*eg* Goudie *et al.* 1979; Whalley *et al.* 1982; Ali & West 1983) this concept is still not generally accepted. However, the production of material finer than sand size from the experiments reported here suggests a possible mechanism for the formation of loess-sized quartz silt in desert environments. Silt production by aeolian rounding could therefore be added to other mechanisms such as deep weathering (Nahon & Trompette 1982), silica dissolution (Pye 1983) and *in situ* salt weathering (Pye & Sperling 1982) which have been demonstrated elsewhere. Some of the silt, from whatever source, may circulate within the deserts, but quantities are also blown beyond desert margins (Whalley & Smith 1981). Although far-travelled desert dust is generally of fine silt (< 10 μm) (*eg* Coudé-Gaussen 1984), nevertheless, Sarnthein & Koopmann (1980) have found terrigenous loess-size silt (modal size < 40 μm) within dust deposited in a zone about 100 km off the W African coast. Observations of wind-blown soils on mainland W Africa also show an abundance of silt-size material with a decrease in grain size towards the coast. Fines produced by aeolian attrition and weathering mechanisms could therefore contribute to both terrestrial (Fig. 6*d*) and offshore silt deposits.

Silica dissolution

Silica for the formation of silcretes can be derived from a number of possible sources and by means of several mechanisms (see *eg* Folk 1978; Smalley *et al.* 1978; Summerfield 1983). These include derivation from abrasion-activated quartz surfaces (Whalley 1978). We would further suggest that high concentrations of silica in solution can be derived from sand grains which have been subjected to aeolian attrition as well as from the resultant dust. Where dust falls with rain, therefore, there is a possibility of silica being effectively and rapidly dissolved and subsequently precipitated in deposits.

An abrasion mechanism for quartz sand grains in deserts has been demonstrated which can be applied to all three areas outlined in the Introduction. The question which remains is the general applicability of this mechanism in real desert situations and its relative significance in the formation of loessic silt.

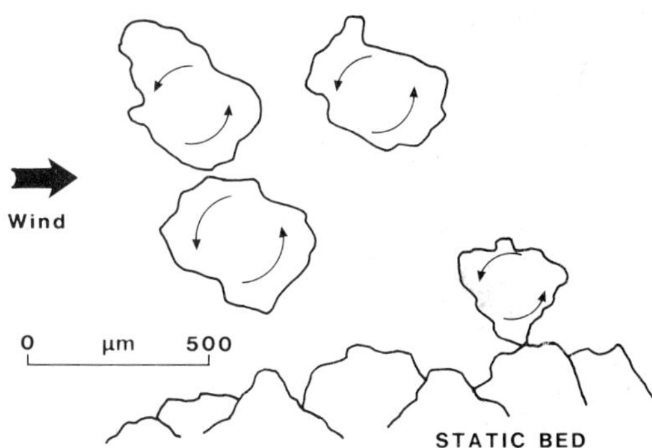

FIG. 7. The effect of rotation on the collision of saltating grains.

ACKNOWLEDGMENTS: We thank the Royal Society for financial help with this project, the Electron Microscopy Unit of Queen's University and, in particular, Stephen McFarland, as well as members of the UMIST Algeria Expedition 1985.

References

ALI, Y. A. & WEST, I. 1983. Relationships of modern gypsum nodules in sabkhas of loess to compositions of brines and sediments in northern Egypt. *Journal of sedimentary Petrology* **53**, 1151–1168.

CHEPIL, W. S. 1945. Dynamics of wind erosion. I Nature of movement of soil by wind. *Soil Science* **60**, 305–320.

COUDÉ-GAUSSEN, G. 1984. Le cycle des poussières éoliennes désertiques actuelles et la sédimentation des loess péridésertiques quaternaires. *Bulletin des Centres de Recherches Exploration–Production Elf Aquitaine* **8**, 167–182.

DE FELICE, P. 1956. Processus de soulèvement des grains de sables par le vent. *Comptes Rendus de l'Academie des Sciences* **242**, 920–923.

FOLK, R. L. 1978. Angularity and silica coatings of Simpson Desert sand grains, Northern Territory, Australia. *Journal of sedimentary Petrology* **48**, 611–624.

GOUDIE, A. S. & WATSON, A. 1981. The shape of desert dune grains. *Journal of Arid Environments* **4**, 185–190.

——, COOKE, R. U. & DOORNKAMP, J. C. 1979. The formation of silt from quartz dune sand by salt weathering in deserts. *Journal of Arid Environments* **2**, 105–112.

KALDI, J., KRINSLEY, D. H. & LAWSON, D. 1978. Experimentally produced aeolian surface textures on quartz sand grains from various environments. *In*: WHALLEY, W. B. (ed.) *Scanning Electron Microscopy in the Study of Sediments*, Geo Books, Norwich, 261–274.

KUENEN, PH. H. 1960. Experimental abrasion 4: Eolian action. *Journal of Geology* **68**, 427–449.

—— & PERDOK, W. G. 1962. Experimental abrasion 5: Frosting and defrosting of quartz grains. *Journal of Geology* **70**, 648–658.

KRINSLEY, D. H., GREELEY, R. & POLLACK, J. B. 1979. Abrasion of windblown particles on Mars—erosion of quartz and basaltic sand under simulated martian conditions. *Icarus* **39**, 364–384.

LINDÉ, K. & MYCIELSKA-DOWGIAŁŁO, E. 1980. Some experimentally produced microtextures on grain surfaces of quartz sand. *Geografiska Annaler* **62A**, 171–184.

NAHON, D. & TROMPETTE, R. 1982. Origin of siltstones: glacial grinding versus weathering. *Sedimentology* **29**, 25–35.

PYE, K. 1983. Formation of quartz silt during humid tropical weathering of dune sands. *Sedimentary Geology* **34**, 267–282.

—— & SPERLING, C. H. B. 1982. Experimental investigation of silt formation by static breakage processes: the effect of temperature, moisture and salt on quartz dune sand and granitic regolith. *Sedimentology* **20**, 49–62.

RUMPEL, D. A. 1985. Successive aeolian saltation: studies of idealized collisions. *Sedimentology* **32**, 267–280.

SARNTHEIN, M. & KOOPMANN, B. 1980. Late Quaternary deep sea record on Northwest African dust supply and wind circulation. *In*: SARNTHEIN, M., SEIBOLD, E. & ROGNON, P. (eds) *Palaeoecology of Africa and surrounding Seas, Sediments and Climatic Changes*. Balkema, Rotterdam, 239–253.

SMALLEY, I. J. & KRINSLEY, D. H. 1978. Eolian sedimentation on earth and Mars: some comparisons. *Icarus* **40**, 276–288.

—— & SMALLEY, V. 1983. Loess material and loess deposits: formation, distribution and consequences. *In*: BROOKFIELD, M. E. & AHLBRANDT, T. S. (eds) *Eolian Sediments and Processes*. Elsevier, Amsterdam, 51–68.

—— KRINSLEY, D. H., MOON, C. F. & BENTLEY, S. P. 1978. Processes of quartz fracture in nature and the formation of clastic sediments. *In*: EASTERLING, K. E. (ed.) *Mechanisms of deformation and fracture*. Pergamon Press, Oxford, 119–127.

SMITH, B. J., MCGREEVY, J. P. & WHALLEY, W. B. 1987. The production of silt-size quartz by experimental salt weathering of a sandstone. *Journal of Arid Environments* **12**, 199–214.

SUMMERFIELD, M. A. 1983. Silcrete. *In*: GOUDIE, A. S. & PYE, K. (eds) *Chemical Sediments and Geomorphology*. Academic Press, London, 59–91.

WAUGH, B. 1970. Petrology, provenance and silica diagenesis of the Penrith Sandstone (Lower Permian) of northwest England. *Journal of sedimentary Petrology* **40**, 1226–1240.

WELLENDORF, W. & KRINSLEY, D. H. 1980. Wind velocities determined from the surface textures of sand grains. *Nature* **283**, 372–373.

WHALLEY, W. B. 1978. Scanning electron microscopic examination of a laboratory-simulated silcrete. *In*: WHALLEY, W. B. (ed.) *Scanning Electron Microscopy in the Study of Sediments*. Geo Books, Norwich, 399–405.

—— & MARSHALL, J. R. 1986. Simulation of quartz grain surface textures: some scanning electron microscopic observations. *In*: SIEVEKING, G. DE G. & HART, M. B. (eds) *The Scientific Study of Flint & Chert*. Cambridge University Press, Cambridge, 227–233.

—— & SMITH, B. J. 1981. Mineral content of Harmattan dust from northern Nigeria examined by scanning electron microscopy. *Journal of Arid Environments* **4**, 21–29.

——, MARSHALL, J. R. & SMITH, B. J. 1982. Origin of desert loess from some experimental observations. *Nature* **300**, 433–435.

WHITE, B. R. & SCHULZ, J. C. 1977. The Magnus effect in saltation. *Journal of Fluid Mechanics* **81**, 497–512.

W. B. WHALLEY, B. J. SMITH, J. J. MCALISTER & A. J. EDWARDS, Department of Geography, The Queen's University, Belfast BT7 1NN, UK.

The mechanics and geological implications of dust transport and deposition in deserts with particular reference to loess formation and dune sand diagenesis in the northern Negev, Israel

K. Pye & H. Tsoar

SUMMARY: Entrainment, transport and deposition of dust in and around deserts has major geological implications. A consideration of the mechanics of dust transport and deposition shows that 'desert loess' can only accumulate in vegetated desert fringe areas which are relatively close to a major source of dust. Dust deposition can also contribute to the stablilization of active sand dunes, and infiltrated dust plays an important role in the early diagenesis of a wide range of desert sediments.

Aeolian dust transport is a process of major importance in arid and semi-arid regions (Jackson *et al.* 1973; Idso 1976; Goudie 1978, 1983; Morales 1979; Péwé 1981; Coudé-Gaussen 1984; Gerson *et al.* 1985). Dust transport and deposition has important geological and geomorphological implications (Fig. 1). Dust contributes to the case-hardening of rocks (Conca & Rossman 1982), desert varnish formation (Potter & Rossman 1977; Allen 1978), duricrust development (Blümel 1980), and reddening and cementation of sediments (Folk 1978; Waugh 1970; Walker 1976, 1979; Bowman 1982). Aeolian transport is also an important process in geochemical cycling from the continents to the oceans (Goldberg 1971; Windom 1975; Prospero *et al.* 1981; Janecek & Rea 1985).

The aims of this paper are twofold: (1) to review briefly the processes of dust transport and deposition in deserts, and (2) to consider some of the geological implications of dust deposition with particular reference to the northern Negev and adjoining areas of southern Israel. A more detailed review of aeolian dust and dust deposits is provided by Pye (1987).

Nature of airborne additions to desert sediments

Airborne additions to desert sediments are of three types: (1) dust, defined here as solid particles which are transported by wind, mainly in suspension; (2) salts dissolved in precipitation; (3) gases which may be adsorbed onto moist surfaces. Types (1) and (2) are the most important. Some common constituents of desert dust are shown in Table 1. The major chemical species in precipitation and their sources are listed in Table 2.

Dust storms are defined by international convention as meteorological events in which visibility at eye level is reduced to less than 1000 m due to active entrainment of dust by the wind. 'Blowing dust' describes the meteorological condition in which dust is entrained locally but visibility at eye level is not reduced to less than 1000 m. 'Dust haze' refers to the residual fine dust which remains in suspension after the cessation of dust entrainment; visibility is usually, though not always, more than 1000 m.

Sources of dust and airborne salts

The main source of airborne salts is sea spray. Atmospheric concentrations of salts are high within about 50 km of many coasts, but decline markedly further inland (Yaalon 1964*a*).

Wind erosion of soils and sediments is by far the most important source of dust (Hidy & Brock 1970; Junge 1979; Prospero 1981). Other sources include volcanic eruptions, gas to particulate conversions in the atmosphere, cosmic dust, fires, direct release of biogenic particles into the atmosphere (spores, pollen, phytoliths), and anthropogenic emissions.

The world's major source areas of dust are the arid regions which extend in a broad belt from West Africa to Central Asia (Fig. 2). Less important sources are the deserts of central Australia, the southwestern United States, SW Africa and South America (Coudé-Gaussen 1984; Middleton 1986*a*, *b*; Middleton *et al.* 1986).

Within deserts, dust deflation takes place mainly from dry lake beds, alluvial fans, wadi beds, bare soils and active dune fields (Coudé-Gaussen 1984; Gerson *et al.* 1985; Pye 1987).

From Frostick, L. & Reid, I. (eds), 1987, *Desert Sediments: Ancient and Modern*, Geological Society Special Publication No. 35, pp. 139–156.

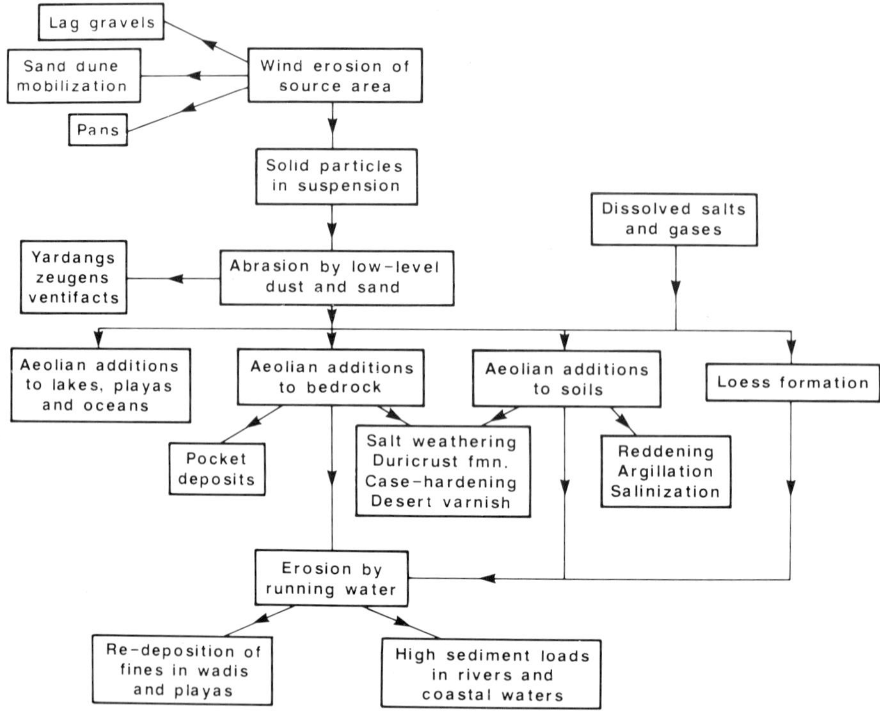

FIG. 1. Some geological and geomorphological implications of dust transport and deposition in deserts.

TABLE 1. *Some common constituents of desert dust and their major origins*

constituent	major origins
quartz, feldspars, heavy minerals	weathering of igneous and metamorphic rocks, sandstones and shales
calcite	limestones, calcareous sandstones, lake beds
dolomite	dolomites, dolomitic limestones, lake beds
smectite, illite, mixed-layer clays, kaolinite, chlorite	soils, shales, lake beds, playas
palygorskite, sepiolite, halite, gypsum	arid soils, playas, lake beds
opal	diatoms, phytoliths
organic carbon	soils, lake beds, fires

Entrainment of dust particles

Dust particles can be entrained by the following mechanisms acting alone or in combination: shear stress exerted by the wind (frictional drag); aerodynamic lift; ballistic impacts; and mechanical disturbance by pedestrian or vehicular traffic.

Frictional drag

The critical drag velocity (threshold entrainment velocity, u_{*t}), required to initiate movement of particles on a dry, well-sorted sand bed was shown by Bagnold (1941) to be dependent on particle size:

$$u_{*t} = A\sqrt{\frac{\rho_p - \rho_a}{\rho_a} \cdot gD} \qquad (1)$$

where ρ_p is the density of the particles, ρ_a is the density of air, g is the acceleration due to gravity, D is the median particle diameter, and A is an empirical coefficient with a value of 0.1 where particle Reynolds number is greater than 3.5 (*ie* for most sands).

Several workers have confirmed Bagnold's finding that u_{*t} attains a minimum value for particles with a size of about 80 μm (Chepil 1945; Iversen & White 1982). With finer particle sizes the sediment surface becomes aerodynamically smooth, and the air drag, instead of being carried by a few, more exposed grains, is borne more or less evenly across the whole surface. Consequently, a relatively greater drag is required to

TABLE 2. *Common chemical species in desert precipitation and their major origins*

constituents	major origins
chlorides (Cl^-)	seawater; dissolution of particulate chlorides
sulphates (SO_4^{2-}, HSO_4^-)	seawater; lake waters; dissolution of sulphate particulates; condensation of gaseous H_2SO_4; sulfite oxidation by O_2, O_3, H_2O_2 and metal catalysts
sulfites (H_2SO_3, HSO_3^-, SO_3^{2-})	SO_2 dissolution (particularly important near sites of industrial emissions)
carbonates (CO_3^{2-}, HCO_3^-, H_2CO_3)	CO_2 dissolution; dissolution of particulate carbonates; seawater; lakewater
nitrites (HNO_2, NO_2^-)	dissolution of NO, NO_2 and HNO_2
nitrates (NO_3^-)	nitrite oxidation by O_3; dissolution of gaseous NO_2^- and HNO_3; dissolution of particulate $NaNO_3$ from soils
ammonium (NH_4OH, NH_4^+)	dissolution of gaseous NH_3; dissolution of particulate $(NH_4)_2SO_4$ and NH_4NO_3
hydroxyls (OH^-)	water dissociation
active hydrogen (H^+)	water dissociation
metallic cations (Na^+, K^+, Ca^{2+}, Mg^{2+}, Fe^{3+})	seawater; lakewater; dissolution of particulates

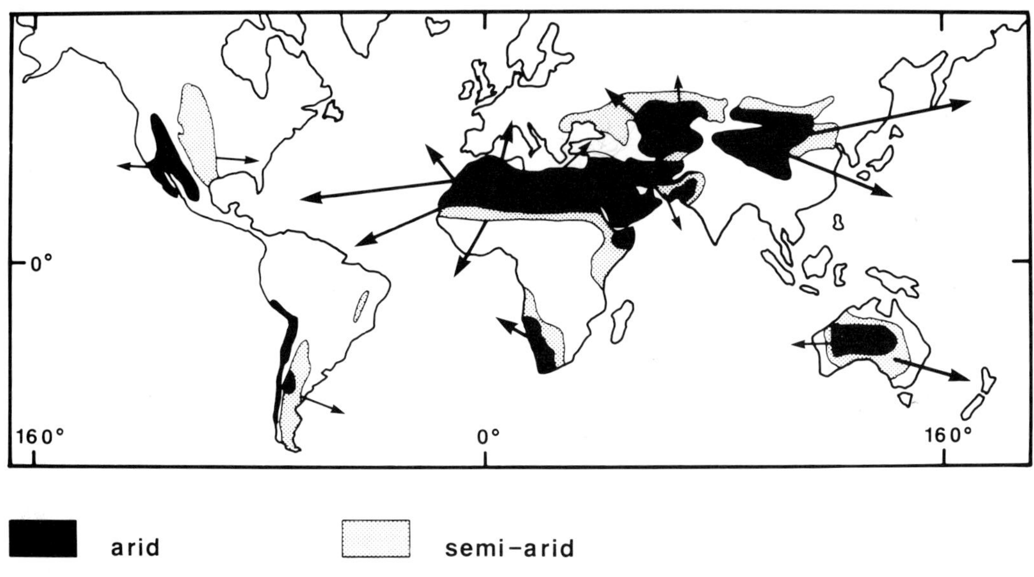

FIG. 2. The major source areas and transport directions of dust (modified after Coudé-Gaussen, 1984).

set the first grains in motion. Interparticle cohesive forces also become important for grains smaller than 80 μm (Iversen *et al.* 1976; Iversen & White 1982). For these reasons, very high drag velocities are required to initiate movement of settled dusts (Bagnold 1960).

The presence of moisture and cementing agents significantly raises the threshold entrainment velocity. Belly (1964) showed that 0.6% moisture (by volume) could double the value of u_{*t} compared with dry sand. Salt precipitates have a similar, though less pronounced effect (Nickling & Ecclestone 1981; Nickling 1984). Algae and fungi can also bind particles in soil crusts (Foster & Nicolson 1981; Van den Ancker *et al.* 1985). Field experiments by Gillette *et al.* (1980, 1982) showed that, for undisturbed desert soils, even a weak crust (modulus of rupture <0.07 MPa) will protect the soil from wind erosion.

The threshold entrainment velocity is also influenced by the presence of non-erodible roughness elements (*eg* gravel particles) on the bed. Low densities of non-erodible roughness elements reduce the threshold velocity and activate erosion by scouring around roughness elements, but high densities increase the threshold (Logie 1982) because a higher proportion of the total shear stress is borne by the non-erodible elements (Lyles *et al.* 1974; Lyles 1977). The critical density at which the effect of the roughness elements changes from one of activation to one of protection (referred to as the inversion point) is dependent on the size and shape of the roughness elements (Logie 1982).

Aerodynamic lift

A steady sheared flow exerts a lift force on particles resting on a surface (Chepil 1945; Chepil & Woodruff 1963). Lift forces strong enough to eject particles from the bed into the flow only occur rarely and are related to strong turbulent eddy activity (Lyles & Krauss 1971). A combination of short-term fluctuations in drag and lift forces is generally responsible for the initiation of grain movement (Chepil 1959).

Higher than average lift forces are associated with strong horizontal vortices and pronounced vertical pressure gradients which accompany dust devils and tornadoes (Greeley *et al.* 1981; Greeley & Iversen 1985, pp. 85–89). However, on Earth these phenomena are localized, of short duration, and of little overall significance as mechanisms of dust entrainment.

Ballistic impact

Bagnold (1941) observed that when particle movement starts, disturbance of the bed by rolling or saltating particles initiates movement of other grains, so that sediment movement can be *maintained* at velocities lower than the fluid threshold. This lower threshold was referred to by Bagnold as the 'impact threshold', and was defined empirically by:

$$u_t = 680\sqrt{D} \cdot \log \frac{30}{D}. \tag{2}$$

Bagnold (1960) noted that, although very high fluid threshold velocities are required to erode fine powders and settled dusts, these particles can be readily ejected into the airflow by the ballistic impacts of saltating sand grains larger than 100 μm. Abrasion of soil crusts by saltating sand grains has also been observed to be an important mechanism by which fine particles are released into the airstream (Chepil 1945; Gillette *et al.* 1974; Hagen 1984).

Mechanical disturbance

Even deposits of very fine, uniform powders can be eroded by the wind where mechanical disturbance has caused sharp edges to protrude into the wind stream. However, once the surface irregularities have been removed by deflation, no further movement takes place (Bagnold 1960). The passage of vehicular or pedestrian traffic may also eject particles from the surface, or indirectly lead to their entrainment by generating local turbulent eddies. Several authors have documented an increase in blowing dust in areas where the desert surface has been disturbed by human activities (Oliver 1945; Clements *et al.* 1963; Nakata *et al.* 1976; Wilshire 1980; Jones *et al.* 1986).

Several local factors can influence the value of u_{*t} that is required to entrain dust, but in the case of dry, non-crusted desert sediments it generally lies in the range 0.2–0.6 m s^{-1} (*eg* Nickling 1983).

Dust dispersion

Once disturbed from rest, silt and clay-size particles are transported mainly in suspension. The height attained and distance travelled by a suspended dust particle (smaller than about 50 μm) in a neutral atmosphere, that is, one in which buoyancy and subsidence effects are negligible, depends on the particle fall velocity (u_f) and the turbulent diffusivity (ε) of the atmospheric boundary layer. The upward component of turbulence represents the force opposing the tendency of a particle to settle. In order to remain suspended for a considerable period of time, a particle must experience more upward than downward movements. The ratio u_f/u_* has been found by several authors to be a suitable criterion for determining the likelihood that a particle will remain in suspension (Sundborg 1955; Chepil & Woodruff 1957; Gillette *et al.* 1974). Gillette *et al.* (1974) found that particles smaller than 15 μm have a sufficiently small u_f/u_* ratio to remain suspended indefinitely. Silt particles larger than 15 μm settle back to the surface at a rate proportional to their u_f/u_* ratios.

The maximum distance and time travelled by a suspended dust particle (smaller than 50 μm) in a neutral atmosphere is inversely proportional to the fourth power of the particle diameter (Tsoar & Pye 1986):

$$L = \frac{\bar{u}2\varepsilon}{K^2 D^4} \quad (3)$$

where L is the maximum distance travelled; $\bar{u}$ is the mean wind velocity; ε is the coefficient of turbulent exchange, given by $\varepsilon = \sqrt{w'^2}l$, where $\sqrt{w'^2}$ is the standard deviation of the vertical fluctuating velocity component and l is the mixing length defined by $l = \alpha\, \kappa z$, where $\alpha = 1$ for neutral atmospheres, κ = the von Karman constant (0.4), and z is height above surface within the lower boundary layer; and K is a constant derived from $K = \rho_p g / 18\,\mu$, where μ is the dynamic viscosity of the air.

Figure 3, based on Equation (3), shows how far different sizes of dust (assumed to be quartz spheres) are likely to be dispersed when $\bar{u} = 15$ m s^{-1} and values of ε lie in the range 10^3–10^7 cm^2 s^{-1}. During typical neutral atmosphere windstorms, when ε lies between 10^4 and 10^5 cm^2 s^{-1}, silt particles larger than 50 μm are unlikely to be transported more than 60 km from source and most will be deposited within about 30 km. Particles corresponding to the typical modal size of loess (20–30 μm) are likely to be transported up to 300 km, while finer grains can be transported thousands of kilometres.

It should be noted that these transport distances assume that settling dust particles are not re-suspended by the wind. The model deployed also assumes no significant convective lifting or significant momentum transfer aloft as would be associated with a jet stream. However, the transport distances predicted in Fig. 3 agree well with the observed distribution, thickness and grain size trends of Pleistocene loess deposits in relation to their source (Tsoar & Pye 1987), suggesting that the model provides a reasonable approximation of long-term silt dispersion.

In nature, many different synoptic conditions are responsible for dust transport. At a very localized scale, dust devils are short-lived columns of dust which result from heating of the air above the desert surface (Sinclair 1969). Haboobs are local dust storms created by cold down-draughts generated ahead of large convective cells (Lawson 1971; Nickling & Brazel 1984). Katabatic winds may also generate local, low-level dust storms in piedmont regions (Bowden *et al.* 1974; Nalivkin 1982). Cyclonic depressions are a major cause of dust storms in North Africa, the Middle East and Central Asia (Yaalon & Ganor 1979; Ganor & Mamane 1982). In cases where such depressions are connected aloft to a jet stream, dust may be dispersed rapidly over very large distances (*eg* Liu Tung Sheng *et al.* 1981). In the southern Sahara and the Sahel, fronts associated with the Harmattan wind system periodically generate plumes of fine dust which extend out over the Atlantic (Junge 1979; Kalu 1979). The characteristics of dust-transporting wind systems are discussed further by Goudie (1983) and Middleton *et al.* (1986).

Dust deposition

Dust particles larger than 15 μm will eventually settle back to the surface due to gravitational forces, but smaller particles remain suspended indefinitely unless they are washed out by

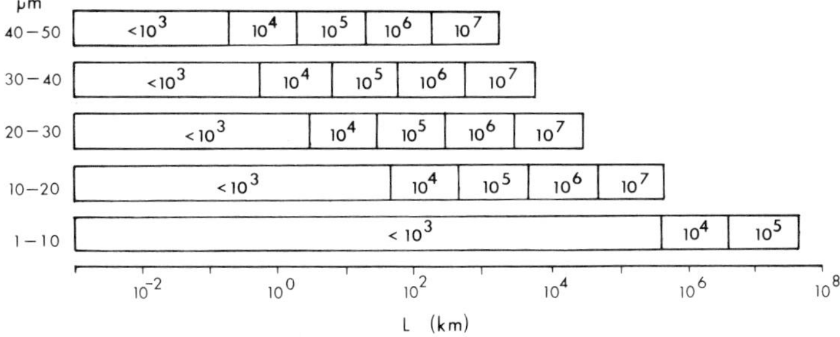

FIG. 3. Predicted maximum transport distances (L) for different size classes of quartz spheres (diameters in μm) in a neutral atmosphere with differing values of the coefficient of turbulent exchange (ε) given in each row segment. After Tsoar & Pye (1986); for explanation see text.

precipitation. Dry dust deposition occurs wherever there is a significant reduction in wind speed and turbulence, as in the lee of topographic obstacles, or where vegetation increases the surface roughness.

Ganor (1975) reported that in Israel, washed out dust contains 50–65% clay size particles, while dry dust contains less than 20% clay. In the dry dust samples, most of the clay particles were found to adhere to the surfaces of larger grains. The grain size of deposited dust reflects three factors: (1) the size distribution of particles in the source area; (2) distance from the source; and (3) frequency and timing of precipitation events. Since the main dust sources (excluding bare cultivated soils) are desert interiors where rain is infrequent, the mean size of deposited dust typically decreases towards the desert margins (*eg* Yaalon & Ganor 1975, 1979).

Reliable data concerning rates of contemporary dust deposition and accumulation are sparse. Most of the available data refer to dustfall rates during individual dust storms or to periods of only a few years. Estimates of longer term dust accumulation rates are based mainly on mineralogical and particle size analyses of soils and marine sediments and are subject to considerable margins of error. Published estimates of short-term dust deposition rates in deserts range from about 50 to 200 tonnes km^{-2} a^{-1} (Yaalon & Ganor 1975; Safar 1980; Péwé *et al.* 1981; McTainsh & Walker 1982). However, the effects of recent human activities mean that modern dust deposition rates are considerably higher than average Holocene rates (Tsoar & Pye 1987). At times during the Pleistocene, when major climatic and environmental changes were taking place, dust deposition rates in some desert margin areas were at least two orders of magnitude higher than at present (Bowler 1978). Thick alternating loess-soil sequences in the desert margin areas of China, Soviet Central Asia and elsewhere clearly show that dust deposition in the Quaternary has been episodic, reflecting cyclical changes in climate and dust availability (Heller & Liu Tung Sheng 1982; Dodonov 1984).

Loess formation in arid and semi-arid areas

Thick deposits of aeolian dust can only accumulate in areas where settled dust is not subsequently resuspended by wind or eroded by surface runoff. Thin dust accumulations often form in the bottom of dry, unvegetated topographic depressions which are sheltered from the wind, but thicker accumulations require either the presence of vegetation or moisture at the ground surface (Cegla 1969; Tsoar & Pye 1987).

In extremely arid parts of the Negev, lack of plant cover prevents accumulation of dust on interfluves (Yaalon & Dan 1974). Dust is eroded by surface wash during periodic rainstorms and deposited in small alluvial fans, in wadi beds and in playas (Issar *et al.* 1984; Fig. 4*a*). 'Pocket' dust deposits are preserved on some jointed bedrock interfluves. In the slightly less arid areas where vegetation cover is denser loess may cover the whole landscape but it is often thicker on N-facing slopes (Dan 1966; Gerson *et al.* 1985; Fig. 4*B*). The soils on these slopes experience lower temperatures and retain more moisture than the S-facing slopes, thereby giving rise to greater vegetation cover (Danin 1970). The thickness of loess on undulating terrain in the Negev Highlands generally increases downslope, indicating that syn- and post-depositional reworking by slopewash is important. Material eroded from the upper, less well vegetated slopes has been trapped on the lower, better-vegetated footslopes. On plateau areas aeolian dust has accumulated without much reworking (Dan *et al.* 1973). These deposits are presently being eroded, suggesting that dust accumulation may have occurred when the vegetation was denser (Gerson *et al.* 1985).

In the northern Negev plains and adjacent hills, loessic deposits have two origins. *In situ* aeolian dust deposits occur on interfluves, while fluvially reworked loessic material fills the valleys and forms the older terraces. In the Besor Basin, the fluvial silts are more than 15 m thick (Gardner 1977; Sneh 1983).

The dust deposited in Israel is derived from the deserts of the Negev, Sinai and north Africa, although the relative magnitudes of contributions from each of these sources have not been determined (Yaalon 1969; Yaalon & Dan 1974; Yaalon & Ganor 1979; Ganor & Mamane 1982). Modern dust storms come most frequently from the SW, with a secondary mode from the E (Yaalon & Ginzbourg 1966; Kastnelson 1970). The mean grain size of *in situ* loess on interfluves in the northern Negev and on the adjoining southern Pleshet Plain decreases towards the N and NE, indicating that southwesterly dust-transporting winds were also dominant in the recent past.

The decrease in mean grain size of loess from S to N is enhanced by greater post-depositional weathering towards the wetter areas in the N. Much of the loess on the northern margins of the Negev and in adjoining semi-arid areas is weathered and displays pronounced pedogenetic horizonation. At Netivot (present rainfall

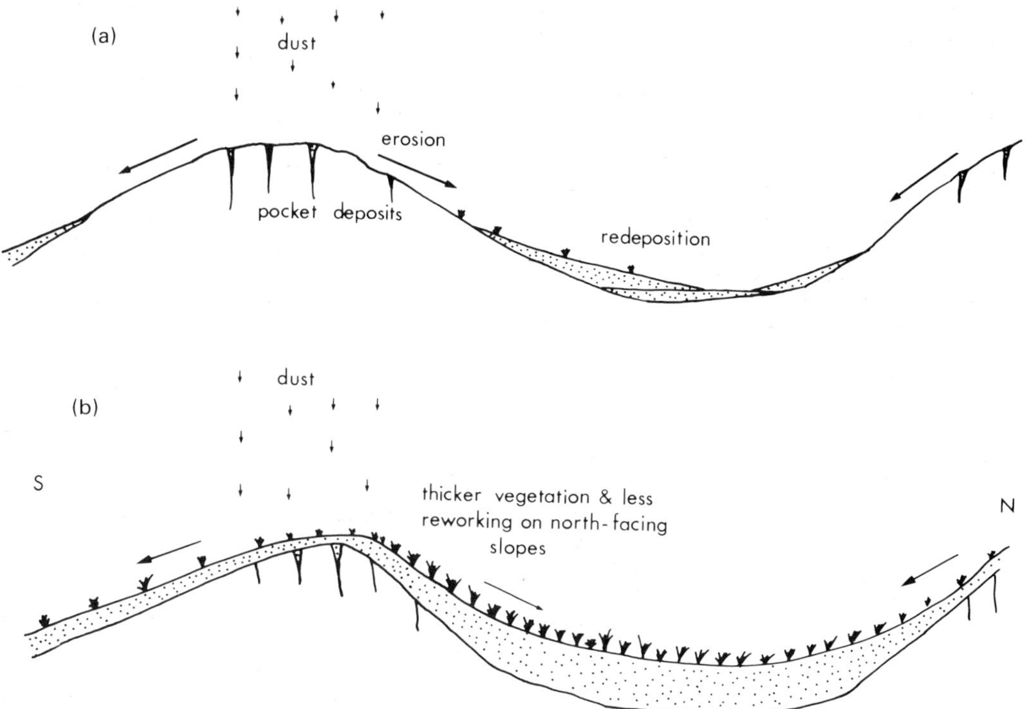

FIG. 4. Schematic representation of the consequences of aeolian dust deposition on hilly terrain in (*a*) a hyper-arid area and (*b*) a semi-desert fringe area. Under very dry conditions vegetation cover is thin and deposited dust is eroded from interfluves and upper slopes during periodic rain-storms. The reworked dust accumulates in small fans on the lower slopes, in wadi beds and in playas. Some dust may be trapped in 'pockets' on bedrock interfluves. In the semi-desert fringe thicker vegetation cover causes more dust to be trapped and loess may blanket the whole landscape. Deposits are often thicker on better-vegetated north-facing slopes and in hollows.

300 mm) NW of Beersheva, the loess sequence consists of a series of superimposed soils which formed contemporaneously with slow dust deposition (Bruins 1976; Bruins & Yaalon 1979). The section exposes up to 8 m of clayey, weathered loess containing a number of nodular carbonate horizons (Fig. 5). In the more arid areas to the SW and W of Beersheva, where the present annual rainfall is between 100 and 200 mm, the loess shows less pedogenic modification (Fig. 6).

A model which attempts to explain the changes in loess character along a hypothetical environmental gradient between a desert and an adjacent semi-arid area is shown in Fig. 7. Both the dust accretion rate (curve A) and the mean size of deposited dust decrease with distance from the desert source. No dust accretion takes place between points W and X where the climate is too arid to support the growth of dust-trapping vegetation. As rainfall increases towards the semi-arid area, so the rate of syn-depositional weathering and leaching (curve B) also increases. At some distance from the source, designated Y in Fig. 7, the weathering rate (indicated notionally by depth of decalcification) becomes equal to the dust accretion rate. Beyond this point the accumulated loess displays significant evidence of weathering and pedogenic horizonation. By comparison, upwind of point Y the loess appears relatively fresh, with limited carbonate leaching and pedogenetic horizonation. Changes in either the dust deposition rate or rainfall would cause a shift in the positions of X and Y. Several such shifts would produce a stratigraphic sequence composed of alternating weathered and relatively unweathered loess horizons. Dust deposition need not cease totally in order to produce a weathered horizon, but the rate of accretion must fall below the weathering rate. Since clay minerals and carbonates are more abundant in the finer size fractions of dust (see below), these minerals are enriched relative to quartz and feldspar in the dust deposited at greater distance from the source. Consequently, the relationship between accretion rate and weathering rate is not linear, and the form of the curves shown in Fig. 7 is schematic.

FIG. 5. Weathered clayey loess (CL) and soil carbonate horizon (SCH) exposed at Netivot, northern Negev.

FIG. 6. Relatively unweathered sandy loess (SL) overlain by recent dune sands (DS), Gaza Strip.

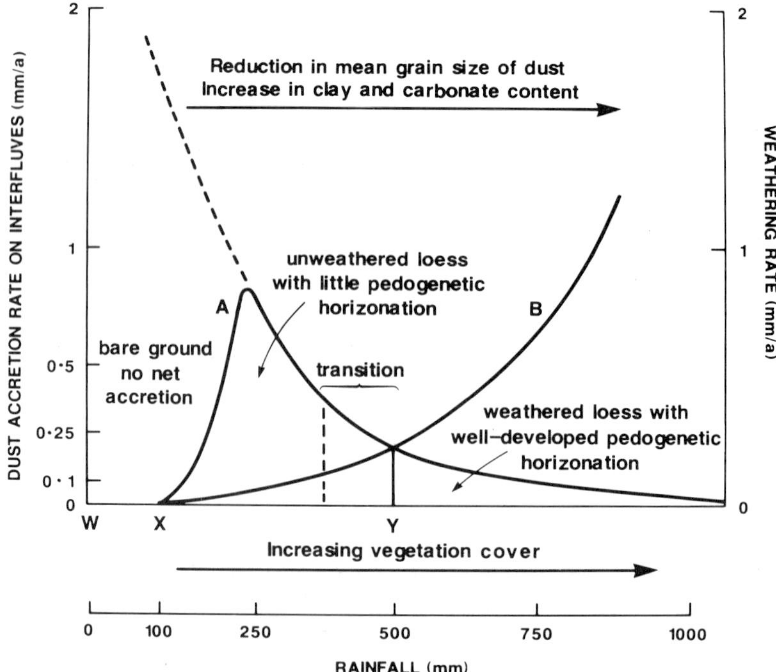

FIG. 7. Schematic model showing expected changes in the character of loess deposits along an environmental gradient between warm desert and sub-humid climatic conditions. For explanation see text.

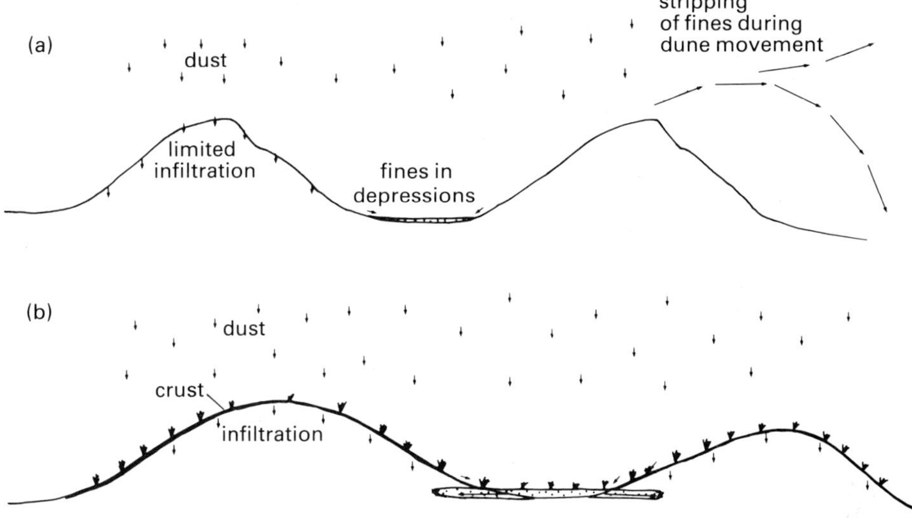

FIG. 8. Schematic representation of the consequences of dust deposition in sand dune terrain: (*a*) an area of thinly vegetated migrating dunes; (*b*) an area of partially vegetated and largely fixed dunes (*eg* the linear dunes of the northern Negev).

Effects of dust deposition on sand dune dynamics and dune sand diagenesis

In sand dune areas, deposition of airborne dust and salts can lead to the formation of a surface crust which tends to stabilize the sand. This in turn may allow colonization of the sand surface by lichens, fungi and vascular plants (Danin 1978; Danin & Yaalon 1982; Van den Ancker *et al.* 1985). Some of the deposited dust particles and salts infiltrate the dunes with rainwater, and can have a significant effect on early diagenesis of the sands.

The effects of deposited dust are to some extent dependent on dune type and mobility. In hyper-arid or very windy deserts, where dunes are unvegetated and highly mobile, dust is usually re-suspended and winnowed during dune migration (Fig. 8*a*). The average silt plus clay content of mobile desert dunes is usually less than 1%. Thin clay crusts and drapes may form, however, in inter-dune hollows where rainwater accumulates and evaporates. In less arid or less windy deserts, where dunes are often partly vegetated and maintain a more or less fixed position, as in the sand-ridge deserts of central Australia and northern Negev, substantial amounts of dust may accumulate in the dune sands over time (Fig. 8*b*).

Tsoar & Møller (1986) found that a stabilized linear dune in the northern Negev contained 10% fines at the surface, and 5–6% fines at a depth of 40 cm (Fig. 9). Vegetated linear dunes in Rice Valley, California contain 4–6.6% fines, and in central Australia up to 5% (Breed & Breed 1979).

The depth of dust infiltration depends on the amount and frequency of rainfall, on the pore size distribution of the dune sand, and on the size of the dust. Orev (1984) found that 1 mm of rain penetrated to a depth of 50 mm in dune sand. Dincer *et al.* (1974) found that 1 mm of rain penetrated to 7 mm in well-graded dune sand with a mean size of 150 μm and to 20 mm in

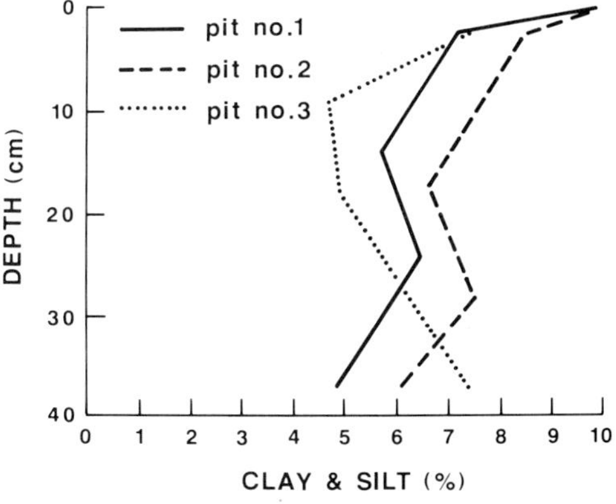

FIG. 9. Variation in silt and clay content with depth in a northern Negev linear dune (after Tsoar & Møller 1986).

300 μm sand. Walther (1973) found that 50 mm of rain wetted the upper 50 cm of a dry sandy soil to field capacity. In arid areas of the northern Negev, where the mean annual rainfall is about 100 mm, rain penetrates the sand dunes to a depth of about 100 cm, compared with 30 cm in neighbouring loessial soils (Tsoar & Zohar 1985).

Leaching experiments by Wright & Foss (1968) showed that fine silt and clay particles are more easily translocated through sand columns than medium and coarse silt particles. Fine silt was completely washed out of a 33 cm long column filled with medium sand by less than 1 litre of water. About 95% of the coarse silt was also partially translocated, with 25–50% passing completely through the sand column. With columns filled with fine sand, appreciable movement of fine silt was observed but there was little movement of coarse silt.

These findings have important implications because, in many natural dusts, different size fractions have different mineral and chemical compositions. Mineralogical stratification of particles infiltrated into dune sand may therefore be expected.

Fig. 10 shows the mineral composition of different size fractions in material collected from a depth of 0–10 cm in a stabilized dune in the northern Negev. The sand fraction (>63 μm) consists principally of quartz, plagioclase and K-feldspar with traces of calcite and heavy minerals (Fig. 10*a*). Calcite and clay minerals are important constituents of the silt and clay fraction (<63 μm), which represents 7.2% of the total sample (Fig. 10*b*). Calcite is the dominant non-clay mineral in the <31 μm fraction, while the proportion of quartz declines progressively with decreasing grain size (Fig. 10*c–f*). Kaolinite, smectite and mixed-layer illite–smectite are the main clays present. The mineral composition of the <63 μm material is similar to that of a 20 cm thick layer of unweathered loess at the top of the Netivot section (Fig. 10*g*), supporting the interpretation that the silt and clay are derived from aeolian dust.

Examination of the dust fraction separated from the dune sands using a scanning electron microscope (SEM) confirmed that carbonate grains are on average smaller than quartz grains (Fig. 11). Calcite occurs mainly as limestone fragments, with some foraminifera. Dolomite is also present, mainly as medium silt-size rhombs or fragments of rhombs. The fine size of the carbonates reflects their low hardness and their susceptibility to disintegrate compared with quartz. After deposition on the dune surface, many of the carbonate grains and clays are translocated in the sand column, whereas the coarser quartz and feldspar components are retained near the surface.

Airborne salts also show differing relative mobilities once deposited with $Cl^- > SO_4^{2-} > HCO_3^-$ (Yaalon 1964*b*). The composition of salts deposited in precipitation at Beer Sheva and Sde Boker in the northern Negev is shown in Table 3. At both locations bicarbonate is present in substantially higher concentrations than sulphate and chloride. When gravity seepage has ceased following rainfall, water is retained in dune sands as films which coat grain surfaces, as menisci at grain contacts, and as moisture adsorbed by colloids clay minerals, and in fine capillaries. In the uppermost 30 cm of the sand column these films gradually dry out due to surface evaporation. Tsoar & Zohar (1985) found that at the end of summer, dune sands in the northern Negev contained moisture equivalent to about half the field capacity (*ie* about 2%) at a depth of 70 cm. During drying the salinity and alkalinity of the moisture films can become very high, leading to solution and reprecipitation of silica. Fine-grained silicate components are most susceptible to dissolution due to their large surface area and frequent structural disordering. The solutions are drawn towards the sand surface, where they eventually evaporate and precipitate amorphous silicate gels (Fig. 12). Figure 10*h* shows the crystalline components present in a typical 2 mm thick dune sand crust. In the crusts studied, halite is frequently present as a cement but gypsum is rare. Some halite may also be detrital, cementing aggregates eroded from soil crusts or evaporites (*cf* Schroeder *et al.* 1985; Eswaran *et al.* 1980). SEM examination showed that most of the

TABLE 3. *Salts deposited annually in precipitation at Beer Sheva and Sde Boker, northern Negev (after Yaalon 1964a)*

Station	distance from sea (km)	elevation (m)	mean annual rainfall (mm)	mean annual raindays	Na^+	K^+	Ca^{2+}	Mg^{2+} (ppm)	Cl^-	SO_4^{2-}	HCO_3^-	conductivity (mmho)
Beer Sheva	46	280	200	34	9	2	43	3	17	33	98	247
Sde Boker	72	470	90	22	2	1	16	1	3	9	42	206

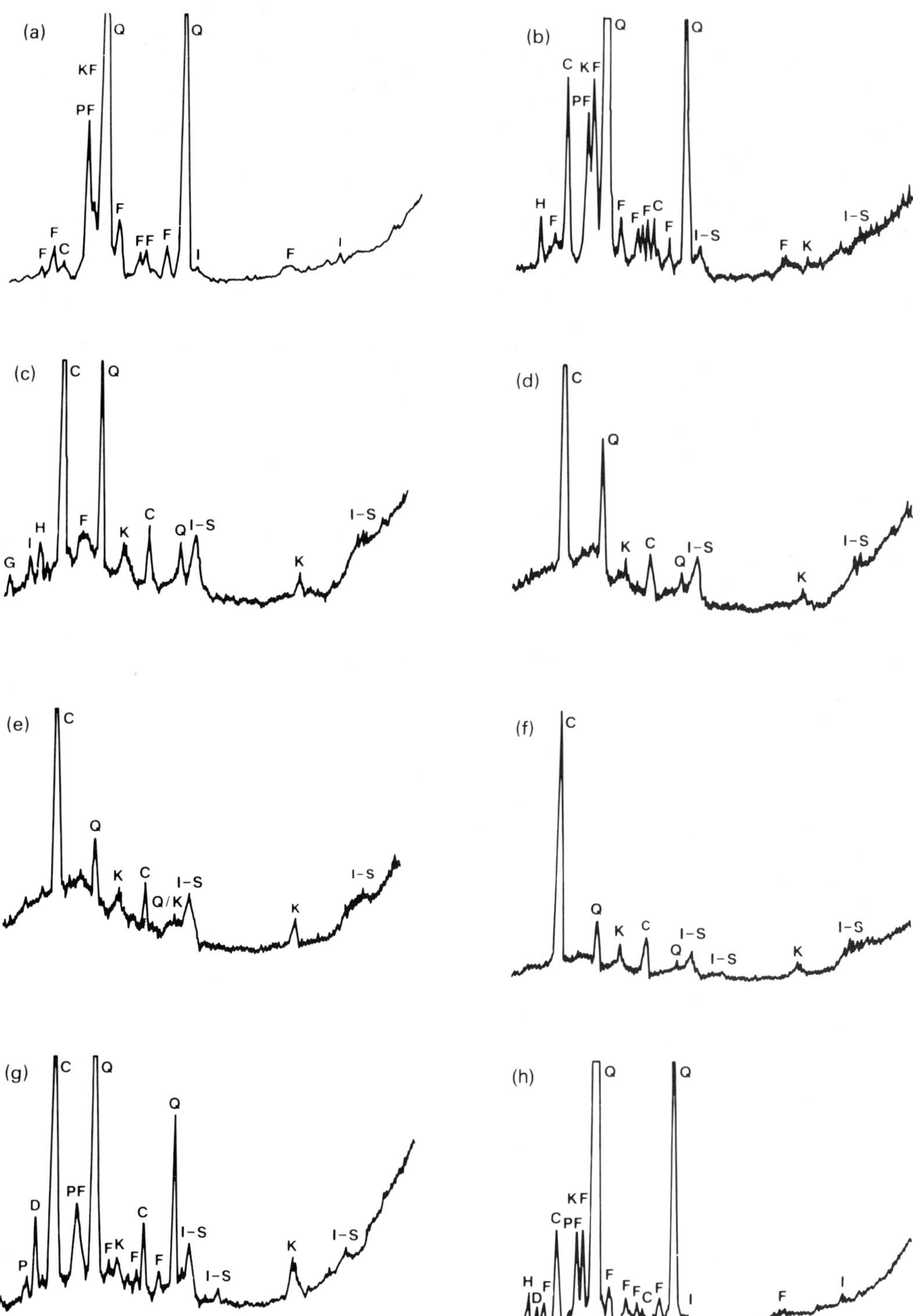

FIG. 10. X-ray powder diffraction traces (CuKα radiation) of different size fractions of dust in the upper 10 cm of a Negev linear dune: (*a*) >63 μm; (*b*) 32–63 μm; (*c*) 16–32 μm; (*d*) 8–16 μm; (*e*) 4–8 μm; (*f*) <4 μm; of a bulk sample of the unweathered 'upper loess' at the Netivot section (*g*), and a bulk sample of dune sand crust (*h*). Q = quartz, PF = plagioclase feldspar, KF = K-feldspar, F = undifferentiated feldspar, C = calcite, D = dolomite, H = halite, K = kaolinite, I–S = mixed layer illite–smectite including some discrete mica (I), illite and smectite, G = gypsum. All samples are air-dry.

FIG. 11. SEM micrograph of the 20–63 μm fraction from a Negev linear dune. Carbonate grains (C) are on average smaller than quartz and feldspar grains (Q). Scale bars = 10 μm.

FIG. 12. The cemented surface crust on a linear dune. Scale bars = 100 μm.

carbonate minerals in the crusts represent detrital dust particles, although some carbonate occurs as intergranular cement. The major cementing agents in the crusts, however, are mixtures of infiltered detrital clays and amorphous silicate precipitates (Figs 13 and 14). Energy-dispersive X-ray microanalysis indicated that the amorphous cements consist principally of Si and Al, with varying amounts of Fe, K and Ca (Fig. 15).

The sands more than 30 cm below the surface generally never dry out, and moisture films containing concentrated salt solutions are in continuous contact with grain surfaces. Under these conditions silica is dissolved and is reprecipitated when the solutions become supersaturated (first with respect to amorphous silica). In this manner 'waxy' or 'scaly' coatings, referred to by Folk (1978) as 'turtle-skin silica coats', are formed. Quartz grains in stabilized dunes rarely show surface textural features indicative of aeolian abrasion (*cf* Krinsley & McCoy 1978) due to the rapidity of chemical alteration of the surface (Figs 16 and 17).

However, the coatings on the grains of stabilized dunes from the Negev do not usually consist only of silica, but include varying amounts of Al, Fe, K and Ca (Fig. 18). Generally, the redder the colour of the grains, the more Fe and Al present. A similar relationship was observed by Walker (1979) in dune sands in Libya. The iron is present as finely disseminated haematite derived from weathering of both infiltered dust and iron-bearing silicate minerals in the sand, principally epidote and hornblende. There is some experimental evidence to suggest that formation of haematite from ferric oxyhydroxide precursors is

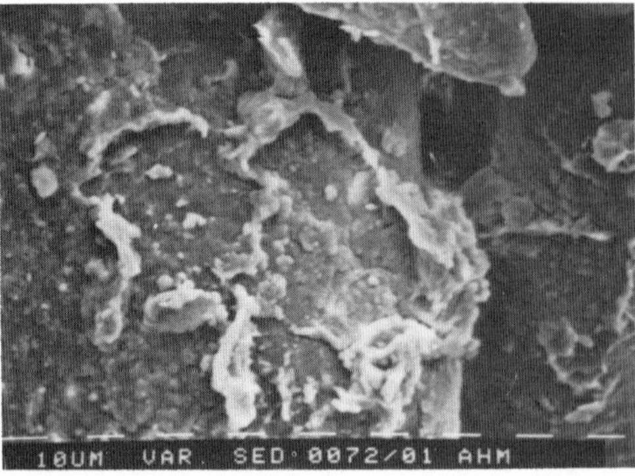

FIG. 13. Amorphous silicate gel and infiltrated dust particles forming meniscus cements around former points of grain contact in a sand dune crust. Scale bars = 10 μm.

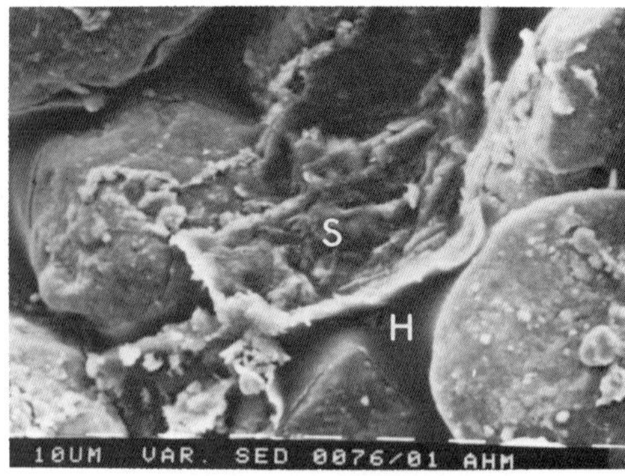

FIG. 14. Sand grains cemented by amorphous silicate material (S) and halite (H) in a dune surface crust. Scale bars = 10 μm.

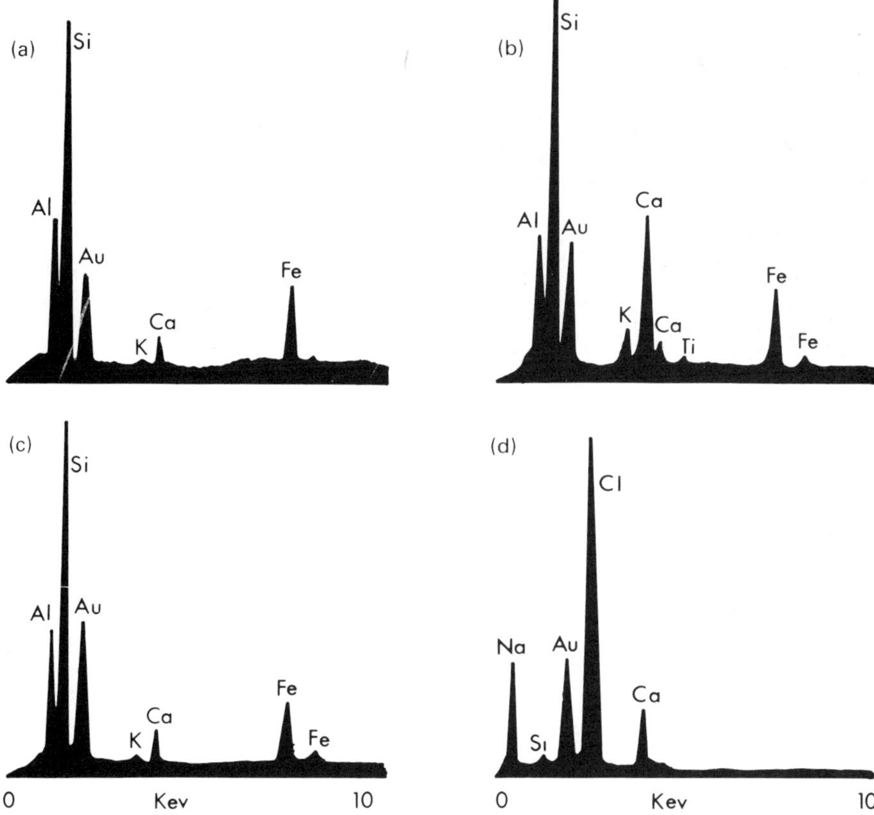

FIG. 15. Energy-dispersive X-ray spectra of the silicate gel (*a–c*) and halite (*d*) cements shown in Fig. 14.

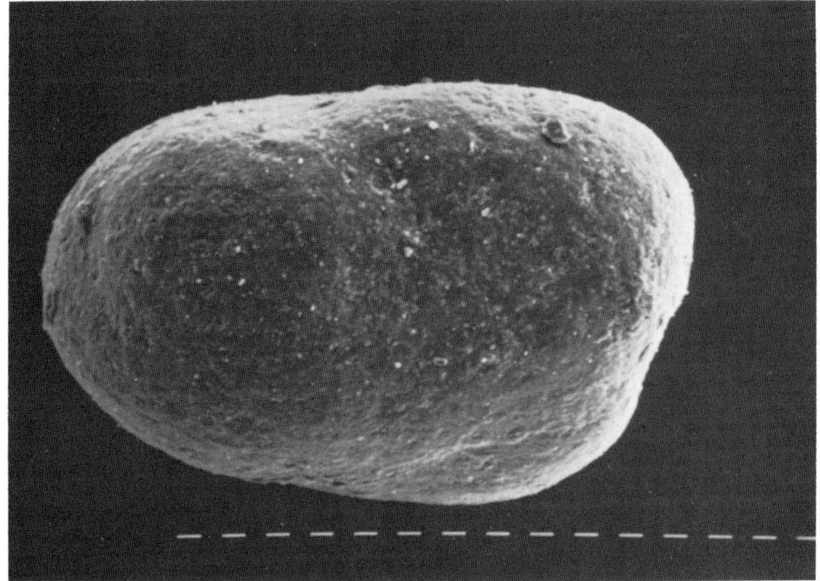

FIG. 16. A reddened sand grain from a depth of 1 m in a stabilized linear dune, northern Negev. Scale bars = 10 μm.

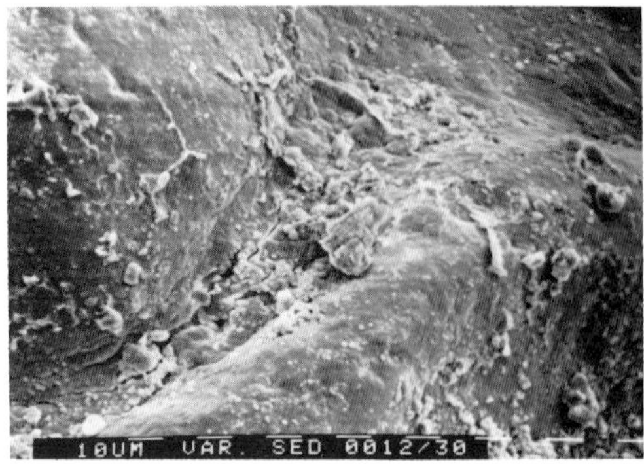

FIG. 17. SEM micrograph showing the surface texture of a typical sand grain from the Negev stabilized linear dunes. The grain surface is coated by amorphous silica, clay minerals, ferric oxide, and carbonate particles. The coatings are thickest in surface pits. Scale bars = 10 μm.

favoured by the presence of chloride ions (Collepardi *et al.* 1973). The Al, K, and Ca in the grain coatings is derived mainly from infiltered dust.

Waugh (1970) suggested that euhedral overgrowths on quartz dune sand grains in the Penrith Sandstone (Lower Permian) of NW England were formed by dissolution of siliceous dusts in the near-surface desert environment. However, a number of factors suggest this interpretation is probably incorrect. Firstly, desert dusts are rarely, if ever, composed of pure silica. Secondly, in polymineralic dusts, quartz occurs mainly in the coarser size fractions and is infiltered only to shallow depths in the sand column. Thirdly, it is inconceivable that large euhedral quartz overgrowths, which require slow crystallization and low concentrations of dissolved silica for their formation (Iler 1979), could develop in a vadose desert environment with discontinuous, highly saline, alkaline fluid films. Under vadose conditions in the Negev, and in central Australia (Folk 1978), impure amorphous silica coatings are produced on sand grains by repeated dissolution and reprecipitation in response to changes in the moisture regime of the sands. It is more likely that the Penrith Sandstone overgrowths were formed below the water table and before significant compaction of the sands occurred.

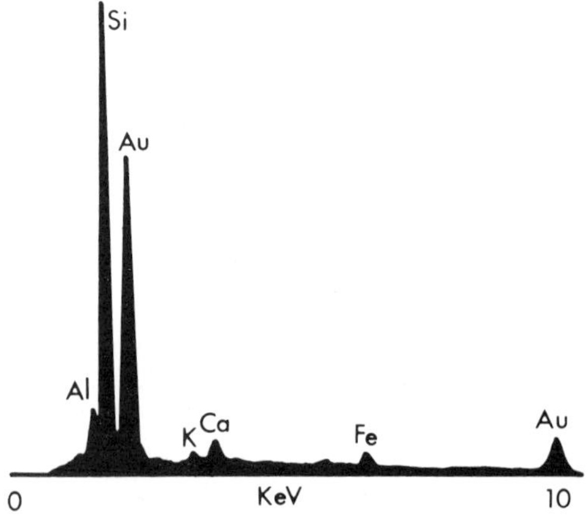

FIG. 18. Energy-dispersive X-ray spectrum (areal analysis) of the grain coating shown in Fig. 17.

Many aeolian sandstones in the geological record, including those of the Rotliegend of the southern North Sea (Pye & Krinsley 1986), show considerable small-scale variability in preserved porosity and authigenic mineralogy. Some of this variability may reflect patterns of early cementation related to infiltration of airborne dust and salts. During slow vertical accretion of aeolian sand bodies, a stacked sequence of cemented surface crusts would be preserved, with alternating layers of less well cemented and more porous sand. The amorphous and poorly crystalline silicate phases formed by dissolution and reprecipitation in the surface crusts and near-surface sand layers might also provide a precursor for later authigenic clay mineral formation.

Conclusion

The grain size distributions, mineral composition and grain surface textures of linear dune sands in the northern Negev have been significantly modified as a result of additions of airborne dust and salts in the Holocene. The precise age of the dunes is not known, but the presence of Late Palaeolithic artefacts on the surface suggests only minor sand movement has occurred in the last 16–20 000 years (Goring-Morris *pers. comm.* 1986). The effects of sub-aerial diagenesis are most marked close to the sand surface, and include reddening, formation of amorphous silica coats on quartz grains, and cementation by mixtures of salts, infiltered dust, and amorphous silicate precipitates. The presence of fine particles and bonding agents helps to maintain the stability of the dunes by raising the threshold velocity for sand movement and by retaining moisture for longer periods, thereby facilitating a more vigorous vegetation.

Although dust is being deposited today in southern Israel, rates of deposition must have been significantly higher earlier in the Quaternary. Stratigraphic relationships indicate that most of the loess in the northern Negev was deposited in the late Pleistocene before invasion of the now-stabilized linear dunes. Pedogenic features of the loess and its topographic distribution indicate that the presence of vegetation played a key role in its accumulation. The main source of coarse and medium silt was probably northern Sinai, particularly the deposits of Wadi el Arish, since it is unlikely that large amounts of such material, which is transported by the wind at relatively low levels above the ground in short-term suspension, could cross the Nile from the north African deserts.

Isotopic data from groundwaters and tufas in Sinai and the Negev indicate wetter conditions during much of the last glacial period (Issar & Bruins 1983). At this time, cyclonic storms moving in from the Atlantic across north Africa appear to have been more frequent than at present. Higher rainfall would have increased the seasonal discharge in Wadi El Arish and other wadis, so increasing the availability of silt during times of low flow. Besides this, strong southwesterly dust-transporting winds would have been more frequent. Higher rainfall would also have encouraged the growth of vegetation, helping to trap deposited dust. Desiccation during the very late Pleistocene and/or Holocene appears to have reduced the dust supply and the frequency of dust storms, and led to a thinning of the vegetation cover with widespread erosion of the loess.

ACKNOWLEDGMENTS: We thank A. Issar, R. Gerson and N. Goring-Morris for discussion in the field, and E. Haines for assistance with analytical work.

References

ALLEN, C. C. 1978. Desert varnish of the Sonoran desert—optical and electron probe microanalysis. *Journal of Geology* **86**, 743–752.

BAGNOLD, R. A. 1941. *The Physics of Blown Sand and Desert Dunes*. Methuen, London, 241pp.

—— 1960. The re-entrainment of settled dusts. *International Journal of Air Pollution* **2**, 357–363.

BELLY, P. Y. 1964. Sand movement by wind. *United States Army Coastal Engineering Research Centre Technical Memorandum* **1**, 1–38.

BLÜMEL, W. D. 1982. Calcretes in Namibia and SE Spain—relations to substratum, soil formation and geomorphic factors. *Catena Supplement* **1**, 67–95.

BOWDEN, L. W., HUNING, J. R., HUTCHINSON, C. F. & JOHNSON, C. W. 1974. Satellite photograph presents first comprehensive view of local wind: the Santa Ana. *Science* **184**, 1077–1078.

BOWLER, J. M. 1978. Glacial age aeolian events at high and low latitudes. *In*: VAN ZINDEREN BAKKER, E. M. (ed.) *Antarctic Glacial History and World Palaeoenvironments*. Balkema, Rotterdam, 149–172.

BOWMAN, D. 1982. Iron coating in recent terrace sequences under extremely arid conditions. *Catena* **9**, 353–359.

BREED, C. S. & BREED, W. J. 1979. Dunes and other windforms of central Australia (and a comparison with linear dunes on the Moenkopi Plateau, Arizona). *In:* EL BAZ, F. & WARNER, D. M. (eds) *Apollo–Soyuz test project summary science report, Volume 2: Earth observations and photography*. NASA, Washington D.C., 319–358.

BRUINS, H. J. 1976. *The origin, nature and stratigraphy of paleosols in the loessial deposits of the N.W. Negev (Netivot, Israel)*. MSc thesis, Hebrew University of Jerusalem.

—— & YAALON, D. H. 1979. Stratigraphy of the Netivot section in the desert löess of the Negev (Israel). *Acta Geologica Academica Scientiarum Hungaricae* **22**, 161–169.

CEGLA, J. 1969. Influence of capillary ground moisture on eolian accumulation of loess. *Bulletin Academica Polonica Science Series Geology Geography* **17**, 25–27.

CHEPIL, W. S. 1945. Dynamics of wind erosion. II. Initiation of soil movement. *Soil Science* **60**, 397–411.

—— 1959. Equilibrium of soil grains at the threshold of movement by wind. *Proceedings of the Soil Science Society of America* **20**, 288–292.

—— & WOODRUFF, N. P. 1957. Sedimentary characteristics of duststorms. II. Visibility and dust concentration. *American Journal of Science* **255**, 104–114.

—— & —— 1963. The physics of wind erosion and its control. *Advances in Agronomy* **15**, 211–302.

CLEMENTS, T., STONE, R. O., MANN, J. F. & EYMANN, J. L. 1963. A study of windborne sand and dust in desert areas. *United States Army Natik Laboratories, Earth Sciences Division, Technical Report* ES-8, 61 pp.

COLLEPARDI, M., MASSIDDA, L. & ROSSI, G. 1973. Effects of pH and salt additions on ageing of ferric iron oxide gels. *Transactions of the Institute of Mining and Metallurgy, Section C* **81**, C42–C46.

CONCA, J. L. & ROSSMAN, G. R. 1982. Case hardening of sandstone. *Geology* **10**, 520–523.

COUDÉ-GAUSSEN, G. 1984. Le cycle des poussières éoliennes désertiques actuelles et la sedimentation des loess peridésertiques quaternaires. *Bulletin des Centre Recherches Exploration Production Elf-Aquitaine* **8**, 167–182.

DAN, J. 1966. *The effects of relief on soil formation and distribution in Israel*. PhD thesis, Hebrew University of Jerusalem, 361pp. (in Hebrew, summary in English).

——, MOSHE, R. & ALPEROVITCH, N. 1973. The soils of Sede Zin. *Israel Journal of Earth Sciences* **26**, 211–227.

DANIN, A. 1970. *A phytosociological–ecological study of the northern Negev of Israel*. PhD thesis, Hebrew

University of Jerusalem, 230 pp. (in Hebrew, summary in English).

—— 1978. Plant species diversity and plant succession in a sandy area in the northern Negev. *Flora* **167**, 409–422.

—— & YAALON, D. H. 1982. Silt and clay sedimentation and decalcification during plant succession in sand of the Mediterranean Coastal Plain, Israel. *Israel Journal of Earth Sciences* **31**, 101–109.

DINCER, T., AL-MUGRIN, A. & ZIMMERMAN, U. 1974. Study of the infiltration and recharge through the sand dune in arid zones with special reference to the stable isotopes and thermonuclear tritium. *Journal of Hydrology* **23**, 79–109.

DODONOV, A. E. 1984. Stratigraphy and correlation of Upper Pliocene–Quaternary deposits of Central Asia. *In:* PECSI, M. (ed.) *Lithology and Stratigraphy of Loess and Paleosols*. Geographical Research Institute, Hungarian Academy of Sciences, Budapest, 201–211.

ESWARAN, H., STOOPS, G. & ABTAHI, A. 1980. SEM morphologies of halite (NaCl) in soils. *Journal of Microscopy* **120**, 342–353.

FOLK, R. L. 1978. Angularity and silica coatings of Simpson Desert sand grains, Northern Territory, Australia. *Journal of sedimentary Petrology* **48**, 611–624.

FOSTER, S. M. & NICOLSON, T. H. 1981. Microbial aggregation of sand in a maritime dune succession. *Soil Biology and Biochemistry* **13**, 205–208.

GANOR, E. 1975. *Atmospheric dust in Israel–sedimentological and meteorological analysis of dust deposition*. PhD thesis, Hebrew University of Jerusalem (in Hebrew, summary in English).

—— & MAMANE, Y. 1982. Transport of Saharan dust across the eastern Mediterranean. *Atmospheric Environment* **16**, 581–587.

GARDNER, R. A. M. 1977. Evidence concerning the existence of loess deposits at Tell Fasa, northern Negev. *Journal of Archaeological Sciences* **4**, 377–386.

GERSON, R., AMIT, R. & GROSSMAN, S. 1985. *Dust availability in desert terrains—a study in the deserts of Sinai and Israel*. Report for the U.S. Army Research, Development and Standardization Group, UK, Contract No. DAJA45-83-C-0041.

GILLETTE, D. A., ADAMS, J., ENDO, A. & SMITH, D. 1980. Threshold velocities for input of soil particles into the air by desert soils. *Journal of Geophysical Research* **85**, 5621–5630.

——, ——, MUHS, D. & KIHL, R. 1982. Threshold velocities and rupture moduli for crushed desert soils for input of soil particles to air. *Journal of Geophysical Research*, **87**, 9003–9015.

——, BLIFFORD, D. A. & FRYEAR, D. W. 1974. The influence of wind velocity on size distributions of soil wind aerosols. *Journal of Geophysical Research* **79**, 4068–4075.

GOLDBERG, E. D. 1971. Atmospheric dust, the sedimentary cycle and man. *Comments Earth Science and Geophysics* **1**, 117–132.

GOUDIE, A. S. 1978. Dust storms and their geomorphological implications. *Journal of Arid Environments* **1**, 291–310.

—— 1983. Dust storms in space and time. *Progress in Physical Geography* **7**, 502–530.

GREELEY, R. & IVERSEN, J. D. 1985. *Wind as a Geological Process*. Cambridge University Press, Cambridge, 333 pp.

——, WHITE, B. R., POLLACK, J. B., IVERSEN, J. D. & LEACH, R. N. 1981. Dust storms on Mars: considerations and simulations. *Geological Society of America Special Paper* **186**, 101–121.

HAGEN, L. 1984. Soil aggregate abrasion by impacting sand and soil particles. *Transactions of the American Society of Agricultural Engineering* **27**, 805–808.

—— & WOODRUFF, N. P. 1973. Air pollution from duststorms in the Great Plains. *Atmospheric Environment* **7**, 323–332.

HELLER, F. & LIU TUNG SHENG 1982. Magnetostratigraphical dating of loess deposits in China. *Nature* **300**, 431–433.

HIDY, G. M. & BROCK, J. R. 1970. An assessment of the global sources of tropospheric aerosols. *Proceedings of the 2nd Clean Air Congress, Washington D.C.*, 1088–1097.

IDSO, S. B. 1976. Dust storms. *Scientific American* **235**, 108–114.

ILER, R. K. 1979. *The Chemistry of Silica*. Wiley, New York.

ISSAR, A. S. & BRUINS, H. J. 1983. Special climatological conditions in the deserts of Sinai and the Negev during the latest Pleistocene. *Palaeogeography, Palaeoclimatology and Palaeoecology*, **43**, 63–72.

——, KARNIELI, A., BRUINS, H. J. & GILEAD, I. 1984. The Quaternary geology and hydrology of Sede Zin, Negev, Israel. *Israel Journal of Earth Sciences* **33**, 34–42.

IVERSEN, J. D. & WHITE, B. R. 1982. Saltation threshold on Earth, Mars and Venus. *Sedimentology* **29**, 111–119.

——, POLLACK, J. B., GREELEY, R. & WHITE, B. R. 1976. Saltation threshold on Mars: the effect of interparticle force, surface roughness, and low density. *Icarus* **29**, 381–393.

JACKSON, M. L., GILLETTE, D. A., DANIELSEN, E. F., BLIFFORD, I. H., BRYSON, R. A. & SYERS, J. K. 1973. Global dustfall during the Quaternary as related to environments. *Soil Science* **116**, 135–145.

JANECEK, T. R. & REA, D. K. 1985. Quaternary fluctuations in the northern hemisphere trade winds and westerlies. *Quaternary Research* **24**, 150–163.

JONES, D. K. C., COOKE, R. U. & WARREN, A. 1986. Geomorphological investigation, for engineering purposes, of blowing sand and dust hazard. *Quarterly Journal of Engineering Geology* **19**, 251–270.

JUNGE, C. 1979. The importance of mineral dust as an atmospheric constituent. *In*: MORALES, C. (ed.) *Saharan Dust*. Wiley, Chichester, 49–60.

KALU, A. E. 1979. The African dust plume: its characteristics and propogation across West Africa in winter. *In:* MORALES, C. (ed.) *Saharan Dust*. Wiley, Chichester, 95–118.

KASTNELSON, J. 1970. Frequency of duststorms at Beer Sheva. *Israel Journal of Earth Sciences* **19**, 69–76.

KRINSLEY, D. H. & MCCOY, F. 1978. Aeolian quartz

sand and silt. *In:* WHALLEY, W. B. (ed.) *Scanning Electron Microscopy in the Study of Sediments.* Geo Abstracts, Norwich, 249–260.

LAWSON, T. J. 1971. Haboob structure at Khartoum. *Weather* **26**, 105–112.

LIU TUNG SHENG, GU XIONG FEI, AN ZHI SHENG & FAN YONG XIANG 1981. The dust fall in Beijing, China, on April 18, 1980. *Geological Society of America Special Paper* **186**, 149–158.

LOGIE, M. 1982. Influence of roughness elements and soil moisture on the resistance of sand to wind erosion. *Catena Supplement* **1**, 161–173.

LYLES, L. 1977. Wind erosion: processes and effect on soil productivity. *Transactions of the American Society of Agricultural Engineering* **20**, 880–884.

—— & KRAUSS, R. K. 1971. Threshold velocities and initial particle motion as influenced by air turbulence. *Transactions of the American Society of Agricultural Engineering* **14**, 134–139.

——, SCHRANDT, N. F. & SCHNEIDLER, N. F. 1974. How aerodynamic roughness elements control sand movement. *Transactions of the American Society of Agricultural Engineering* **17**, 134–139.

MCTAINSH, G. H. & WALKER, P. H. 1982. Nature and distribution of Harmattan dust. *Zeitschrift für Geomorphologie* **26**, 417–435.

MIDDLETON, N. J. 1986*a*. A geography of dust storms in southwest Asia. *Journal of Climatology* **6**, 183–196.

—— 1986*b*. Dust storms in the Middle East. *Journal of Arid Environments* **10**, 80–86.

——, GOUDIE, A. S. & WELLS, G. L. 1986. The frequency and source areas of dust storms. *In:* NICKLING, W. G. (ed.) *Aeolian Geomorphology.* Allen & Unwin, London, 237–259.

MORALES, C. (ed.) 1979. *Saharan Dust—Mobilization, Transport, Deposition.* Wiley, Chichester, 297 pp.

NAKATA, J. K., WILSHIRE, H. G. & BARNES, C. G. 1976. Origin of Mojave Desert dust plumes photographed from space. *Geology* **4**, 644–648.

NALIVKIN, D. V. 1982. *Hurricanes, Storms and Tornadoes.* Amerind, New Delhi.

NICKLING, W. G. 1983. Grain size characteristics of sediment transported during dust storms. *Journal of sedimentary Petrology* **53**, 1011–1024.

—— 1984. The stabilizing role of bonding agents on the entrainment of sediment by wind. *Sedimentology* **31**, 111–117.

—— & BRAZEL, A. J. 1984. Temporal and spatial characteristics of Arizona dust storms (1965–1980). *Journal of Climatology* **4**, 645–660.

—— & ECCLESTONE, M. 1981. The effect of soluble salts on the threshold shear velocity of fine sand. *Sedimentology* **28**, 505–510.

OLIVER, F. W. 1945. Dust storms in Egypt and their relation to the war period, as noted in Maryut, 1939–45. *Geographical Journal* **106**, 26–49.

OREV, Y. 1984. Sand is greener. *Teva va-Aretz* **26**, 15–16 (in Hebrew).

PÉWÉ, T. L. (ed.) 1981. Desert dust: origin, characteristics and effect on man. *Geological Society of America Special Paper* **186**, 303 pp.

——, PÉWÉ, E. A., PÉWÉ, R. H., JOURNAUX, A. & SLATT, R. M. 1981. Desert dust: characteristics and rate of deposition in central Arizona. *Geological Society of America Special Paper* **186**, 169–190.

POTTER, R. M. & ROSSMAN, G. R. 1977. Desert varnish: the importance of clay minerals. *Science* **196**, 1446–1448.

PROSPERO, J. M. 1981. Arid regions as sources of mineral aerosols in the marine atmosphere. *Geological Society of America Special Paper* **186**, 71–86.

——, GLACCUM, R. A. & NEES, R. T. 1981. Atmospheric transport of soil dust from Africa to South America. *Nature* **289**, 570–572.

PYE, K. 1987. *Aeolian Dust and Dust Deposits.* Academic Press, London, in press.

—— & KRINSLEY, D. H. 1986. Diagenetic carbonate and evaporite minerals in Rotliegend aeolian sandstones from the southern North Sea: their nature and relationship to secondary porosity development. *Clay Minerals* **21**.

SAFAR, M. I. 1980. *Frequency of dust in day-time summer in Kuwait.* Meteorological Department, State of Kuwait, 107 pp.

SCHROEDER, J. H., KACHHOLZ, K-D. & HEUER, M. 1985. Eolian dust in the coastal desert of Sudan: aggregates cemented by evaporites. *Geo-Marine Letters* **4**, 139–144.

SINCLAIR, P. C. 1969. General characteristics of dust devils. *Journal of Applied Meteorology* **8**, 32–45.

SNEH, A. 1983. Redeposited löess from the Quaternary Besor Basin, Israel. *Israel Journal of Earth Sciences* **32**, 62–69.

SUNDBORG, A. 1955. Meteorological and climatological conditions for the genesis of aeolian sediments. *Geografiska Annaler* **37**, 94–111.

TSOAR, H. & MØLLER, J. T. 1986. The role of vegetation in the formation of linear sand dunes (analysis of the case of the Negev–Sinai border line). *In:* NICKLING, W. G. (ed.) *Aeolian Geomorphology.* Allen & Unwin, London.

—— & PYE, K. 1987. Dust deposition and the question of desert loess formation. *Sedimentology* **34**, 139–153.

—— & ZOHAR, Y. 1985. Desert dune sand and its potential for modern agricultural development. *In:* GRADUS, Y. (ed.) *Desert Development.* Reidel, Dordrecht, 184–200.

VAN DEN ANCKER, J. A. M., JUNGERIUS, P. D. & MUR, L. R. 1985. The role of algae in the stabilization of coastal dune blowouts. *Earth Surface Processes of Landforms* **10**, 189–192.

WALKER, T. R. 1976. Diagenetic origin of continental red beds. *In:* FALKE, H. (ed.) *The Continental Permian in Central, West, and South Europe.* Reidel, Dordrecht, 240–282.

—— 1979. Red color in dune sand. *U.S. Geological Survey Professional Paper* 1052, 61–81.

WALTHER, H. 1973. *Vegetation of the Earth.* Springer, New York,

WAUGH, B. 1970. Petrology, provenance and silica diagenesis of the Penrith Sandstone (Lower Permian) of northwest England. *Journal of sedimentary Petrology* **40**, 1226–1240.

WILSHIRE, H. G. 1980. Human causes of accelerated wind erosion in California's deserts. *In:* COATES, D. R. & VITEK, J. D. (eds) *Thresholds in Geomorphology.* Allen & Unwin, London, 415–434.

Windom, H. L. 1975. Eolian contributions to marine sediments. *Journal of Sedimentary Petrology* **45**, 520–529.

Wright, W. R. & Foss, J. E. 1968. Movement of silt-sized sediment in sand columns. *Proceedings of the Soil Science Society of America* **32**, 446–448.

Yaalon, D. H. 1964*a*. Airborne salts as an active agent in pedogenetic processes. *8th International Congress on Soil Science, Bucharest, 1964*, 997–1000.

—— 1964*b*. Downward movement and distribution of anions in soil profiles with limited wetting. *In*: Hallsworth, E. G. & Crawford, D. V. (eds) *Experimental Pedology*. Butterworth, London, 157–164.

—— 1969. Origin of desert loess. *Abstracts of the 8th INQUA Congress Paris*, 2, **755**.

—— & Dan, J. 1974. Accumulation and distribution of loess-derived deposits in the semi-desert and desert-fringe areas of Israel. *Zeitschrift für Geomorphologie Supplement Bd.* **20**, 91–105.

—— & Ganor, E. 1973. The influence of dust on soils during the Quaternary. *Soil Science* **116**, 146–155.

—— & —— 1975. Rates of aeolian dust accretion in the Mediterranean and desert fringe environments in Israel. *9th International Congress on Sedimentology, Nice, 1975, Theme 2*, 169–174.

—— & —— 1979. East Mediterranean trajectories of dust carrying storms from the Sahara to Sinai. *In*: Morales, C. (ed.) *Saharan Dust*. Wiley, Chichester, 187–193.

—— & Ginzbourg, D. 1966. Sedimentary characteristics and climatic analysis of easterly dust storms in the Negev (Israel). *Sedimentology* **6**, 315–332.

K. Pye, Department of Earth Sciences, University of Cambridge, Downing Street, Cambridge CB2 3EQ, UK.

H. Tsoar, Department of Geography, Ben-Gurion University of the Negev, Beer Sheva, Israel.

Rates and modes of dust accretion and deposition in an arid region—the Negev, Israel

R. Gerson & R. Amit

SUMMARY: Rates of dust accretion and deposition are dependent on the amount of available dust and the trap efficiency of a particular site. Several types of dust-trapping terrains are widespread in deserts: (1) Gravelly (Serir) surfaces that turn with time into Reg soils; (2) vegetated surfaces in the desert fringe that may turn into löessial terrains; (3) stabilized sand dunes; (4) playa surfaces. Loessial terrains exhibit a high rate of dust accretion during the late Pleistocene—0.07–0.15 mm a^{-1} on the interfluves and ≤ 0.5 mm a^{-1} along the flood plains. Gravelly surfaces usually trap about 0.1 mm a^{-1} of dust initially but the rates decrease to several μm a^{-1} due to plugging with dust and salts and may ultimately remain constant as a gravel-free B horizon develops. The amounts of imported dust, from both local and distant sources, have changed during the Quaternary due to climatic fluctuations. Roofless ancient building—most efficient dust traps—show that although large amounts of dust were available (much of it from local sources) during the late Holocene, there was not intensive dust accretion during this period due to increasing aridity and decreasing trap efficiency. Wash and gullying led to destruction of the once widespread efficient trapping terrains.

Dust as defined here is particulate material of fine-grained size—including silt and clay, and sometimes very fine sand—which may be carried long distances in an airborne fashion. Dust both originates in, and is being deposited and trapped for long periods of time in deserts. Desert terrains can be subdivided into two types:

(1) Terrains which serve mostly as sources of dust. These are areas where active chemical and mechanical weathering produces dust-sized particles (Goudie *et al.* 1979; Pye & Sperling 1983). The dust produced by weathering is mobilized and transported by wash to lowland areas, such as flood plains and playas, from which it is transported by wind and spread over vast areas (Yaalon & Dan 1974; Morales 1979; Péwé 1981; Goudie 1983).

(2) Terrains which act as 'dust traps'. Here more dust is deposited than is mobilized and carried away; thus, dust accretion is the dominant geomorphic process. Soils develop mainly through the addition of airborne dust (Yaalon & Ganor 1973; Gile *et al.* 1981). Dust is trapped for long periods of time in gravelly terrains, in playas and in vegetated loessial terrains along the desert fringe.

The present article deals mainly with the latter type of terrain, where dust accretion is a prevailing process. The objective of the paper is to assess the rates of accretion in the various dust sinks and relate these rates both to the operative mechanisms and to the palaeoenvironments existing during the periods of dust deposition.

The Negev—environmental setting

The Negev is currently under an extremely arid to moderately arid–dry semi-arid climate (Fig. 1). Rainfall occurs during the winter and the mean annual rainfall ranges from 30 mm in the S to 250–300 mm in the NW (*Atlas of Israel* 1985). Hot summers and warm to cool winters characterize the region (*Atlas of Israel* 1985) as a part of a wider latitudinal belt of similar climate (UNESCO 1977).

The relief consists of hills, mountains and plateaux underlain mostly by limestones, dolomites, chalks, marls and shales. Small areas are underlain by igneous and metamorphic crystalline rocks, sandstones and chert. Neighbouring areas, such as the Sinai, southern Jordan, the Arabian Desert and the eastern Sahara make up a vast area of similar environments.

The vegetation is sparse, restricted mostly to depressions and runnels fed by runoff, to wadi beds, and to soil pockets and colluvium at the foot of rock outcrops (Evenari *et al.* 1971; Danin *et al.* 1975; Danin 1983). Scattered annuals and bushes characterize most of these areas while bushes and trees are found along the wadi beds. The moderately arid to dry semi-arid plains in the northwestern Negev, characterized by grass and shrub, have a denser vegetation. Scattered trees may also be found. However, this area is currently under various kinds of cultivation—dry farming or grazing—and its natural state has been greatly disturbed by human activities.

From Frostick, L. & Reid, I. (eds), 1987, *Desert Sediments: Ancient and Modern*, Geological Society Special Publication No. 35, pp. 157–169.

The soils are typical of those that form under an aridic climate (Fig. 1). Various loess-rich soils cover c. 5000–6000 km^2, dune sand and sandy soils cover c. 1600 km^2 and gravelly soils in the lowlands cover c. 2300 km^2. Rock outcrops and desert lithosols occur typically in the hills and mountains.

Dust accretion in gravelly (Reg) soils

Coarse gravelly alluvium serves as an efficient trap for atmospheric dust. Most commonly, the original coarse gravelly alluvium does not contain much dust-sized material. The fine-soil fraction may reach 2% to 5%, but the silt content is usually

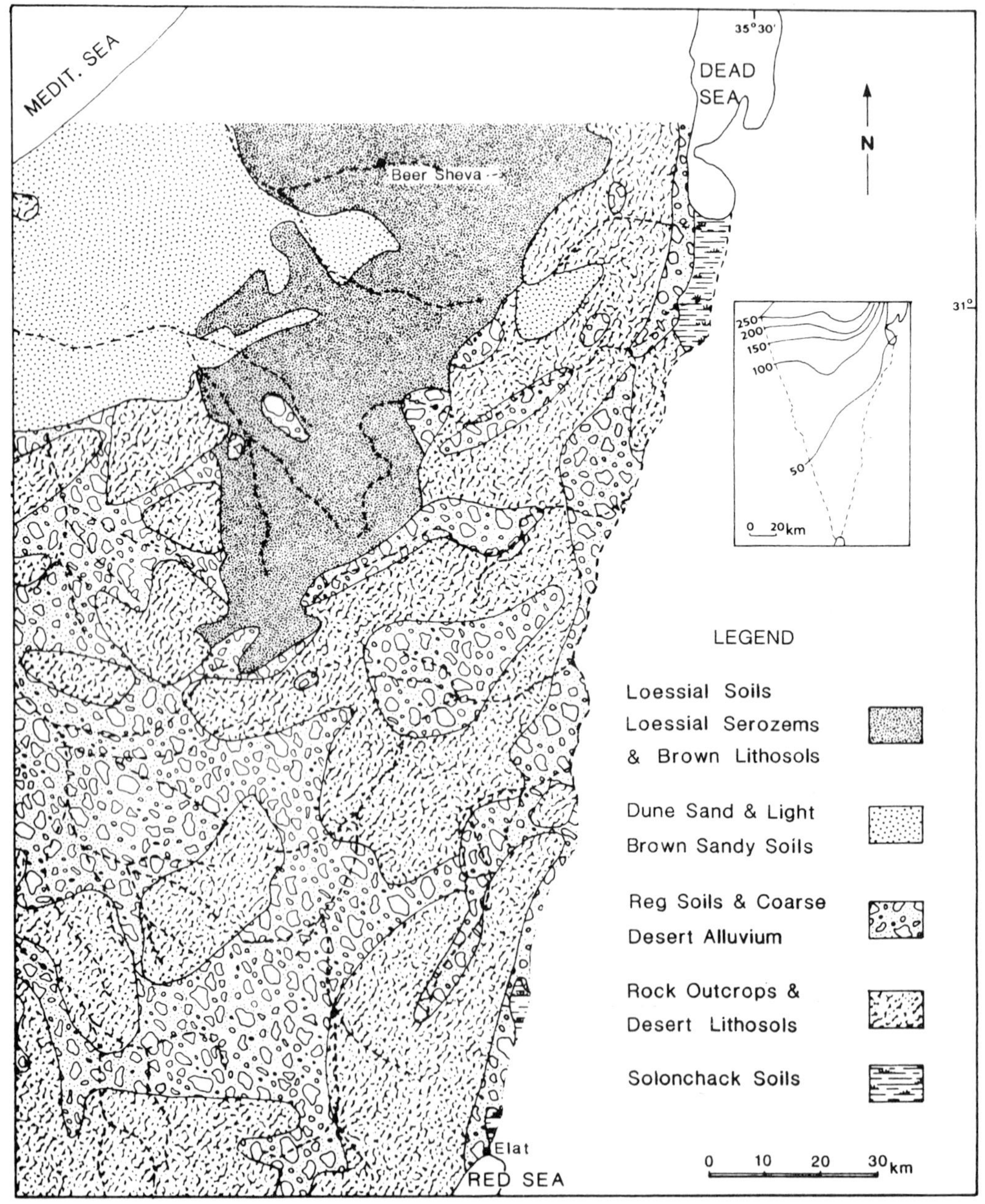

FIG. 1. The Negev soil associations and 30-year (1950–80) mean annual precipitation, mm (inset).

less than 1%. No clay is to be found (Gerson *et al.* 1985). The high effectiveness of dust trapping is due to two major factors: (1) There is a surficial roughness, consisting of gravel-bar and swale topography and a micromorphology formed by the exposed gravel (Fig. 2*a*); the high friction and the resulting low boundary wind speed favours dust settlement; the same factor inhibits effective surficial wash and, consequently, water movement across such coarse gravelly surfaces is very limited. (2) There is high initial porosity (20%–40%) and a high level of permeability. The pores are frequently large, ranging from 0.5 to 10 mm in diameter. Rates of rainfall infiltration of 60 to 80 mm hr^{-1} and higher characterize recent coarse gravelly surfaces (Greenbaum 1986). Run-off is not produced even by intense storms falling on very young terraces.

Reg soils on Holocene gravelly terraces in the Negev and the Sinai usually have the following characteristics: (1) The C horizon occupies most of the soil profile (Fig. 3*a*). The C horizon considered here is the gravelly layer slightly affected by pedogenic processes—*ie* some addition of fines and salts and the shattering of large clasts by mechanical weathering. It is 35 to 50 cm

FIG. 2. Properties of coarse gravelly alluvial surfaces: (*a*) characteristic surface of a late Holocene terrace; (*b*) developed desert pavement on a late Pleistocene surface.

FIG. 3. Profiles of Reg soils: (*a*) recent (< 1000 years old) gravelly alluvium; (*b*) mature Reg soil with a distinct gravel-free B horizon.

thick, very gravelly and the fine fraction makes up less than 10%. The fine fraction consists of sand, silt and clay. The average composition is: sand (60%) ≫ silt (35%) ≫ clay (5%). The size of the sand ranges from coarse to fine. (2) The B horizon is usually thin (≤5 cm thick) and gravelly. The average composition of the fine fraction is: silt (45%) > sand (40%) ≫ clay (15%). The sand is usually fine to very fine. (3) The A horizon is vesicular (A_v). It is thin (≤5 mm) and its average composition is: silt (50%) > sand (40%) ≫ clay (10%). The texture of the sand is fine to very fine. A gravel-bar and swale topography typifies the surface. Salinity is not very high, with 9% to 10% gypsum and 1% to 2% salts in the fine-soil fraction of the C horizon. Some vegetation may occur with annuals after rainstorms and some scattered shrubs.

Mature Reg soils on alluvial surfaces of Pleistocene age older than 50 000 years differ from the Holocene Reg soils in several respects (Fig. 3*b*):

(1) The soil profile is thicker, ranging from 80 to 160 cm.
(2) The soil horizons are well defined and thicker. For instance, the A horizon is 1 to 5 cm thick; the B horizon is 5 to 40 cm thick; and the C horizon is 70 to 120 cm thick.
(3) The B horizon is entirely or almost entirely gravel-free.
(4) The salinity of the C horizon and the lower parts of the B horizon is very high: the fine fraction contains 10% to 30% gypsum and 3% to 10% salts.
(5) 60% to 80% of the gravel is mechanically shattered by salts.
(6) The texture of the non-gravelly fraction is finer than in the Holocene Reg soils: in the A_v horizon the composition is typically silt (55%) > sand (30%) ≫ clay (15%); while in the B horizon it is silt (50%–60%) ≫ sand (20%–25%) ≥ clay (15%–25%); and in the C horizon it is sand (60%) ≫ silt (30%) ≫ clay (10%). The texture of the sand in the B and A_v horizons is fine to very fine.
(7) The ground surface usually consists of a well-developed desert pavement that covers >85% of the surface and is composed of small stones, most of which are 1 to 5 cm in size (Fig. 2*b*). Bar and swale topography is not discernible. The surface is barren of vegetation.

The dating of the Reg soils for use in the derivation of the rates of dust accretion is based on the studies of alluvial terraces along the Aqev-Zin Valleys in the central Negev (Goldberg 1976; Goldberg & Brimer 1983) and in the Dead Sea area, where alluvial surfaces are related to the levels of Lake Lisan (Begin *et al.* 1985; Amit & Gerson 1986). Relative age dating methods were used in areas where radiometric or archaeological ages were not available. The relative age dating methods included use of the degree of soil development, the preservation of bar and swale morphology, the degree of mechanical weathering and desert pavement evolution, and the degree of chemical weathering of surficial rocks. All were calibrated in areas where radiometric dates are available and where comparison can be made with the building stones of dated archaeological sites.

Dust penetration into a gravelly mantle is fast in the initial stages. Some dust in a dry state may penetrate to a depth of several centimetres or more in newly exposed porous gravel (Fig. 4*A*). However, during a period of 500–1000 years, there develops a thin loessial (and sometimes biological—algae and lichen) crust on exposed fines (Fig. 4*B*). With time, soil horizons form. A thin, continuous vesicular A horizon takes about 1000 to 5000 years to develop, while a silt–loam cambic B horizon is discernible after several thousands of years (Fig. 4*C*). The formation of patchy desert pavement takes some 5000 to 10 000 years. Only after several 10^4 years is there a smooth continuous desert pavement with a 1 to 5 cm thick A_v horizon. A gravel-free B horizon occurs in soils older than 50 000 years (Fig. 4*D*). During the formation of this horizon there is an abundant accumulation of gypsum and salts.

All these processes involve the accretion of extraneous dust-sized material, including silt, clay, very fine sand and salt crystals. The surficial crust, the cambic and gravel-free B horizons and the salts are composed mostly of materials from atmospheric sources (Gerson *et al.* 1985; Amit & Gerson 1986). Examination of large clasts in the soil profile shows that there is no appreciable contribution of fine soil particles from the weathering of gravel. The fines modify the rates of water infiltration and dust penetration by changing the texture, density and arrangement of the soil. There is also a concomitant change in the mechanical properties of the surface. The evolution of a smooth desert pavement of low roughness decreases boundary friction, and hence dust settlement and entrapment. It also leads to an increase in surficial runoff and wash erosion.

Fig. 5 presents the average changes in the content of dust in the soil profiles and in the rate of dust accretion with time. Several trends may be distinguished. There is an increase in the content of dust in the soil profile with time. This increase of dust content approximates a power function. Also, the average rate of dust accretion

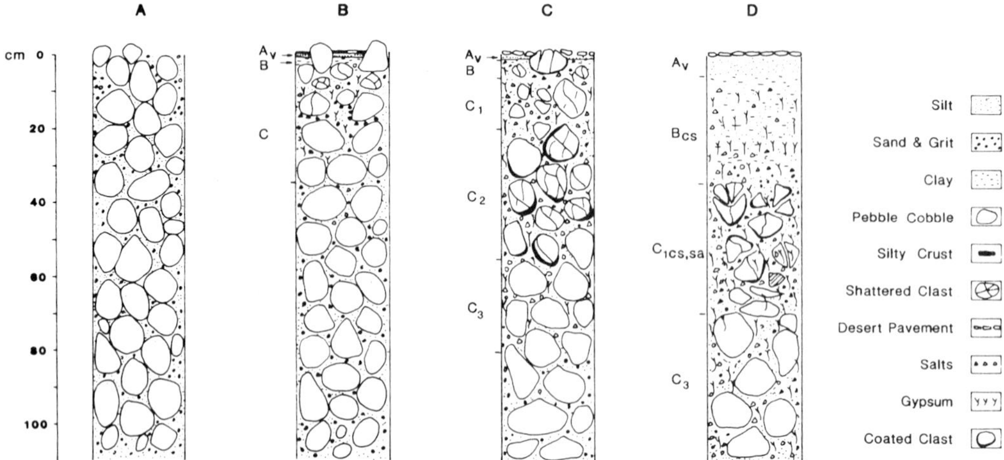

FIG. 4. Stages in the development of a Reg soil profile with accumulation of dust and salts: (*A*) recently deposited coarse gravelly alluvium (< 1000 years old); (*B*) late Holocene Reg soil (2000–4000 years old); (*C*) early Holocene Reg soil (10 000–14 000 years old); (*D*) Middle to late Pleistocene Reg soil (> 100 000 years old; *cf* Fig. 3*b*).

decreases with time from between 0.02 and 0.05 mm a^{-1} at the beginning of soil evolution to between 0.001 and 0.003 mm a^{-1} *c*. 100 000 years. The change from whole soil-profile penetration of dust to accretion in a gravel-free B horizon is brought about by low and constant rates of dust addition to the profile. This is related to the plugging of the C and B horizons with dust and salts and possibly to the development of the smooth desert pavement armour.

The filtering (*ie* sieving) effect of the dust as it accumulates in the soil profile is indicated by the tendency of the soil texture to become finer with time; the ratio of coarse to fine sand in the soil matrix decreases as the soil becomes older and dust content increases.

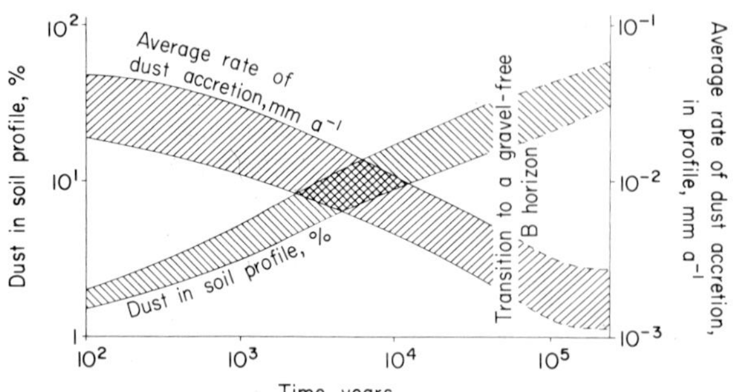

FIG. 5. Dust content and the rates of dust accretion with time in Reg soils.

Dust accretion in loessial terrains

The northern and northwestern Negev has been subjected to favourable conditions for dust accretion for a long period of time for several reasons:

(1) The region is located at the desert fringe downwind of vast source areas of dust, including the deserts of the Negev and the Sinai, N Africa and Arabia (Yaalon & Dan 1974).

(2) The area is moister than the deserts to the SW, S and SE of it. Both humidity of the air and rainfall change significantly as one passes from the desert to this less arid area. For example, rainfall changes from less than 100 mm a^{-1} to more than 200 mm a^{-1} over distances of a few tens of kilometres and the average relative humidity of the air changes from less than 45% in the deserts to more than 55% in the northwestern Negev. Both factors encourage the deposition of airborne dust.

(3) The moderately arid to semi-arid climate encourages the growth of relatively dense grass and shrub vegetation, which in turn moderates boundary wind speeds and constitutes a most efficient dust trap.

This combination of factors has made the region one in which loess deposition and accumulation are a major geological and pedological process. It is important to note that the aridic and semi-aridic soil-forming processes have not changed the general nature of the original airborne dust significantly. Some translocation of clays, some weathering and some precipitation of pedogenic carbonate and gypsum have occurred (Bruins 1976; Eisenberg 1980), but the general properties have not changed. The soil texture is generally similar to the original wind-deposited parent material (Gerson *et al.* 1985).

The northern and northwestern Negev are mantled by a continuous cover of reworked loess between 1 and 15 m thick in the Be'er Sheva Basin and in the southern coastal plain. Hills and flood plains alike are veneered by the loess cover. Usually, north-facing hillslopes have a thicker löess mantle than S-facing ones, due to higher humidity and denser vegetation on the former (Dan 1966; Dan *et al.* 1981; Gerson *et al.* 1985). S of this area, currently under a desert climate, there are small to large patches of loessial soils on high plateaux, at the base of rocky hill slopes, and along the flood plains of most stream channels. Such patches are remnants of a past continuous cover which used to stretch S of the northwestern Negev. The area marked in Fig. 1 as 'Loessial Soils, Loessial Serozems & Brown Lithosols' indicates the extent of this past cover. Furthermore, there is a continuous loessial mantle underlying the present sand-dune terrain of the northwestern Negev. The textural composition of the loess and the loessial soils is usually silty clay loam [silt (60%) $\gg$ clay (30%) $\gg$ sand (10%)] to clay loam [clay (40%–50%) $\approx$ silt (40%–50%) $\gg$ sand (2%–10%)], respectively (Bruins 1976; Gerson *et al.* 1985).

The calculation of the average rates of dust accretion in the northern Negev is based on data from several sites:

(1) Loessial palaeosols in the Netivot region, WNW of Be'er Sheva (Bruins 1976), and in the Ramat Hovav area, some 10 km S of Be'er Sheva (Enzel 1984).
(2) Loessial deposits along the Nahal (wadi) Besor flood plain, S and W of Be'er Sheva (Fig. 6), and in the Sede Zin plain, some 35 km S of Be'er Sheva.

The first set of sites are a product of loess accumulation whereas, the last includes areas of reworked loess.

The late Pleistocene was a major period of loess deposition over the Negev. However, certain areas, such as the Sede Zin Plain, and possibly the Besor Valley, had also experienced the deposition of reworked loess during the Middle Pleistocene.

The long-term average rates of loess accretion during the late Pleistocene in the Be'er Sheva region were 0.07 to 0.15 mm a^{-1}. It is likely that there were considerable fluctuations in accretion rates during this long period. However, the accretion was uninterrupted.

Extensive fluviatile reworking of loess occurred at the same time as there was aeolian accretion over the interfluves and the hill slopes. The dust that was not trapped on the drainage divides and slopes was transported and deposited in the broad flood plains that have at times constituted some 10% to 30% of the area. The rates of deposition along some of the flood plains have been quite high. Along the lower Besor Valley, W of Be'er Sheva, some 10 to 15 m of mostly reworked loess (Sneh 1983) was deposited during a period of 30 000 to 40 000 years during the Late Palaeolithic period, *ie* from 50 000–40 000 to 12 000 years BP (dates from P. Goldberg and N. Goring-Morris 1985, pers. comm.). Average rates of accumulation of 0.2–0.5 mm a^{-1} have been calculated for the lower Nahal Besor flood plain. Along the upper Besor Valley, 8 to 10 m of reworked loess was deposited during the late Pleistocene. This area is more arid than the Be'er Sheva–Lower Besor Valley region to the N and NW. It was also more arid during the generally wetter late Pleistocene (Magaritz 1986). As a result, the rates of dust accretion on both interfluves and flood plains were lower. Rates of 0.2 to 0.3 mm a^{-1} have been calculated for the upper Besor flood plain, based on the dating of calcareous palaeosols and the thickness of the loessial deposits (Magaritz 1986, unpublished). In the Sede Zin plain some 4 to 6 m of fluvial silts have been deposited since the early Pleistocene. Average deposition rates of 0.01 to 0.06 mm a^{-1} have been calculated for this area.

Discussion

Two types of desert terrain have been assessed with respect to dust accretion:

(1) Terrains in which the dust is added to the surface of a sediment or soil profile. These are the loessial terrains, where airborne or reworked dust is accreted at the surface for long periods of time and only small amounts penetrate into subsurface layers. Vegetation is an efficient dust-trapping agent in this environment and much of the deposited dust adheres to the surface and remains there.

(2) Terrains such as *serir* and gravelly colluvia in which dust finds its way mainly to shallow subsurface horizons. Here the gravelly mantle is the dust trap. Dust from both direct airborne sources and local wash is translocated by illuvial processes into the coarse gravel. The properties of the soil change with time as the voids are

plugged with both fines and salts that are added to the matrix.

The trap efficiency of the two types of terrain is different even over long periods of dust accretion because of three factors: (1) Grass and bush vegetation in a moderately arid to semi-arid environment is far more efficient than a gravelly surface with no vegetation in moderating the wind field close to the surface, and so trapping dust and hindering further dust transportation. (2) Gravelly surfaces in arid to extremely arid environments are usually devoid of vegetation due to the climatic aridity and the initially low water-holding capacity. Once dust has been trapped in the gravelly mantle, salinity often prevents growth of vegetation. (3) The weathering of surficial gravel and the formation of a smooth desert pavement composed of small stones leads to decreasing surface roughness and less deposition of dust, to lower rates of water infiltration and dust penetration, and to more effective overland flow and wash erosion. All these lead to a marked decrease in dust penetration with time.

A compilation of the average rates of dust accretion in various arid regions is presented in Fig. 6. The highest rates of dust accretion under natural conditions are found in loessial terrains. Both the northern Negev and the much larger loess region in north-central China have experienced high rates of aeolian dust accretion over long periods of time—0.07 to 0.15 mm a^{-1} are average rates for the late Pleistocene. In both areas the loess deposition was uninterrupted; there are no major disconformities in the loessial sections. However, the presence of palaeosols with different degrees of development indicates varying degrees of loess accretion under fluctuating climatic regimes.

The data from the northern Negev indicate that during the period between 50 000 to 40 000 and 12 000 years BP the rates of dust importation were especially high. In addition to aeolian accretion along the drainage divides, there was extensive deposition along the flood plains. The streams were not carrying any appreciable amounts of gravel and reworked loess was being deposited at a rate of several hundred micrometres per year. Some of the loess made its way

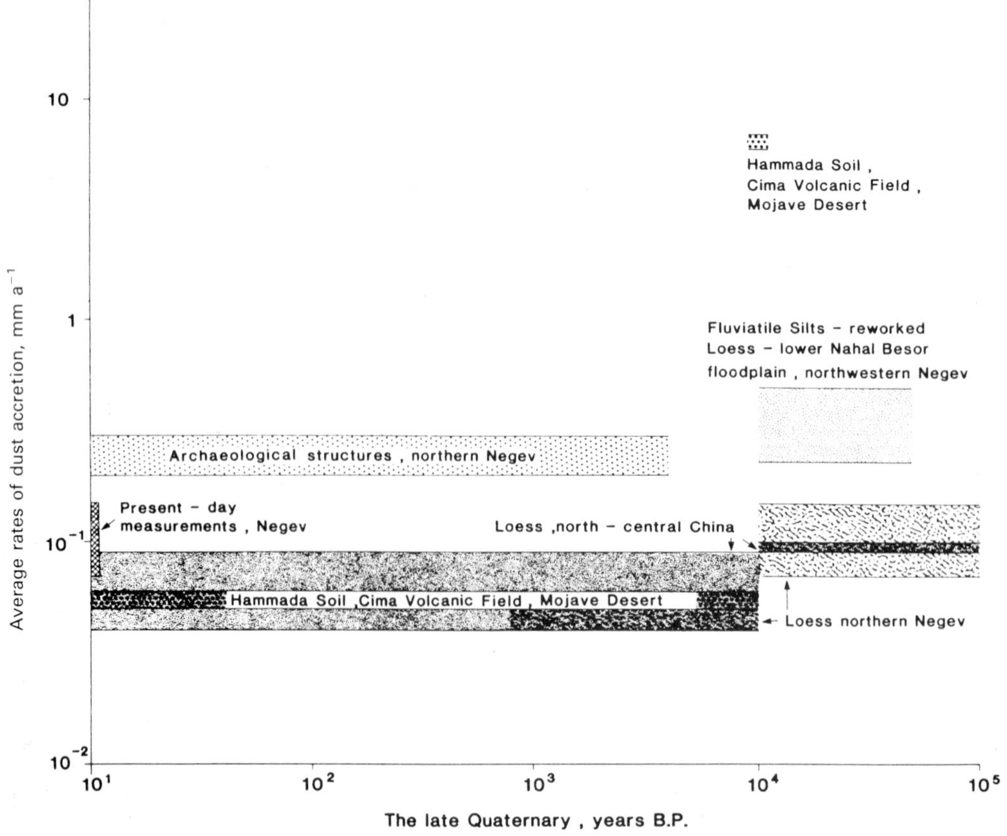

FIG. 6. Average rates of dust accretion in arid regions and related areas during the late Quaternary.

to the Mediterranean Sea and is not accounted for in the northern Negev. Hence, for the period 50 000 to 12 000 years BP the rate of dust accretion, averaged over the vast area of the northern Negev, may have been between 0.2 to 0.4 mm a^{-1}. Trap efficiency for that period must have been very high. A dense cover of vegetation, composed of grass, bushes and trees dominated the landscape. Indications (including palynological evidence) of climates wetter than the present during the late Quaternary are numerous (Begin *et al.* 1974; Horowitz 1979; Gat & Magaritz 1980; Goldberg 1981; Gerson 1982; Magaritz 1986).

Dust accretion declined during the Holocene. No appreciable reworked loess of this period is found in the flood plains and additions to the loess mantle on the interfluves is thin. Intense surface wash and gullying have been the prevailing processes. The effects of man may also have been important during the last 4000 years.

In the Reg soils a clear trend of decreasing rates of dust accretion with time is demonstrated. The rates presented in Fig. 5 are based on the amounts of dust in the coarse gravelly alluvium and the approximate age of the alluvial surface on which each soil has developed. Rates of 0.02 to 0.05 mm a^{-1} are implied for the Holocene, and the rates decrease as the soil becomes older—several micrometres per year appear to be characteristic of an old Reg soil. However, as shown for the loessial soils, the importation of dust to the region may have varied greatly during the Quaternary. Reg soils, with their low trap efficiency, may have trapped comparatively large quantities of dust if their initial development took place during periods of high rates of airborne dust importation and a relatively wet climate. However, our studies of alluvial surfaces of different age are unable to distinguish such occurrences in the gravelly plains in the Negev (Gerson *et al.* 1985). Dust accretion in these soils must have been a slow and declining process. The low trap efficiency and the rapidly changing soil properties undoubtedly account for this situation (Amit & Gerson 1986).

The relatively high rates of dust accretion in the basaltic hammada surfaces of the Cima Volcanic Field, Mojave Desert, California (Wells *et al.* 1984) deserve attention. Here a gravel-free horizon of about 1 m has developed during a period of less than 20 000 years. The rates of sub-pavement redistribution of material must have been rather high. The area is located downwind of a dust-rich dried lake or playa and the receiving surface is very flat and rough. These conditions, together with the area having a more humid climate than the latest Pleistocene–Holocene coarse alluvial surfaces in the Negev, may explain the different rates between the two regions. The rates suggested by Wells *et al.* (1984) for a brief period around 10 000 years BP (Fig. 6, Cima Volcanic Field) appear to be extremely high. At the beginning, when the lava-flow surface was very rough and highly porous, large rates of dust trapping must have occurred. However, as the hammada soil with its gravel-free horizon developed, so the rate of dust accretion must have decreased appreciably.

The behaviour of both the loessial terrains and the gravelly plains during the Holocene may be further illustrated by an examination of dust accretion at archaeological sites. Some 50 ancient buildings which date back 1200–5000 years have been studied in the Negev. Several trends have been observed:

(1) Roofless round and rectangular buildings have trapped as much as 1.0 m of airborne dust if the walls were originally or have remained at least 1.2 to 1.5 m above the floor. Most structures have captured 0.5 to 0.8 m of dust (Fig. 7*a, b*).
(2) Based on stratigraphical and chronological evidence, it took 1000–2000 years for the accumulation of these thicknesses of dust.
(3) As the buildings fill with dust, their trap efficiency decreases due to the decrease in their effective depth and their roughness. As a rate less dust is trapped in them with time.
(4) No vegetation interferes with the process when the rates of dust accretion are at their highest.

The rate of dust accretion in archaeological structures may be assessed in two ways: either as a rapidly decreasing rate as the buildings fill up, or as an average rate of accretion since the building was abandoned and became roofless. High rates were calculated using both methods—0.8 to 0.3 mm a^{-1} by the former, and 0.2 to 0.3 mm a^{-1} based on the latter (Fig. 6). It is obvious that accumulations in man-made structures do not represent the amounts and the rates of dust accretion in natural environments. However, the quantities trapped in these structures may serve at least as an approximation of the amount of dust that has passed across a given site at 1 to 2 m above the ground, whether the source of the dust is local or distant.

The most efficient natural traps might have accumulated similar amounts of dust during the late Holocene; but none have done so. Very little dust has been trapped in recent times in natural receptacles. The major reason for this situation must be the low trap efficiency of most natural terrains. In fact, the increasing presence of sand in the northwestern Negev in the last 15–25 000

FIG. 7. Dust accretion in ancient buildings: (*a*) A round structure in Be'er Ressisim, Western Negev, abandoned *c*. 4000 years ago; (*b*) dust fill in a building in Tel Arad, northeastern Negev, abandoned *c*. 4500 years ago. Most of the fill is dust, with some gravel.

years (N. Goring-Morris 1985, pers. comm.) may have had a significant effect. The abrasion of loessial crusts by the impact of saltating sand grains could have destroyed the protective surface and helped initiate erosion through gullying and dust storms.

The representativeness of measurements of dust deposition in modern settings is debatable. Questions such as: how much of the dust measured is derived from natural sources? how efficient are the measuring devices? and how much of the measured dust would be deposited and added to the soil? are still unanswered. Fig. 6 also presents results from Ganor (1975) and Yaalon & Ganor (1975). Yaalon (1986, pers. comm.) maintains that about half of the measured amount settles and is added to the natural soil (about 0.04 to 0.08 mm a^{-1}). Such rates—about half those of the late Pleistocene—seem to be too high considering the overall poor development of the Holocene soils and the intense wash erosion and gullying characteristic of the late Holocene.

Conclusions

It is evident that the best combination of conditions for the occurrence of high dust accretion is large amounts of dust importation and high trap efficiency. Such conditions prevailed during much of the late Pleistocene in the highlands of the central and northern Negev as well as in the lowlands of the northern Negev, the northwestern Negev and northeastern Sinai. An area of about 6000 km^2 was covered by an almost continuous loess mantle under a moderately arid to semi-arid climate. The gravelly plains to the S and E of the Negev highlands developed Reg soils, which initially incorporated large amounts of dust. In total, an estimated 5–15×10^9 m^3 of dust was deposited during the late Pleistocene in the Negev. The composition of the dust (30% to 50% quartz and feldspar) implies that most of it was derived from distant, allochthonous sources because the Negev is underlain mostly by carbonate rocks. The Sinai,

the Sahara, the Arabian Desert, the previously exposed Mediterranean shelf and the Gulf of Suez were the major source areas.

The favourable conditions changed during the latest Pleistocene and Holocene. Less dust from distant sources has been (and is being) trapped in the Negev region. On the other hand, large amounts of dust from distant as well as local sources were (and still are) available to the region. In fact, the high rates of dust deposition in the Atlantic Ocean (Dubief 1979; Jaenicke 1979; Péwé 1981; Schütz *et al.* 1981) as well as in the archaeological structures of the Negev bear witness to this availability. However, increasing aridity has caused a deterioration of vegetation, and a decline in the trap efficiency of both the Negev highlands and the north-northwestern Negev. Wash erosion and gullying have become the prevailing processes. Only coarse gravelly surfaces and playas still act as efficient dust traps.

Rates of dust accretion in the Negev have changed greatly and vary according to the conditions of the terrain. These rates are similar in some cases to those calculated for the extensive loessial terrains in north-central China, that derived dust mostly from the Gobi Desert (Fig. 6; Liu 1985; Sadao & Wang 1984). However, the rates calculated for both the Negev and China are far less than those found for past proglacial regions such as the Great Plains and the Mississippi Valley in the United States. There, late Wisconsin loess of up to 20 m thick accumulated during a period of 4000 to 12 000 years. Rates were between 0.1 to 4 mm a^{-1} during that period; the highest rates were experienced close to densely vegetated flood plains, and the lower ones at distances greater than 30 km from them (Leonard & Frye 1954; Ruhe 1954, 1983; Ruhe *et al.* 1967; Frazee *et al.* 1970; Kleiss & Fehrenbacher 1973).

ACKNOWLEDGMENTS: The authors would like to thank Michael Kidron for drawing the text figures and D. H. Yaalon and three anonymous referees for their constructive comments.

References

1985. *Atlas of Israel.* Survey of Israel, Tel Aviv, 3rd edition.

AMIT, R. & GERSON, R. 1986. The evolution of Holocene Reg (Gravelly) soils in deserts—an example from the Dead Sea region. *Catena* **13**, 59–79.

BEGIN, Z. B., BROECKER, W., BUCHBINDER, B., DRUCKMAN, Y., KAUFMAN, A., MAGARITZ, M. & NEEV, A. 1985. Dead Sea and Lake Lisan levels in the last 30 000 years. *Geological Survey of Israel Preliminary Report*, 18 pp.

——, EHRLICH, A. & NATHAN, Y. 1974. Lake Lisan, the Pleistocene precursor of the Dead Sea. *Israel Geological Survey Bulletin* **63**, 30 pp.

BRUINS, H. J. 1976. *The origin, nature and stratigraphy of paleosols in loessial deposits of the NW Negev, Netivot, Israel.* MSc thesis, The Hebrew University, Jerusalem, 155 pp.

DAN, J. 1966. *The effect of relief on soil formation and distribution in Israel.* PhD thesis (in Hebrew), The Hebrew University, Jerusalem, 361 pp.

——, GERSON, R., KOYUMDJISKY, H. & YAALON, D. H. 1981. *Aridic soils of Israel, properties, genesis and management.* Division of Scientific Publications, The Volcani Center, Israel 190, 353 pp.

DANIN, A. 1983. *Desert vegetation of Israel and Sinai.* Cana Publishing House, Jerusalem, 148 pp.

——, ORSHAN, G. & ZOHARI, M. 1975. The vegetation of the northern Negev and the Judean Desert of Israel. *Israel Journal of Botany* **24**, 118–172.

DUBIEF, J. 1979. Review of the North African climate with particular emphasis on the production of aeolian dust in the Sahara. *In*: MORALES, C. (ed.) *Saharan dust.* John Wiley and Sons, Chichester, 27–48.

EISENBERG, J. 1980. *The effect of parent material, exposure and relief on soil and vegetative characteristics in the Be'eri badlands of the Northern Negev.* MA thesis (in Hebrew), Tel Aviv University, Ramat Aviv.

ENZEL, Y. 1984. *The geomorphology of lower basin of Nahal Sekher.* MSc thesis (in Hebrew), The Hebrew University, Jerusalem, 106 pp.

EVENARI, M., SHANNAN, L. & TADMOR, N. H. 1971. *The Negev: The challenge of a desert.* Harvard University Press, Cambridge, Massachusetts, 344 pp.

FRAZEE, C. J., FEHRENBACHER, J. B. & KRUMBEIN, W. C. 1970. Loess distribution from a source. *Soil Science Society of America Proceedings* **34**, 296–301.

GANOR, E. 1975. *Atomspheric dust in Israel—sedimentological and meteorological analysis of dust deposition.* PhD thesis (in Hebrew), The Hebrew University, Jerusalem, 224 pp.

GAT, J. R. & MAGARITZ, M. 1980. Climatic variation in the eastern Mediterranean Sea area. *Naturwissenchaften* **67**, 80–87.

GERSON R. 1982. The Middle East: landforms of a planetary desert through environmental changes. *Striae* **17**, 52–78.

——, AMIT, R. & GROSSMAN, S. 1985. *Dust availability in desert terrains. A study in the deserts of Israel and the Sinai.* Institute of Earth Sciences, The Hebrew University of Jerusalem, Israel, 220 pp.

GILE, L. H., HAWLEY, J. W. & GROSSMAN, R. B. 1981. Soils and geomorphology in the basin and range area of southern New Mexico—guidebook to the desert project. *New Mexico Bureau of Mines and Mineral Resources Memoir* **39**, 222 pp.

GOLDBERG, P. 1976. Upper Pleistocene geology of the Avdat/Aqev area. *In*: MARKS, A. E. (ed.) *Prehistory and Paleoenvironments in Central Negev, Israel, Vol. I.* Southern Methodist University Press, Dallas, 25–51.

—— 1981. Late Quaternary stratigraphy of Israel: an eclectic view. *C.N.R.S. Colloque No. 598: Prehistoire du Levant*, Lyons, 1980, 55–66.

—— & BRIMER, B. 1983. Late Pleistocene geomorphic surfaces and environmental history of Avdat/Havarim area, Nahal Zin. *In*: MARKS, A. E. (ed.) *Prehistory and Paleoenvironments in the central Negev, Israel, Vol. III.* Southern Methodist University Press, Dallas, 1–13.

GOUDIE, A. S. 1983. Dust storms in space and time. *Progress in Physical Geography* **7**, 502–530.

——, COOKE, R. U. & DOORNKAMP, J. C. 1979. The formation of silt from quartz dune sand by salt weathering processes in deserts. *Journal of Arid Environments* **2**, 105–112.

GREENBAUM, N. 1986. *Infiltration losses and their effect on runoff production in an extremely arid climate.* MSc thesis (in Hebrew), The Hebrew University, Jerusalem.

HOROWITZ, A. 1979. *The Quaternary of Israel.* Academic Press, New York, 394 pp.

JAENICKE, R. 1979. Monitoring and critical review of the estimated source strength of mineral dust from the Sahara. *In*: MORALES, C. (ed.) *Saharan dust.* John Wiley and Sons, Chichester, 233–242.

KLEISS, H. J. & FEHRENBACHER, J. B. 1973. Löess distribution as revealed by mineral variations. *Soil Science Society of America Proceedings* **37**, 291–295.

LEONARD, A. B. & FRYE, J. C. 1954. Ecological conditions accompanying loess deposition in the Great Plains region of the U.S. *Journal of Geology* **62**, 399–404.

LIU, T. 1985. *Loess and the environment.* China Ocean Press, Beijing, 251 pp.

MAGARITZ, M. 1986. Environmental changes recorded in the upper Pleistocence along the desert boundary, southern Israel. *Palaeogeography, Palaeoclimatology, Palaeoecology* **53**, 213–229.

MORALES, C. (ed.) 1979. *Saharan dust—mobilization, transport, deposition.* John Wiley & Sons, Chichester, 316 pp.

PÉWÉ, T. L. 1981. Desert dust: an overview. *In*: PÉWÉ, T. L. (ed.) *Desert dust: origin, characteristics and effect on man.* Geological Society of America, Special Paper **186**, 1–10.

PYE, K. & SPERLING, C. H. B. 1983. Experimental investigation of silt formation by static breakage processes: the effect of temperature, moisture and salt on quartz dune sand and granitic regolith. *Sedimentology* **30**, 49–62.

RUHE, R. V. 1954. Relations of the properties of Wisconsin loess to topography in western Iowa. *American Journal of Science* **252**, 663–672.

—— 1983. Depositional environment of Late Wisconsin löess in the Midcontinental United States. *In*: WRIGHT, H. E. JR (ed.) *Late Quaternary Environments of the United States Vol. 1: The late Pleistocene*, 130–138. University of Minnesota Press, Minneapolis.

——, RAYMOND, B. D. & CADY, J. G. 1967. Landscape evolution and soil formation in southwestern Iowa. *Soil Conservation Service, U.S. Department of Agriculture, Technical Bulletin* **1349**, 241 pp.

SADAO, S. & WANG, Y. (eds) 1984. *The recent research of loess in China.* Kyoto University and Northwest University, 243 pp.

SCHÜTZ, L., JAENICKE, R. & PIETREK, H. 1981. Saharan dust transport over the North Atlantic Ocean. *Geological Society of America Special Paper* **186**, 87–100.

SNEH, A. 1983. Redeposited loess from the Quaternary Besor Basin, Israel. *Israel Journal of Earth Sciences* **32**, 63–69.

UNESCO 1977. Map of the world distribution of arid regions with explanatory note. *UNESCO M.A.B Technical Notes* **7**, Paris.

WELLS, S. G., MCFADDEN, L. D. & MAHRER, K. D. 1984. Types and rates of late Cenozoic geomorphic processes on lava flows of the Cima Volcanic Field, Mojave Desert. *1984 Annual Meeting of the Geological Society of America, Reno.* Western Geological Excursions **1**, 191–208.

YAALON, D. H. & DAN, J. 1974. Accumulation and distribution of loess-derived deposits in the semiarid and desert fringe of Israel. *Zeitschrift für Geomorphologie* **29**, 91–105.

—— & GANOR, E. 1973. The influence of dust on soils during the Quaternary. *Soil Science* **116**, 146–155.

—— & —— 1975. Rates of aeolian dust accretion in the Mediterranean and desert fringe of Israel. *IXeme Congress Internationale Sedimentologie* 169–174.

R. GERSON & R. AMIT, Institute of Earth Sciences, The Hebrew University of Jerusalem, Jerusalem 91904, Israel.

Wind-blown materials and W African soils: an explanation of the 'ferrallitic soil over loose sandy sediments' profile

H. Vine

SUMMARY: It is suggested that incorporation of dust from the hinterland during episodes in the Pleistocene when the Sahara encroached far S of its present limits accounts for much of the clay and Fe_2O_3 content of soils of southern Nigeria. The amount of pelagic deposition of clay reported for DSDP Site 366 on the Sierra Leone Rise in the last 1–3 million years is shown to correspond to the anomalous accumulation of clay in deep uniform subsoils of the 'red ferrallitic soils over loose sandy sediments'. Particle-size analyses of soils at seven sites (showing 10% to 50% clay in the subsoil) are given, with determinations of Fe_2O_3 in the clay fraction (5% to 9%). Silt-size quartz in the dust is thought to have been lost through solution, and feldspars and clay minerals have weathered to kaolin. Wind systems displaced 600–800 km to the S compared with the situation today would have deposited dust similar to present-day Harmattan fall-out in the N of Nigeria, in the latitude of southern Nigeria and of the Sierra Leone Rise. Creep, mass-movement, faunal activity and eluviation of clay and Fe_2O_3 have affected the incorporation of dust, in addition to mixing coarser and finer layers of sediments together.

A significant factor in the formation of lateritic ironstone on old peneplain surfaces in W and E Africa may have been deposition of clayey and ferruginous dust (Vine 1949). The addition of large amounts of such dust possibly contributed to the formation of the 'Acid Sands' soils of southern Nigeria (Vine 1956)—soils included in the more general classes of Red and Yellowish-Brown Ferrallitic Soils on Loose Sandy Sediments of the CCTA *Soil Map of Africa* (D'Hoore 1964). These ideas arose from consideration of the very sandy soils, termed the 'Northern Drift' type by Doyne *et al.* (1938), which prevail in a wide zone in the N of Nigeria and in adjoining territory. The Northern Drift soils occur partly in the form of fixed dunes, developed during arid periods in the past; present-day mean annual rainfall in this zone is approximately 150 to 800 mm. The writer regarded the sand as the residue left after the fine fractions of soils had been blown away southwards, a process that would have been most vigorous during episodes when the Sahara was advancing over areas where there had been sufficiently humid conditions for the formation of red clayey soils.

A well-known feature of the interior of W Africa at the present day is the very dry and dusty Harmattan wind which blows by day from the NE during the period October to March (Dubief 1979, pp. 29–31). The main source area of the Harmattan dust and haze was considered by Hamilton & Archbold (1945) to be the desert region around Faya Largeau 700 km NE of Lake Chad which had a particularly 'light and dusty' soil. They showed that convective uplift and gusty surface winds in this area are caused by disturbances which penetrate across the desert from the Mediterranean, and that there are severe dust storms when the convection is most intense. Kalu (1979) developed this explanation. He regarded the whole of the 'alluvial plain of Bilma and Faya Largeau' as the source of the dust carried away southwestwards by winds at altitudes of about 600 to 1200 m. The dust forms a 'plume', from which there is fall-out in progressively decreasing amounts and with a decreasing mean particle size. McTainsh & Walker (1982) accounted for the replenishment of the alluvial deposits in the region of Faya Largeau, balancing the loss of material by wind action, as the result of fluvial transportation of material from the high-rainfall regions S of Lake Chad, with intermittent overspill from the present lake area to the Bodele Depression further N. In 1976–77 they recorded 13 dust plumes with an average duration of six days passing over northern Nigeria. Most of the heavy deposition of dust tended to occur between the source area and Kano.

Clay content of Harmattan dust fall-out

Dust deposited at distances of 1000 to 1700 km downwind from Faya Largeau has been found to contain large quantities of clay and iron oxide. Doyne *et al.* (1938) collected a sample at Samaru, Zaria (not at Ibadan in southern Nigeria as McTainsh & Walker supposed), and determined the particle-size distribution by a sedimentation

From FROSTICK, L. & REID, I. (eds), 1987, *Desert Sediments: Ancient and Modern*, Geological Society Special Publication No. 35, pp. 171–183.

method, after thorough disaggregation. The results were: 64.7% clay (<2 μm), 24.5% 'silt' (2–20 μm), 10.8% 'fine sand' (20–200 μm). The reported 'fine sand' was mostly coarse silt (20–50 μm). There was about 10% $CaCO_3$, mainly in the 'silt' size fraction. The clay fraction was analysed for SiO_2, Al_2O_3 and Fe_2O_3 (see Table 4 below).

McTainsh & Walker (1982) carried out particle-size analyses of 39 carefully collected samples of dust fall-out, thoroughly disaggregated. The results are given on a weight basis. The amount of clay fraction (<2 μm)was found to range from 2.3% to 32%; in 31 of the 39 samples it was between 10% and 30%. In four samples collected at Kano during periods of poor visibility the clay fraction ranged from 21% to 28% (average 26%), and in four collected when the visibility was better the range was 7% to 15% (average 11%). In these eight Kano samples fine sand (50–200 μm) was 4% to 32% (average 10%), coarse silt (20–50 μm) was 28% to 42% (average 35%) and fine silt (2–20 μm) was 20% to 38% (average 30%).

X-ray diffraction analysis of the clay fraction from some of McTainsh & Walker's samples collected at Kano showed about 60% kaolinite, 10% illite, and 25% smectite, with a small percentage of quartz. In the whole sediment, quartz was dominant, with varying amounts of feldspar and calcite as well as the clay minerals. The percentage of clay size material in fall-out samples collected in the same way at other places was:

Maiduguri	12%
Jos	16%
Zaria	24% (two samples)
Sokoto	28%

Wilke *et al.* (1984) analysed the minerals of the clay fractions in fall-out collected at four stations during the 1978–79 Harmattan season, with the following results:

Maiduguri	Kaolinite (60%), Illite
Jos	Kaolinite (40%), Illite
Kano	Kaolinite (40%), Mixed-layer (Illite–Smectite) (30%), Illite
Sokoto	Kaolinite (40%), Mixed-layer (Illite–Smectite) (30%), Illite.

The very fine dust-haze material

Dubief (1979) described 'dense dry haze' as a layer which persists over W Africa downwind of the stations where samples of fall-out were collected for the studies detailed above. After a particular dust storm at Bilma in early March he showed how the front of the dust haze progressed, taking 3½ to 4 days to reach the coast between Lagos and Monrovia.

The dust present in the atmosphere at Ibadan during the 1955–56 dry season was investigated by McKeown (1958) by trapping air in a settlement counter and examining the deposit under a microscope at times corresponding to settlement of certain sizes of particle (*eg* 53 min for 1.0 μm, 5 h for 0.3 μm, 97 h for 0.06 μm). Air was sampled at 1600 local time, when the lower layers of the atmosphere were well mixed by turbulence. Size analyses were carried out on eight occasions. Six samples showed about 80% of the particles in the 0.2–0.3 μm diameter range and two showed about 75% in the 0.3–0.4 μm range.

McKeown's observations indicate that, at a great enough distance from the source, deposition can be entirely of clay-size mineral material. The deposition of such fine material is achieved by rainstorms (Dubief 1979) and may take place over the ocean.

It may also be noted that, if the rates of fall of particles 1 μm or less in diameter are as slow as indicated, then much of the clay fraction found by McTainsh & Walker (1982) in dust fall-out must have been carried as fragmented soil (silt-size aggregates) and not as single particles, because, in the three clay-fractions which they analysed in more detail, 40% to 70% of the material was less than 0.2 μm in diameter.

How much wind-blown dust has been incorporated in the 'Acid Sands' (ferrallitic soils over loose sandy sediments) and how have the deep uniform subsoils been formed?

These soils are extensive in Nigeria (see Fig. 1). In the southern part of the country all soils except some in valley bottoms show an increase in clay content with depth. The thickness of the 'eluvial horizons' is generally between 20 and 80 cm. In the Acid Sands the total amount of clay in the very thick subsoils ('illuvial' horizons) seemed to the writer to be anomalous, as in many places there was little clay in the parent material of intensely weathered Tertiary and Upper Cretaceous sandstones (Vine 1949). It was envisaged that there must have been a relative accumulation of clay over a considerable time period. This clay would be derived from the sporadic clay beds in the sandstones by leaching and eluviation whilst sandy surface soil was gradually lost by erosion. Grove (1951, p. 12) also indicated that this could be an effective process, though extremely slow under forest. The idea that deposition of clayey and ferruginous dust was a major factor (Vine 1956) offered a better explanation; but there was no substantial evidence that there was a sufficient quantity of dust being deposited. Subsequently, cores drilled in the sea bed in various parts of the world have revealed the widespread presence of considerable amounts of material finer than loess that have been transported long distances and deposited as fall-out or by rain-wash.

Even more clearly than in the case of the clay content, there seems to be far more iron oxide in the profile than could be derived from the underlying sandstone and clay layers.

The clay and iron oxide may quite possibly, therefore, have been contributed both by the addition of wind-blown dust and by the break-up and dispersal of the fine layers that occur in the underlying sediments. Even so, it is remarkable that there has been thorough mixing to such great depths. Grove (1951) found that the red subsoil of the Acid Sands in the rolling landscapes which he was studying could be 12 to 18 m thick, but was less on valley slopes (*eg* 8 to 10 m in the sides of a particular gully).

There are several problems involved in explaining the origin of these soils which will be considered further in this paper:

(1) Whether the rate at which clay was deposited by dust fall-out on the Sierra Leone Rise at DSDP Site 366 during the Pleistocene can explain the amounts of clay in Acid Sands profiles at a number of sites—these sites being at different elevations and thought to be of differing age.

(2) Whether the ratio of Fe_2O_3 to clay in these soils is similar to that in Harmattan dust and could be accounted for by Pleistocene fall-out.

(3) Whether 'slope pedimentation' (Rohdenburg 1969), 'breakaway retreat' (Moss 1965), and other processes involving creep and mass movement, which might explain the landscapes where these soils occur and the depths of subsoil, can yet be fully defined.

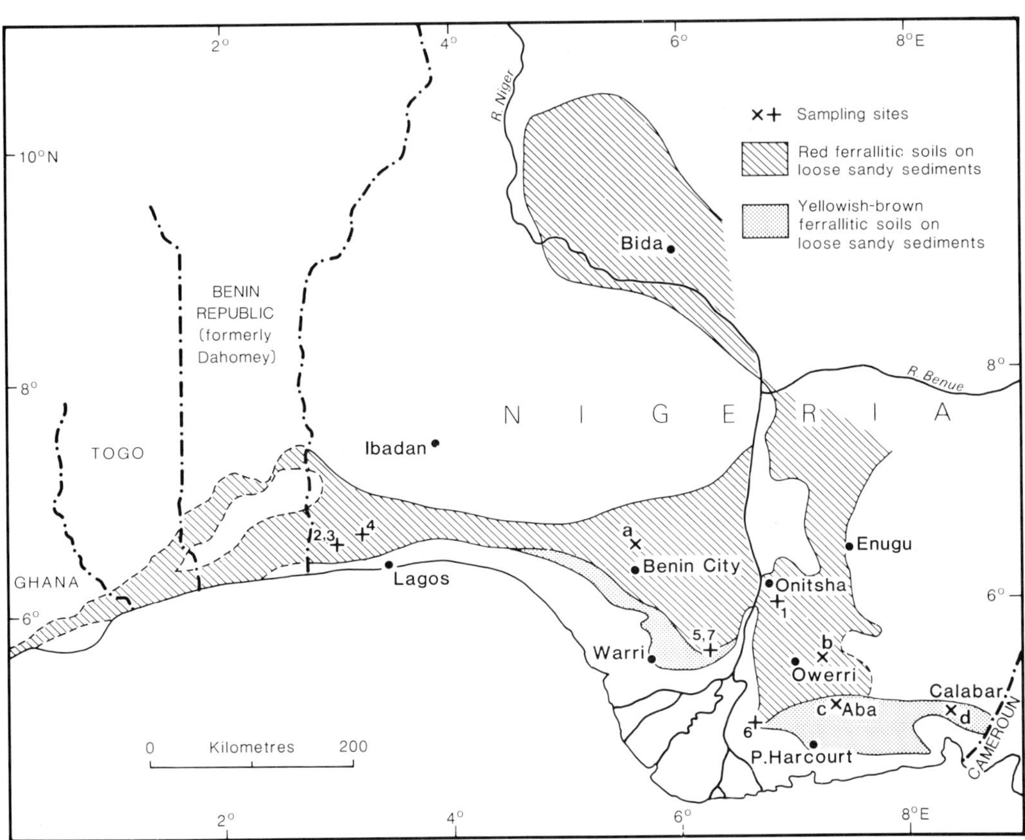

FIG. 1. Map showing occurrence of ferrallitic soils on loose sandy sediments (Tertiary and Upper Cretaceous) in the southern half of Nigeria, Benin Republic, Togo and Ghana, as on CCTA *Soil Map of Africa* 1964. The region of red soils includes three land form-related sub-types: areas almost wholly of 'Acid Sands', areas of mottled soils and 'Acid Sands', and areas of 'Acid Sands' and soils with hardened layers.

A further problem that arises is that neither silt-size quartz nor clay minerals other than kaolin, which should accumulate in aeolian dust, have been recorded. Relatively rapid weathering of these materials in humid tropical conditions appears to provide the explanation for their absence.

The soils in relation to geology

An easily accessible section of 'loose sandy sediments' overlain by red ferrallitic soil is provided by a large pit near Oba, 8 km from Onitsha on the modern highway to the S, on a broad rolling summit about 130 m above sea level and 110 m above the nearby Niger flood plain (Site 1, Fig. 1). An increased demand for sand for use in concrete buildings meant that by 1980 there was an excavation 20 to 25 m deep exploiting this extensive supply of unconsolidated sandstone. The sediments in this locality are believed to be Oligocene–Miocene, and have been regarded as part of the Lignite Series or Ogwashi–Asaba Formation (Grove 1951; Reyment 1965). They could now be placed in the diachronous Benin Formation as defined by Short & Stäuble (1967) and Avbovbo (1978); the cross-bedded ('false bedded') sand in the pit is typical of what Parkinson (1907) first termed Benin Sands.

Highly weathered unconsolidated sandstones similar to those in the Oba pit are exposed in gullies in several areas. The clay layers that occur sporadically are typically about 20 cm thick. Where there are thicker clay beds they tend to prevent drainage, and lignites immediately above them are preserved. With this exception the sediments are very much altered, and in particular, originally plentiful plant remains have been oxidized and disappeared (A. M. J. de Swardt pers. comm.). It can be suggested that there would probably have been pyrites in these deltaic deposits, as a result of which acid-sulphate weathering would occur upon exposure to oxygen as the groundwater receded during uplift (in older formations and in the northern part of the Benin Formation), or during eustatic falls in sea level. Information on the Benin Formation in the unweathered state is scarce. Avbovbo (1978) notes that this formation attains a thickness of almost 2000 m at the middle of its area. He describes the deposits as having been laid down in braided streams on sandy alluvial plains and as being highly porous and freshwater-bearing. He stated: 'Mineralogically the sandstones consist dominantly of quartz and potash feldspar with minor amounts of plagioclase'.

The problematic relationship between soil and geology was very clearly seen in the pit at Oba where a random sample of subsoil at about 120 cm depth was collected as well as a piece of very weakly coherent cross-bedded sandstone from about 20 m. The section deserves a fuller study. Analyses of the two samples are given in Table 1 and X-ray difffractograms of the 0.2–0.5 mm and 0.05–0.2 mm grain size fractions are given in Fig. 2. The soil profile consists of a metre

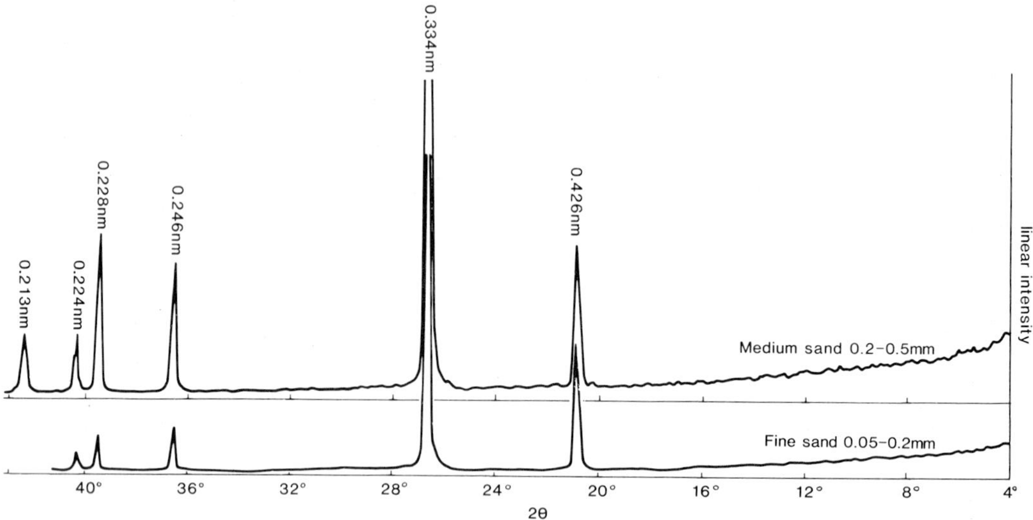

FIG. 2. X-ray diffractograms (CuKα radiation, Ni filter) of two sand fractions of weathered cross-bedded sandstone of the Benin Formation, from approximately 20 m depth at Site 1, indicating quartz alone, with no trace of feldspar or mica.

TABLE 1. *Details and analyses of soil samples from Sites 1 to 7 and deep sample of weathered cross-bedded sandstone from Site 1*

Site	Height above sea (m)	Soil series	Material sampled	Depth (cm)	Munsell colour (moist)	Particle size						Fe_2O_3*		XRD† of clay
						Gravel <2.0 mm	Coarse sand 2.0–0.2 mm	Fine sand 0.2–0.05 mm	Coarse silt 50–20 μm	Medium & fine silt 20–2 μm	Clay <2 μm	in clay fraction (%)	in whole specimen (%)	
1	130	Kulfo	Subsoil	120	10R 4/8 red	nil	74.6	9.2	1.4	1.0	13.8	6.1	—	(ii)
			False-bedded sandstone	2000	2.5Y 8/4 pale yellow	nil	90.8	4.3	0.2	1.0	3.7	—	0.083	(i)
2	40	Alagba	Soil	0–5	2.5YR 3/2 dusky red	0.1	71.1	14.1	1.3	2.6	10.9	8.1	—	—
				18–23	2.5YR 2/4 dark red	0.3	71.9	11.9	1.3	2.6	12.3	8.6	—	—
				35–50	2.5YR 3/6 dark red	0.2	56.5	8.1	0.8	1.8	32.8	8.2	—	—
			Subsoil	150–180	2.5YR 4/6 red	0.4	35.4	8.3	1.0	2.2	53.1	6.4	—	(ii)
3	40	Alagba	Soil	0–5	2.5YR 3/2 dusky red	nil	74.1	9.8	1.2	2.4	12.5	5.2	—	—
				13–20	2.5YR 2/4 dark red	nil	67.2	10.3	1.3	3.1	18.1	6.6	—	—
				35–48	2.5YR 4/6 red	nil	55.2	7.8	0.8	2.1	34.1	7.3	—	(ii)
			Subsoil	150–175	2.5YR 4/8 red	nil	34.6	7.9	1.2	2.7	53.6	7.6	—	—
4	120	Alagba–Owode	Soil	0–5	2.5YR 3/2 dusky red	nil	38.6	34.0	4.8	8.0	14.6	8.2	—	—
				18–28	2.5YR 3/4 dark reddish-brown	nil	52.6	11.7	2.1	3.3	30.3	7.3	—	—
				28–50	2.5YR 3/6 dark red	nil	39.9	8.9	1.7	2.9	46.6	7.3	—	(ii)
			Subsoil	140–175	2.5YR 3/6 dark red	nil	33.0	10.0	2.1	3.1	51.8	6.3	—	(ii)
5	30?	Kulfo	Soil	45	5YR 4/8 yellowish-red	nil	42.5	34.8	2.7	2.0	18.0	6.6	—	—
			Subsoil	90	2.5YR 4/8 red	nil	46.3	27.5	2.3	1.7	22.2	6.0	—	(iii)
6	20?	Ahiara	Soil	0–20	7.5YR 3/4 dark brown	nil	86.7	4.7	0.2	1.1	7.3	6.3	—	—
				45	10YR 5/3 brown	nil	93.1	2.8	0.7	0.8	2.7	6.6	—	—
			Subsoil	200	5YR 5/8 yellowish-red	nil	81.6	4.2	1.9	1.4	10.9	6.0	—	(iii)
7	20?	Ahiara	Soil	0–2	5YR 2/3 v. dark reddish-brown	nil	73.4	17.5	1.2	1.4	6.5	4.8	—	—
				2–22	5YR 3/3 dark reddish-brown	nil	82.9	11.4	0.5	0.6	4.6	5.2	—	—
			Subsoil	82–135	5YR 5/8 yellowish-red	nil	64.1	18.1	1.7	1.7	14.4	5.7	—	(iii)

* Expressed on oven-dry basis. Extraction with dithionite–citrate–bicarbonate (Mehra & Jackson 1960).

† (i) Very well ordered kaolinite with sharp intense peaks.
(ii) Kaolin with fairly narrow 12.3° and 24.95° peaks (CuKα radiation) and low 19.9–20.4 –21.4° and 37.8–38.3–38.5° peaks.
(iii) Kaolin with broader 12.3° and 24.95° peaks and still lower bands at 19.9–21.4° and 37.8–38.5° than in (ii).

of loose sandy brownish layers merging downwards into red coherent but friable subsoil to 5 m. Pale yellow cross-bedded sandstone could be seen in much of the pit face from 5 m downwards. Some of it was obscured by rain-wash, but a remarkably sharp uneven line could be seen around the pit separating the red subsoil from the sandstone. The sandstone is clearly the source of the sand fractions in the soil profile. On the other hand, the analyses show that it contains very little clay, almost no Fe_2O_3, and no feldspar which might decompose to form clay.

The type of soil at the Oba pit is commonly known as the *Kulfo Series*. Similar soils with higher percentages of clay material form the *Alagba Series* (including soils that were termed *Orlu Series*). Soils with the same type of profile but lower clay percentages have been termed the *Ahiara Series*.

Areas almost wholly composed of Acid Sands

There are large rolling to gently undulating areas where only the non-mottled, non-gravelly Acid Sands soils occur, with sandy hydromorphic soils confined to incised valleys. The two main areas of the Benin Formation fall into this category and they form the immediate hinterland of the Niger Delta. They include the sampling sites 1, 5, 6 and 7 and a, b, c and d shown in Fig. 1, E and W of the Niger. Observations in these areas suggest that the amount of clay in the subsoil is related to elevation and to the age of the surface, but that the proportion of clay beds intercalated in the sands is also a factor.

At the Institute for Oil Palm Research near Benin City a rolling upland surface is occupied by soils of the Alagba and Kulfo Series, whilst the northern part of the Institute land is on the irregular slopes of an incised valley about 130 m deep, where the soil changes to the very loose sandy material of the Ahiara Series. Geomorphologically the valley slopes steeply down to an uneven terrace or bench about 1 km wide, then drops steeply again to a perennial stream. Similarly, at Ahiara, where Site b (Fig. 1) is on a rolling upland clothed with the Alagba Series soil, there is a steep drop to an uneven area where a Government Farm Centre was established on the Ahiara Series and then a further drop to a stream.

Two surfaces are clearly seen on a broader scale over the Benin Formation on the western side of the Niger. The main road E of Benin City is on an undulating surface at 150 to 250 m above sea level deeply incised by wide-spaced streams. Approximately 12 km from the Niger it drops to a lower surface which slopes down gradually to the river. A continuous scarp (of smoothed form in the unconsolidated soil and sandstone) runs many kilometres southwestwards. The upper surface rises gradually to 450 m above sea level. The soil is of the Alagba Series, in some parts having more than the average amount of clay, which is possibly due to thick beds of clay associated with the lignite. The lower surface is an undulating plain falling from about 90 m above sea level to about 20 m, and soils identified as Kulfo Series were found at several sites examined in this area. Samples used in the present study were collected in 1980 at the Oleh junction on the Ughelli–Kwale highway (Site 5) and Emakporo's palm plot (Site 7), near the lowest part of the undulating plain. The profiles were identified as the Kulfo and Ahiara Series respectively.

East of the Niger, Site 6 (Ahiara Series) is on a low undulating surface interrupted by swampy areas. The broad summit where Site 1 (Kulfo Series) is located may be a continuation of an intermediate surface which widens around and beyond Owerri (Fig. 1). E and NE from the locality of Site 1, Acid Sands soils occur over two extensive uplands, with summit levels rising to 350 m (the Awka–Orlu Uplands, formed of Eocene strata) and 450 m (the Udi Plateau, formed of Upper Cretaceous strata). Of these, the former appears to have somewhat more clayey soils and subsoils (Grove 1951), but a few figures given by Obihara *et al.* (1964) for the main Acid Sands soil of the Udi Plateau, termed the Nkpologu Series, indicate even there, there is considerably more clay in the soil than in the examples from the Kulfo Series at Sites 1 and 5.

Areas of mottled soils and Acid Sands

These are large areas immediately NE and NW of Lagos, correlated stratigraphically with the Benin Formation of the Niger Delta region, but with a larger proportion of clay or sandy clay. Profiles either show red or orange-brown non-mottled soil (in the Owode and Agege Series) over red-mottled sandy clay at approximately 1 to 2.5 m depth or are of the Acid Sands type (mainly the Alagba Series) as at Sites 2 and 3. It will be seen in Table 1 that the subsoils at these two sites contained 53% clay-sized material. A large proportion of this could have been derived from the varied sedimentary provenance but it is suggested that deposition of dust, as elsewhere, is a major source—in particular, it accounts for the iron oxide. A borehole sunk 8 km WSW of here recorded 10 m of 'earthy red sandy clay', underlain by alternating 'variegated sand and clay' and 'brown fine-grained sand', in equal amounts, down to 30 m (Jones & Hockey 1964). The altitude of about 40 m does not imply correlation in age with the low undulating plain described above (*ie* with Site 5 at about 30 m). The Niger Delta region is obviously one of relatively rapid subsidence, with a gentle folding of surfaces.

Areas of Acid Sands and soils with hardened layers

These are areas of broken topography where short steep slopes are associated with resistant layers in the subsoil. Moss (1965) depicted the formation of soil profiles containing great amounts of ferruginous gravel (Asaba Series) in the talus below a lateritic ironstone, which merged laterally into deep non-gravelly Acid Sands soil (typically, the Alagba Series). He envisaged that, whilst the scarp retreated slowly, the ferruginous gravel of the Asaba Series dissolved away and the fine (< 2 mm) residue became Alagba Series. In course of time this soil would become more extensive and the hardened upper land surface would eventually disappear. He suggested that such a process might account generally for the formation of Acid Sands soils involving the mixing together of materials from finer and coarser layers. It can be objected that the ferruginous rubble and gravel would not dissolve away and that the clayey material would contain dispersed iron oxide. The problem of the formation of the Alagba Series in this situation is probably the same as it is in simpler landscapes—the sand fractions are derived directly from more or less unconsolidated sediments, together with some of the clay and a smaller proportion of the Fe_2O_3. The suggestion that deposition of Pleistocene wind-blown dust accounts for the rest of the clay and Fe_2O_3 also applies.

The problem of the great depth of the soils

An intensive pedological study of soils of this group was made by Fauck (1972) (for localities see Fig. 4). He considered the red subsoil to be a very thick illuvial horizon (B horizon). He mentioned that clay skins were reported in these soils in Benin Republic (Dahomey). Ogunwale *et al*. (1975) also reported thin clay coatings in root channels from 80 cm downwards in a 3 m deep

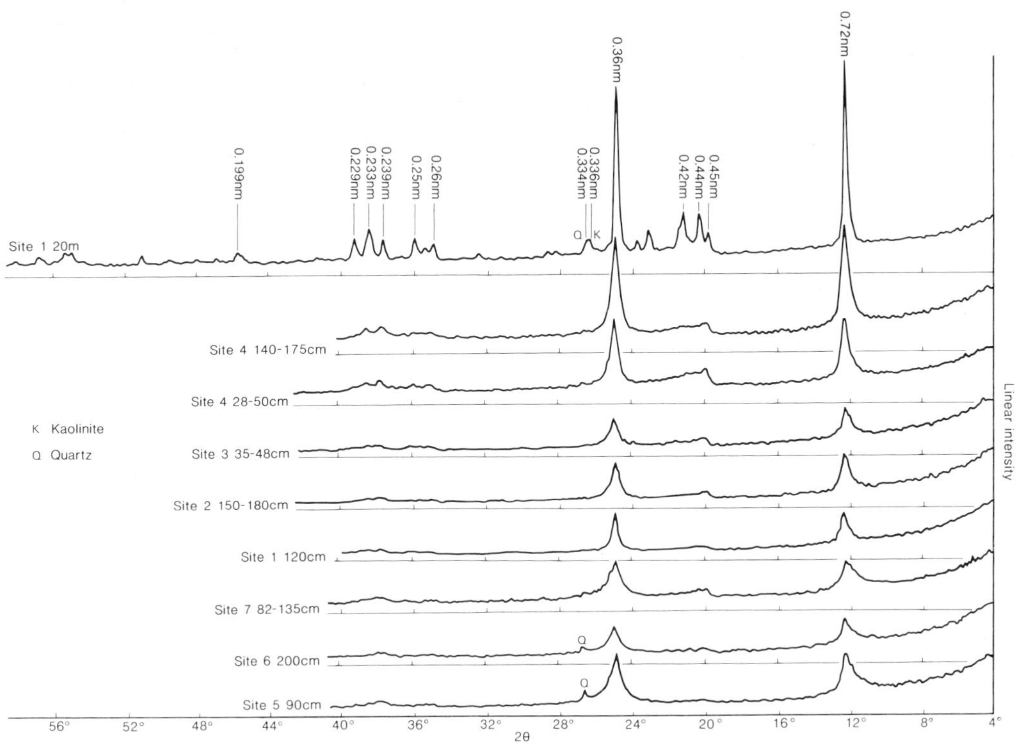

FIG. 3. X-ray diffractograms (CuKα radiation, Ni filter) of clay fractions. Sample No. 121 is (top) clay <2 μm, with Al-exchangeable cation, from weathered false-bedded sandstone at approximately 20 m depth at Site 1. The others are clay < 1.6 μm, with Mg-exchangeable cation, after cleaning with dithionite, from soil profiles at Sites 1 to 7. No trace of any clay mineral other than kaolin can be seen.

pit in Alagba Series dug close to Site a (at the Oil Palm Research Institute, Fig. 1). He cited this as evidence of the process of illuviation from the upper layers.

Fauck remarked that this red illuvial layer could be over 6 m thick. He elaborated a very interesting theory in which the whole profile was developed *in situ* from the parent rocks. He did not envisage incorporation of wind-blown material, but suggested accumulation of clay and Fe_2O_3 by dissolution of quartz and removal of dissolved silica in drainage waters. He stressed the importance of organic compounds ('fulvic acids') as complexing agents, effecting the downward movement of clay and Fe_2O_3, and reported analyses showing some organic matter at the base of the profile. Soil microbes are particularly important in the formation of such complexing agents in the topsoil.

It seems that the processes described by Fauck could lead to as much as 12 to 18 m of soil in southern Nigeria if the addition of wind-blown material was sufficiently prolonged. It is doubtful, however, if the thorough incorporation of material of the varying parent rock strata is brought about without considerable mass-movement. The 'slope pedimentation' process which Rohdenburg (1969) considered in relation to the sedimentary rock region of southern Nigeria, involving alternating stability and instability due to climatic change, is mainly concerned with surface movement and the accumulation of a gravelly layer. Grove (1951) showed that collapse of the sides of deep gullies tended to be brought about by sliding on fine-grained layers where water collects after very heavy rains.

Soil analyses

Samples representing profiles of red ferrallitic soils over loose sandy sediments at Sites 1 to 7 (see Fig. 1) were selected for the examination of particles from different size fractions. Separation had to be complete, but with a minimum of mechanical attrition of sand. Because of the low percentages of fine sand and silt, the procedure was first to remove the coarse sand, and then as much as possible of the clay-sized fraction including finely disseminated iron oxide. The amount of clay-sized material obtained was increased by heating with hydrogen peroxide to decompose organic components. Clay in suspension in dilute alkali was repeatedly decanted, after allowing a settling time of 24 hours. This gave a separated clay fraction with a nominal maximum diameter of 1.6 μm.

The remaining sediment was treated with citrate, bicarbonate and dithionite (Mehra & Jackson 1960) to dissolve Fe_2O_3. The cleaned fine sand (50–200 μm) was removed by sieving and the coarse silt (20–50 μm) and fine and medium silt (2–20 μm) were separated by sedimentation.

Examination of clay fraction

In approximately 0.1 g subsamples of the <1.6 μm clay fraction, Na was replaced by Mg as the exchangeable cation. The specimen was

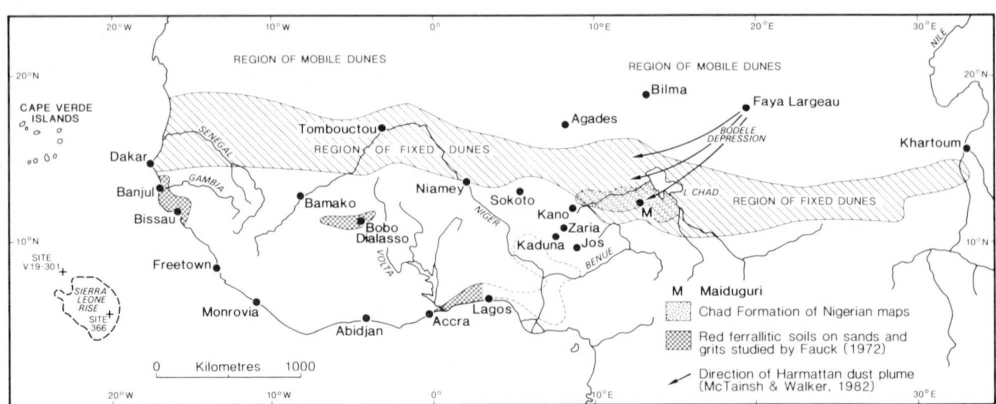

FIG. 4. Map based on Grove (1958), showing limits of occurrence of late Quaternary fixed dunes in parts of present-day Sahel and Sudan and of active dune formation in regions of the present-day Sahara, and of the area of the Plio–Pleistocene Chad Formation recognized in Nigeria but not mapped separately elsewhere. Also shown are: Site V19–301 studied by Parkin (1974) and Parkin & Padgham (1975) and DSDP Site 366; the areas of red ferrallitic soils and loose sandy sediments in which are marked the sites studied by Fauck (1972), from the CCTA *Soil Map of Africa*, 1964; is the main origin of the present-day Harmattan, are taken from McTainsh & Walker (1982).

then mixed to a paste with water on a glass slide, smeared with a spatula and air-dried. Diffractograms were obtained with CuKα radiation and Ni filter.

Iron was determined colorimetrically by the thiocyanate method.

All the clay specimens are dominated by 'kaolin' (Fig. 3). No other clay mineral is indicated by any discernible peak or band. There is notably no sign of illite. A trace of quartz is present in three specimens. The clays from the red subsoils at the Oba pit (Site 1) and the other six sites all appear to consist of disordered kaolin giving poorly-defined peaks. The sharpest peaks are given by those from 28–50 and 140–175 cm at Site 4, where the profile is intermediate between the Alagba Series and the Owode Series. In contrast, the diffractogram for the clay present in small amounts in the weathered sandstone at Site 1 is remarkable. It is almost identical in the positions, sharpness and relative heights of peaks to that reproduced in Brindley & Brown (1980) for 'the very well-ordered geode kaolinite from Keokuk, Iowa', which is regarded as well crystallized.

Previously (Vine 1949), SiO_2/Al_2O_3 ratios very close to the value 2.00 for pure kaolinite were obtained in analyses of clay fractions of samples from a depth of about 120 cm in Acid Sands profiles at Sites a, b, c and d (Fig. 1). The clay content was about 30% in all of these samples. The results are shown in Table 2.

Jungerius & Levelt (1964) reported X-ray diffraction studies of soils of southeastern Nigeria, including five profiles which seem fairly typical of Acid Sands soils. Kaolin was dominant, but traces of mixed-layer clays were found in three of the five profiles, and one also contained more than a trace of gibbsite. Ogunwale *et al.* (1975) also found a trace of a mixed-layer clay (mica–smectite) in their Alagba Series profile, close to Site a.

Comparison between soils and deep-sea sediments

The southeastern part of the Sierra Leone Rise, where DSDP Site 366 was located, is a 15 000 km^2 plateau under 2700–2900 m of water. It was chosen to provide a complete record of pelagic Tertiary and Upper Cretaceous deposits in the eastern North Atlantic, away from turbidity currents and above the Carbonate Compensation Depth (Lancelot *et al.* 1977). Over 800 m of sediment have accumulated since the Palaeocene. The top 120 m were assigned to the Pliocene and Pleistocene, a period of 6 million years. Sediments of this part of the column have between 64% and 70% porosity. The whole of 'Unit 1' (the top 136 m) is mottled, the top 63 m are yellowish-brown in colour (10YR 5/4–5/8), and the next 73 m are very light grey to light olive-grey (N8 to 5Y 6/1). The $CaCO_3$ content increases from around 60% in the top 63 m (marls and oozes), to 75% in the next 73 m (nanno-fossil oozes). Clay interbeds 5 to 15 cm thick occur at places in the top 136 m. These could represent very fine dust material.

Mélières (1977) carried out X-ray diffraction studies on samples from Site 366. He stated that quantitative estimates were made using an internal standard so that absolute percentages could be given, the crystalline components summed, and a rough estimate made of amorphous material. The amount of amorphous material was found to be correlated with terrigenous quartz, feldspar and clays, and was therefore thought to be terrigenous also. Its chemical nature is unknown, but presumably some part of it was formed from siliceous microfossils. According to fig. 2 in Mélière's paper, the top 25 m of deposits are Pleistocene, with Pliocene reaching down to 100 m. No date is indicated for the division between these two. Three specimens taken within

TABLE 2. *Clay-fraction analyses (Vine 1949) and derived molar ratios*

Site	Rainfall (mm)	Colour	% in ignited clay			Total Fe_2O_3 % calculated on oven-dry basis	SiO_2/Al_2O_3	Fe_2O_3/Al_2O_3
			SiO_2	Al_2O_3	Total Fe_2O_3			
a	1800	Red	46.4	40.7	12.6	11.0	1.94	0.198
b	2200	Red	47.9	41.6	12.2	10.6	1.96	0.187
c	2500	Orange-yellow	48.5	41.6	10.0	8.7	1.98	0.152
d	3300	Yellow	48.4	41.8	10.1	8.8	1.97	0.154

Samples from depth of 150 cm, clay content of each approximately 30%.

the top 25 m contain 15.7% of clay on average (Table 3). The 25 m would become 17 m if compressed from about 65% porosity to 45%, which is the average porosity measured in subsoils of Alagba Series. Thus there is a Pleistocene accumulation of the equivalent of 3 m of 100% clay at 45% porosity—1 m each of kaolin, smectite, and illite plus mixed-layer minerals.

In the Kulfo Series profile at the Oba pit (Site 1) it is estimated that there is the equivalent of 4.5 m of soil with 22% clay, or 1.0 m of 100% clay. For a typical Alagba Series soil there is an average clay content of 33% over a thickness of 15 m. In this case the total clay accumulation would be the equivalent of 5.0 m of 100% clay, at 45% porosity. Ogunwale *et al.* (1975) recorded 33% clay between depths of 1.5 and 3.0 m in their pit near Site a. The full thickness of the red subsoil around that site has not been recorded but only a few hundred metres away a pit was dug to 9.5 m without change in the material for which figures of about 31% of clay and 2–20 μm silt combined were recorded (Oil Palm Research Station 1948).

There is therefore fair agreement between the amount of clay (including kaolin, smectite, illite and interstratified minerals) deposited during the Pleistocene at Site 366 and that in soils of the Alagba Series. Clay accumulation probably occurred during only about the last quarter of that time around Site 1. The low surfaces, of Ahiara Series soils, might be about one-tenth the age of the older soils.

The simplifying assumption has been made that clay minerals other than kaolin have been *altered* to kaolin by weathering in hot moist and seasonally moist conditions. This would be consistent with the very high degree of disorder indicated in the diffractograms.

There is a major discrepancy in silt-size quartz between the Pleistocene deep-sea sediment and the soils. The average figures in Table 3 show almost as much quartz as kaolinite in the sediment samples. Fütterer (1977) examined the silt-size material in detail. Qualitatively there was no doubt, from the fossil content and the size-range of mineral particles, that the non-calcareous mineral part of the silt was wind-blown terrigenous material. Angular quartz grains appear in SEM photographs of 2–20 μm fractions, together with various microfossils. Some flaky particles were considered to be mica. Few quartz grains exceeded 20 μm. If an attempt is made to put some of the data from chemical analysis and grain counts together, it appears that the quartz makes up on average 75% (by weight) in seven samples from Pleistocene layers.

Greenwood (1933) first reported analyses of Acid Sands soils, and remarked that the figures showed that the coarse sand and the clay fraction were predominant, and the comparative absence of fine sand and silt gave the wet soil a peculiar gritty feel. For the Ogba Experimental Farm near Benin City he found 1.5% 2–20 μm silt at 0–22 cm, 1.8% at 85 cm, and 2.2% at 120–210 cm. For Site a (Fig. 1) an average of 1.3% of 2–20 μm silt was found from 5 to 180 cm, with 3.0% in the 0–5 cm layer (Vine 1949). Ogunwale *et al.* (1975) gave very similar values for their pit close to Site a. It must be noted that Ogunwale *et al.* reported a small amount of feldspar (1% to 3%) in the silt from this profile, which appears to contradict what has been said earlier about the complete decomposition of feldspar. However, the writer has found titanium minerals in the 2–20 μm fraction of Acid Sands soils, with peaks in diffractograms at 0.325 nm (rutile) and 0.352 nm (anatase). Small peaks indistinguishable from these can be seen in the diffractogram given by Ogunwale *et al.*, and it is virtually certain that their identification of feldspar from the peak labelled 0.323 nm was mistaken.

Larger amounts of silt have been found in numerous analyses of soils from adjoining areas in Nigeria over Basement Complex rocks, particularly gneisses. A number of soils of the Basement Complex areas have also been fractionated by the methods described in this paper and the silts examined by X-ray diffraction. Surprisingly, the greater amounts of silt proved not to be due to much more quartz, but to silt-size particles of kaolin and sometimes mica. This result has removed an apparent obstacle to explaining the very low silt percentages, compared with those in

TABLE 3. *X-ray mineralogy of Pleistocene samples from Site 366 (Mélières 1977)*

Sample	Depth (m)	Quartz (%)	Feldspar (%)	Calcite (%)	Halite (%)	Illite (%)	Mixed layer (%)	Smectite (%)	Kaolinite (%)	Amorphous (%)
366 1–2, 6–7	1.9	3.0	—	53	2.0	1.8	1.0	2.2	4.2	32
366 2–2, 19–20	11.5	7.1	—	29	1.8	4.8	2.0	6.3	7.3	40
366A 2–3, 63–64	13.5	5.2	0.8	51	1.6	2.0	2.5	6.6	6.5	23
Mean of three		5.1	0.3	44	2.0	2.9	1.8	5.0	6.0	32

the sea-floor sediments, as due to the dissolution of fine quartz in percolating rainwater.

Silt-size particles, identified by X-ray diffraction as almost entirely of quartz, generally make up about 1% to 3% in the Acid Sands soils. Table 1 shows one surface soil with 4.8% coarse silt and 8.0% 2–20 μm silt; there is also three times as much fine sand as in the 18–28 cm depth sample, and the explanation lies in the sifting of particles by worms, ants, *etc*. Some unpublished SEM studies of fine sand, coarse silt and 2–20 μm silt of some of the Acid Sands soils show clearly that, in the hot humid conditions of southern Nigeria, solution weathering is producing silt by disintegration and flaking of pitted and fissured sand grains whilst smaller particles disappear. This is in accordance with Fauck's indications of the dissolution of quartz taking place at a much greater rate than hitherto recognized.

It is interesting to note that approximately equal amounts of quartz and clay minerals were found by Parkin (1974) and Parkin & Padgham (1975) in samples from a core of the top 12 m of sea-bed deposits at Site V19-301. This site is under 4724 m of water, close to the slope down from the Sierra Leone Rise (see Fig. 4). There appears to be a possibility of the deposits being partly derived by downslope movement. The average rate of accumulation at this site in the last 700 000 years is clearly identified from the magnetic reversal at 11.25 m depth in the core. The average $CaCO_3$ content was about 20%. On a carbonate-free basis the accumulation rate was 9 m in 0.7 million years. This is twice the average rate at Site 366 during the 6 million years of the Plio–Pleistocene record; the apparent difference in the rates of deposition could be real, and not the result of downslope movement. Samples taken at close intervals in the core were studied by X-ray diffraction. Ignoring the possible presence of amorphous matter, the averages of carbonate-free material are: 52% quartz, 33% kaolin plus chlorite, 13% illite, 3% smectite. From all the evidence it seems likely that these deposits consist mainly of Harmattan dust washed out by rain in the Doldrums.

The source of dust at the present time and during the Pleistocene

McTainsh & Walker (1982) presented the chemical analyses of a 'lake deposit' near Faya Largeau and of two samples of Harmattan dust, pointing to their similarity and supporting the idea of the Faya Largeau area as a main source of the dust material. One of the dust samples was collected by McTainsh & Walker at Kano; this was one of the three samples in which it was possible to determine particle size fractions below 2 μm, showing 40–70% of the <2 μm clay to be less than 0.2 μm. The other dust sample was collected by the Geological Survey, presumably at their headquarters at Kaduna, and was examined microscopically and analysed at the Imperial Institute in London (Jones 1938). Table 4 shows the percentages of SiO_2, Al_2O_3 and Fe_2O_3 in these samples, together with those found by Doyne *et al.* (1938) in the clay fraction of dust fall-out at Samaru, Zaria. For comparison, the percentages are on the basis of the ignited weight (*ie* weight after heating to redness, driving off H_2O).

The iron content is best considered in terms of the ratio of Fe_2O_3 to Al_2O_3 (Tables 2 and 4) because this would not change in a material at first containing feldspars, *etc.* which had become weathered to clays after transportation or deposition. Comparing the soils with Harmattan dust it is remarkable that there is more than enough iron in modern Harmattan dust to account for the iron in the Red and Yellowish-brown Soils over Loose Sandy Sediments. The total amounts of Fe_2O_3 in the clay fractions of soils from Sites 1–7 were not determined. The dithionite-extractable Fe_2O_3 in the clay fractions of Alagba and Alagba–Owode soils at Sites 2, 3 and 4 (Table 1) was somewhat lower than the total Fe_2O_3 for Sites a, b, c and d. This is probably because some Fe_2O_3 may not have dissolved in the buffered dithionite extraction process.

McTainsh & Walker (1982) suggested that the system of deposition of material into low ground

TABLE 4. *Analytical data for three dust samples and a specimen of a suggested source deposit*

Specimen	% in specimen (ignited weight basis)			Molar ratios	
	SiO_2	Al_2O_3	Fe_2O_3	SiO_2/Al_2O_3	Fe_2O_3/Al_2O_3
Harmattan dust—Kano (McTainsh & Walker)	75.7	12.7	5.1	10.1	0.256
Harmattan dust—Geological Survey	66.5	13.9	5.6	7.9	0.256
Clay fraction of Harmattan dust—Zaria (Doyne *et al.*)	67.5	22.9	10.8	5.0	0.302
Lake deposit—Borkou, near Faya Largeau	52.5	8.4	5.0	10.6	0.381

and removal from there in dust storms of the Harmattan could be taken as a contemporary model 'against which to test hypotheses about aeolian systems of the Quaternary'. It seems that the same system may have functioned during most of the Quaternary, when lake deposits were being laid down in various parts of the Chad Basin. Exposed to arid conditions, they would supply the finer part of the dust mobilized by the moving sand which also supplied the quartz in the dust cloud.

The source area of the dust storms would certainly have shifted many times through several degrees of latitude. The time scale of climatic changes is indicated by the recognition in the vicinity of Lake Chad of three periods of dune activity during southward advances of the Sahara in the last 60 000 years, and a considerable number before then, possibly as far back in time as 2 or 3 Ma (Durand & Mathieu 1980; Servant & Servant-Vildary 1980). It is realistic to visualize that there were 'Harmattan' dust plumes across the middle and southern parts of Nigeria when there were source areas much further S than they are now.

A wider question is how much dust may have been produced in regions to the E and W of the Chad Basin during the Pleistocene and been deposited in soils, as well as in marine sediments. For example, Prospero (1981) shows that even nowadays the course taken by dust outbreaks from the central Sahara can be as far S as the 10° N line.

ACKNOWLEDGMENTS: I wish to thank the Geology Department of Leicester University for allowing me the use of the XRD apparatus and R. Wilson for his guidance. I was helped by many people in the past in the field and the laboratory when I was employed in Nigeria, first in the Government Agricultural Service and later at the University of Ibadan.

References

AVBOVBO, A. A. 1978. Tertiary lithostratigraphy of Niger Delta. *The American Association of Petroleum Geologists Bulletin* **62**, 295–306.

BRINDLEY, G. W. & BROWN, G. (eds) 1980. *Crystal Structures of Clay Minerals and their X-ray Identification.* Mineralogical Society, London.

D'HOORE, J. L. 1964. *Soil Map of Africa, Scale 1 to 5,000,000. Explanatory Monograph.* Commission for Technical Cooperation in Africa, Publication No. 93, Lagos.

DOYNE, H. C., HARTLEY, K. T. & WATSON, W. A. 1938. Soil types and manurial experiments in Nigeria. *Third West African Agricultural Conference: Papers (Nigeria).* Government Printer, Lagos, pp. 227–298 and map.

DUBIEF, J. 1979. Review of the North African climate with particular emphasis on the production of eolian dust in the Sahel Zone and in the Sahara. *In*: MORALES, C. (ed.) *Saharan Dust—Mobilization, Transport, Deposition.* John Wiley & Sons, Chichester, 27–48.

DURAND, A. & MATHIEU, P. 1980. Evolution paléogéographique et paléoclimatique du bassin tchadien au Pléistocène supérieur. *Revue de Géologie Dynamique et de Géographie Physique* **22**, 329–341.

FAUCK, R. 1972. *Les Sols Rouges sur Sables et sur Grès d'Afrique Occidentale.* Memoires O.R.S.T.O.M., Paris, No. 61.

FÜTTERER, D. 1977. Late Neogene silt at the Sierra Leone Rise (Leg 41 Site 366): terrigenous and biogenous components. *In*: LANCELOT, Y., *et al.* (eds) *Initial Reports of the Deep Sea Drilling Project, Volume 41.* US Government Printing Office, Washington, 1049–1059.

GREENWOOD, M. 1933. Some cocoa soils of the eastern plains in Nigeria. *Tenth Annual Bulletin, Department of Agriculture, Nigeria.* Government Printer, Lagos, 36–49; 221–223 and Tables.

GROVE, A. T. 1951. Land Use and Soil Conservation in parts of Onitsha and Owerri Provinces. *Bulletin No. 21, Geological Survey, Nigeria.*

—— 1958. The ancient erg of Hausaland, and similar formations on the south side of the Sahara. *Geographical Journal* **124**, 528–533.

HAMILTON, R. A. & ARCHBOLD, J. W. 1945. Meteorology of Nigeria and adjacent territory. *Quarterly Journal of the Royal Meteorological Society* **71**, 231–265.

JONES, B. 1938. Desiccation and the West African Colonies. *Geographical Journal* **91**, 401–423.

JONES, H. A. & HOCKEY, R. D. 1964. The Geology of part of South-Western Nigeria. *Bulletin No. 31, Geological Survey, Nigeria.*

JUNGERIUS, P. D. & LEVELT, T. W. M. 1964. Clay mineralogy of soils over sedimentary rocks in Eastern Nigeria. *Soil Science* **97**, 89–95.

KALU, A. E. 1979. The African dust plume: its characteristics and propagation across West Africa in winter. *In*: MORALES, C. (ed.) *Saharan Dust—Mobilization, Transport, Deposition.* John Wiley & Sons, Chichester, 95–118.

LANCELOT, Y., SEIBOLD, E. & GARDNER, J. V. 1977. Introduction. *In*: LANCELOT, Y., *et al.* (eds) 1977. *Initial Reports of the Deep Sea Drilling Project, Volume 41.* US Government Printing Office, Washington, 7–18.

McKeown, H. D. J. 1958. Dust concentrations in the Harmattan. *Quarterly Journal of the Royal Meteorological Society* **84**, 280–282.

McTainsh, G. H. & Walker, P. H. 1982. Nature and distribution of Harmattan dust. *Zeitschrift für Geomorphologie* N.F. **26**, 417–435.

Mehra, O. P. & Jackson, M. L. 1960. Iron oxide removal from soils and clays by a dithionite–citrate system buffered with sodium bicarbonate. *Clays and Clay Minerals* **7**, 317–327.

Mélières, F. 1977. X-ray mineralogy studies, Leg 41, Deep Sea Drilling Project, eastern North Atlantic Ocean. *In*: Lancelot, Y. *et al.* (eds) 1977. *Initial Reports of the Deep Sea Drilling Project, Volume 41*. US Government Printing Office, Washington, 1065–1086.

Moss, R. P. 1965. Slope development and soil morphology in a part of South-West Nigeria. *Journal of Soil Science* **16**, 192–209.

Obihara, C. H., Bawden, M. G. & Jungerius, P. D. 1964. *Soil Survey Memoir No. 1. The Anambra-Do Rivers Area*. Government Printer, Enugu.

Ogunwale, J. A., Ashaye, T. I., Odu, C. T. I. & Fayemi, A. A. A. 1975. Characterization of selected sandstone-derived soils in the ecological zones of Nigeria. *Geoderma* **13**, 331–347.

Oil Palm Research Station, Nigeria. 1948. *Seventh Annual Report 1946–47*. Department of Agriculture, Nigeria.

Parkin, D. W. 1974. Trade-winds during the glacial cycles. *Proceedings of the Royal Society of London* **A337**, 73–100.

—— & Padgham, R. C. 1975. Further studies on trade winds during the glacial cycles. *Proceedings of the Royal Society of London* **A346**, 245–260.

Parkinson, J. 1907. The post-Cretaceous stratigraphy of Southern Nigeria. *Quarterly Journal of the Geological Society of London* **63**, 311–320.

Prospero, J. M. 1981. Arid regions as sources of mineral aerosols in the marine atmosphere. *In*: Péwé, T. L. (ed.) *Desert Dust: Origin, Characteristics, and Effect on Man*. Geological Society of America Special Paper **186**, 71–86.

Reyment, R. A. 1965. *Aspects of the Geology of Nigeria: The Stratigraphy of the Cretaceous and Cenozoic Deposits*. Ibadan University Press, Ibadan.

Rohdenburg, H. 1969. Hangpedimentation und Klimawechsel als wichtigste Faktoren der Flächen—und Stufenbildung in den wechselfeuchten Tropen an Beispielen aus Westafrika, besonders aus dem Schichtstufenland Südost-Nigerias. *Gottinger Bodenkundliche Berichte* **10**, 57–152.

Servant, M. & Servant-Vildary, S. 1980. L'environment quaternaire du bassin du Tchad. *In*: Williams, M. A. J. & Faure, H. 1980. *The Sahara and the Nile*. Balkema, Rotterdam, 133–162.

Short, K. C. & Stäuble, A. J. 1967. Outline of geology of Niger Delta. *The American Association of Petroleum Geologists Bulletin* **51**, 761–779.

Vine, H. 1949. Nigerian soils in relation to parent materials. *Proceedings of the First Commonwealth Conference on Tropical and Sub-Tropical Soils, 1948*. Commonwealth Bureau of Soil Science, Technical Communication No. 46, 22–29.

—— 1956. Studies of soil profiles at the W.A.I.F.O.R. Main Station and at some other sites of oil palm experiments. *Journal of the West African Institute for Oil Palm Research* **1**, 8–59.

Wilke, B. M., Duke, B. J. & Jimoh, W. L. O. 1984. Mineralogy and chemistry of Harmattan dust in northern Nigeria. *Catena* **11**, 91–96.

H. Vine, 2 Swale Close, Oadby, Leicester LE2 4GF, UK.

Dune dynamics and deposits

Continental sabkha pans and associated nebkhas in southern Kuwait, Arabian Gulf

A. Gunatilaka & S. Mwango

SUMMARY: Small evaporitic pans within a sand-sheet/gravel-plain terrain characterize the coastal margins of southern Kuwait. Sedimentological studies indicate that the pans commenced as ephemeral lakes during a Pleistocene pluvial phase, and later evolved into sabkhas during the Holocene with the changing climatic and hydrological regimes. The lacustrine episode is represented by laminated calcite micrite in the lower parts of the section, grading upward into a diagenetic gypsum facies of the sabkhas, with diagnostic evaporite textures. Dunes are represented only by nebkhas, which occur within the sabkha pans. Their initial build-up is controlled by a halophyte which colonizes the sabkha surface, trapping any moving sand. The nebkhas pass through phases of growth, stabilization and decay that are controlled by fluctuations in the water table. A rise of the water table can waterlog the roots and kill the plants—in which case the trapped sand starts drifting. A continued fall in the water table can lead to deflation of the entire complex. The nebkhas have a poor preservation potential. In this area the aeolian system is highly starved of sand in relation to the available wind energy. Stable isotope data indicate that the groundwaters are derived from old formation waters, with only minor meteoric recharge.

During the last two decades, the recognition of modern sabkha environments as major depositional systems has introduced a new dimension into palaeoenvironmental analysis (Evans 1966; Shearman 1966; Hardie 1968; Butler 1969; Evans *et al.* 1969; Kinsman 1969; Glennie 1970; Amiel & Friedman 1971; Purser 1973; Handford 1981, 1982). Sabkhas or saltflats are relatively flat geomorphic surfaces in deflational–depositional equilibrium whose existence is controlled by the local level of the groundwater table. Both coastal and continental sabkhas have been recognized, and in the southern part of the Arabian Gulf one passes into the other without any noticeable geomorphic discontinuity (Kinsman 1969). In contrast, in the northwestern Arabian Gulf there is a distinct topographic barrier between the two types of sabkhas. While there is a continuous spectrum of sabkha environments from coastal to continental type, there are also great variations within each in terms of processes and/or products (Purser 1985).

Extensive continental sabkhas are lacking in Kuwait. Instead, the desert terrain in the S of the country is studded with isolated, very small sabkha pans. The large dune forms typical of the southern Arabian peninsula are totally absent. The only recognizable aeolian dunes are *nebkhas* (an Arabic word used to describe small, rounded sand bodies trapped by vegetation; also spelt *nebkhas*) which are confined to the sabkha pans.

The purpose of this study is to describe the evolution of a Quaternary continental sabkha pan and nebkha complex from an area of limited sand supply in relation to the available wind energy. Nebkhas may be typical of areas of net deflation and low sand supply.

Regional setting, physiography and climate

The study area is situated in the S of Kuwait, near the border with Saudi Arabia and within the former neutral zone between the two States (Fig. 1). The desert regions of southern Kuwait are characterized by very flat or gently undulating erosional surfaces. There are no significant rock outcrops within a radius of more than 100 km. The land surface rises gently westward from the coast to an elevation of 100 m above sea level 40 km inland (Fig. 2).

Much of this study is confined to an area between 6 and 20 km from the coast, where the sabkha–nebkha complexes are best developed. Extensive sand sheets and gravel plains occupy the relatively high areas in the aeolian terrain. Sand sheets consist of broad swells (0.5–2 km in width) with the intervening troughs forming the sabkha pans (Fig. 2). Fryberger *et al.* (1983) describes this as a 'ridge and swale' topography. This is the dominant landform of the coastal sabkhas all the way to Saudi Arabia, covering a very extensive area.

The sedimentary cover is thin. The sand sheets are no more than 1.5–2 m thick and in many places the wind has eroded down to Tertiary bedrock. The gravels occur as a lag deposit and

From FROSTICK, L. & REID, I. (eds), 1987, *Desert Sediments: Ancient and Modern*, Geological Society Special Publication No. 35, pp. 187–203.

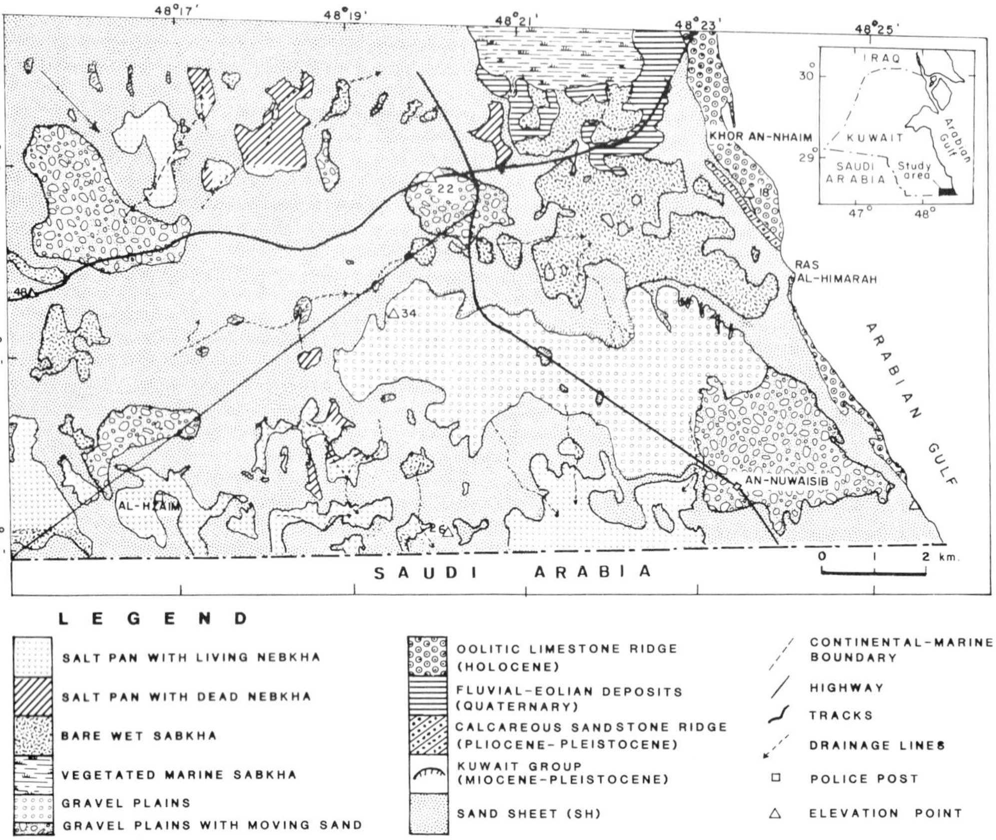

48°17'
48°19'
48°21'
48°23'
48°25'
28° 36'
28° 34'
28° 32'
KHOR AN-NHAIM
RAS AL-HIMARAH
ARABIAN GULF
AN-NUWAISIB
AL-HAIM
SAUDI ARABIA
0 1 2 km.
IRAQ
KUWAIT
SAUDI ARABIA
Study area
Arabian Gulf
30°
29°
47°
48°
LEGEND
SALT PAN WITH LIVING NEBKHA
SALT PAN WITH DEAD NEBKHA
BARE WET SABKHA
VEGETATED MARINE SABKHA
GRAVEL PLAINS
GRAVEL PLAINS WITH MOVING SAND
OOLITIC LIMESTONE RIDGE (HOLOCENE)
FLUVIAL-EOLIAN DEPOSITS (QUATERNARY)
CALCAREOUS SANDSTONE RIDGE (PLIOCENE-PLEISTOCENE)
KUWAIT GROUP (MIOCENE-PLEISTOCENE)
SAND SHEET (SH)
CONTINENTAL-MARINE BOUNDARY
HIGHWAY
TRACKS
DRAINAGE LINES
POLICE POST
ELEVATION POINT

FIG. 1. Location of study area and map of the aeolian terrains of southern Kuwait.

Wafra
Continental Sabkha – Desert sands
Coastal (Marine) Sabkha
m.
S.S
P
G.W.T
P : Pan
S.S : Sand Sheet
G.W.T : Ground-Water Table
offshore
Km.

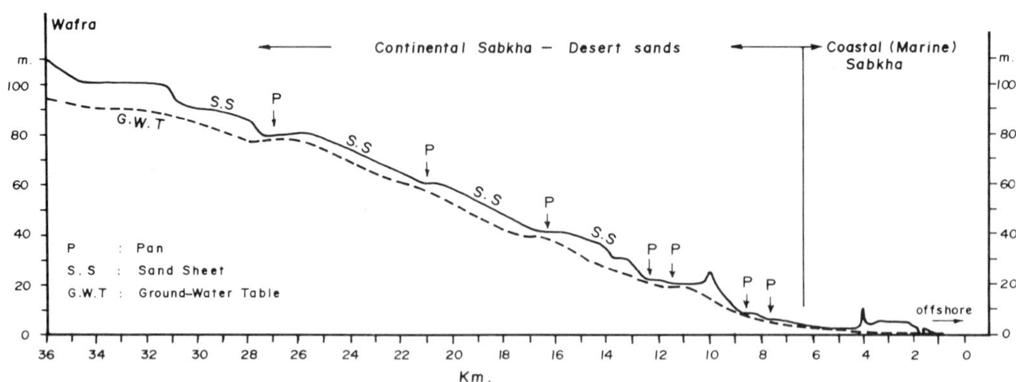

FIG. 2. Section across the study area from Wafra oasis to coast showing the sand sheet–sabkha pan complexes with ridge and swale topography. The first 14 km from the coast to desert corresponds to the E–W distance represented by the map of the study area (Fig. 1).

are part of the extensive Dibdibbah gravels which originated in Saudi Arabia during a Pleistocene pluvial episode (Al-Sayari & Zotl 1978). These are probably sheet-flood or braid-plain deposits and mark the eastern limits of the Dibdibbah gravel plain. This continental complex is separated from the Holocene marine sabkhas to the E by a distinct topographic barrier (6–7 m above sea level) made up of coarse sand and gravel of fluvial-aeolian origin. To the N and S, the sand sheets spread to the coast and completely surround the marine sabkhas.

The arid climate is characterized by high average summer temperatures (45°–48°C) and relatively cold winter temperatures (12°–15°C) with night-time temperatures occasionally dropping to 0°C and below. The relative humidity is variable and averages 60%–70%. Rainfall ranges from 25–334 mm a^{-1}, but averages 115 mm a^{-1}. Evaporation far exceeds precipitation although no absolute values are available for the pan areas. The main wind vector is from the NW ('the Shamal') for most of the time (Fig. 3). Seasonal southeasterly winds blowing from the Gulf can cause very high humidity. Sand storms are rare but dust storms (locally called toz) are common in the summer.

There are no previous regional studies on the aeolian terrains of southern Kuwait. The coastal zone and the evaporite minerals have been studied by several authors (Gunatilaka *et al.* 1980, 1984, 1985). Khalaf *et al.* (1984) made a reconnaissance study of surface deposits in the desert regions of Kuwait. The most relevant study is one on the hydrological characteristics of the coastal and continental areas which has a direct bearing on the present investigation (Robinson *et al.* 1987). However, there is much related research in Saudi Arabia and Israel that has important implications for the present study (McClure 1976; Al-Sayari & Zotl 1978; Fryberger *et al.* 1983; Evenari *et al.* 1985).

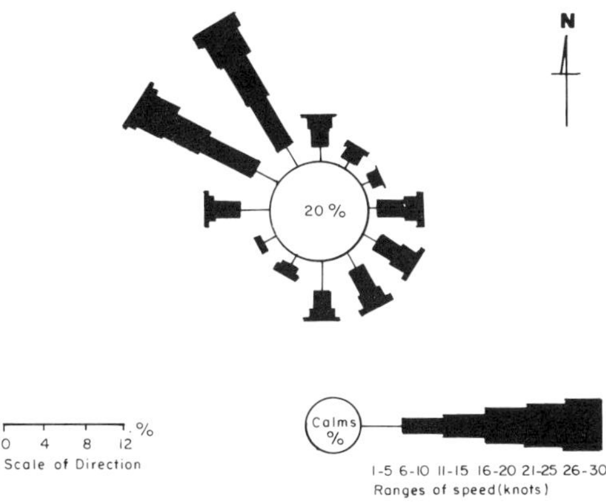

FIG. 3. Wind rose showing wind speed and direction at Kuwait International Airport (25 years data).

The sedimentary framework

In this paper the two main geomorphic forms, the sabkha pans and nebkhas and the processes responsible for them, are described separately. However, it needs to be emphasized at the outset that they are not forms that just happen to lie alongside each other in the landscape, but are in fact intimately (and genetically) associated, effectively constituting a single geomorphic entity. The link between the sabkha pans and nebkhas is provided by the dynamics of the local and/or regional groundwater tables. Climatic (especially wind and evaporation) and hydrological (especially water-table fluctuations, composition and capillarity) factors are the dominant and common controls responsible for the evolution of these complexes in this desert terrain with limited sand supply.

Inland sabkha pans

In contrast to the extensive continental sabkhas of the southern Arabian Gulf, the sabkhas of southern Kuwait are small and isolated geomorphic features. They have irregular outlines and are 'pan-like' in character, generally occurring 2–5 m below the elevation of the sand-sheet/gravel-plain terrain (Figs 2, 4). These pans are very unlike the continental sabkha plains of Kinsman (1969), an example of which is the Sabkha Mutti (~5000 km^2 in area) in the United Arab Emirates (UAE), or the sabkhas of the Dharan area of Saudi Arabia (Fryberger *et al.* 1983).

Morphology of the pans

The pans are ideally described by the term *mamlahat*, (singular, *mamlahah*) an Arabic word which refers to rounded to irregular pan-like depressions containing evaporitic salts which may be covered temporarily by a layer of rainwater. If the depressions are filled up with silt, but without salts, the term *khabrats* is used. The term playa is not very suitable for these small continental sabkha pans, since they have open hydrological systems and lack the most soluble evaporite minerals.

The long dimension of the pans varies from about 0.8–1.5 km and their width from about 0.2–0.8 km. They appear to be oriented in the direction of the prevailing winds (NW–SE).

FIG. 4. View looking SW from sand sheet to sabkha pan with nebkha dunes. The elevation difference between the sand sheet and this pan is ~5.5 m.

Aerial photographs and topographic maps indicate that several of these pans were once connected to each other down-gradient. Along the upwind margins of the pans the sand sheets are slowly encroaching upon the pan surface. On the downwind side the margins have been stripped of sand, exposing the Tertiary bedrock surface and/or gravels. These exposed areas are sometimes characterized by a concrete-like duricrust in which coarse sand and gravels have been cemented by calcite. The stripped sand is transferred to the adjacent sand-sheet area.

The pans can be called sand-sheet sabkhas within the desert terrain. Most of them are probably in a depositional–deflational equilibrium with respect to the water table.

Composition of the sediments

The Quaternary sediments of the pans are made up essentially of quartz, calcite and gypsum. The proportions of these minerals (as determined by X-ray diffraction analysis) vary from the bottom to top in the cores studied (2.5–3.5 m in length). The dominant mineral at a given depth in the cores is a reflection of the environmental conditions operating at the time of deposition. The lithology of the sediments and the inferred environmental processes are summarized in Fig. 5.

Quartz and related siliciclastics. These are present throughout the sediment column. The granules and pebbles at the bottom of the section (basal 0.5 m) are composed of quartzite, igneous and metamorphic rocks. These materials were derived from a distant source area and deposited on the erosional surface of the Tertiary Kuwait Group (Fig. 5). The prolate and oblate shape of the majority of these pebbles suggests that they were transported by shallow sheet floods. The dominance of rounded sand grains throughout the rest of the section indicates that aeolian conditions prevailed throughout the entire history of these pans (*ie* the pans post-date the fluvial phase).

Gypsum. The gypsum content is highly variable within the section (Fig. 5). Two or three major morphologies are present. The most common is found in the silt and fine sand size fraction which is present throughout the section. Both microscopic and SEM examination indicates that this material is not composed of euhedral grains, but of abraded and cleaved fragments of detrital-aeolian origin. These grains were probably derived from deflating pans further inland and by deflation of the ubiquitous gypsum 'Gatch' layers exposed on the upper surface of the Kuwait Group. Gatch is a thick gypsum crust found all over Kuwait. In contrast to this fine-grained gypsum, there are also very large crystals (up to 5 cm long) that have grown displacively within the sediments. This later evaporitic phase contains quartz and older gypsum grains enclosed poikilotopically within it and is the typical sabkha habit of gypsum. This intrasedimentary growth of gypsum disrupts any layering that may be present in the sediments.

The top 0.30 m of the pans is characterized by one or two hard gypsum crusts, which probably reduce evaporative water losses from the surface. Their origin in desert environments had been studied in detail by Watson (1979). Immediately above or below the crust are sugary white, irregular layers of primary gypsum nodules which

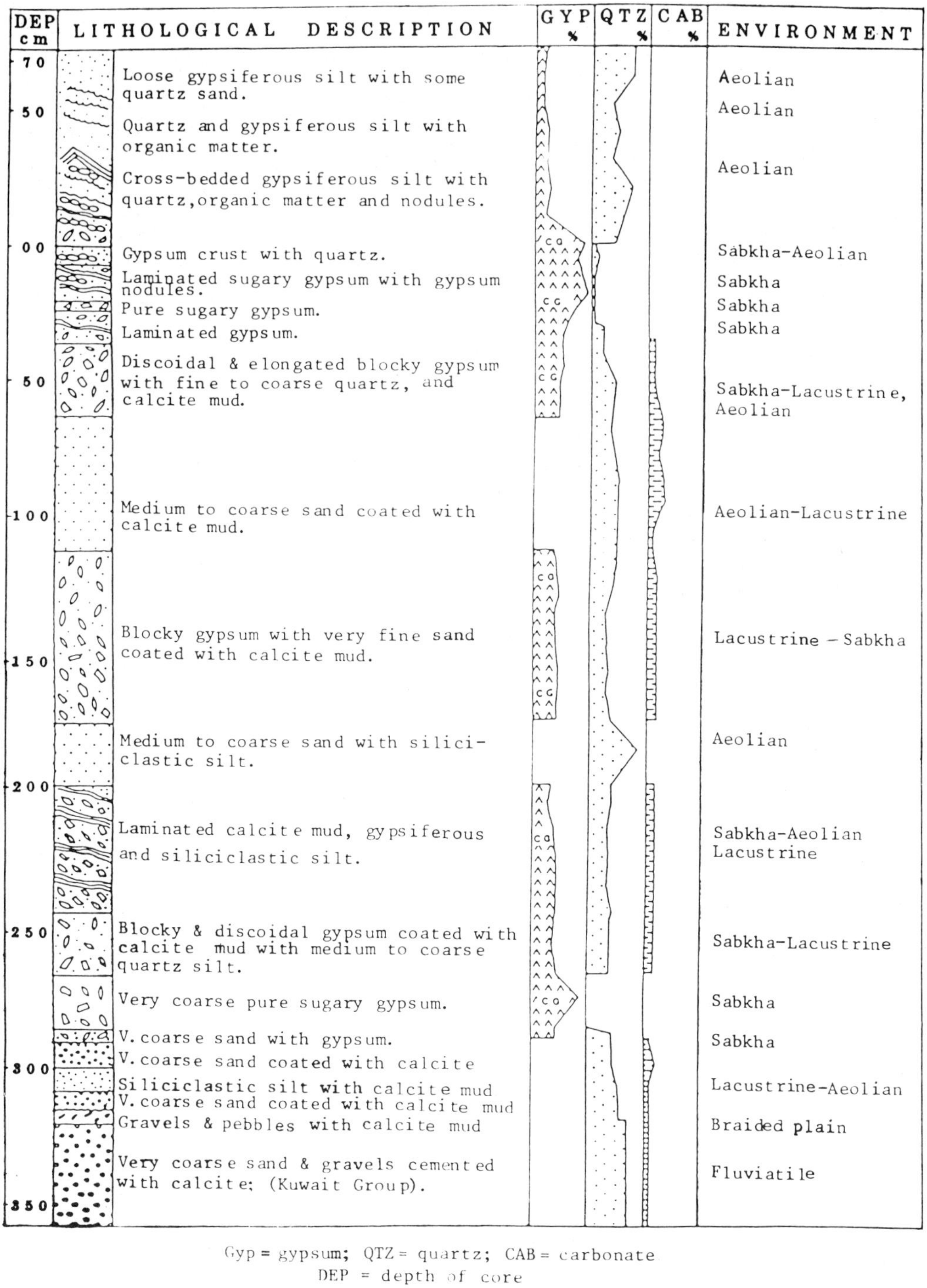

DEP cm	LITHOLOGICAL DESCRIPTION	GYP %	QTZ %	CAB %	ENVIRONMENT
70	Loose gypsiferous silt with some quartz sand.				Aeolian
50	Quartz and gypsiferous silt with organic matter.				Aeolian
	Cross-bedded gypsiferous silt with quartz, organic matter and nodules.				Aeolian
00	Gypsum crust with quartz.				Sabkha-Aeolian
	Laminated sugary gypsum with gypsum nodules.				Sabkha
	Pure sugary gypsum.				Sabkha
	Laminated gypsum.				Sabkha
50	Discoidal & elongated blocky gypsum with fine to coarse quartz, and calcite mud.				Sabkha-Lacustrine, Aeolian
100	Medium to coarse sand coated with calcite mud.				Aeolian-Lacustrine
150	Blocky gypsum with very fine sand coated with calcite mud.				Lacustrine – Sabkha
	Medium to coarse sand with siliciclastic silt.				Aeolian
200	Laminated calcite mud, gypsiferous and siliciclastic silt.				Sabkha-Aeolian Lacustrine
250	Blocky & discoidal gypsum coated with calcite mud with medium to coarse quartz silt.				Sabkha-Lacustrine
	Very coarse pure sugary gypsum.				Sabkha
	V. coarse sand with gypsum.				Sabkha
300	V. coarse sand coated with calcite				
	Siliciclastic silt with calcite mud				Lacustrine-Aeolian
	V. coarse sand coated with calcite mud				
	Gravels & pebbles with calcite mud				Braided plain
	Very coarse sand & gravels cemented with calcite; (Kuwait Group).				Fluviatile
350					

Gyp = gypsum; QTZ = quartz; CAB = carbonate
DEP = depth of core

FIG. 5. Generalized lithologic section from cores showing vertical distribution of the main minerals in sabkha pans, and the inferred environments.

are diagnostic of these pans and discussed in detail later. Between the groundwater table and the nodular layer, the sediments of some pans are dark grey in colour, changing to a reddish-brown near to the nodules, indicating different degrees of oxidation–reduction. The veneer of sediment covering the pan surface is white in colour due to the dehydration of the gypsum crust to bassanite ($CaSO_4 \cdot \frac{1}{2}H_2O$). This is a very unstable and transitory phase in these environments and changes to gypsum between summer and winter seasons and from dry to humid spells. Even overnight, bassanite may change to gypsum (Gunatilaka *et al.* 1985).

Calcite. This is generally confined to the middle two-thirds of the section and is totally absent in places (Fig. 5). It is microcrystalline and can be studied only using the SEM (Fig. 6). The euhedral, scalenohedral (?) shape and very small size (~20 μm in length) classifies it as a micrite and indicates *in situ* precipitation from an aqueous phase. At some levels it forms more than 30%–40% of the sediment and displays a laminated appearance; in other levels it completely coats gypsum and quartz. Most of the crystals have undergone some dissolution at their edges. Its laminated nature, fine grain size and euhedral morphology suggests that an ephemeral or seasonally ponded, aqueous phase prevailed within the pans from time to time. Very minor amounts of dispersed halite are found within the upper 0.1 m of the sabkha.

FIG. 6. Scanning electron photomicrograph of calcite micrite from a core. Note euhedral crystallites with slight dissolution of edges.

Nebkhas (dune facies)

Large-scale dune forms of the type typical of other parts of Arabia are absent from Kuwait (Holm 1960; Glennie 1970; Chapman 1971; Fryberger *et al.* 1983). However, the sabkha pans with shallow water tables develop a system of vegetated dunes which have been called shrub coppice dunes or nebkhas by some authors (Military Engineering Experimental Establishment 1969; Cooke & Warren 1973). Similar coastal dunes have been described as wind-shadow dunes by Goldsmith (1973).

Morphology

Looking down from the sand sheet into the pan areas, the nebkhas appear as a series of semicircular mounds vegetated on their windward side (Fig. 4). Individual dunes are 1–3 m in height, up to 3 m across and slightly elongated downwind in the direction of the prevailing 'shamal' (in contrast to lunettes which are oriented transverse to the wind). Dune crests are rounded. The plant initiating these dunes is exclusively the halophyte species *Nitraria retusa*. Although rounded crests are the norm, it is not unusual to find conical- or triangular-shaped nebkhas within some pans. Their external morphology is dependent very much on the pattern of branching of the plant and its stage of growth, as will be apparent later. Nebkhas are the only dunes of any significance in southern Kuwait and they are always genetically associated with the sabkha pans. The absence of large dune forms (barchans, seifs, *etc.*) is probably because of the lack of a continuous sand supply, as much of the region has been deflated down to near bedrock levels.

Nebkhas have not been reported from the inland sabkhas of the southern part of the Arabian Peninsula. This is a region of net sand deposition, where the aeolian system is highly sand-charged in relation to the available wind energy. Nebkhas have also been reported from northeastern Saudi Arabia (Al-Sayari & Zotl 1978). The hydrohalophyte *Nitraria retusa* and its relationship to nebkhas, 'rebkas' and 'phytogenic hillocks' in the Sinai desert has been discussed by Evenari *et al.* (1985).

Growth and development of nebkhas and shadow dunes

Because the salinity of the groundwater is relatively low (15‰–24‰), the salt-resistant plant *Nitraria* is able to colonize the pan surface. It acts as an obstacle or baffle and traps the sand moving across the pan surface. As the sand develops bedforms, the plant keeps pace, growing above the active dune surface while at the same time maintaining root contact with the capillary zone of the pan.

Nebkhas commence as small, barchan-shaped dunes (~0.5 m in length) around very young plants. By the time the dunes are 0.5 m in height, the rounded form has been reached. In a given pan all sizes of nebkhas can be seen, but their maximum height is about 3 m and their average height is ~2 m. The decay of the halophyte is accompanied by deflation of the sediments on the windward side of the plant, transport downwind and deposition to the lee of the dying plant. This results in the development of a pyramidal, wind-shadow dune at the expense of the decaying nebkha (Fig. 7).

Shadow dunes typically have a quasi-triangular ground plan, whose base is approximately normal to the main wind vector (NW–SE; ~300°). In contrast to the maximum dimensions of the nebkhas, the maximum dimensions of the shadow dunes are height 1.5 m, width 2–3 m and length 4–5 m (measured from the lee-side of plants). A central ridge tapers from a high point immediately in front of the *Nitraria* bush to zero height downwind along the axis of the shadow dune (Fig. 7). Sloping away on either side of the ridge axis are steep (20°–26°), convex-up slip faces, which give the bedform the appearance of a pyramid. Small current-ripples cover the entire surface of the dune. Bagnold (1954) recognized lee-trending shadow dunes as a distinct dune type, as they form in the wind shadow behind a discrete, roughness element (a halophyte bush in this case).

Sediments

In contrast to the pan sediments, the nebkhas and shadow dunes are composed only of quartz sand and gypsum. The nebkhas are rich in gypsum (at least 40%) which is derived from deflation of the pans and from local diagenetic precipitation within the mound due to capillary concentration of groundwater. Gypsum rosettes and nodules are very common within the nebkhas. The quartz sand within the mounds is derived from the nearby sand-sheets which are migrating over the pans.

In the shadow dunes, where the sediments are derived from the immediate upwind nebkha, aerodynamic processes cause some grain sorting. Generally, the lower parts of these dunes are gypsum rich, whereas in the upper parts quartz exceeds gypsum. However, cross-strata within these dunes show some mineral and size segregation, as will be discussed later.

Flow conditions around dunes

The theoretical basis of the dynamics of fluid-flow over discrete objects is well-established (Prandtl 1952; Thwaites 1960). However, there have been very few detailed aerodynamic studies of three-dimensional flow around discrete vegetation reported in the geological literature (Hesp 1981). The flow structure as discussed here applies to a much later stage when the nebkhas are undergoing decay and deflation, with transport

FIG. 7. A wind-shadow dune forming behind a decaying nebkha. Note the orientation of ripples on the dune surface on either side of the photograph. The highest point on the dune is where flow reversals are greatest (maximum deposition). Footprints for scale.

of the previously trapped sediments and their subsequent deposition immediately downwind, forming a shadow dune. In this study visualization of flow over the pyramidal bedform was accomplished by a crude field method of attaching thin coloured ribbons to drinking straws and inserting them into the surface of dunes during strong northwesterly winds. This showed that horizontal flow separation caused by two opposing eddies behind the plants dominates the flow pattern over the dune (Fig. 8).

Experimental studies show that a low-pressure region or 'cavity zone' exists behind the obstacles, which extends downwind and which tends to suck in air into a lee eddy or vortex (Hesp 1981). The exact three-dimensional flow pattern over the plants is dependent on the width and 'porosity' of the foliage (defined as density of foliage/unit area). Most studies have confirmed that the flow-structure in the lee of vegetation is dominated by horizontal separation with only a very minor vertical separation component (Hesp 1981). The shadow dune itself is the result of the reverse flows that occur within the horizontally-separated flow-cell (Fig. 8).

The basal shape of these dunes is nearly symmetrical. The eddies start from the sides of the bush and meet at a centre-line represented by the ridge. The ridge or dune crest is the zone of maximum sediment accumulation, where the two opposing eddies meet and where flow reversals are highest (well within the cavity zone where the sand grains are 'sucked in'). Sand grains transported from the upwind side and past the vegetation follow the flow paths within the two eddies and are deposited about the centre-line where the flow velocity is suddenly reduced. The sand is also transported close to the bed surface of the dune, as evidenced by the wind-ripples covering it (Fig. 7). The orientation of the central ridge can therefore be correlated with the prevailing azimuth of the wind (NW–SE).

The flow pattern envisaged for the *Nitraria* dunes is only a crude approximation based on field observations using the straw flags. The flow structure as illustrated by Hesp (1981) is, however, generally applicable to all vegetation morphologies that are: (1) discrete; (2) semicircular to circular; and somewhat porous. In *Nitraria* nebkhas the foliage is mostly above the surface and hence the morphology of the bedform is not always pyramidal. The presence of erosional

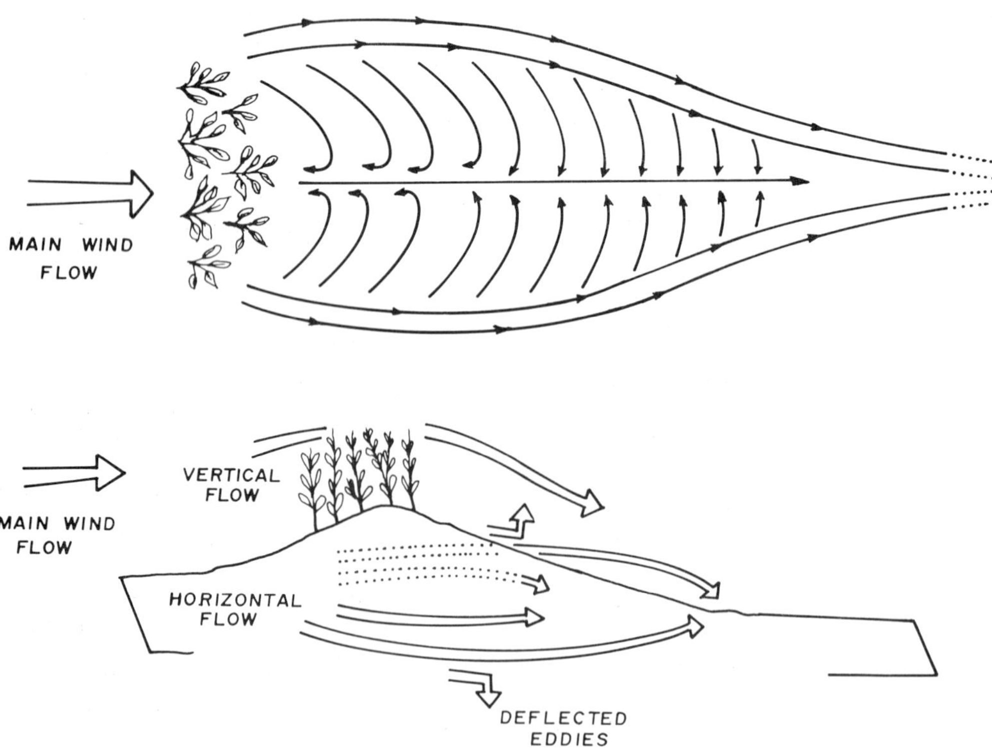

FIG. 8. Inferred flow pattern across *Nitraria* plants showing horizontal flow separation behind an obstacle. Note flow reversals in the central area (plan view); *cf* Fig. 7.

blowouts on the windward side of some dunes suggests that a more complex flow structure may occur on this side. The shapes of the shadow dunes in the pans are dependent to a large extent on the 'plan shape' and porosity of the plant body, and are therefore variable. Actively growing *Nitraria* plants are not always semicircular, and they have a very low porosity (Fig. 4). The nebkha is essentially a sediment-trapping and binding stage in dune development, and cannot strictly be considered as an aeolian bedform as will be seen later when their internal sedimentary structures are described.

The side slopes are not symmetrical about the central ridge. The NE slopes are much less steep than those on the SW (orientated downwind). In some of the dunes studied, the central ridge shows a slight sinuosity, suggesting that the eddy pattern is slightly asymmetric. This may be due to the fact that the plant base itself is asymmetric with respect to the wind vector. Hence, either the two opposing eddies are unequal, or more likely, there are very minor but systematic wind shifts about a vector mean (Fig. 3). In addition, there appears to be a very small slope at the downwind toe of the dune, where it flattens out on to the pan surface. This flattened area is probably the zone of reattachment of the flow. The three-dimensional flow envelope defined by the horizontal and vertical configuration of the flow streamlines determines the morphology of the shadow dunes.

The dimensions of these dunes with respect to the flow conditions were not investigated. This aspect has been thoroughly studied by Bagnold (1954). He concluded that the maximum dune height is a function of the width of the 'obstacle' (plants in the present study) and the angle of repose of the sand-grain population. The wider the base of the vegetation, the greater the height to which a dune grows before the limiting angle of repose is reached. The width of the vegetation barrier dominates over wind velocity as the major variable controlling shadow-dune length. The higher the wind velocity, the shorter the dune length for a given vegetation width. There is no apparent relationship between dune length and plant height (Hesp 1981). If the sediment supply is substantial, long shadow dunes can form.

Sedimentary structures

Diagnostic sedimentary structures in nebkhas and shadow dunes

The internal geometry of these 'dunes' is dependent on their stage of growth. In a mature nebkha with actively growing *Nitraria* plants, the bedform is dome-shaped. At this stage the nebkha can be up to 3 m in height. However, much of the wind energy above the base is dissipated within the vegetation and most of the sand in the flow is trapped within the plants. The sediments within the nebkha are very moist and cohesive because of capillary rise. Only in the uppermost 0.2 m are the sediments dry and loose, with the sand at the angle of repose and avalanching down the sides of the nebkha. Imperfectly developed avalanche cross-strata may be present in the upper layers (Fig. 9). In the rest of the nebkha, bioturbation by roots and animals destroys any sedimentary structures that may have been present. However, isolated primary gypsum nodules may be present within the moist areas of the dune. Thick masses or thin horizons of dark organic matter or dried leaves are also present at various levels in the nebkha. The most prominent feature, however, consists of irregular layers of sugary gypsum at the base of the dune (up to 1 m long and 5 cm thick). These probably result from differential growth and displacement of gypsum crystals (Fig. 9). They resemble ptygmatic layers in gneissic rocks.

The general lack of an internal structure at this stage is due to the baffling and trapping action of the plants where the sands fall through the vegetation. The accompanying bioturbation by roots and burrowing animals destroys laminations.

If the flow conditions envisaged earlier generally hold true for all shadow dunes, then the resulting flow patterns should be reflected in their internal sedimentary structures. Twenty-two dunes from four pans were sliced transversely (normal to the wind direction) and longitudinally (parallel to wind) and their sedimentary structures were examined.

Internal structures normal to wind direction

A highly variable pattern of internal sedimentary structures was observed in this orientation, depending on the spatial distribution of the dunes in a given pan. A common internal structure can be deciphered for shadow dunes whose nearest neighbours are more than ~ 7 m away from each other (yet no two shadow dunes have identical structures). The most distinctive feature in these sections is the bidirectional azimuths and bimodal dip of the cross-stratification, dipping away to either side from the central ridge of the shadow dune (Fig. 10). The dip of the cross-beds are variable, with values of $18°–28°$ in the central area (Fig. 11), decreasing to values of $\sim 3°$ at the toes of the dip slopes on the northeasterly side (Fig. 10). The dips are steeper on the SW side ($>20°$) than on the NE side ($<18°$) when

FIG. 9. Section of a mature nebkha cut parallel to the wind direction. Note irregular layer (arrow) formed by differential growth and displacement of sugary gypsum crystals at the bottom and immediately above the gypsum crust (c) of the pan. Poorly developed avalanche (?) cross-strata at the top. This is an early transitional stage between nebkhas and shadow dunes.

observed in the downwind direction. If dips of <6° are omitted from the measurements they fall into two distinct modes; one of 11°–13° (mostly on the NE side) and the other of 23°–25° (mostly on the SW side). The lower angle dips (<18°) are generally more abundant in all dunes. The bipolar azimuth distributions appear to correlate very well with the prevailing northwesterly 'shamal' winds of this region, and secondary wind vectors (Fig. 3) have a negligible influence on this distribution. The bimodality of the dips probably reflects a slight asymmetry in the flow pattern.

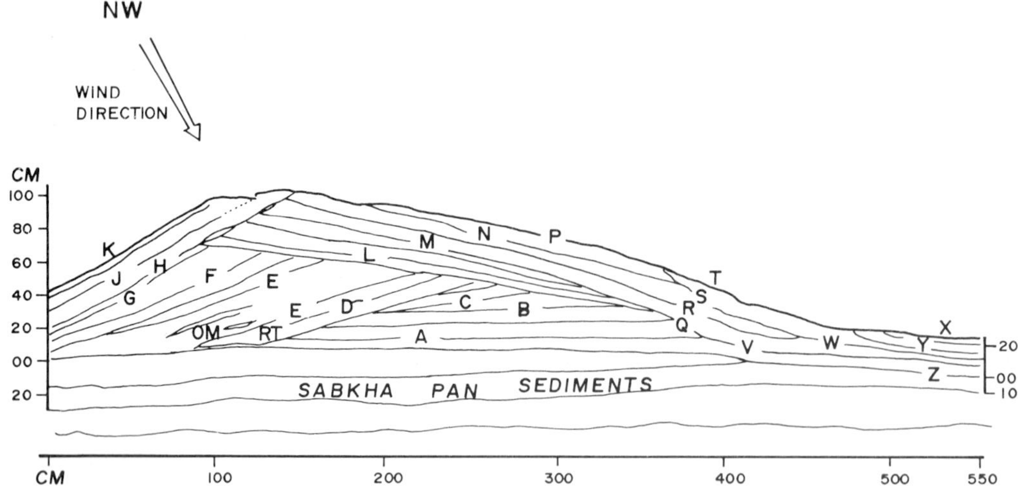

FIG. 10. Composite sketch made from several sections showing bi-directional azimuths and bimodal dip of cross-strata in shadow dunes. Sections normal to the prevailing wind vector. The letters denote various dips of the cross-strata

FIG. 11. Photograph corresponding to central area of composite sketch (Fig. 10). Note alternating light/dark layers of grainfall cross-strata. The dark linear patch is organic matter; (R) is a root. Gypsum crust of the pan is at the bottom of the photograph.

On the other hand, it is not known if the relict morphology of the nebkhas had a role in controlling the bimodal dip distribution in shadow dunes.

When dunes occur in closer proximity to each other (<4–5 m apart), the internal structures are more complex. Two shadow dunes occurring side by side (albeit rarely) are often separated by a narrow deflation corridor with the adjacent bipolar cross-strata truncated and disturbed somewhat at their lateral toes. Obviously, the proximity of so many dunes on the wind field affecting any one individual is an extremely important consideration. Perhaps this is one of the reasons for the asymmetry of these dunes alluded to earlier, although this aspect is beyond the scope of the present study.

In the leeward side of the dunes, where horizontal flow separation is dominant (the 'calm' zones), sand grains lose much of their momentum and fall onto the surface. This is grainfall deposition and the corresponding internal structures produced are grainfall 'laminations' (Hunter 1976). We suspect that much of the bidirectional or winged cross-strata in these dunes are grainfall laminations (Fig. 10). There is very little evidence for avalanching down lee slopes in the zone of flow separation, because the slopes are rarely at the angles of initial yield of the sand grains. Sandflows and slumps were not observed and sandflow cross-strata are unlikely to be present. However, we are not certain of the exact mechanism of formation of these grainfall strata as they can form by a complex of processes including pure grainfall, and processes transitional between traction and grainfall. Also, winds that are too weak to cause traction on a flat surface may cause traction on slopes. Such tractional transport at wind velocities too weak to cause ripples is common on lee slopes in the secondary circulation zone. Weak traction results in smooth slopes, with net deposition giving rise to grainfall laminae (Hunter 1976).

The lee slopes on which grainfall laminations occur are much steeper than the gentler slopes of the rippled surface, but not steep enough for sand avalanching. Preservation in ancient dunes will therefore depend upon the angle of the original leeward depositional surface. The grainfall laminae in the shadow dunes are tabular (Fig. 11). The suggestion of Hunter (1976) that much of the cross-strata dipping at angles of 20° to 28° originate from grainfall and that they can form on any surface in a zone of flow separation appears to be the case for shadow dunes as well. In the composite sketch of sedimentary structures contructed parallel to the dip direction, it can be seen that the grainfall strata have highly variable dips (Fig. 10). The laminae alternate dark and light due to size grading and the concentration of fine, leaf organic matter between the layers. The lighter layers are relatively rich in gypsum and the darker layers are rich in quartz. Each grainfall layer is well sorted and packed (Fig. 11).

Careful field monitoring of grainfall deposition confirms that wind-speed fluctuations are responsible for grain segregation and the resulting grainfall laminations (Hunter 1976). In the shadow dunes investigated it appears that *within* a given grainfall layer, size grading is negligible, which attests to the fairly uniform flow conditions during the time interval represented by the layer (Fig. 11). We are not certain if each layer represents a single episode of grainfall deposition or not, but the boundaries between the layers are very sharp. Shadow dunes with bimodal cross-strata have also been reported by McBride & Hayes (1962); Land (1964) and Goldsmith (1973), but their origin was explained as due to bimodal or changing wind directions, rather than by horizontal flow separation in a unidirectional wind system.

In most of the dunes studied, the *Nitraria* bush is tallest in the middle and tapers to the sides. Flow separation past such an obstacle may result in slightly inclined, opposing eddies in the horizontal plane. This may be partly responsible for the 'winged' nature of the internal geometry of shadow dunes (Figs 10, 11). In dunes sectioned parallel to the wind, the internal structures were not as diagnostic. Wind ripples are common on lee slopes that are inclined less steeply than the angle of repose and where traction transport occurs (Fig. 7). The internal structures produced by net deposition are a type of climbing-ripple cross-lamination (Hunter 1976; Allen 1982). Ripples climb in the direction of the prevailing wind even though they are within the zone of flow-separation. Because the two opposing lee slopes in the shadow dunes are transverse to the wind, the secondary circulation in the zone of flow separation is very weak. If the slopes are oblique to the wind, part of the free air stream enters the flow-separation zone creating vortices, and the flow above the slopes then also tends to have an along-slope component (Allen 1982).

In the dunes studied, most of the ephemeral ripples that trend parallel to the slopes or along slope can be accounted for by lee vortices. This relationship appears to be a common one (McKee 1966; Goldsmith 1973; Hunter 1976). However the preservation potential of these ripples within the dunes is uncertain, as we have not observed any ripple-foreset cross-laminae within transverse sections of the dunes.

In contrast to the nebkha phase, there are neither primary gypsum nodules in the shadow dunes nor any intrasedimentary diagenetic growth of secondary gypsum crystals to disrupt the internal structures. Hence, the grainfall layering remains intact (Fig. 11). Microscopic examination of the grainfall sediments show minor corrosion of gypsum grains, probably because of the annual rainfall which seeps through these porous and permeable sediments. The long-term effects of this intrastratal dissolution could be significant.

Sedimentary structures within the sabkhas and their origin

The diagnostic sedimentary structures of the sabkha sediments are the irregular layers of nodular gypsum which occur as thin sheets confined within the upper 0.1–0.25 m of the pan sediments (Fig. 12). These layers are invariably associated with a gypsum crust and occur where the ground surface is within the capillary fringe (usually within 0.7–0.8 m above the groundwater table). The white nodules contrast with the brown-coloured host sediment. They vary in diameter from a few millimetres to several centimetres and are elliptical to near-spherical in shape. Several of these nodules may join together, forming continuous layers, displacing the host sediment with upward-bulging diapir-like structures (Fig. 12). Such structures are not unlike the incipient enterolithic gypsum veins in the loess sabkhas of northern Egypt (Ali & West 1983). Under the SEM, the nodules are seen to be composed of microcrystalline (10–60 μm), tabular to lenticular gypsum.

The formation of the gypsum nodules and layers are easily explained by the upward movement of groundwaters by capillary rise and evaporation at the surface according to the *per ascensum* model (Watson 1979; Ali & West 1983). Ali & West (1983) suggest that if the interstitial waters are deficient in Na^+ and Cl^- ions, gypsum could form a very thick crust close to the surface. The upward increase in salinity of the interstitial water with the crystallization of halite above the gypsum is evidence for the model.

The gypsum nodules and layers described

FIG. 12. Three-dimensional section of a sabkha pan showing three white layers of coalesced primary gypsum nodules with enterolithic veins and diapirs (arrow). Roots growing horizontally are seen in the pan sediments. Note completely structureless nebkha dune sediments above the pan due to bioturbation by roots. The boundary between the two facies is very sharp.

above are very similar to the anhydrite of marine sabkhas along the Arabian Gulf (Shearman 1966; Butler 1970; Kinsman 1974; Gunatilaka *et al.* 1980). In all cases these are confined to the capillary zone within 1 m of the water table. The absence of anhydrite in the continental sabkha pans of Kuwait is due to the very low chlorinity of the interstitial brines in the nodular gypsum zone (~ 2.6 **M** $Cl^- \ kg^{-1}$), which is insufficient to form anhydrite at the prevailing temperatures. The low salinity of the pan groundwater (Table 1) is due to the greater winter rainfall and the large influx of very dilute waters by subsurface lateral flow from the continental interior (Robinson *et al.* 1987). According to Butler (1969) and Kinsman (1974) a chlorinity of 4–5 **M** $Cl^- \ kg^{-1}$ is required for anhydrite stability in the sabkha environment.

Several lines of evidence suggest that these gypsum nodules were not originally anhydrite. The criteria used have been adapted from Shearman (1966) and Ali & West (1983):

(1) Microscopic (including SEM) and X-ray studies of nodules have not revealed the presence of any traces of relic anhydrite.
(2) The volume increase when anhydrite → gypsum ($\sim 66\%$) should disrupt the host sediments far more than is observed within the pan sediments (Fig. 12).
(3) The nodules are soft, friable and uncemented. Nodules of small crystals overlie nodules of larger crystals.

Detailed descriptions of modern environments containing primary gypsum nodules are rare in the literature. An exception is the excellent study on the Egyptian sabkhas by Ali & West (1983). These authors also cite at least 50 other worldwide occurrences of calcium sulphate facies nodules. The nodular gypsum of the sabkha pans of Kuwait are strikingly similar to those of the Egyptian sabkhas, the playas of Bristol Dry Lake, California and the sabkhas of the Laguna Mormona in Baja California—all of which are associated with halophytes and nebkhas (Pierre & Ortlieb 1980; Handford 1982; Ali & West 1983). In all these areas, evaporation far exceeds rainfall, the sabkha 'soils' are close to the water table and contain capillary brines rich in Ca^{2+}, SO_4^{2-}, Na^+ and Cl^- ions. Interestingly, the climatic conditions of all these areas are closely comparable.

Biological aspects

Sometimes, the entire upper 0.5 m of the dunes is composed of layered organic matter derived from several years of winter shedding of *Nitraria* leaves. In the upper layers, dried leaves are still recognizable, whereas the lower layers are of moist peat. The layering is due to thin horizons of sand that alternate with the leaf organic matter.

The organic material provides a cool, sheltered niche for two species of small beetle. A white-spotted, carnivorous, 'head-hunting' black beetle, 4–5 cm in length occupies large burrows within the sand (up to 5 cm in diameter) in the lower part of the nebkha. This beetle (*Thermophilum duodecimguttatum*; Cicindelidae or tiger beetles) squirts a foul-smelling dark secretion when disturbed. Immediately outside its burrow are dozens of heads and carapaces of the smaller beetles. It is not known whether the predator forays into the organic matter zone in search of the prey, immobilizing it and bringing it back to its own burrow or whether it entices the prey into the burrow. The carnivore leaves a distinct walking trail on the sands (like the imprint of a bicycle tyre). In addition at least one species of lizard (*Diplometopon zarudnyi*) takes refuge in the dunes.

The groundwater table in sabkha pans

The groundwater table has a direct influence on the origin and evolution of the pan-nebkha complexes. The water table is within 1 m of the surface in most sabkha pans and appears perched in relation to that in the surrounding sand sheet and gravel plain areas, where it is more than 3–5 m deep (Fig. 2). The annual water-table fluctuation between the dry summer and wet winter seasons can be between 0.25 to 0.47 m (5 years of observations). When the rainfall data (~ 115 mm a^{-1}) are considered, this variation can only be accounted for by significant lateral, subsurface flow from the continental interior.

The best developed sabkha–nebkha complexes with *Nitraria* plants and dunes in a state of active growth are in those pans within 5–7 km of the coast (Fig. 4). Here, the standing groundwater table in the summer is within 0.75–1.2 m of the surface, and the entire complex is in a state of depositional–deflational equilibrium. In some pans further inland, where the water table is much shallower (<0.7 m), the halophytes are dead or dying and the nebkhas are presently undergoing deflation with the development of pyramidal, wind-shadow dunes to the lee of the decaying plants. Some of the deflated sediment is also redistributed on the pan surfaces. The decay of the halophyte is probably due to the waterlogging of its roots. The most distant pan studied is

TABLE 1. *Chemical and isotopic composition of selected groundwater from the study area*

Pan no.	Water table depth (m)	Ca^{2+}	Mg^{2+}	Na^{+}	K^{+}	SO_4^{2-}	Cl^{-}	Total alkalinity	IAP_G	Water		Sulphate	
										$\delta^{18}O$	δD	$\delta^{34}S$	$\delta^{18}O$
CS-1	0.74–1.0	2400	1060	9600	740	3615	18570	148	3.6×10^{-5}	−0.5 to −0.8	−15.2 to −16.7	13.4 to 13.7	13.4 to 14.6
CS-2	0.88–1.1	2200	1170	9190	700	3815	18220	163	3.5×10^{-5}	−0.9	−13.8	13.8	13.3
CS-3	0.48–0.96	1250	800	7000	875	3010	17750	88	1.3×10^{-5}	−0.8 to −1.0	−14.8 to −16.1	13.5 to 13.8	13.7 to 13.9
CS-4	0.5–0.6	1100	560	7350	440	2100	18450	118	9.4×10^{-6}	0.9	−3.3	12.9	12.9
CS-5	0.5–0.6	610	320	1800	150	1600	13100	110	3.9×10^{-6}	−0.3	−8.2	—	13.4
CS-6	Deep aquifer	390	115	1100	54	900	2195	140	1.4×10^{-6}	−1.9	−14.1	14.6	14.3
CS-7	Water pumping	405	115	940	26	945	2085	138	1.7×10^{-6}	−1.9	−13.9	14.9	14.5
Rainwater										−2.2(w)	−7.9	—	—
										−0.9(s)	2.9	—	—
Gulf seawater		425	1510	10875	875	3495	26625	200	6×10^{-6}	2.4	14.7	20.9	10.4

$K_{SP}^{Gyp} = 2.5 \times 10^{-5}$; IAP_G = Ion activity product; Isotopic standards V—SMOW and CDT; Chemical data in ppm.

22 km from the coast; here, the entire complex has been deflated down to the bedrock, exposing fluvial gravels. The early stages of calcrete/gypcrete development can be seen in this location. Here it is inferred that the water table has fallen drastically.

It appears that the sabkha–nebkha complex passes through phases of growth, stabilization and decay that are controlled by the hydrological characteristics of the local and regional groundwater tables. As the nebkha builds up in height, the roots keep pace, so that contact with the moist capillary zone is retained.

It has been suggested that the nebkhas grow to a maximum height of ~5 m, after which growth is minimal because of loss of root contact with 'soil' moisture (Military Engineering Experimental Establishment 1969; Cook & Warren 1973). In the present study area nebkhas rarely develop to heights greater than 3 m. After a period of stabilization, decay sets in, probably due to a rising water table. The maximum height to which a dune grows must be dependent on the type of plant responsible for trapping the available sediments. Waterlogging may be more important in destroying a nebkha than loss of root contact with the moist soil layers, as evidenced by the observation that lateral growth of roots are more common (up to 7 m long) than roots growing vertically downwards. In this connection the development of the root system of *Nitraria* (a typical Saharo–Arabian halophyte) along the same lines as that of the plant responsible for a 'rebkha' in the Sinai, and its relationship to moderate-salinity waters at between 0.3 and 1 m depth is of relevance (Evenari *et al.* 1985).

Chemistry of the pan groundwaters

Table 1 summarizes the groundwater chemistry of some of the pans. For comparison, the Dammam limestone aquifer of Eocene age, 45 km W of the coast, was also sampled. Salinity of the groundwater increases from the desert to the coastal pan areas (from 5‰–6‰ to 24‰). In most pans, the groundwater table and/or interstitial water is supersaturated with respect to gypsum which is precipitating diagenetically within the host sediments. The product of $^{m}Ca^{2+} \cdot {}^{m}S_4^{2-}$ in relation to K_{SP}^{Gyp} (2.5×10^{-5}) indicates the degree of supersaturation of gypsum (Table 1). But periods of undersaturation are indicated by the leached outlines of the gypsum crystals seen in thin section. Active nebkhas containing growing *Nitraria* plants are never found in pans where the salinity exceeds 26‰–28‰.

The very high Ca^{2+} and SO_4^{2-} values of the pans relative to aquifer water or seawater are noteworthy. There is a steady evaporative concentration of these waters from the inland areas to the coast. Pore waters from the uppermost moist layers of sediment indicate that the salinity reaches values of 135‰–145‰ (~2.5 **M** Cl^- kg^{-1}), sufficient to precipitate gypsum but not anhydrite. The composition and behaviour of the water table can explain the precipitation of the gypsum nodules according to the *per ascensum* mechanism. The overall chemistry of the different pans (Table 1) suggests that they have open hydrological systems and therefore are not 'playas'.

The stable isotopic composition (^{34}S, ^{18}O, D) of the groundwaters of five of these pans was determined as part of a regional study of groundwaters in the State of Kuwait (Robinson *et al.* 1987). In summary, most of these groundwaters are old formation waters with low δD and $\delta^{18}O$ values. Sabkha pan groundwaters are derived from these formation waters but are affected by evaporative concentration. The inland sabkha samples tend to have very low sulphate isotope values and their isotopic history is more complex. However, they have similar $\delta^{34}S$ values to those of the Neogene aquifers of Arabia (Shampine *et al.* 1978). ^{14}C and tritium analyses of Kuwait aquifer waters suggest ages of up to 40 000 years (Abusada 1982). Most of the recharge of similar aquifers in Saudi Arabia is thought to have taken place approximately 20 000 years ago. Several studies of palaeowaters in desert areas show similar 'light' isotopic signatures (Gat & Issar 1974; Fleischer *et al.* 1977). They presumably reflect the meteoric water cycle of a former climatic period. The lighter isotope values of the Kuwait palaeowaters are consistent with a cooler and wetter climate than today. Chapman (1971) proposes a cool, moist pluvial period in eastern Arabia between 11 000 and 60 000 years BP.

Preservation potential and the geological record

The pre-Holocene geological history of the study area was dominated by a Pleistocene pluvial episode that shed vast amounts of coarse clastic debris originating from interior Arabia onto a Tertiary erosional land surface. These deposits are represented today by the sheet-flood deposits of a very extensive braid-plain called the Dibdibbah. The sediments were transported to eastern Arabia through the Wadi Al-Rimah–Al-Batin system, over 500 km in length. Large deflation hollows on the erosional surface (the eventual location of the sabkha pans) received some of the fluvial debris in their early stages. It appears that this earlier fluvial phase was followed by one characterized by alternating, ephemeral lacustrine (or ponded) and aeolian conditions. The laminated, calcite micrite horizons probably represent these lacustrine phases. Similar lacustrine mud deposits of Quaternary times have been described in the Rub' Al-Khali (the Empty Quarter) by McClure (1976). With the filling up of the hollows, the environment gradually changed in the Holocene to one dominated by aeolian and sabkha-pan conditions (Fig. 5). Much of the aeolian material was probably derived from the reworking of the thin, sheet-flood deposits. Intrasedimentary, diagenetic gypsum characterizes the sabkha facies. In the absence of a significant volume of sand, large dune bedforms could not form. Instead the available sand was trapped by the halophytes into nebkhas. The very thin sand sheets (0.1–0.7 m thick) and gravel of the plains marginal to the pans attests to a starved sand supply. The changing conditions within the pans from the Pleistocene to the present reflects the changing climate and hydrology of these areas. Observations of both the sediments and the groundwater point to the very important role Pleistocene pluvial events have played in shaping the evolution of these terrains.

The preservation of the sabkhas is very much dependent on a rising water table, which allows the accumulation of sand-sheet sediments moving across them. However, an identifiable sequence would be produced only if the thickness of the sand sheet was significant. A rising water table can also destroy the nebkhas. A combination of rapidly rising water table and much larger sand supply could cover the entire pan and its shadow dunes so that the sand sheets merge across the pans. In the geological record such a situation would be represented by regionally extensive sheet sandstones with very low angle climbing-ripple cross-laminations, containing very large (up to 2 km long) lenses rich in evaporites (gypsum and/or anhydrite). The evaporites would pinch out sharply on either side. Lateral facies changes and upward lithological transitions would be very abrupt. If the shadow dunes were also preserved in ancient sediments, their asymmetric, bimodal cross-bedding with bipolar azimuths would indicate: (1) the azimuth of the prevailing or dominant regional wind vector; and (2) the presence of nebkhas with halophytes.

On the other hand, a drastic drop in the water table could lead to deflation of the entire evaporite pan complex. In this case, if later sand migration occurs, then the base of the sheet would have a lag layer of pebbles and a calcrete of duricrust horizon. This situation occurs in the furthest inland pan studied. The above features reflect indirectly the palaeohydrological regime of the environment.

Nitraria retusa is an angiosperm. This group appears in the rock record only from the Cretaceous period onwards. Any recognition of shadow-dune sedimentary structures in pre-Cretaceous rocks would suggest the existence of gymnosperm-controlled nebkhas. A possible candidate for this search is the Triassic rocks of SW England.

ACKNOWLEDGMENTS: We thank Professors D. J. Shearman and David Almond for reading the manuscript critically and for making valuable suggestions. Dr Ian West's comments at the refereeing stage have considerably improved the manuscript.

References

ABUSADA, S. M. 1982. Interpretation of environmental isotope data of Kuwait groundwater. *Gulf and Arabian Peninsula Studies* **2**, 80–95.

ALI, Y. A. & WEST, I. 1983. Relationships of modern gypsum nodules in sabkhas of lӧess to compositions of brines and sediments in northern Egypt. *Journal of sedimentary Petrology* **53**, 1151–1168.

ALLEN, J. R. L. 1982. *Sedimentary structures: Developments in Sedimentology*, 30A, 30B, Vols I & II. Elsevier, Amsterdam.

AL-SAYARI, S. S. & ZOTL, J. G. 1978. *The Quaternary Period in Saudi Arabia*. Springer-Verlag, Vienna.

AMIEL, A. J. & FRIEDMAN, G. M. 1971. Continental sabkhas in Arava Valley between Dead Sea and Red Sea. Significance for the origin of evaporites. *American Association of Petroleum Geologists Bulletin* **55**, 581–592.

BAGNOLD, R. A. 1954. *The Physics of Blown Sand and Desert Dunes*. Chapman & Hall, London.

BUTLER, G. P. 1969. Modern evaporite deposition and geochemistry of coexisting brines, the sabkha, Trucial coast, Arabian Gulf. *Journal of sedimentary Petrology* **39**, 70–89.

—— 1970. Holocene gypsum and anhydrite of the Abu Dhabi sabkha, Trucial Coast: an alternative explanation of origin. *In*: RAU, J. L. & DELLWIG, L. F. (eds) *3rd Symposium on Salt*. Northern Ohio Geological Society, Cleveland, Ohio, 120–152.

CHAPMAN, R. W. 1971. Climatic changes and evolution of landforms in the eastern province of Saudi Arabia. *Geological Society of America Bulletin* **82**, 2713–2728.

COOKE, R. U. & WARREN, A. 1973. *Geomorphology in Deserts*. Batsford Ltd, London.

EVANS, B., SCHMIDT, V., BUSH, P. & NELSON, H. 1969. Stratigraphy and geologic history of the sabkha, Abu Dhabi, Persian Gulf. *Sedimentology* **12**, 145–159.

EVANS, G. 1966. Recent sedimentary facies of the Persian Gulf region. *Philosophical Transactions of the Royal Society of London, Series A* **259**, 291–298.

EVENARI, M., GUTTERMAN, Y. & GAVISH, E. 1985. Botanical studies on coastal salinas and sabkhas of the Sinai. *In*: FRIEDMAN, G. M. & KRUMBEIN, W. E. (eds) *Hypersaline Ecosystems: The Gavish Sabkha*. Ecological Studies **53**, Springer Verlag, Berlin, 145–182.

FLEISCHER, E., GOLDBERG, M., GAT, J. R. & MARGARITZ, M. 1977. Isotopic composition of formation waters from deep drillings in southern Israel. *Geochimica et Cosmochimica Acta* **41**, 511–525.

FRYBERGER, S. G., AL-SARI, A. M. & CLISHAM, T. J. 1983. Eolian dune, interdune, sandsheet, and siliciclastic sabkha sediments of an offshore prograding sand sea, Dharan area, Saudi Arabia. *American Association of Petroleum Geologists Bulletin* **67**, 280–312.

GAT, J. R. & ISSAR, A. 1974. Desert isotope hydrology: Water sources of the Sinai Desert. *Geochimica et Cosmochimica Acta* **38**, 1117–1131.

GLENNIE, K. W. 1970. *Desert Sedimentary Environments*. Developments in Sedimentology **14**, Elsevier, Amsterdam.

GOLDSMITH, V. 1973. Internal geometry and origin of vegetated coastal sand dunes. *Journal of sedimentary Petrology* **43**, 1128–1142.

GUNATILAKA, A., SALEH, A. & AL-TEMEEMI, A. 1980. Plant controlled supratidal anhydrite from Al-Khiran, Kuwait. *Nature* **288**, 257–260.

——, AL-TEMEEMI, A., SALEH, A. & NASSAR, N. 1985. A new occurrence of bassanite from recent evaporitic environments, Kuwait. *Journal of Kuwait University (Science)* **12**, 157–166.

——, SALEH, A., AL-TEMEEMI, A. & NASSAR, N. 1984. Occurrence of subtidal dolomite in a hypersaline lagoon, Kuwait. *Nature* **311**, 450–452.

HANDFORD, C. R. 1981. A process sedimentary framework for characterizing recent and ancient sabkhas. *Sedimentary Geology* **30**, 255–265.

—— 1982. Sedimentology and evaporite genesis in a Holocene continental-sabkha playa basin—Bristol Dry Lake, California. *Sedimentology* **29**, 239–253.

HARDIE, L. A. 1968. The origin of the Recent nonmarine evaporite deposit of Saline Valley, Inyo County, California. *Geochimica et Cosmochimica Acta* **32**, 1279–1301.

HESP, P. A. 1981. Formation of shadow dunes. *Journal of sedimentary Petrology* **51**, 101–112.

HOLM, D. A. 1960. Desert geomorphology in the Arabian Peninsula. *Science* **132**, 1369–1379.

HUNTER, R. E. 1976. Basic types of stratification in small eolian dunes. *Sedimentology* **26**, 361–387.

KHALAF, F., GHARIB, I. M. & AL-HASHASH, M. Z. 1984. Types and characteristics of the recent surface deposits of Kuwait, Arabian Gulf. *Journal of Arid Environments* **7**, 9–33.

KINSMAN, D. J. J. 1969. Modes of formation, sedimentary associations and diagnostic features of shallow-water and supratidal evaporites. *American Association of Petroleum Geologists Bulletin* **53**, 830–840.

KINSMAN, D. J. J. 1974. Calcium sulphate minerals of evaporite deposits: their primary mineralogy. *In: 4th Symposium on Salt*. Cleveland, Northern Ohio Geological Society **1**, 343–348.

LAND, L. S. 1964. Eolian cross-bedding in the beach-dune environment, Sapelo Island, Georgia. *Journal of sedimentary Petrology* **34**, 289–294.

MCBRIDE, E. F. & HAYES, M. O. 1962. Dune cross-bedding on Mustang Island, Texas. *American Association of Petroleum Geologists Bulletin* **46**, 546–551.

MCCLURE, H. A. 1976. Radiocarbon chronology of late Quaternary lakes in the Arabian Desert. *Nature* **263**, 755–756.

MCKEE, E. D., 1966. Structures of dunes at White Sands National Monument, New Mexico. *Sedimentology* **7**, 1–69.

MILITARY ENGINEERING EXPERIMENTAL ESTABLISHMENT 1969. Glossary of local physiographic terms about world hot deserts. *M.E.X.E. Report 1124*, Christchurch, Hampshire.

PIERRE, C. & ORTLIEB, L 1980. Sedimentation et diagenese dans trois lagunes evaporitiques de Basse Calfornie (Mexique): données geochimiques et isotopiques sur les sediments et les saumures interstitielles. *Science de la Terre* **24**, 129–156.

PRANDTL, I. 1952. *Essentials of Fluid Dynamics*. Blackie and Sons Ltd, London.

PURSER, B. H. (ed.) 1973. *The Persian Gulf*. Springer-Verlag, Berlin.

—— 1985. Coastal Evaporite Systems. *In*: FRIEDMAN, G. M. & KRUMBEIN, W. E. (eds) *Ecological Studies 53: Hypersaline Ecosystems*. Springer-Verlag, Vienna, 72–102.

ROBINSON, B., STEWART, M. & GUNATILAKA, A. 1987. Use of sulphur and other stable isotopes in environmental studies of regional groundwater flow and sulphate mineral formation in Kuwait: *In:* KROUSE, H. & GRINENKO, V. (eds) *Use of Isotopes in Sulfur Cycle Studies*. SCOPE Report, Wiley, New York, in press.

SHAMPINE, W. J., DINCER, T. & NOONS, M. 1978. An evaluation of isotope concentrations in the groundwater of Arabia. *In: Isotope Hydrology 1978*. Vol. 2. Proceedings of the International Symposium of Isotope Hydrology, Meulenberg, I.A.E.A., Vienna, 443–463.

SHEARMAN, D. J. 1966. Origin of marine evaporites by diagenesis. *Transactions of the Institute of Mining and Metallurgy Bulletin* **77**, 208–215.

THWAITES, B. (ed.) 1960. *Incompressible Aerodynamics*. Clarendon Press, Oxford.

WATSON, A. 1979. Gypsum crusts in deserts. *Journal of Arid Environments* **2**, 3–20.

A. GUNATILAKA & S. MWANGO, Department of Geology, Kuwait University, Box 5969, Kuwait 13060.

Dune networks

A. Warren & S. Kay

SUMMARY: Dune networks are one of the commonest of dune types, and yet the least understood. A nomenclature is developed for describing the components of these systems. It uses the idea of dune memory: dune systems with little sand have little memory, those with more sand 'remember' the effect of the most dominant wind throughout the year; some dunes may have 'mega-memories', remembering events from the Pleistocene. The nomenclature defines the components of dune networks according to both relative dominance and whether the components are transverse, seif-like or nodal elements. Using this nomenclature, the dynamics of two dune networks in the Wahiba Sands of Oman are described. In one low-memory dune system, where individual dunes are 1 to 2 m high, a plot of 150 by 50 m was surveyed. In a second, higher memory dune system, where individual dunes are 5 to 6 m high, a 200 by 200 m plot was surveyed and an intensive study of sand movement was conducted. Both areas were surveyed in January and March 1986. A further survey was conducted in mid-July. Dune movement is related to topographic parameters such as altitude, slope angle and slope azimuth, slope curvature and proximity to other slopes. An analysis of sand loss or gain at one site is also presented. The dunes were found to be highly mobile. Some crests moved up to 80 cm in 9 hr. Movement was restricted to the upper flanks of ridges. However, adjustment to new wind conditions is very quick.

Dune networks are one of the commonest types of dune terrain. A dune network is defined as one in which transverse and longitudinal elements, each aligned to a number of winds, are superimposed. Dune networks are sometimes referred to by the Mauritanian term 'akle' (Monod 1958). Dune networks correspond roughly to the category of 'complex crescentic dunes' in the survey of Breed *et al.* (1979), and this category covers about 40% of the sand seas they surveyed. Surveys of dunes on Mars show that they are very common there too. Despite a wide distribution, they are the least understood and least studied of all dune types. The purpose of the present study is to develop a terminology for describing these systems and to record the dynamics of two dune networks.

Early empirical work on aeolian dune dynamics is summarized by Cooke & Warren (1973); more recent work is reviewed in Warren (1984). However, detailed empirical observation of the dynamics of desert dunes are few. There have been only two close studies of dune dynamics in the last two decades: Howard *et al.* (1977) examined a barchan dune; and Tsoar (1983) reported a number of observations made on a longitudinal or seif dune. Some of the techniques used in these two studies have been adopted in the present research.

Study areas

The Wahiba Sands lie in the eastern part of the Sultanate of Oman (Fig. 1). They contain a great range of dune forms (Warren in press; Goudie *et al.* in press). The dominant forms are mega-ridges with superimposed smaller ridges; there are also large and small transverse dunes, mega-barchans, and a variety of dune networks. Some of the dunes, notably the mega-ridges that dominate the northern part of the Sands, appear to be relics of a former, perhaps stronger wind regime (Glennie 1970). However, in the S and E of the Sands, very active, white sands are formed into mobile, modern dune forms. Most of these white sands are in the form of dune networks.

The Sands experience two distinct wind seasons (Warren in press). Between December and March, winds are light and generally easterly. Directions and speeds are very variable as winds are drawn in to cyclonic depressions tracking to the N. For example, the winds in spring 1986 changed daily and even hourly in both speed and direction. In the summer or kharif season on the other hand, strong onshore southwesterlies or southerlies blow over the Sands, and leave a distinct imprint on the dune forms, as will be shown.

The dynamics of two mobile dunes in the modern active dune area are reported here. The first and smaller site is at Ras Dhabdhub (21° 32′ N, 58° 49′ E; Fig. 1). This site is at the northern end of a large salient of mobile dunes extending from the coast into the central Sands, known as Dhabdhub. The salient of active dunes is apparently advancing slowly northward over the more stable ancient dunes to the N. The active dunes at this site were transverse to the easterlies at the time of our study, but appear to become nearly

From FROSTICK, L. & REID, I. (eds), 1987, *Desert Sediments: Ancient and Modern*, Geological Society Special Publication No. 35, pp. 205–212.

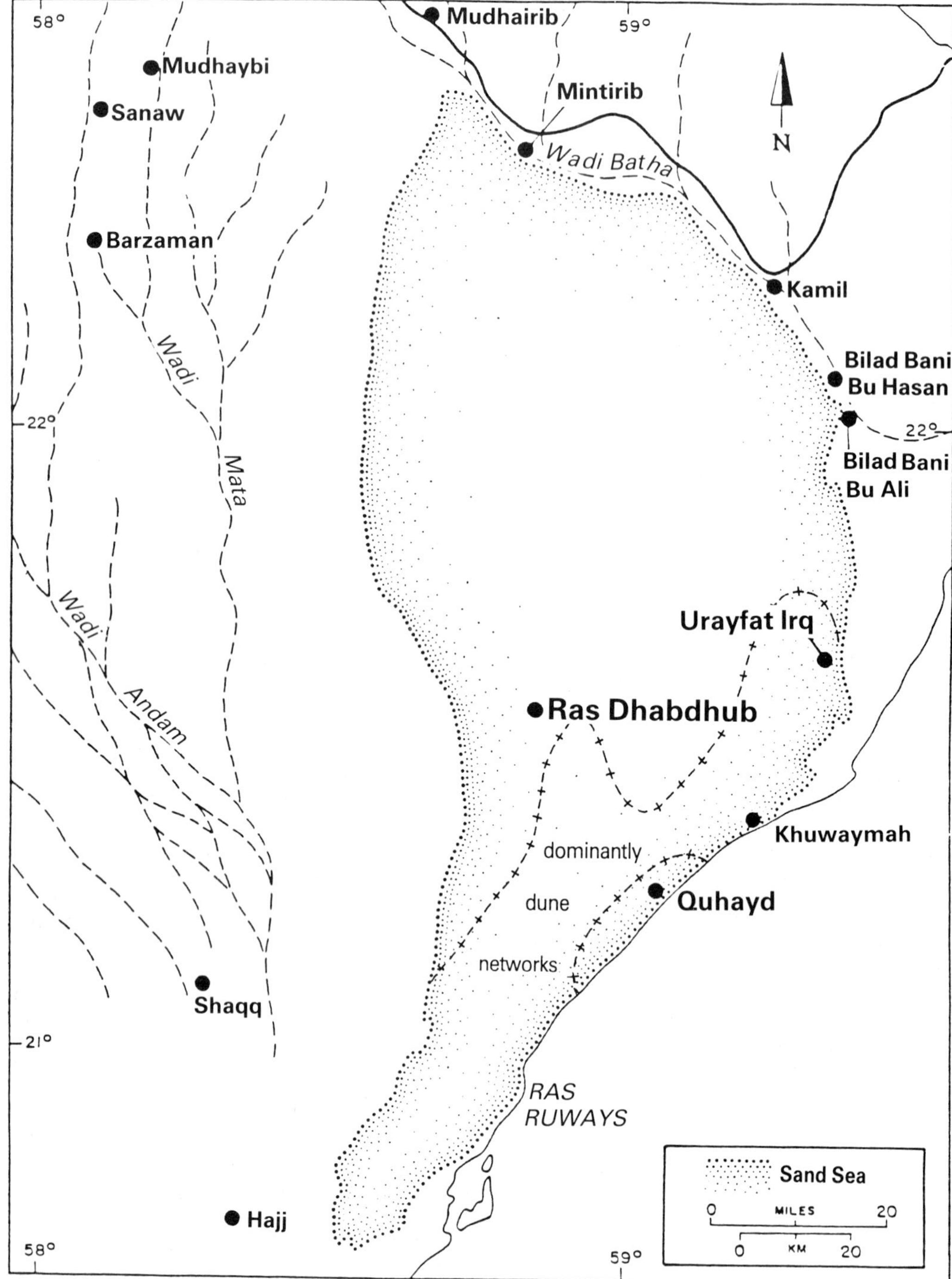

FIG. 1. Location of study areas in eastern Oman.

parallel to the strong south-southwesterly winds of summer. The dunes are only 1 to 2 m high. The sand contains a considerable proportion of coarse, red, cherty grains. A site 100 by 50 m was chosen for observations.

The second site is the 'Urayfat Irq (21° 40′ N, 59° 19′ E; Fig. 1). The dominant dune form here is a series of ridges, 5 to 10 m high and spaced about 30 m apart, transverse (roughly orthogonal) to the monsoon winds which blow from a SSW direction in this area. These ridges are overlain by others, as will be explained below. The sand is fine and white and dominated by calcareous material, but with a small proportion of fine, dark minerals (as yet unidentified). Dunes such as these are typical of a very large area in the eastern and southern part of the Sands. A plot 200 by 200 m was surveyed, and a range of close monitoring techniques was applied.

Methods

Survey and graphical analysis

Two methods were used to map the movement of dunes. First, two large-scale land surveys were made of the 'Urayfat and Ras Dhabdhub sites. At 'Urayfat, six stations were fixed and used to make tachymetric observations using a theodolite and electronic distance measurement. The conceptual scale was around 1:200 with a 0.5 m contour interval. Spot heights were obtained along ridgelines, gullies, bottoms of slip faces, *etc.* Two complete surveys were made at both 'Urayfat and Ras Dhabdhub, one in late January and one in late March. A further survey has been undertaken in mid-July 1986. At 'Urayfat some 1500 spot heights were recorded each time, giving a density equivalent to a 5 m sample grid.

The surveys, after initial processing, have been used to construct Digital Terrain Models (DTM). One of these—that for the Ras Dhabdhub site—is presented here.

Graduated pegs

To complement the land survey, a network of round dowling pegs was installed. The pegs, mostly of 90 cm length, were graduated in centimetres from the top. Loss or gain of sand could thus be recorded. Disturbance of the site was restricted to periods when pegs had to be replaced. This disturbance of the sand around the pegs created a much more mobile environment, but only for two or three days after installation if the winds were light, and for much less time if they were strong. The form of the surface on which the pegs sat was recorded on five occasions by land-based photogrammetry and once by land survey.

Because the pegs were round in cross-section, there was only slight scour where they emerged from the sand, but they did create wake trains even in gentle winds. The directions of these trains were recorded to show the direction of movement of sand on the dune. Howard *et al.* (1977) used a similar technique.

One hundred and twenty-one pegs were installed on a 2 by 2 m grid of irregular shape. The maximum dimensions of the grid were 44 by 22 m, the long dimension being transverse to the dune ridge, from one swale to the next. Observations were generally made on a daily or twice daily basis, although on days when there had only been light winds, no observations were made. This simple technique has yielded a large database. Observations from one line are given here. The daily rates of change will later be correlated with altitudinal position on the dune, slope curvature and slope aspect and with the data from anemometry.

A notation system for dune network systems

The studies of the dynamics of the 'Urayfat Irq and the dunes at Ras Dhabdhub have highlighted the need for a nomenclature for dune networks. The following notation was tested in the Wahiba Sands.

The first part of the notation refers to whether a dune system can or cannot 'hold', 'fix' or 'remember' the results of wind action. The ability of a dune system to 'remember' a wind refers to its capacity to hold the effects of one wind when the wind changes. Active dunes occur in a spectrum between two extremes: low memory (LM); and high memory (HM). LM (low-memory) dunes have little sand, and are small (usually less than 3 m high). They are re-orientated to new winds with ease, and successive winds leave little trace of their precursors. The dunes at Ras Dhabdhub appear to be such a system. HM (high-memory) patterns occur on deeper sand as at the 'Urayfat Irq where the dunes are about 6 m high. They 'remember' the effects of the dominant annual event for the rest of year. In all networks it is of course the stronger winds that are remembered, and the weaker that are forgotten.

The larger the dune, the longer its memory. The highest memory systems in the Wahiba Sands are the mega-ridges of the main body, whose outer form (and probably also internal structure) apparently 'remembers' the high winds of the last glacial period. These might be termed mega-memory (MM) systems.

To survive, a dune ridge must be larger than can be modified by a succeeding wind. In most climates, winds from one compass quarter usually dominate at least for a season, and the winds that blow from other quarters are much less frequent and usually of low velocity. In seasons with a very dominant wind, the effects of all earlier winds on the outer form of the dune field may be obliterated at that season, but the effects of the dominant wind will be retained in the seasons of lighter winds. Different annual wind regimes will create different combinations of ridges at any one time, and, depending on the relative strengths of the winds, there will be different combinations of survivors.

The notation refers to a specific time and to the ridges retained by the network at that time, including short-lived ridges created by the last wind. It is useful therefore for describing a network as recorded on an aerial photograph.

The obliteration of the effects of light winds on the outer form of the dune does not necessarily extend to the sedimentary structure. Our experience in the 'Urayfat Irq shows that, even when a new wind has obliterated the surface expression of an earlier, lighter wind from a dune network, the inner structure still preserves some of the bedding attributable to the earlier winds. Another notation system, independent of the first, would be needed to refer to that part of the memory held in the internal structure of the network; but that is beyond the scope of this paper.

The next part of the notation refers to the *patterns* that are retained by the network (see Fig. 2). The strongest winds create the primary transverse dune ridges: System T1; a sub-notation can be used to identify the wind that is responsible—*eg* T1-SSW is a dominant ridge system transverse to a SSW wind. The next strongest wind creates a secondary, crossing set of transverse ridges: System T2; this too can be annotated with its wind, *eg* T2-NE. Successive sets of diminishing importance are T3, T4, *etc*.

In other types of dune, such as in the mega-ridge system of the Wahiba Sands, there are dunes that are longitudinal to various winds; these would be given the notation L1, L2, *etc*.

The final two parts of the notation refer to secondary forms that result from the interference of two or more winds. Like the transverse ridges, these too may appear and disappear seasonally. The first of these is what could be termed a nodal dune, formed at the crossing of different transverse systems (see Fig. 2). They can be termed N (1, 2) if they form at the nodes of T1 and T2; N(2, 3) if they form at the nodes of T2 and T3; and so on.

The second form of 'interference' dunes is the seif-type dunes formed by the interference of two winds. The dynamics and form of free-standing seif have been explained by Bagnold (1941) and examined in detail by Tsoar (1983). Seif dunes are also found within networks. They can be denoted by the letter S and given the following notation: S(1–2) if they form by the interference of the winds that form T1 and T2 ridges; and so on. Seif dunes have not been shown on Fig. 2 to avoid complication.

The 'Urayfat Irq in the winter of 1985–86 was a HM system with T1-SSW and T2-NE, and a T3 whose orientation depended on the day on which it was seen. The system had a well-developed set of N(1–2) and occasional S(1–2). This approximates to the pattern shown in Fig. 2. The Ras Dhabdhub system seems to be a LM system but may have T1-E and S(1–2) (where T2 is minor and lies across winds from the SSW).

Preliminary analysis of network dune dynamics

Fig. 3 shows contour maps of the two successive surveys at Ras Dhabdhub. Changes are apparent. The outline of the main dune ridge is much less indented on the later survey, because colls appear to have been filled in. These two maps, nevertheless, are not in themselves very useful and must be subject to further analysis. One of these analyses is shown in Fig. 4 where the losses and gains are plotted on a grid basis. This is much more useful. It can be seen that, as a whole, this particular dune seems to have gained sand in the period between the two surveys, and to have gained it most markedly in the colls of the ridge.

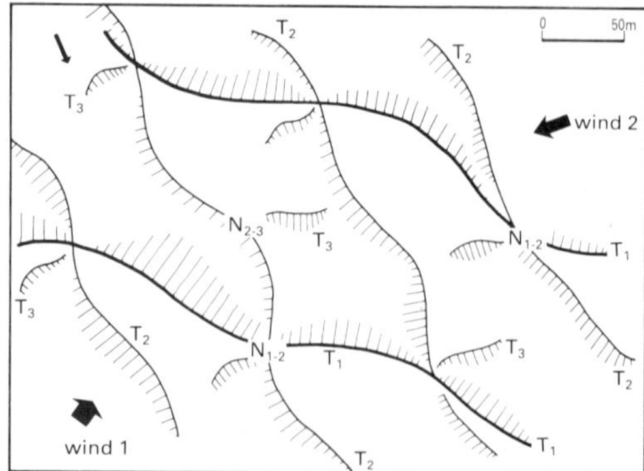

FIG. 2. A notation system for dune networks. The network shown is diagrammatic.

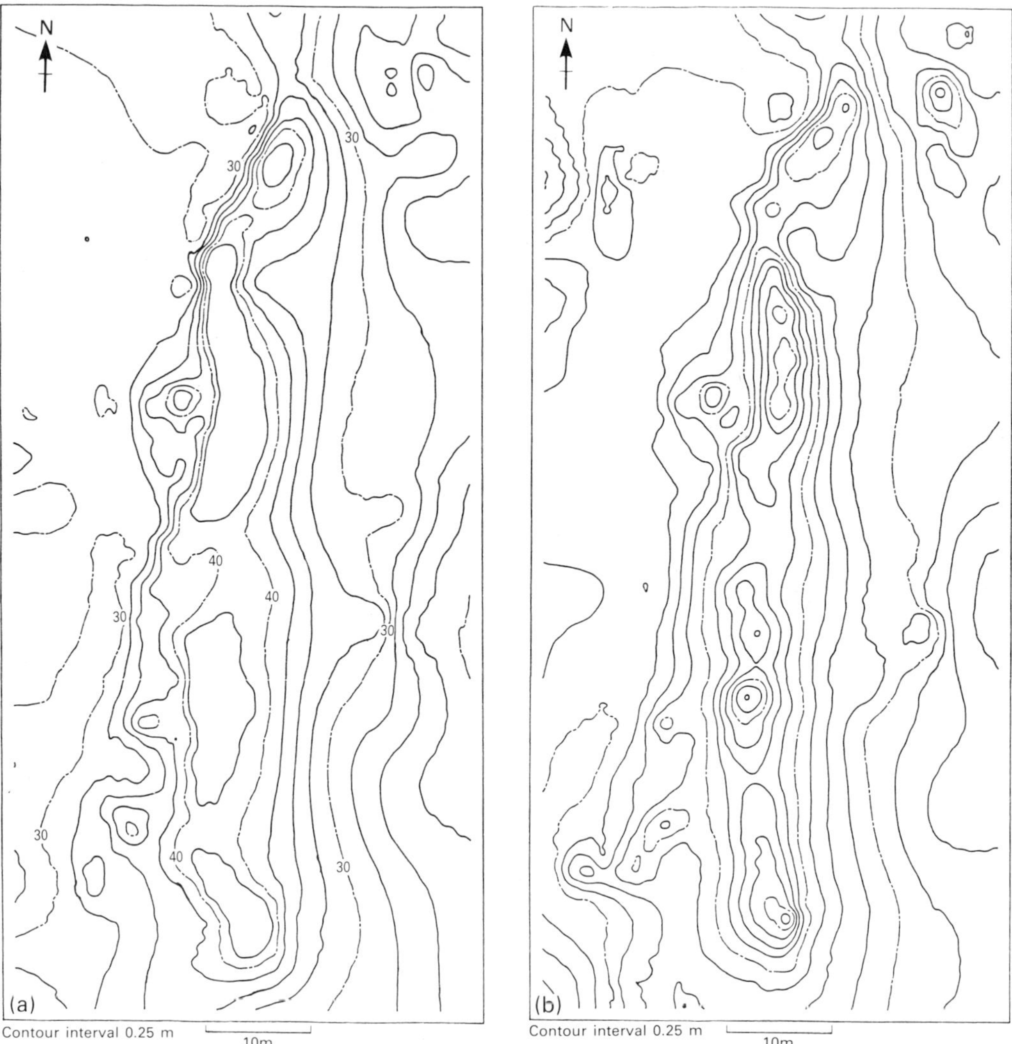

FIG. 3. (*a*) Contour plot of land-survey data, Ras Dhabdhub site, first survey—14 January 1986. (*b*) Contour plot of land-survey data, Ras Dhabdhub site, second survey—28 March 1986.

There were small losses on the western side, but the slip faces here seem to have remained virtually in the same place. This raises many questions: Where has the sand come from? Which kinds of site attract sand? Some of these questions will be answered when further analysis of the data for Ras Dhabdhub has been undertaken. This will allow a quantitative comparison of gains and losses, and an explanation in terms of slope angle, slope curvature, and position. The same will be possible for the 'Urayfat Irq.

Changes in surface elevation and dune morphology

The pegs at the 'Urayfat Irq were installed on a T1-SSW ridge onto which a T2-ENE ridge was occasionally superimposed. There were even more occasional T3 ridges aligned across daily winds of different orientation.

The most striking conclusion to be drawn from the peg observation is that the dunes are very mobile. At one fixed point the crest of a slip face

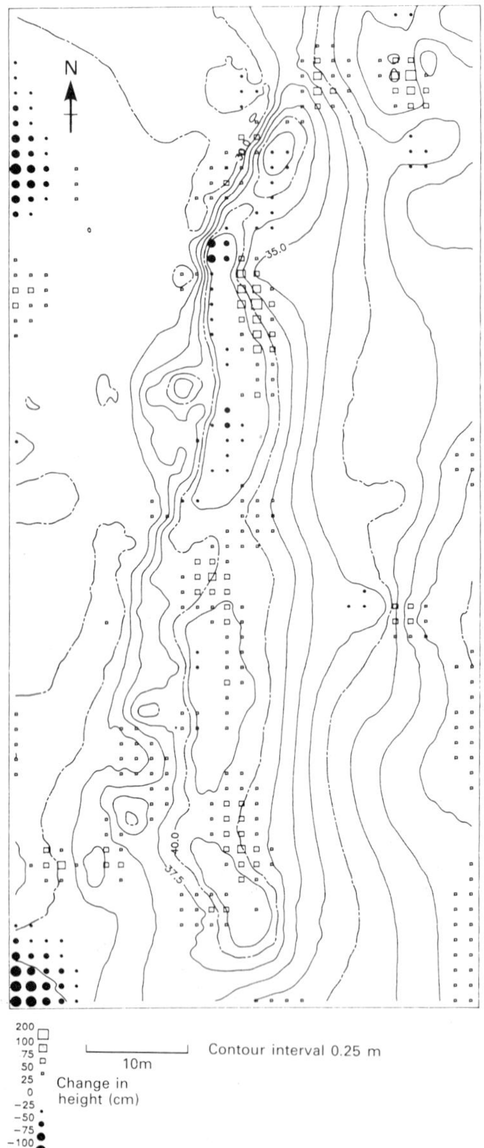

FIG. 4. Analysis of losses and gains between surveys of 14 January and 28 March, Ras Dhabdhub site.

moved 80 cm in a 9-hr period. The secondary (T2-ENE) and tertiary ridges (T3-various) were shown to be very short-lived: they could be eliminated or reformed within a day. There are no comparative data from other areas on daily rates of movement to suggest whether or not these rates are unusual. When they are compared to the data on seasonal or annual rates of movement (the only reliable figures available), the daily rates recorded at 'Urayfat appear extraordinarily high. Much of the daily movement would be masked in a longer-term study, because much of it is back and forth, but it may also be the case that our longer-term observations will show a high rate of seasonal and annual movement.

The second observation is that this mobility is narrowly restricted in area, at least for the period of observation (when wind velocities were not as high as they can be in the summer). The upper flanks of the dunes were by far the most mobile zone. The hollows were by far the most stable. This can be seen on Fig. 5. The ordinate gives millimetres sand loss or gain per hour. This is a conservative statement of accretion and erosion rates since many of the periods between observations include nights when there was very little wind. Pegs W19 to W17 on the windward flank were eroded and the sand went to create a new low slip face near W16. Further up the western flank, W13 and W12 also suffered erosion, and the sand from these points appears to have contributed to the marked upward growth at W11. The crest of the dune is seen to have grown slightly at 01 and been eroded slightly at E11. The eastern (usually leeward) upper flank was not as mobile as the western. The gains at E12 relate to the growth of a slip face, and the losses at E13, E14 and E15, took place in easterly winds.

The restriction of sand mobility to the upper flanks can be explained in three ways:

Higher wind velocities

The upper parts of dunes are the ones that experience the highest wind velocities and, therefore, the greatest number of sand-moving events. Most relationships between wind velocity and sand-carrying capacity are cubic *eg* $q_s = au^3_c$, where q_s is unit sand discharge, u_c is wind velocity above the sand entrainment threshold, and a is a constant. If this relationship is combined with the logarithmic relationship between wind velocity and height above the base of small hills derived by Jackson & Hunt (1975), it can be seen that the upper parts of dunes are more vulnerable to sand movement. Wind shear is higher on the steep windward upper slopes than on the more level summits (Bagnold 1941), so that it is the former rather than the latter that experience greatest erosion. Sand is moved from these upper flanks and deposited either on crests as flat-bedded crestal sands or beyond on the slip structures of the lee face where there are inevitable changes in level. At higher wind speeds, the erosion extends down to the windward middle slopes.

Oblique and vortex flow

In a period of very changeable winds, such as during the period of our observations, many

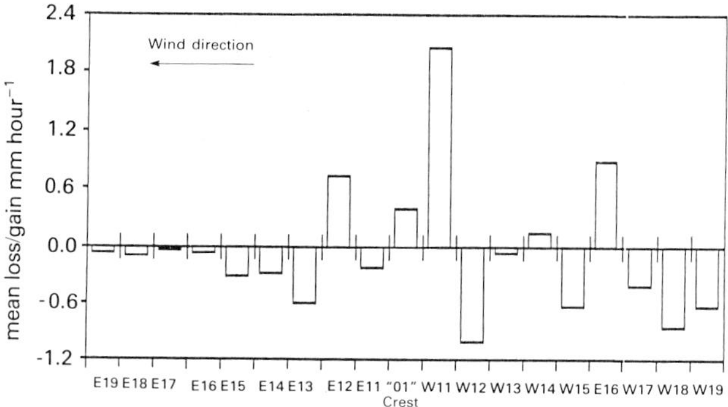

FIG. 5. Losses and gains of sand on a dune network as recorded by graduated pegs. The letters on the abscissa refer to peg numbers, E and W of the summit. The losses and gains are mean figures for hourly rates of change in sand level derived from 27 observations over 32 days in February and March 1986.

winds approach dune ridges obliquely. Tsoar (1983) found that these winds create strong lee flows, and that they carry a heavy load of sand parallel to the ridge and, therefore, oblique to the approaching wind. The winds in the lee of Tsoar's oblique ridge had higher velocities than the approaching wind. When a wind approached the ridges in the 'Urayfat Irq obliquely, there were indeed strong oblique lee winds, at least as demonstrated by the orientation and size of the wake trains of the measuring pegs. These lee winds were only active on the upper slopes of the dunes. Tsoar found that, in many cases, the strong lee flows were helical, and this too is demonstrated by the orientation of wake trains behind the pegs in the present study. This type of flow actually appears to bring sand up the lee of the ridge against the direction of the approaching wind (unlike the flow in Tsoar's observations). Sand was thus removed from some of the shallower hollows and accumulated in the upper flanks and crests.

Adjustment to new wind conditions

Because of the very changeable nature of the wind speed and direction at the time of our observations, winds commonly approached dune slopes which had yet to adjust to their sand-carrying capacities. Slopes that were steeper than equilibrium probably accelerated the flow, particularly near the ground, so steepening the velocity–height relationship. This increases the drag of the wind, and so increases its sand-carrying capacity. There was therefore erosion of the slope.

This effect is particularly marked at the top of a slip face when the wind reverses. The top of the face is regraded very quickly indeed (the actual time depending on the strength of the reversed wind). It usually begins with 'fluting' of the uppermost part of the face which is then rapidly rounded. Regrading then extends to the rest of the face. The new windward slope ceases to be unstable and is no longer a slip face.

Slopes lower down the dune, which present less of a contrast in slope angle and curvature to the incoming wind than do the slip faces, are, nevertheless, also regraded. The upper western and southern slopes of the larger dunes in the study area were repeatedly eroded and buried by the changeable winds of the study period.

This erosion led to yet another type of crestal instability. The gentler winds of the winter season created secondary dunes with smaller wavelength than the stronger winds of summer. The wavelength of these dunes was of the order of 5 m, and they built slip faces of no more than 20 cm high. They were larger on the higher ridges. These low dunes appeared on the southern flanks of almost all the main ridges. The growth of such a low dune on the pegged site is shown at peg W16 (Fig. 5). One feature of these low dunes is that their slip faces propagate laterally. This can be explained in terms of oblique flow: if there is even slight obliqueness of the approaching wind, helical lee-side airflow helped to extend the new slip face laterally.

Conclusions

The nomenclature developed in this paper proved valuable in the description and understanding of the network dunes that were studied. In particular, the idea of dune memory is helpful in separating dune systems that can and cannot be used as indicators of winds of the annual regime.

The division of dunes into transverse, and seif-like ridges and nodes appears to cover most of the dune types that were encountered.

The dunes have been shown to be highly mobile, but the mobility is restricted to the upper flanks. The explanation for the actual changes observed must lie in spatial differentiation in the power of the wind across the dune surface. The next stage in this research is therefore to measure wind velocities and wind velocity profiles and to trap sand at different points on a dune slope as it adjusts to new wind conditions.

ACKNOWLEDGMENTS: Acknowledgment must first be made of the generous help of many sponsors of the Royal Geographical Society's Wahiba Sands Project. The corporate sponsors include Gulf Air, Assarain Enterprise, Land Rover Ltd, Mohsin Haider Darwish, Racal Electronics, Suhail and Saud Bahwan, Taylor Woodrow Towell, and Zubair Enterprises. Particular thanks are due to the Sultan of Oman's Armed Forces; we should especially like to thank Naqib Chris Griffiths for his quick response to our many requests for help and for his interest and enthusiasm, and Mul/2 Said Mubarak and his group of soldiers from the Coast Security Force for their cheerful help in erecting and dismantling equipment at the dune sites, and for many other services including the domestic arrangements at our field base. We also wish to thank the Natural Environment Research Council of the United Kingdom for the loan of an automatic weather station, Wild Heerbrug (UK) for the loan of survey equipment, the Department of Land Survey, North-East London Polytechnic for the loan of a photogrammetric camera, and the Central Research Fund of the University of London for their help in purchasing equipment. Apart from general thanks to our colleagues in the Eastern Sands Project, we are particularly indebted to James Cutler for his land survey work.

References

BAGNOLD, R. A. 1941. *The Physics of Blown Sand and Desert Dunes.* Methuen, London.

BREED, C. S., FRYBERGER, S. G., ANDREWS, S., MCCAULEY, C., LENNARTZ, F., GEBEL, D. & HORSTMAN, K. 1979. Regional studies of sand seas using Landsat (ERTS) imagery. *In*: MCKEE, E. D. (ed.) *A Study of Global Sand Seas.* United States Geological Survey Professional Paper **1052**, 305–397.

COOKE, R. U. & WARREN, A. 1973. *Geomorphology in Deserts.* Batsford, London.

GLENNIE, K. W. 1970. *Desert Sedimentary Environments.* Elsevier, Amsterdam.

GOUDIE, A. S., WARREN, A., JONES, D. K. C. & COOKE, R. U. 1987. The sediments of the Wahiba Sand Sea Oman. *Geographical Journal*, in press.

HOWARD, A. D., MORTON, J. B., GAD-EL-HAK, M. & PIERCE, D. B. 1977. Simulation model of erosion and deposition on a barchan dune. *NASA Contractor Report, NASA CR-2838*, Washington DC.

JACKSON, P. S. & HUNT, J. C. R. 1975. Turbulent windflow over a low hill. *Quarterly Journal of the Royal Meteorological Society* **101**, 929–955.

MONOD, T. 1958. Majabat al Koubra. *Memoires de l'Institute Francais d'Afrique Noire* **52**.

TSOAR, H. 1983. Dynamic Processes acting on a longitudinal (seif) dune. *Sedimentology* **30**, 567–578.

WARREN, A. 1984. Desert Geomorphology. *Progress in Physical Geography* **8**, 399–420.

—— 1987. The dunes of the Wahiba Sands. *Journal of Oman Studies, Special Report*, **3**, in press.

A. WARREN & S. KAY, Department of Geography, University College London, Gower Street, London WC1E 6BT, UK.

Complex star dunes and associated aeolian bedforms, Hopeman Sandstone (Permo–Triassic), Moray Firth Basin, Scotland

L. B. Clemmensen

SUMMARY: The Hopeman Sandstone, which is exposed at the southern edge of the Moray Firth Basin, is composed of a number of aeolian sandstone units. These units, which have been mapped continuously for 4.5 km, represent part of an ancient sand sea in which complex star dunes alternated with sand-filled star-dune corridors. The palaeowind regime was complex with prevailing winds from the NNE, secondary winds from the SSE and subordinate winds from the NW. The complex star dunes are represented by two types of deposits: Type I—large-scale or giant-scale mainly trough-formed high- to medium-angle cross-bedding, which constitute the slip-face deposits of actively migrating star dunes; and Type II—bimodally dipping more wedge-shaped large-scale or giant-scale sets, which apparently formed on the opposed flanks of relatively stationary star arm segments. The exposed star arms trend in three main directions being roughly parallel with the palaeowind directions. The dune-dune corridors contain small- and medium-sized crescentic or reversing dunes which frequently overlap the star-dune flanks, interdune flats, and rare ephemeral streams. The palaeowind pattern inferred from the Hopeman Sandstone is very similar to that inferred from early Triassic reversing dune deposits in E Greenland, suggesting the existence of a regional Trade Wind system in the period near the Permo–Triassic boundary.

Modern sand seas are commonly very complex as revealed by recent studies, *eg* Breed & Grow (1979), Breed *et al.* (1979) and Lancaster (1983). Compound and complex dunes (also termed draas by Wilson (1972)) of linear, star and crescentic geometry are the dominant bedforms, and all types frequently occur within the same sand sea though spatially separated. This is particularly well illustrated by the Namib Desert in which compound and complex linear, crescentic and star dunes cover more than 80% of the area (Lancaster 1983). These large dunes are commonly associated with smaller simple dunes. Star dunes in the Namib Desert, for example, are frequently associated with smaller barchanoid and reversing dunes (Lancaster 1983).

From studies of modern sand seas it is therefore to be expected that ancient sand seas also should be composed mainly of various compound or complex dunes, and a number of closely interbedded complex and simple dune types. Ancient compound crescentic dunes (draa) have also been recognized at several localities in recent years (*eg* Kocurek 1981; Clemmensen & Abrahamsen 1983; Blakey & Middleton 1983; Ross 1983) and linear draas have been described from the Permian Yellow Sands of NE England by Steele (1983) and Clemmensen (1985).

Complex and simple star dunes, in contrast, seem to be rare in the geological record. The present paper describes in some detail an ancient system of complex star dunes and associated aeolian bedforms from the Permo–Triassic Hopeman Sandstone Formation, Moray Firth, Scotland (Fig. 1). As few recent star dunes have been trenched (*eg* McKee 1966, 1982; Nielson & Kocurek, in press) and then only in their uppermost portions (which in any case are unlikely to be preserved), this paper adds to the fund of knowledge of the internal structures of star dunes.

Geological setting

The Hopeman Sandstone Formation (Peacock *et al.* 1968; Warrington *et al.* 1980; Benton & Walker 1985) is exposed for almost 10 km along the coast of Moray Firth between Burghead and Lossiemouth (Fig. 1). A large portion of the coast is composed of cliffs 10 to 30 m high revealing mainly the lateral building of a number of closely associated aeolian sandstone units (Figs 2, 3). Additional information is gained from studies of abandoned quarries, which occur near the cliff edge E of Hopeman, and from foreshore exposures (Fig. 2).

The general strike and dip of the formation is difficult to estimate because dune sandstone is the only facies at most localities, but the dip appears to vary between a few degrees and 8°. The sandstones contain frequent joint zones and a few minor faults. Deformed bedding occurs at several localities along the coast as described in detail by Peacock (1966) and Glennie & Buller (1983).

The age of the Hopeman Sandstone has been the subject of some debate. Walker (1973) dates

From Frostick, L. & Reid, I. (eds), 1987, *Desert Sediments: Ancient and Modern*, Geological Society Special Publication No. 35, pp. 213–231.

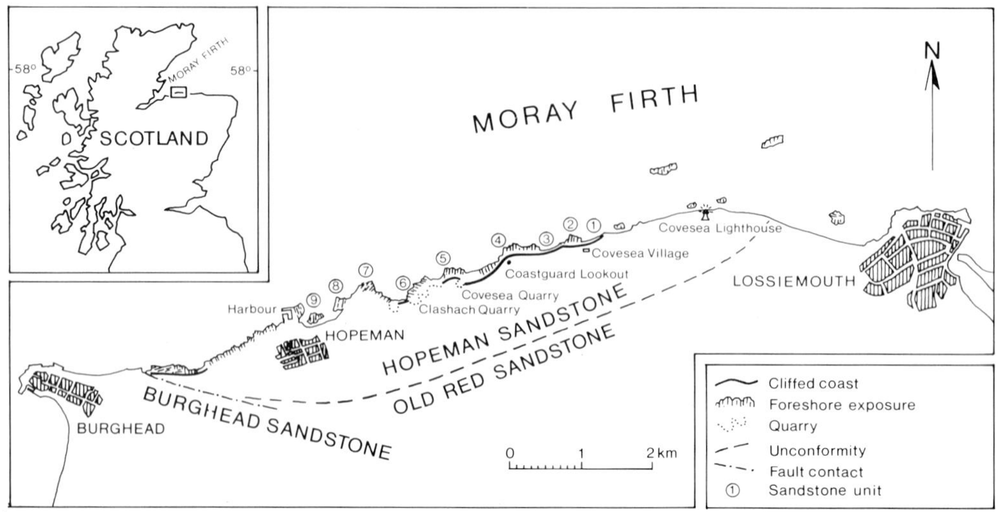

FIG. 1. Location map. Geology based on Peacock *et al.* (1968) and Glennie & Buller (1983). The Cutties Hillock locality occurs 7 km inland in a small fault-bounded block.

the Hopeman Sandstone Formation as earliest Triassic (*Lystrosaurus* Zone) on the basis of the reptilian fauna from the Cutties Hillock locality. Benton & Walker (1985) distinguish between the Hopeman Sandstone Formation (coastal exposures) and Cutties Hillock Sandstone Formation (isolated exposures 7 km inland). The Cutties Hillock Sandstone Formation contains reptiles indicating a late Permian age near the Permo–Triassic boundary (Benton & Walker 1985). The tracks in the Hopeman Sandstone Formation are believed to have been made by dicynodont reptiles, but this is a group which ranges in time from the middle of the Permian until late in the Triassic (Benton & Walker 1985). Glennie & Buller (1983) and Glennie (1985) presume that the deformation structures in the Hopeman Sandstone lie at the same stratigraphical horizon and owe their origin to the Zechstein transgression, and they define a lower unit of early Permian ('Weissliegend') age and an upper unit of late Permian–early Triassic age. This latter subdivision was questioned by Benton & Walker (1985) and is not supported by the present fieldwork, which has indicated that the deformations occur randomly in the whole sequence and do not define

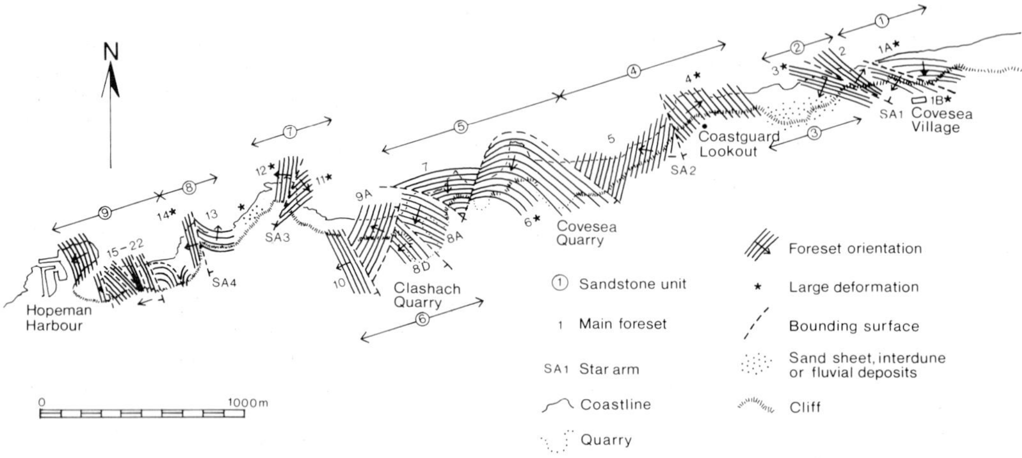

FIG. 2. Sandstone units and main dune foresets (numbered) between Covesea Village and Hopeman Harbour.

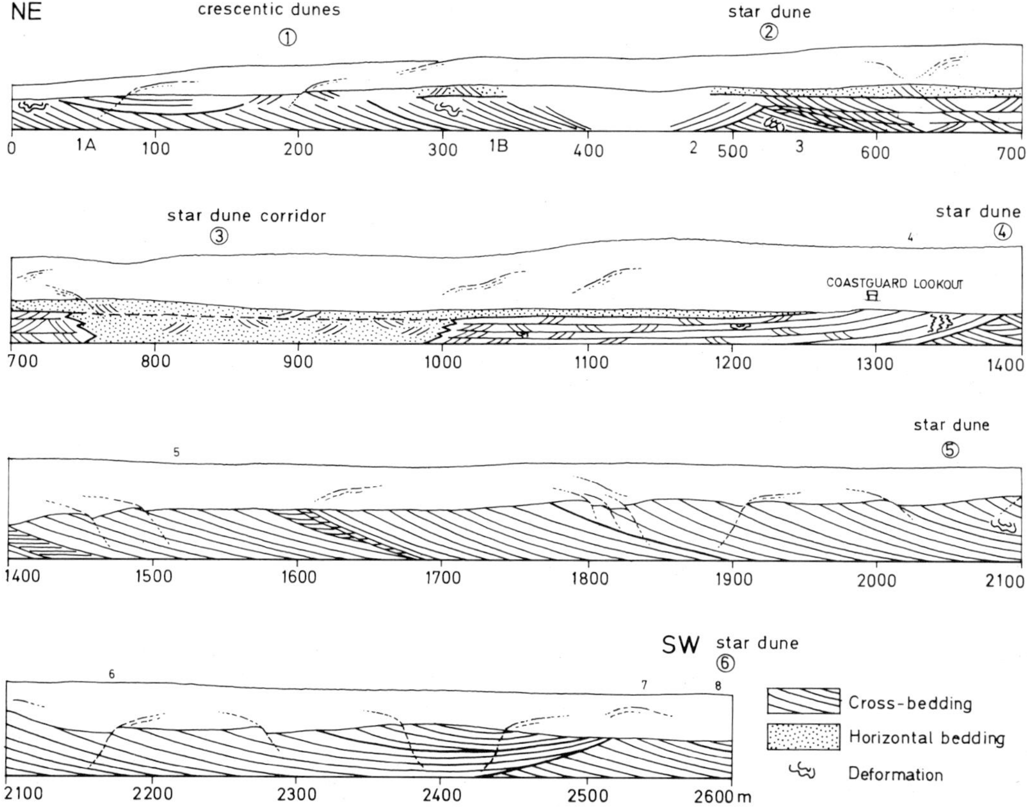

FIG. 3. Generalized appearance of the coastal cliff between Covesea Village and Clashach Quarry showing the lateral relationships of the nearly time-equivalent sandstone units 1–6. Main dune foresets are numbered. Note occurrence of deformed bedding in dune foresets 1A, 1B, 3, 4 and 6 and the capping aeolian sand sheet deposit.

any stratigraphic markers, and that the sandstone units are closely associated genetically (Figs 2, 3). The distribution of the deformation structures and their stratigraphical significance will be commented on in more detail after the individual sandstone units have been described.

Outline description and interpretation of the aeolian sandstones

An aeolian origin of the Hopeman Sandstone is generally accepted (*eg* Shotton 1956; Peacock 1966; Glennie & Buller 1983; Trommestad 1982). The sandstones are well-sorted, quartz-rich, and fine- to medium-grained (Peacock *et al.* 1968). Granule-rich layers and sand-covered clay curls occur in interdune and sand-sheet deposits, and small pebbles and intraformational clay clasts appear in the rare fluvial deposits. The colour of the weathered sandstones varies from light grey to yellow and orange-brown reflecting the main cementing materials, which are quartz, iron oxide and leucoxene (Peacock *et al.* 1968).

Two main types of aeolian sandstone units are distinguished. Type I is composed of small-scale (< 1 m thick), large-scale (1–10 m thick) and giant-scale (> 10 m thick) mainly trough-formed, high- to medium-angle cross-bedding of dune slip face and associated lee-side deposits. Type II consists of oppositely dipping more wedge-shaped large-scale or giant-scale, mainly medium-angle cross-bedding. These are interpreted as flank deposits of linear dune segments. The foresets of Type I frequently contain sandflow strata, while the foresets of Type II are dominated by wind-ripple and grainfall strata.

The Hopeman Sandstone is thought to contain two different types of deposit related to one main form of complex star dune (Fig. 4). The basic shape of the star dunes are here compared to the complex star dunes in the Rub' al Khali, Saudi Arabia (Breed & Grow 1979, fig. 179C). The fossil star dunes are associated with smaller

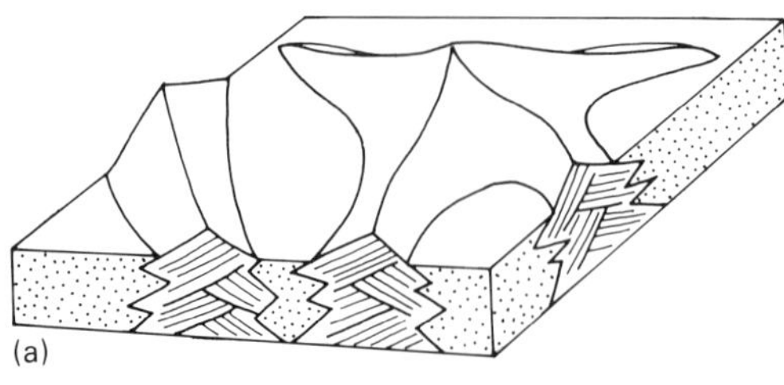

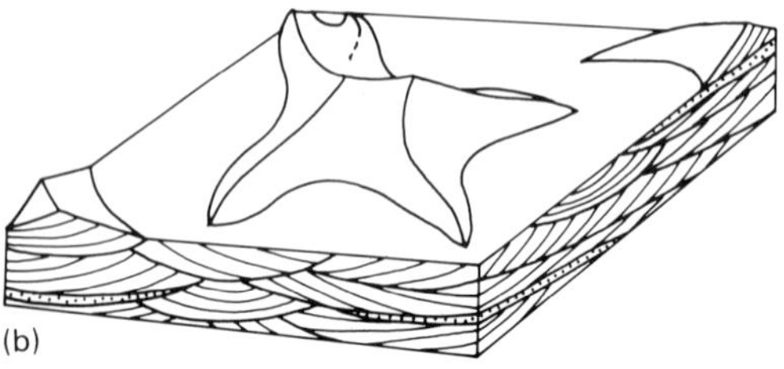

FIG. 4. Depositional models for complex star dunes with short radiating arms and crescentic curvatures. (*a*) Relatively stationary star dunes. Vertical accretion dominates and medium- to high-angle dune-flank deposits are separated by thick interdune deposits (see also Nielson & Kocurek, in press). (*b*) Laterally migrating star dunes. Trough-formed lee-side deposits with a large range of foreset dip angles are separated by thin interdune deposits.

crescentic and reversing dunes, and the following features contribute to this interpretation:

(1) Individual aeolian sandstone units in Type I display a unimodal (crescentic dunes), bimodal (reversing dunes) or polymodal orientation of their foresets (complex star dunes with crescentic segments, Fig. 4*b*). The dominance of slip-face deposits suggests actively migrating bedforms. The overall foreset dip distribution is polymodal and can be explained by assuming that the palaeowind regime was complex with winds blowing from three main directions (NE, SSE and NW). Recent star dunes form under a complex wind regime with winds from three or more directions (McKee 1979, 1982; Fryberger & Dean 1979; Lancaster 1983; Wasson & Hyde 1983).

(2) The linear dune segments in Type II units trend in at least three directions suggesting the existence of complex star dunes with radiating arms. The preservation of oppositely dipping flank deposits suggests that deposition here was dominated by vertical accretion (Fig. 4*a*). The exposed dune arms are roughly parallel to the palaeowind directions that have been deduced.

(3) The great thickness of many sets (10–35 m) and the occurrence of thick sand flows in Type I foresets (up to 14 cm) suggest that original dune heights were considerable. Recent star-dune complexes are commonly of great height; in modern deserts they reach heights of 200–300 m (Breed *et al.* 1979; Lancaster 1983).

(4) Second- and third-order bounding surfaces are numerous and display a great variation in dip angle and dip orientation suggesting a complex dune topography. First-order bounding surfaces are only developed between the star-dune system and an overlying aeolian sand sheet unit.

(5) Lateral facies studies indicate close association between star dunes and other aeolian bedforms (here smaller crescentic or reversing dunes) very similar to those in recent deserts (*eg* Breed & Grow 1979; McKee 1982; Lancaster 1983; Nielson & Kocurek, in press).

(6) Vertical facies studies indicate that the star-dune system is overlain by aeolian sand sheet deposits. According to Nielson & Kocurek (in press) star-dune deposits are highly susceptible to modification by sand-sheet processes.

Aeolian sandstone units

The coastal exposures of the Hopeman Sandstone are composed of a number of closely interrelated genetic aeolian sandstone units. The best exposures are between Covesea Village and Hopeman Harbour and over this distance (*c.* 4.5 km) it is possible to obtain a nearly complete lateral record of nine main sandstone units (Figs 2, 3). W of Hopeman Harbour the discontinuous nature of good coastal cliff exposures and the occurrence of considerable soft-sediment deformation makes it impossible to make safe lateral correlations. The general characteristics of the sandstone units here are, however, not significantly different from those seen E of Hopeman Harbour. E of Covesea Village there is only one good coastal exposure, at the lighthouse, but here deformation structures totally overprint the original bedding. In the following, therefore, only the well-exposed sandstone units between Covesea Village and Hopeman Harbour are described, starting from the E.

Sandstone unit 1 (crescentic dunes)

Description

This sandstone unit occurs in the coastal cliffs and foreshore immediately N and NE of Covesea Village (Figs 2, 3). The major part of the unit is composed of two large- or giant-scale sets (sets 1A and 1B) with an exposed thickness of 6 m. They dip up to 27° and are inclined towards the S (175° and 215°). Large-scale deformation structures occur in the upper part of both sets. Towards the top of the unit several large-scale southward-dipping trough-formed sets occur, as well as a few eastward-dipping medium-scale sets. The lower boundary of the set is unexposed. The upper contact is a well-defined first-order bounding surface overlain by a minimum 5 m thick sequence composed of small-scale bi-directionally dipping trough-formed or wedge-shaped cross-bedding, and horizontal lamination.

The eastward continuation of the unit is covered by modern beach sand. Towards the W the unit is replaced by the northeastward-dipping flank of a presumed star-dune arm (set 2, sandstone unit 3, Fig. 2). The exact relationship between sets 1B and 2 is unfortunately obscured by sand and vegetation.

Interpretation

Sandstone unit 1 apparently represents the slip-face deposits of southward-migrating crescentic dunes or incipient star dunes. With time, the crescentic dunes were eroded and overlain by sand-sheet deposits. While evidence for reversed winds during deposition of the crescentic dune is not strong, the overlying sand-sheet deposits were formed under alternating northeasterly and southerly winds.

Sandstone unit 2 (star dune arm)

Description

This sandstone unit occurs on the coastal cliffs and foreshore immediately N and NW of Covesea Village (Figs 2, 3). It comprises two main sets (sets 2 and 3) overlain by a number of medium-scale and large-scale sets which form a wedge-shaped unit thickening to the SW (Fig. 5). Set 2, which dips steeply (30°) towards the NE, has an exposed thickness of only 2 m. Set 3, which dips steeply (up to 32°) towards the S (185°–195°) and displays several deformation structures, has an exposed thickness of *c*. 5 m and contains sand-flow strata. Although the contact between the two sets is not exposed, it is inferred from the overall field relationship that set 3 is erosively overlain by set 2.

Set 2 is overlain by a southwestward-dipping second-order bounding surface and wedge-shaped co-sets of small-scale cross-bedding. The basal co-set (A_1) displays ascending sets with foresets dipping towards the E, while the overlying co-set (B_1) shows descending sets with foresets dipping towards the SW.

Set 3 is also overlain by a southwestward-dipping second-order bounding surface. This southwestward-dipping bounding surface, which also overlies set 2, is overlain by a number of ascending and descending small- and large-scale sets (co-set A_2) finally cut by a nearly flat-lying second-order bounding surface (Fig. 5). Co-set A_2 is a lateral equivalent of co-set A_1. The overlying sediments, which are more sheet-like, contain two southwestward-dipping sets (co-set B_2) and are eroded by a nearby flat-lying first-order bounding surface upon which *c*. 3 m of sand-sheet deposits occur (Fig. 5). Co-set B_2 is a lateral equivalent of co-set B_1. All these overlying

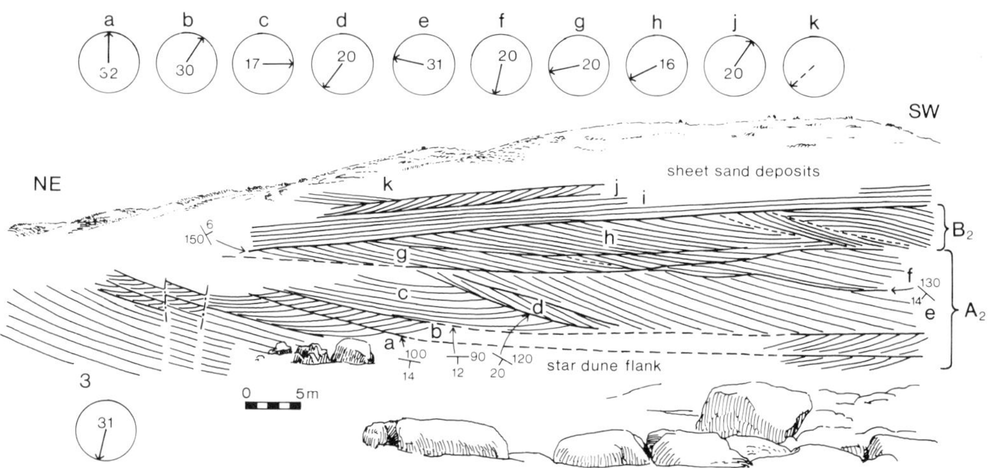

FIG. 5. Dune set 3 (lower left corner) of sandstone unit 2 overlain by a number of small- to large-scale trough-formed or wedge-shaped cross-bedded sets in co-set A_2 (set a–f) and co-set B_2 (set g–h). These sets formed on the flank of a complex star-dune arm and indicate shifting winds. The star-dune deposits are overlain by sand-sheet deposits (i, j, k) dominated by horizontal or low-angle stratification. Coastal cliff below Covesea Village. Small circles give dip angle in degrees (figures) and dip direction (arrows) of each set.

sets have medium- to high-angle foresets (16–32°) and frequently contain sand-flow strata.

The lower boundary of the unit is not exposed. The upper boundary is formed by the nearly flat-lying first-order bounding surface at the base of the sand-sheet deposits. The westward continuation of the unit is characterized by a gradual decrease in the dip angle of the second-order bounding surfaces in co-set A_2. Co-sets A_2 and B_2 thereby become gradually indistinguishable. In the same direction the two units are gradually replaced by a transition zone to sandstone unit 3 in which multidirectional trough-formed and wedge-shaped large-scale and small-scale cross-bedded sets are intercalated with evenly laminated sandstones. The exposed width of the transition zone is approximately 100 m.

Interpretation

The overlapping sets (sets 2 and 3) are thought to represent deposition on the opposing flanks of a star dune arm. The overlapping sets much resemble structures described from recent longitudinal dunes (Tsoar 1982). The arm was aligned in a NW–SE direction, and from the orientation of the foresets, a closure towards the SE is suggested. The northeastern flank of the presumed dune arm appears rather simple (although the exposure is very limited), while the southern flank had a more complex morphology with superimposed reversing and crescentic dunes on the lower slip face and dune apron (Fig. 5). The dunes in this superimposed system (units A_2 and B_2) migrated towards the N, E or SW. This superimposed dune complex was finally eroded and overlain by sand-sheet deposits in which bidirectionally dipping dune foresets indicate the occurrence of opposed winds.

Sandstone unit 3 (interdune flat)

Description

The unit occurs immediately SW of sandstone unit 2 and can be followed for *c.* 175 m in the coastal cliffs (Fig. 3). It is dominated by *c.* 10 m of relatively well-sorted horizontally or very low angle stratified sandstone (mainly wind-ripple strata). The general dip of the whole sequence is *c.* 5° towards the N. High-angle cross-stratified sets are very rare in the central part of the unit, but become increasingly more common towards the sides of the body. Wide transition zones (25–100 m) characterized by interbedded, evenly laminated sandstone and multidirectionally dipping small-scale and large-scale cross-stratified sandstone occur at the contacts with sandstone units 2 and 4.

Sandstone unit 3 forms a lateral equivalent of sandstone unit 2, but unlike the latter unit there is no well-defined upwards boundary to the sand-sheet deposits. The unit infills a low-lying area between sandstone units 2 and 4 (Fig. 6).

Interpretation

Low-angle aeolian stratification occurs in a number of environments including interdune areas, sands sheets and dune aprons and lower plinths (Kocurek 1986). The interfingering with the complex star dunes on both sides indicates that these sediments coexisted with the dunes and should be envisaged as interdune deposits. The great thickness of the unit is noteworthy and indicates that the nearby star dunes were stationary or only experiencing slow migration (*cf* Ahlbrandt & Fryberger 1981; Nielsen & Kocurek, in press). With time, the interdune environment was replaced by the uppermost sand sheet unit, which can be traced continuously for *c.* 900 m along the cliffs.

Sandstone unit 4 (star dune arm)

Description

Sandstone unit 4, which occurs in the cliffs below the Coastguard Lookout, is composed of two giant-scale sets (sets 4 and 5) and associated intrasets (Figs 2 and 3). Set 4, which has an exposed thickness of nearly 15 m, dips towards the N and NE (350–64°). The dip angle of the set gradually decreases from *c.* 30° to only a few degrees in this direction. The giant-scale set possesses numerous third-order bounding surfaces that divide the set into *c.* 1–3 m thick northeastward-dipping subsets (bundles, Fig. 7). Some of the subsets wedge out up-dip, others wedge out down-dip (Fig. 7). Several of the subsets contain ascending and descending intrasets (Fig. 7).

Ascending intrasets dominate in the more steeply dipping part of the set, while descending (northward-migrating) intrasets dominate in the low-angle dipping 'bottomsets' and 'toesets'. These intrasets are small-scale or large-scale, and of trough-formed, wedge-shaped and occasionally tubular geometries. The intrasets become increasingly more common in a northeastward direction and finally completely overprint the low-angle 'toesets' (Fig. 6). A very large deformation structure occurs in the more steeply dipping part of the set (Fig. 7; see also Glennie & Buller 1983, fig. 11).

FIG. 6. Northward-dipping sets in the distal zone ('toesets') of set 4 in sandstone unit 4 (star-dune arm), overlain by aeolian sand sheet deposits (ss). Note small deformation structure at the base (*). In the background is the southward-dipping dune set 3 of sandstone unit 2 (star-dune arm), and in between are the thick interdune deposits (sandstone unit 3). View towards NE.

FIG. 7. Northward-dipping giant-scale dune-flank deposits (set 4) in sandstone unit 4, a star-dune arm exposed in the coastal cliff below the Coastguard Lookout. Note large deformation structures (*) and small-scale ascending sets (arrows). Cliff height *c*. 20 m.

Set 4 is bounded below by a steeply northeastward-dipping second-order bounding surface which erodes dune set 5 (Fig. 8). Upwards, to the NE, the sandstone unit is bounded by a well-developed first-order bounding surface and overlying sand-sheet deposits. Laterally, to the NE, there is a gradual transition from the flat-lying strata in the ('toesets') to the interdune deposits in sandstone unit 3.

Set 5 dips to the W (242°–300°) for *c*. 400 m and has an exposed thickness of *c*. 30 m. The set has well-developed tangential foresets and a maximum dip angle of 30°. A prominent third-order bounding surface divides the set in two (Fig. 9). Of particular interest is the occurrence of large-scale ascending intrasets. Near the overlying bounding surface to set 4, two relatively large sets ascend the giant-scale foresets (Fig. 8). The lower of these ascending sets is *c*. 4 m thick and low-angle. The upper set is more steeply dipping and *c*. 2 m thick. Approximately 250 m further W, a second horizon with ascending cross-stratification appears (Fig. 9). This horizon is composed of seven small-scale trough-formed or wedge-shaped sets forming co-sets up to 3.5 m thick. The horizon is traceable in the exposure for *c*. 100 m and defines a prominent bounding surface (Fig. 9).

Set 5 is bounded above by a steeply dipping second-order bounding surface and is followed towards the SW by set 6.

Interpretation

Sandstone unit 4 is interpreted as part of a star dune. Dune sets 4 and 5 define one of the arms of this dune. The arm was trending NNW–SSE and the cliff section reveals a cross-section of this arm. The orientation of the foresets on the two flanks suggest a closure of the arm towards the NNW. The northeastward-facing flank gradually merged into a complex field of relatively small crescentic and reversing dunes in the dune apron ('bottomset' and 'toeset' zones). The numerous third-order bounding surfaces in set 4 and the associated intrasets seem to indicate that the bundles of this set were deposited during alternating winds from the NE and the S and that the flank repeatedly changed from lee-side to stoss-side.

Set 5 forms the westward-facing flank of the dune arm. This flank also shows evidence of reversing winds as indicated by the ascending intrasets, but there is no distal field of small crescentic of reversing dunes on this side. Instead the flank is eroded by a south-southwestward-

FIG. 8. Contact between set 4 and set 5 in sandstone unit 4 in the coastal cliff below the Coastguard Lookout. Note the large ascending intraset in set 5 and the deformation structure (*) in set 4. Cliff height *c*. 20 m.

FIG. 9. Westward-dipping foresets in dune set 5 of sandstone unit 4, a laterally migrating star dune arm exposed in the coastal cliff between the Coastguard Lookout and Covesea Quarry. Note zone with ascending (northward-migrating) small-scale cross-bedding. Cliff height *c.* 35 m.

dipping giant-scale set (set 6) probably representing a crescentic segment of a large complex star dune.

The different nature of the two flanks of the dune arm is probably related to the wind regime with prevailing winds from the NE and subordinate winds from the SSE and the NW. The relatively small crescentic and reversing dunes were formed on the more exposed northeastern side of the dune arm. The opposite flank was more sheltered and classic lee-side deposits were formed during at least two episodes of lateral migration of the dune arm in a westerly direction (*ie* the resultant wind vector).

Sandstone unit 5 (migrating complex star dune)

Description

Sandstone unit 5 is exposed in the foreshore and coastal cliffs between Covesea Quarry and Clashach Quarry (Figs 2, 3). It is composed of two giant-scale sets (sets 6 and 7). The sand body is bounded below by a steeply southwestward-dipping second-order bounding surface and on top by a newer steeply southwestward-dipping bounding surface at the contact with sandstone unit 6. Set 6 is clearly trough-formed (Fig. 10) as revealed by the strikingly curved foresets seen on the wave-cut platform and by a change in foreset dip directions from SW to SE. A large deformation structure occurs in the upper half of the set (Fig. 10). Set 6, which dips up to 32°, has well-developed granule-bearing tangential toesets and overlies set 7 from which it is separated by a well-exposed trough-shaped bounding surface. A curved bounding surface in the uppermost part of set 6 divides the set into two subsets. Set 7, which dips at up to 22° is weakly trough-formed to planar as revealed by the nearly straight foresets in the foreshore, and dips towards the SSE. Sets 6 and 7 are of considerable dimensions: set 6 has a width of *c.* 600 m and an exposed height of *c.* 35 m; set 7 has a width of *c.* 300 m and an exposed height of *c.* 20 m.

Interpretation

Sandstone unit 5 represents the slip-face deposits of crescentic dune segments. Because of the close association with star-dune deposits in sandstone unit 6, it is thought also that unit 5 was formed by a complex star dune that migrated actively to the SW. The lack of a pronounced erosion surface between units 5 and 6 indicates that the two dune deposits were nearly contemporaneous.

FIG. 10. Giant-scale southwestward-dipping trough-formed cross-bedding in set 6 of sandstone unit 5, a crescentic segment of a migrating star dune exposed in the coastal cliff at Covesea Quarry. Deformation in upper part of set (*). Cliff height *c.* 30 m.

Sandstone unit 6 (migrating complex star dune)

Description

This sandstone unit is exposed in Clashach Quarry and nearby abandoned quarries and in the associated foreshore (Fig. 2). The sandstone unit is separated from sandstone unit 5 by a weakly developed curved bounding surface dipping towards the SW. It is of complex build and is composed of several subunits that overlap each other to the S and W. The first subunit is composed of set 8A, which constitutes a number of subsets all dipping towards the S or SSW with medium- to high-angle foresets (14°–26°). The best exposed subset is seen in the westward-facing wall of a small abandoned quarry and consists of *c.* 25 m of weakly trough-formed cross-stratification (Fig. 11). This set contains numerous well-sorted sand-flow layers up to 12 cm thick. The following subunit is composed of three sets (sets 8B, 8C, 8D, Fig. 12). The basal set (set B) is wedge-shaped, has a preserved thickness of 1 m and dips steeply (25°) towards the NW (Fig. 12). Set C (*c.* 5 m thick) on top of set B is trough-formed, dips steeply (up to 28°) towards the W, contains sand-flow strata and seems to be composed of three subsets displaying slight variations in dip direction (265°–300°). The second-order bounding surfaces separating sets A, B and C are curved and moderately to steeply dipping in the same direction as the overlying sets. Finally set C is cut by set D (*c.* 10 m exposed) with foreset dips (up to 28°) towards the SW (220°–235°). Set D contains sand-flow layers up to 10 cm thick and possesses several descending intrasets.

Set 8D is overlain by a large trough-formed set (set 9A) dipping with intermediate angles (16°–23°) towards the NW (300°–335°; Fig. 13). This set is finally overlain by another trough-formed set—set 10 (*c.* 10 m thick) dipping steeply (up to 30°) towards the SW (220°–265°; Fig. 13). Set 10, which contains sand-flow strata, is overlain by a nearly horizontal first-order bounding surface and several metres of low-angle stratified sand-sheet or dune-apron deposits (Fig. 13). It is possible that these sand-sheet deposits can be correlated with those overlying sandstone units 2, 3 and 4. Locally, relatively thin wedge-shaped sets are intercalated between sets 8D and 9A. At a small headland near Clashach Cove, a 2–4 m thick westward-dipping set (set 9B, Fig. 13) occurs at this interval and, in a cove *c.* 100 m towards the E, a *c.* 1 m thick northeastward-dipping set is seen.

FIG. 11. Giant-scale weakly trough-formed southeastward-dipping dune set 7 of sandstone unit 5 overlain by the basal trough-formed set 8 of sandstone unit 6 (a migrating complex star dune) in an abandoned quarry E of Clashach Quarry. Cliff height *c.* 35 m.

FIG. 12. Sandstone unit 6, a migrating complex star dune. Note the well-developed jointing, parallel to the fault, which strongly overprint primary structures in dune set 9B. Dune foresets in set 10 are overlain by a first-order bounding surface and (presumed) sand-sheet deposits (ss). The exposure occurs in an abandoned quarry W of Clashach Quarry. Cliff height *c.* 20 m.

FIG. 13. Polymodally orientated foresets in the upper part of sandstone unit 6, the central part of a migrating complex star dune. Exposure in an abandoned quarry E of Clashach Quarry.

Interpretation

The large spread in dip directions (S, SW, W and NW) seems difficult to explain by the migration of simple crescentic dunes. It is suggested therefore that this sandstone unit represents a complex star dune composed of crescentic segments. Competitive migration of individual segments caused the observed pattern of cross-strata. The lack of large sets dipping in an easterly direction is explained by an overall migration of the star dune towards the SW causing erosion on the opposing parts of the dune. The complex star dune was initiated apparently in sandstone unit 5, and unit 6 records a continued migration of the dune to the SW. The star dune is thought to be comparable in geometry to recent complex star dunes in the Rub' al Khali, Saudi Arabia, which are characterized by crescentic segments (Breed & Grow 1979, fig. 179C).

Sandstone unit 7 (star dune arm)

Description

Unit 7 is exposed on a headland near the Hopeman Golf Course, *c.* 300 m W of Clashach Cove (Fig. 2). It is composed of bimodally-dipping large- to giant-scale sets (sets 11 and 12; Fig. 2). The easternmost extension of the unit is covered by the sea, but it is still quite obvious that this unit is overlain by the westernmost set (set 10) in sandstone unit 6. Towards the W the continuation of the unit is obscured by deformation structures. Further westwards low-angle granule-bearing interdune facies and sandy fluvial channel deposits rich in intraformational clay clasts occur. Small current ripples indicate that the interdunes were ephemerally wetted in rainy periods. The lowermost set (11), which dips towards the SE (130°–142° with medium-angle foreset (14°–22°)), has an exposed thickness of *c.* 4 m. The overlying set 12, which is separated from set 11 by a well-developed erosive second-order bounding surface, dips steeply (up to 26°) towards the WNW (265°–285°). It has an exposed thickness of *c.* 6 m. Set 12 contains thick sandflow strata, while set 11 is dominated by grainfall and relatively coarse-grained wind-ripple strata.

Interpretation

The overlapping structures are thought to represent deposition on the opposed flanks of a star dune arm trending NE–SW. The exposed structures only represent a small portion of the former dune arm, which was probably of considerable size. The structures that are preserved resemble those described from recent sinuous longitudinal dunes (Tsoar 1982), although in the present case only one unit with overlapping sets occurs. The orientation of the foresets suggests a closure of the dune arm towards the SW, and it is thought that the most persistent winds, those from the NE and blowing slightly oblique to the crestline, caused the extension of the dune arm towards the SW. The orientation of the dune arm facilitated the development of opposed slip faces by winds blowing from the SE or the NW.

Sandstone unit 8 (star dune arm)

Description

This sandstone unit, which is exposed at a headland *c.* 500 m E of Hopeman Harbour (Fig. 2), is composed of two bimodally-dipping large-scale or giant-scale sets. The extension of the unit towards the E is obscured by areas with numerous joints and deformed bedding, but *c.* 300 m eastwards, the sequence of low-angle interdune deposits and associated fluvial channel sandstones already described occur.

Towards the W the sandstone unit is replaced by a series of trough-formed large-scale sets (sandstone unit 9; Fig. 2). Unfortunately the contact between units 8 and 9 is covered by modern beach sand. The easternmost set (set 13) has low- to high-angle foresets (12°–28°) that dip towards the N (322°–70°) and are strikingly curved and concave in a downwind direction indicating a trough-formed set. It overlies set 14, which dips steeply (up to 26°) towards the W (255°–265°). The bounding surfaces between the two sets dip towards the E. Much of the contact between the two sets is obscured by a highly jointed zone. Set 13 has an exposed width of *c.* 150 m and an exposed thickness of *c.* 3 m; set 14, which contains small deformation structures, can be followed downwind for *c.* 80 m and has an exposed thickness of *c.* 5 m.

Interpretation

It is suggested that the overlapping sets formed on the opposed flanks of a sinuous-crested star-dune arm. The dune arm was orientated NNW–SSE and, judged from the orientation of the foresets, the dune arm diminished in size in a northwesterly direction. The trough-formed nature of set 13 is striking and suggests that the dune arm (at least locally) had a 'tear drop' morphology with peaks and saddles at the crest line of the dune (*cf* Tsoar 1983). The trough-formed set was probably formed by northwards migration of the crescentic slip face of the saddle under the influence of southerly winds.

Sandstone unit 9 (crescentic dunes)

Description

This sandstone unit is composed of 8–10 overlapping large-scale trough-formed cross-stratified sets. The sand body is exposed for *c.* 350 m in the foreshore E of Hopeman Harbour (Fig. 2). It cannot be traced on the western side of the harbour. The sets, which mainly have foresets dipping towards the SW at high angles (20°–28°), have widths up to 100 m. The trough-formed geometry of the sets is very well displayed in the foreshore exposures, and the curved foresets of one very large set possess a 140° variation in dip direction. The sets are separated by second-order bounding surfaces that mostly dip 12°–14° towards the SE.

Interpretation

The unit is interpreted as a number of crescentic dunes mainly migrating towards the SW. The medium-angle dip of the second-order bounding surfaces suggests that these dunes migrated on the lower slope of the westward-facing flank of the star-dune arm immediately to the E (set 14, sandstone unit 9).

Deformation structures

Deformation structures characterize a large portion of the Hopeman Sandstone (Peacock 1966; Glennie & Buller 1983). During the present study they have been observed in sandstone units 1, 2, 4, 5, 7 and 8 (Figs 2, 3), and they are also frequent W of Hopeman Harbour. The deformation structures occur at various levels in the foreshore and in the overlying sediments of the coastal cliffs up to altitudes of 20–30 m (Figs 7, 8, 10). They have been observed in star dunes, in crescentic dunes, and in close association with the ephemeral stream deposits between sandstone units 7 and 8. Deformation structures are on the other hand absent in the thick interdune deposit of sandstone unit 3.

Individual dunes may be free of deformation structures or deformed to varying degrees. When developed, the deformation structures frequently occur in isolated parts of the dune foresets separated by non-deformed strata. In dune set 4, for example, deformation structures occur at random intervals from the steepest angled (Figs 7, 8) to the lowest angled part (Fig. 6) of the flank deposits.

On a larger scale, deformed and non-deformed dune sets alternate in a random way, and the only apparent trend is a decline in deformation structures away from the present shoreline.

There is a strong spatial association between large deformation structures and joint zones (*eg* in sandstone units 4, 5, 7 and 8; Figs 7, 8, 10).

The deformation structures probably formed by liquefaction. However, it does not seem necessary to postulate a marine transgression to explain their genesis (Glennie & Buller 1983) as almost similar structures occur in many ancient inland dune sequences (Doe & Dott 1980; Horowitz 1982).

There seems to exist a gradual upwards transition from the more stationary star dunes (units 2, 4, 7, 8) to more actively migrating star dunes (units 5, 6). Deformation structures occur in both dune types, although they are more frequent in connection with the stationary star dunes. The two dune associations are only separated by a second-order bounding surface and are both capped by a first-order bounding surface and associated aeolian sand sheet deposits. All star dunes clearly belong to the same draa system.

It seems, therefore, difficult to divide the dune deposits of the Hopeman Sandstone into two time-stratigraphic intervals as proposed by Glennie & Buller (1983). Instead it is suggested that all the sandstones belong to the same stratigraphic interval, the age of which is probably late Permian (Benton & Walker 1985), or perhaps early Triassic (Benton pers. comm.).

Discussion

Dune types and depositional model

The Hopeman Sandstone was deposited in a number of closely related aeolian sandstone units. Most of the units are thought to represent large complex star dunes (Fig. 14). These dunes were associated with smaller crescentic or reversing dunes, which frequently overlapped the star-dune flanks (Fig. 14). It seems that both relatively stationary as well as more actively migrating star dunes were present, but it is suggested that all star dunes were of the same general morphology with crescentic curvatures and short radiating arms. The stationary star dunes are represented by the oppositely dipping wedge-shaped sets (Type II), while the more actively migrating star dunes are represented by trough-formed high- to medium-angle cross-bedding (Type I). The dune arms were orientated in at least three main directions (Fig. 15) as judged from the exposures.

There is much lateral variation (Fig. 14). Generally speaking, large star dunes occupied most of the coastline that has been studied (units 2, 4, 5, 6, 7, 8; Fig. 14). This system of star dunes was flanked on both sides by fields of medium-sized crescentic dunes (units 1, 9). It appears that the most actively migrating star dunes formed in the central part of the section (units 5, 6), but these dunes could be slightly younger than the more stationary star dunes (units 2, 4, 7, 8). The stationary star dunes were separated by interdune flats, which locally contained ephemeral streams (unit 3 and sediments between unit 7 and 8). Smaller crescentic and reversing dunes frequently lapped onto the star-dune arms.

The average spacing of the reconstructed star dunes (Fig. 14) is *c.* 950 m when the more actively migrating star dunes of sandstone units 5 and 6 are considered as one landform. Recent star dunes in large sand seas have wavelengths around 2000 m (*eg* the Grand Erg Oriental, Algeria, 2070 m; and Rub' al Khali, Saudi Arabia, 2060 m; Breed & Grow 1979). In somewhat less extensive sand seas, such as the Namib Desert, the average wavelength is 1332 m (Lancaster

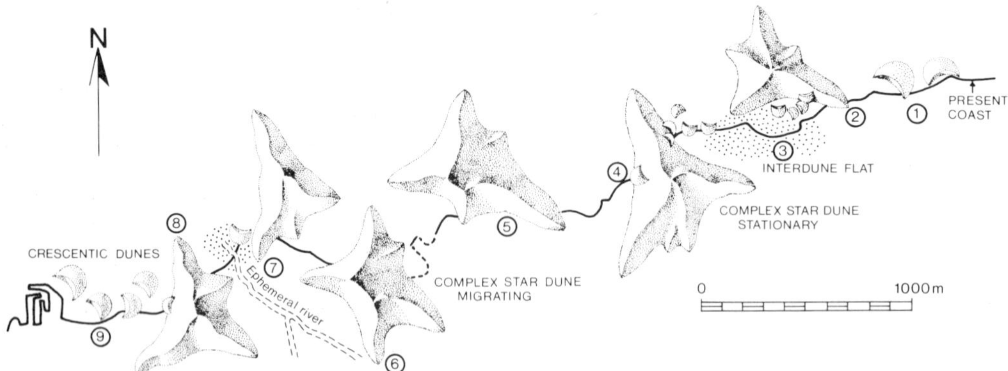

FIG. 14. Tentative reconstruction of dune morphology of the Hopeman Sandstone E of Hopeman Harbour.

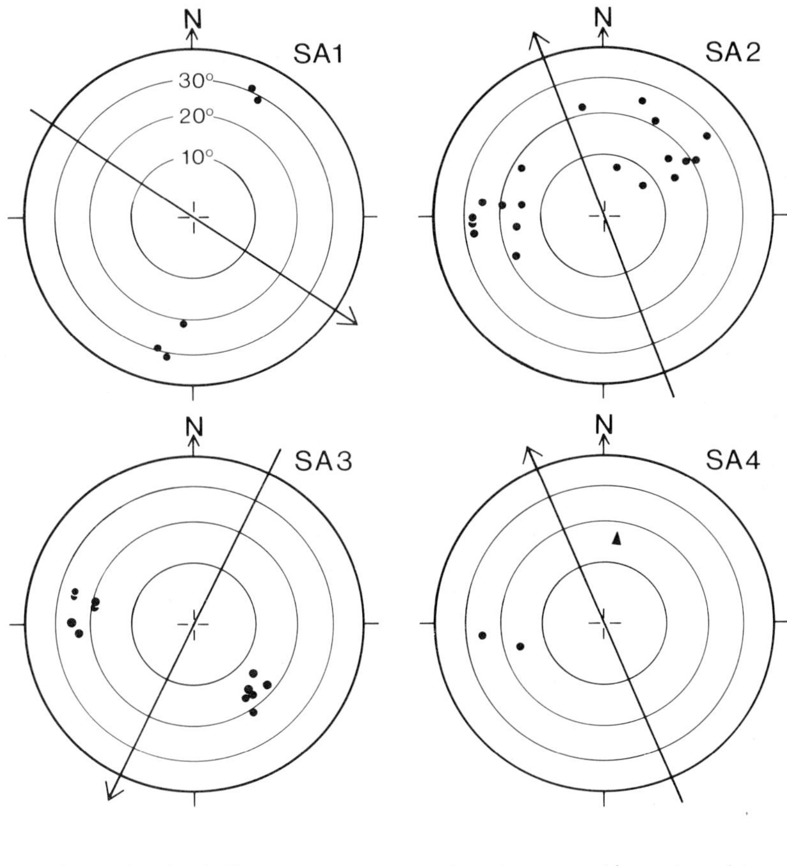

FIG. 15. Trends of exposed star arms (SA1, *etc.*). For location of arms, see Fig. 2. Arrows are vector means.

1983). The size of the ancient star dunes is therefore best compared with those from the Namib Desert, which should indicate that their average heights were approximately 100 m (*cf* Lancaster 1983, p. 268).

Although the exposed thickness of the Hopeman Sandstone is less than 45 m there is some evidence to suggest that stationary star dunes were gradually replaced with time by more actively migrating star dunes. This trend was terminated apparently by the development of the capping aeolian sand sheet deposits. However, the time taken for the deposition of the star-dune system must have been brief as suggested by the limited thickness of the exposed sandstone units and by their close genetic association.

Fossil star dunes have not previously been identified in the geological record, although they are very common in modern deserts. Their apparent scarcity in ancient sequences could be related to the geometrical similarities of many star dune arms to crescentic dunes. In fact, the deposits of ancient migrating complex star dunes actually resembles very closely those of crescentic dunes, but they have a larger range of foreset dips. Some star-dune arms are similar in form to longitundinal dunes, but these arms would probably in many cases experience some long-term lateral migration and each arm produce a cross-bedding of unidirectional dip (*cf* Rubin & Hunter 1985). In the present case, oppositely dipping dune flank deposits are preserved indicating the dominance of vertical accretion during formation of these star-dune arms. However, the occurrence of medium to high-angle cross-bedding is worth noting, since Nielsen & Kocurek (in press) suggest that star dune deposits consist mainly of low-angle stratification.

Palaeowinds

It is common practice to deduce palaeowind directions from the foreset orientation of dune deposits. In most previous work, the foreset orientation of ancient dune deposits has been unimodal (*eg* Bigarella 1962; McKee 1979; Gradzinski *et al.* 1979; Kocurek 1981; Marzolf 1983) reflecting the existence of only one significant palaeowind direction and the deposition of crescentic dunes. In a few cases, a bimodal distribution of dune foresets has been recorded and related to two different palaeowind directions (*eg* Laming 1966; Clemmensen 1978).

The measurements of foreset dip direction in the Hopeman Sandstone have been divided into two groups. The first group comprises dunes with well-developed lee-slope deposits (crescentic dunes migrating which include star dunes); the foreset dip distribution is clearly polymodal (Fig. 16). Inspection of the data indicates that the foresets can be classified in three groups. Most of the data fall into the SSW sector (46%); a second group lies in the NNW sector (29%); and a third group lies in the SE quadrant (14%). This classification, however, still leaves a number of foreset azimuths (11%) unclassified. The distribution of the foresets is most easily explained by suggesting three significant palaeowind directions: prevailing winds from the NE, secondary winds from the SSE and less frequent but neverthless significant winds from the NW (Fig. 16).

In recent deserts, star dunes form in connection with a highly variable wind regime (Fryberger & Dean 1979; Wasson & Hyde 1983; Nielson & Kocurek, in press). They are frequently associated with small crescentic dunes and form by the modification of or merging of individual crescentic dunes (Nielson & Kocurek, in press).

The second group of cross-strata comprises the more stable star dune arms. The dune flank cross-bedding azimuths of the dune arms are plotted along with the foresets of the 'crescentic' dunes; they fall mainly at the margins of the fields delineated already in Fig. 16. Measurements of foreset orientation in this group are also used to define the trends of star arms by assuming that deposition took place on opposed flanks. The data indicate that the star arms (at least those that appear in exposures) were trending in three main directions—NE–SW, NNW–SSE, and NW–SE (Fig. 15) but most of the star dunes probably had more arms than those that can be deduced (Fig. 14). To judge from the exposed dune arms, they had trends roughly parallel to

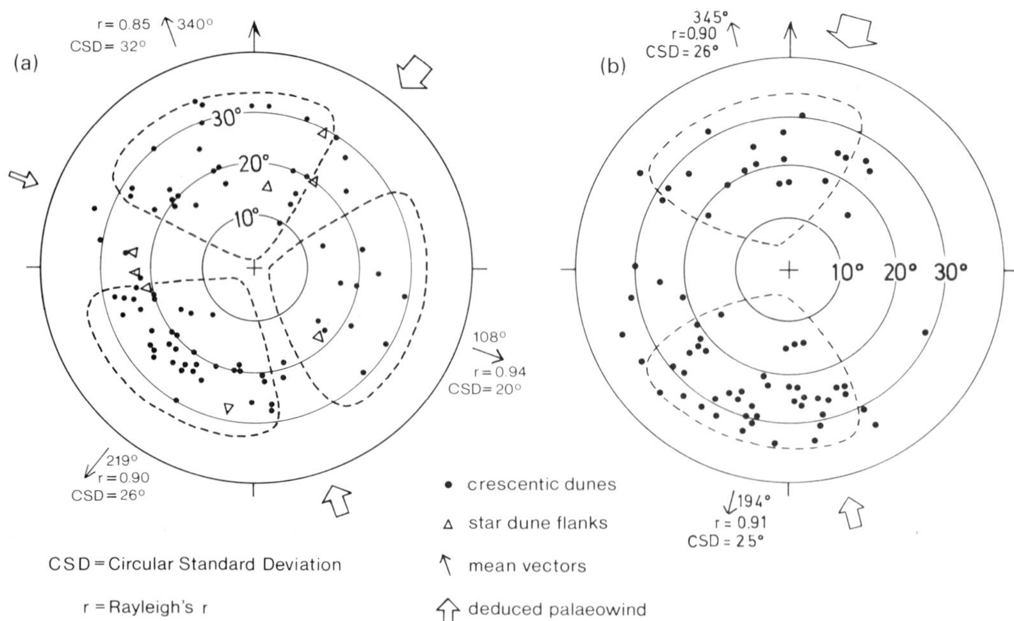

FIG. 16. Palaeowind roses. (*a*) Distribution of dune foresets in the Permo–Triassic Hopeman Sandstone. Each dune foreset is represented by only one point which, in most cases, is the vector mean of several (2–7) measurements. The grouping of data is based on visual inspection; high *r*-values indicate that each mode is statistically significant. (*b*) Distribution of dune foresets in the early Triassic Kolledalen Member, E Greenland (source: Clemmensen 1978).

the palaeowinds, while the associated slip faces were oriented, in most cases, oblique to these winds.

The resultant wind (*ie* sand-flow) direction has been calculated to be 250°, which is in reasonably good agreement with the evidence from sandstone units 5 and 6 that suggests that this particular complex star dune migrated towards the SW.

The palaeowinds could have been regional winds, or local winds, or even secondary circulations caused by the topographical irregularities of the dune system itself.

If all the aeolian sandstones of the region were to belong to the same stratigraphical unit, it should be of interest to compare the palaeowind data of the Hopeman Sandstone with that of other nearby Permo–Triassic localities. Lower Permian (Rotliegend) sandstones occur in the Moray Firth and in the Northern Permian Basins, but there is an obvious difference between the prevailing NE palaeowinds observed in the Hopeman Sandstone and the SW to NW palaeowinds deduced from core material at these other localities (Glennie 1982). Lower–Middle Triassic dune sandstones of similar light grey to yellow colours as the Hopeman Sandstone occur in E Greenland (*cf* Clemmensen 1978). These aeolian sandstones were deposited as reversing (barchanoid) dunes under the influence of alternating palaeowinds from the NNE (prevailing) and SSE (Fig. 16). The alternating palaeowinds were interpreted as seasonal Trade Winds by Clemmensen (1978) and were related to the annual migration of the Intertropical Convergence Zone. The North Sea has rotated *c.* 10° clockwise since the Triassic, while the rotation of E Greenland has been negligible since the Triassic (Habicht 1979; Smith & Briden 1977). Thus the prevailing northeastern wind that formed the Hopeman Sandstone was *c.* 10° more northerly than the present dip azimuths would indicate.

The striking similarity between the wind system in E Greenland and in Scotland, at least with regard to the two most dominant wind directions (Fig. 16) would indicate that both sequences were of comparable age (*ie* early Triassic) and that the north-northeastern and south-southeastern palaeowinds were of a regional (*ie* Trade Winds) rather than a local nature. The NW palaeowind direction is not clearly seen in the E Greenland data, which seems to suggest that its presence in Scotland indicates a more local (?seasonal) wind pattern. The relatively large number of unclassified foreset azimuths at the Moray Firth localities (11%) indicates that secondary winds within the dune system itself were indeed active in forming lee faces.

In support of the idea of a regional Trade Wind system near the Permo–Triassic boundary is the evidence of prevailing palaeowinds from the NE during deposition of the early Triassic Frodsham Sandstone in England (Thompson 1969) and the early Triassic dune sandstones on the western side of Arran in Scotland (Hegner pers. comm.).

Recognition of star dunes in cores

To judge from the sedimentary structures of the Hopeman Sandstone, star-dune systems would show the following characteristics in cores:

(1) Individual aeolian units are characterized by unimodal, bimodal or polymodal foreset dip orientation; a number of observations from superimposed units would produce a polymodal distribution.

(2) Set size is highly variable (up to 35 m or more), and intrasets are common. There is a large variation in the direction and angle of dip of second-order bounding surfaces.

(3) Both high-angle slip-face deposits with sandflow strata and less steep dune-flank deposits of grainfall and wind-ripple strata occur.

(4) Dune deposits are associated with interdune deposits and may be capped by aeolian sand sheet deposits.

(5) The sequential pattern of sediments is dependent on the dynamics of the star-dune system (*cf* Nielson & Kocurek, in press). If vertical accretion dominates, uniform sequences of star dunes are deposited, separated by thick interdune deposits. If lateral migration dominates, thinner units of star-dune and interdune facies form and there is greater facies repetition.

Conclusions

Star dunes are common in recent deserts, but seem not to have been reported from ancient desert deposits. The complex star dunes in the Hopeman Sandstone alternate with star-dune corridors, which contain fields of smaller crescentic or reversing dunes, or interdune flats, or (rare) ephemeral streams. Some of the complex star dunes seem to have been comparatively stationary, while others were more actively migrating. The stationary star dunes are characterized by medium-angle dune flank deposits, while the laterally migrating star dunes have steeper dipping lee-side deposits. The thickness of the Hopeman Sandstone is limited, but it appears that stationary star dunes were gradually replaced by laterally migrating star dunes and that the dune system was finally capped by aeolian sand sheet deposits. The star dune system was deposited during a complex wind regime with prevailing winds from the NNE, secondary winds from the SSE and subordinate winds from the NW.

ACKNOWLEDGMENTS: The author would like to thank Gary Kocurek, Mike Talbot and Henrik Olsen for critically reading the manuscript, and Kjell Trommestad for showing him the field localities. René Madsen very skilfully prepared the drawings and Pernille Andersen typed the manuscript. The fieldwork was funded by a grant from the Danish Ministry of Energy.

References

AHLBRANDT, T. S. & FRYBERGER, S. G. 1981. Sedimentary features and significance of interdune deposits. *In*: ETHRIDGE, F. G. & FLORES, R. M. (eds) *Recent and ancient nonmarine depositional environments: models for exploration*. Society of Economic Paleontologists and Mineralogists, Special Publication **31**, 293–314.

BENTON, M. J. & WALKER, A. D. 1985. Palaeoecology, taphonomy, and dating of Permo–Triassic reptiles from Elgin, North-East Scotland. *Palaeontology* **28**, 207–234.

BIGARELLA, J. J. 1962. Eolian environments: their characteristics, recognition and importance. *In*: RIGBY, J. K. & HAMBLIN, W. K. (eds) *Recognition of Sedimentary Environments*. Society of Economic Palaeontologists and Mineralogists, Special Publication **16**, 12–62.

BLAKEY, R. C. & MIDDLETON, L. 1983. Permian shoreline eolian complex in central Arizona: Dune changes in response to cyclic sealevel changes. *In*: BROOKFIELD , M. E. & AHLBRANDT, T. S. (eds) *Eolian Sediments and Processes*. Elsevier, Amsterdam, 551–581.

BREED, C. S. & GROW, T. 1979. Morphology and distribution of dunes in sand seas observed by remote sensing. *In*: MCKEE, E. D. (ed.) *A Study of Global Sand Seas*. United States Geological Survey, Professional Paper **1052**, 253–302.

——, FRYBERGER, S. G., ANDREWS, S. E., MCCAULEY, C., LENNARTZ, F., GEBEL, DL. & HORSTMANN, K. 1979. Regional studies of sand seas using Landsat (ERTS) imagery. *In*: MCKEE, E. D. (ed.) *A Study of Global Sand Seas*. United States Geological Survey, Professional Paper **1052**, 305–398.

CLEMMENSEN, L. B. 1978. Alternating aeolian, sabkha and shallow-lake deposits from the Middle Triassic Gipsdalen Formation, Scoresby Land, East Greenland. *Palaeogeography, Palaeoclimatology, Palaeoecology* **24**, 111–135.

—— 1985. Ancient aeolian bedforms and wind regime—three examples from the Permian–Triassic of the UK. *International Association of Sedimentologists, 6th European Meeting, Lleida*, 92–95.

—— & ABRAHAMSEN, K. 1983. Aeolian stratification and facies association in desert sediments, Arran basin (Permian), Scotland. *Sedimentology* **30**, 311–339.

DOE, T. W. & DOTT, J. R. 1980. Genetic significance of deformed cross bedding—with examples from the Navajo and Weber sandstones of Utah. *Journal of sedimentary Petrology* **50**, 793–812.

FRYBERGER, S. G. & DEAN, G. 1979. Dune forms and wind regime. *In*: MCKEE, E. D. (ed.) *A Study of Global Sand seas*. United States Geological Survey, Professional Paper **1052**, 137–169.

GLENNIE, K. W. 1982. Early Permian (Rotliegendes) palaeowinds of the North Sea. *Sedimentary Geology* **34**, 245–265.

—— 1985. Early Permian palaeowinds of the North Sea—reply. *Sedimentary Geology* **45**, 297–313.

—— & BULLER, A. T. 1983. The Permian Weissliegend of NW Europe. The partial deformation of aeolian dune sands caused by the Zechstein transgression. *Sedimentary Geology* **35**, 43–81.

GRADZINSKI, R., GAGOL, J. & SLACZKA, A. 1979. The Tumlin Sandstone (Holy Cross Mts, central Poland): Lower Triassic deposits of aeolian dunes and interdune areas. *Acta Geologia Polonica* **29**, 151–175.

HABICHT, J. K. A. 1979. Palaeoclimate, palaeomagnetism and continental drift. *Bulletin of the American Association of Petroleum Geologists, Studies in Geology* **9**, 1–31.

HOROWITZ, P. H. 1982. Geometry and origin of large-scale deformation structures in some ancient wind-blown sand deposits. *Sedimentology* **29**, 155–180.

KOCUREK, G. 1981. Erg reconstruction: The Estrada Sandstone (Jurassic) of northern Utah and Colorado. *Palaeogeography, Palaeoclimatology, Palaeoecology* **36**, 125–153.

—— 1986. Origins of low-angle stratification in eolian deposits. *In:* NICKLING, K. G. (ed.) *Aeolian Geomorphology. Proceedings of the 16th Annual Binghamton Geomorphology Symposium*, pp. 177–193.

LAMING, D. J. C. 1966. Imbrication, palaeocurrents and other sedimentary features in the Lower New Red Sandstone, Devonshire, England. *Journal of sedimentary Petrology* **36**, 940–959.

LANCASTER, N. 1983. Controls on dune morphology in the Namib Sand Sea. *In*: BROOKFIELD, M. & AHLBRANDT, T. S. (eds) *Eolian Sediments and Processes*. Elsevier, Amsterdam, 261–289.

MARZOLF, J. E. 1983. Changing wind and hydrologic regimes during deposition of the Navajo and Aztec Sandstones, Jurassic (?), southwestern United States. *In*: BROOKFIELD, M. & AHLBRANDT, T. S. (eds) *Eolian Sediments and Processes*. Elsevier, Amsterdam, 635–660.

MCKEE, E. D. 1966. Structures of dunes at White Sands National Monument, New Mexico. *Sedimentology* **7**, 3–69.

—— 1979. Sedimentary structures in dunes. *In*: McKee, E. D. (ed.) *A Study of Global Sand Seas*. United States Geological Survey, Professional Paper **1052**, 83–291.

—— 1982. Sedimentary structure in dunes of the Namib Desert, Southwest Africa. *Geological Society of America Bulletin* **188**, 1–64.

Nielson, J. & Kocurek, G. Development, processes, migration and deposits of star dunes. *Geological Society of America Bulletin*, (in press).

Peacock, J. D. 1966. Contorted beds in the Permo-Triassic aeolian sandstones of Morayshire. *Geological Survey of Great Britain Bulletin* **24**, 157–162.

——, Berridge, N. G., Harris, A. L. & May, F. 1968. Geology of the Elgin district. *Geological Survey of Great Britain Memoir*, 165 pp.

Ross, G. M. 1983. Bigbear erg: a Proterozoic intermontane eolian sand sea in the Hornby Bay Group, Northwest Territories, Canada. *In*: Brookfield, M. E. & Ahlbrandt, T. S. (eds) *Eolian Sediments and Processes*. Elsevier, Amsterdam, 482–519.

Rubin, D. M. & Hunter, R. E. 1985. Why deposits of longitudinal dunes are rarely recognized in the geological record. *Sedimentology* **32**, 147–157.

Shotton, F. W. 1956. Some aspects of the New Red Sandstone Desert in Britain. *The Liverpool and Manchester Geological Journal* **1**, 450–465.

Smith, A. G. & Briden, J. C. 1977. *Mesozoic and Cenozoic maps*. Cambridge University Press, 63pp.

Steele, R. P. 1983. Longitudinal draa in the Permian Yellow Sands of North-East England. *In*: Brookfield, M. E. & Ahlbrandt, T. S. (eds) *Eolian Sediments and Processes*, Elsevier, Amsterdam, 543–550.

Thompson, D. B. 1969. Dome-shaped dunes in the Frodsham Member of the so-called 'Keuper' Sandstone Formation (Scythian–?Anisian: Triassic) at Frodsham, Cheshire (England). *Sedimentary Geology* **3**, 263–289.

Trommestad, K. E. 1982. *Examples of fluvial, eolian and caliche deposits in the southern part of the Permo-Triassic Moray Firth Basin*. MSc thesis, University of Bergen.

Tsoar, H. 1982. Internal structures and surface geometry of longitudinal (seif) dunes. *Journal of sedimentary Petrology* **52**, 823–831.

—— 1983. Dynamic processes acting on a longitudinal (seif) sand dune. *Sedimentology* **30**, 567–578.

Walker, A. D. 1973. The age of the Cuttie's Hillock Sandstone (Permo-Triassic) of the Elgin area. *Scottish Journal of Geology* **9**, 177–183.

Warrington, G., Audley-Charles, M. G., Elliott, R. E., Evans, W. B., Ivimey-Cook, H. C., Kent, P. E., Robinson, P. L., Shotton, F. W. & Taylor, F. M. 1980. A correlation of the Triassic rocks in the British Isles. *Geological Society of London, Special Report* **13**, 1–78.

Wasson, R. J. & Hyde, R. 1983. Factors determining desert type. *Nature* **304**, 337–339.

Wilson, I. G. 1972. Aeolian bedforms—their development and origin. *Sedimentology* **19**, 173–210.

L. B. Clemmensen, Institute of General Geology, Øster Voldgade 10, DK-1350 Copenhagen K, Denmark.

Late Palaeozoic cold-climate aeolianites, southern Cooper Basin, South Australia

B. P. J. Williams, E. K. Wild & R. J. Suttill

SUMMARY: Core facies analysis of the Permo–Carboniferous Gidgealpa Group of the southern Cooper Basin, South Australia has revealed sandstones of aeolian origin within the glacigenic Merrimelia Formation. The aeolianites comprise exclusively porous, poorly cemented, medium- and fine-grained sandstones dominated by parallel lamination. Sharp, planar truncation surfaces divide the suite into units composed mainly of translatent wind-ripple lamination and minor amounts of sandflow bedding with primary depositional dip (post-compaction) varying between 0° and 25°. Distinctive compressional, tensional and dissipation structures are preserved in deformed lee-slope deposits and the inclusion of low-angle cross-strata and wedge-bedded units locally indicate deposition on the basal aprons of dunes or in contiguous sand sheets. Wet interdune sediments with adhesion ripples are also exhibited in the cores. The aeolianites, up to 232 m thick, have been traced subsurface by integrating core facies, wireline logs and seismic reflection analysis. A low-interval transit time on sonic logs is characteristic of the porous aeolianite sand and has facilitated mapping of the dune field using seismic data. The sandstone body trends SE–NW, is 6 to 8 km wide and passes laterally into outwash fan and fluvioglacial, braidplain deposits. Potential hydrocarbon traps may occur at the top of the aeolianites, or within them beneath intraformational seals formed by possible muddy interdune facies. The overall high-latitude, cold-climate nature of the Merrimelia Formation, deposited during the extensive Gondwanaland glaciation, is well known but this discovery not only records the first Gondwanan aeolianite suite but may also detail the first pre-Quaternary example of paraglacial aeolianites.

The Gidgealpa Group of the southern Cooper Basin, South Australia, attains its maximum thickness in the Nappamerri Trough (Fig. 1) where in excess of 1370 m of non-marine clastic sediments were deposited during the Late Carboniferous–Permian.

The stratigraphy of the Cooper Basin sequence has been reviewed by a number of workers (Kapel 1966, 1972; Martin 1967; and Gatehouse 1972), and the subdivision of the Gidgealpa Group into its constituent formations has been established using both lithostratigraphic and palynostratigraphic criteria (Paten 1969; Price 1973; Thornton 1979). A recent revision of the sequence has been the formal incorporation of the Merrimelia Formation into the basal Gidgealpa Group (Williams & Wild 1984*a*, *b*). This stratigraphic and sedimentological readjustment enables a more comprehensive overview to be taken of the evolution of the Late Carboniferous–Permian clastic wedge, and its emplacement in the overall tectonic development of the Basin.

The present study stems from a regional sedimentary facies analysis of the Gidgealpa Group currently being undertaken by the authors for South Australian Oil & Gas Corporation Party Limited (Williams 1982, 1984) which is concerned with the sedimentological evolution of the Permo–Carboniferous succession and its potential for additional hydrocarbon reserves.

Merrimelia Formation

The Merrimelia Formation comprises a varied assemblage of conglomerates, conglomeratic mudstones, sandstones and fine- to coarse-grained mudrocks which have been assigned a palynological age of Stage 2 (Late Carboniferous–Early Permian). Grund (1966) was the first to describe the Merrimelia Formation as a glacigenic sequence incorporating terrestrial and sub-aqueous tillites, glaciofluvial, glaciolacustrine, interglacial and periglacial sediments resulting from two phases of glaciation. Correlation of the Formation has been made with sediments of glacial origin containing a flora of a similar age that occur widely throughout Australia (Paten 1969; Harris 1981) and although the glacigenic nature of the sequence was verified by Martin (1967) and Battersby (1976), only recently has a more detailed appraisal of the glacial sedimentology of the Merrimelia Formation been attempted (Williams & Wild 1984*a*, *b*).

The Merrimelia Formation attains a maximum thickness in excess of 350 m, but its distribution over the southern half of the basin is quite irregular, being thickest in coeval depressions and fault-bounded graben and half-graben. Glacigenic and fluvioglacial conglomerates, diamictites and sandstones are common along early structural 'highs', whereas thick, glaciolacustrine

From FROSTICK, L. & REID, I. (eds), 1987, *Desert Sediments: Ancient and Modern*, Geological Society Special Publication No. 35, pp. 233–249.

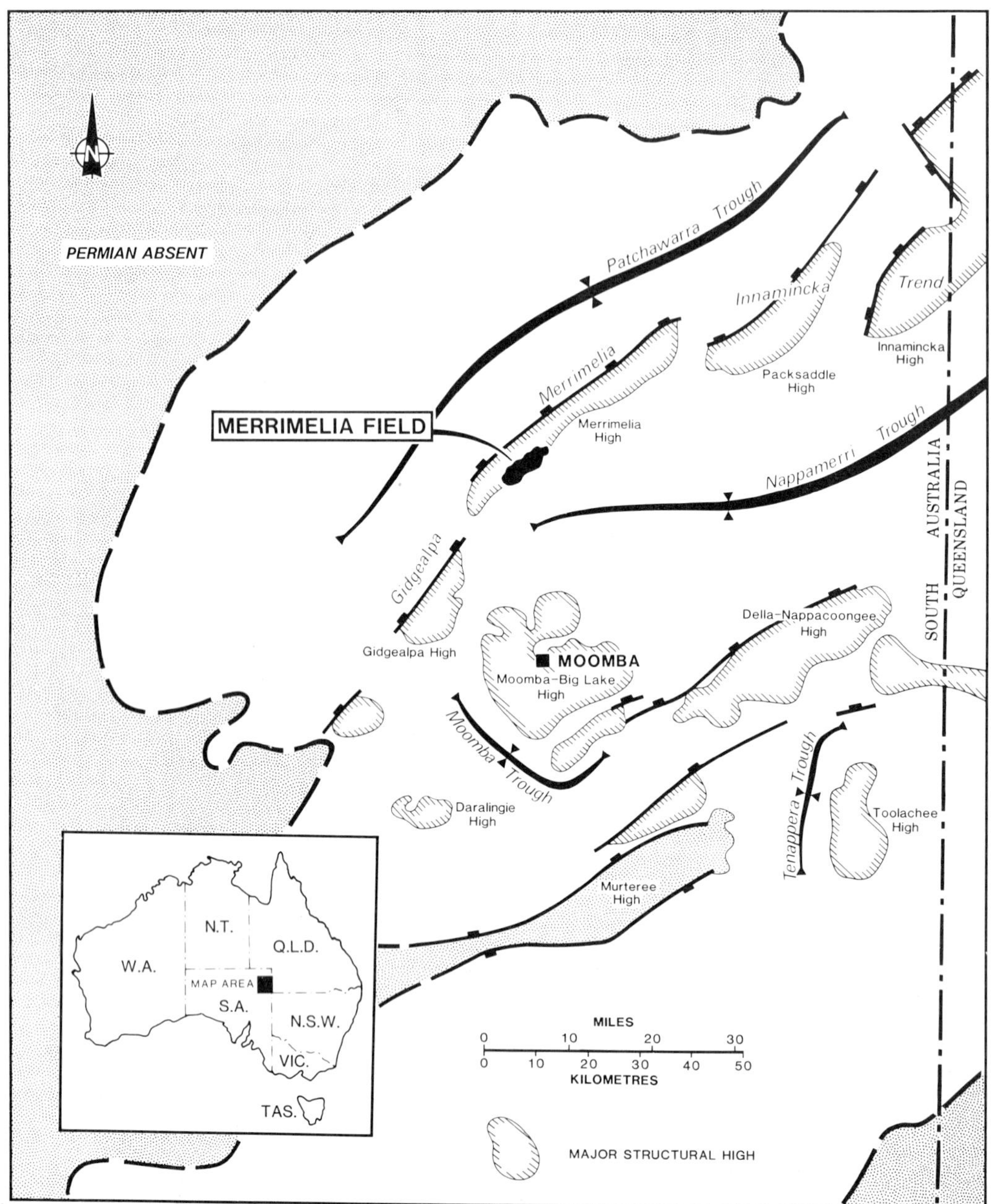

FIG. 1. Southern Cooper Basin, South Australia, indicating major structural features and location of Merrimelia Field.

mudrocks are restricted to the major troughs. To date, some 37 cored wells, of Merrimelia-type facies or Merrimelia/Tirrawarra-type facies, have been subjected to detailed facies analysis. Typical examples of the glacigenic suite, assigned to the Merrimelia Formation (*sensu stricto*), have been identified by Williams & Wild (1984*a*).

The Tirrawarra Sandstone is intimately associated with the Merrimelia Formation and is also of variable thickness throughout the southern Cooper Basin, attaining a maximum of some 75 m (Thornton 1979). The Tirrawarra Sandstone is dominated by thick multistorey sandstones with subsidiary developments of conglomerate,

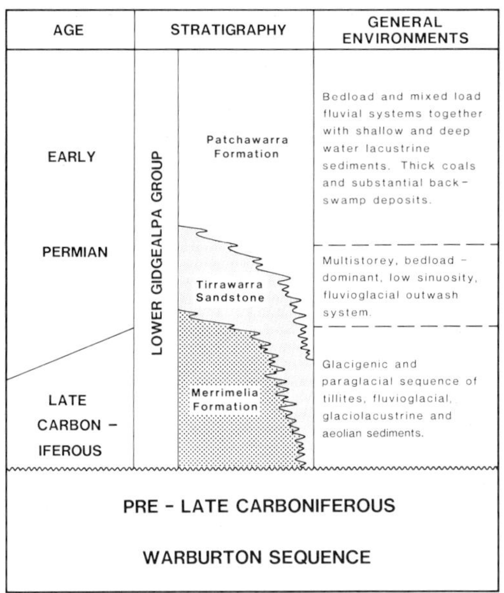

FIG. 2. Southern Cooper Basin: stratigraphy of lower Gidgealpa Group and generalized depositional environments.

mudrock and coal (Williams 1982, 1984). The sandstones are essentially litharenitic in composition but display a varied diagenetic mineral assemblage which generally reduces reservoir quality. The overall sedimentological attributes of the Tirrawarra Sandstone point to its origin mainly in low-sinuosity, bedload-dominant fluvial channels with marked vertical and lateral variations in sedimentary style (Wild & Williams 1984).

It has been clearly shown (Williams & Wild 1984*a*, *b*) that the Merrimelia Formation and Tirrawarra Sandstone comprise a diachronous, interfaced sequence with no intervening unconformity. The sedimentary history of these formations depicts the evolution of a predominantly glacigenic system into one in which fluvial processes dominate both temporally and spatially (Fig. 2). It is in this context of interfaced paraglacial sedimentation of Merrimelia/Tirrawarra style that the discovery of aeolian sediments has been made within the basal sequences of the Gidgealpa Group in the Merrimelia Field. The aeolianites are very closely related to bedload-dominated, fluvial sediments which had their origins as medial bars on large, humid (wet) low slope, braided outwash fans.

No hydrocarbons have yet been produced from the Merrimelia Formation and, due to the pervasive oxidation by meltwater underflows, the glaciolacustrine sequences in the Formation are thought to be low in organic matter. Although the Formation has been generally assumed to be non-prospective for hydrocarbons, Battersby (1976) has suggested that the glaciolacustrine mudrocks might have supported a rich, algal flora and, as such, a thick sequence of source rocks might occur within the Formation in as yet undrilled parts of the Cooper Basin. To test this hypothesis the South Australia Oil & Gas Corporation (SAOGC) are currently undertaking a detailed source rock study on the glaciolacustrine facies identified within the Basin (Williams 1982, 1984). This, together with the discovery of new, sedimentologically distinctive, potential reservoir bodies within the Merrimelia Formation point to the need for re-evaluation of the hydrocarbon potential of basal Gidgealpa Group sediments.

Merrimelia Formation of the Merrimelia Field

The Merrimelia Field is located some 38 km N of Moomba, South Australia (Fig. 1). Eighteen wells have been drilled and tested on the important Merrimelia structural complex which forms part of the Gidgealpa–Merrimelia–Innamincka (GMI) Trend. This Trend is regarded as having been a positive structural element during most of Permo–Triassic time (Battersby 1976; Thornton 1979), separating two main areas of Permian sedimentation in the Cooper Basin, namely the Patchawarra and Nappamerri Troughs (Fig. 1).

The Merrimelia structural complex is bounded by major faults on both northwestern and southwestern flanks (Fig. 3) with a number of subsidiary faults transecting the axial traces of the major folds. The crest of the structure has no Permian sedimentary cover, due in part to non-deposition, but due mainly to two periods of major uplift and erosion during the Permian. Stratigraphic analysis of the Gidgealpa Group across the structure reveals a rapid thinning onto the Merrimelia 'high' and confirms its absence on the highest parts of the structure (Martin 1967; Battersby 1976).

The thickness of the Gidgealpa Group reduces markedly when traced updip between Merrimelia 1 and 5 (Fig. 3). However, the Tirrawarra Sandstone (*sensu stricto*) is present in Merrimelia 1, 4 and 5 and exhibits minimal attenuation of

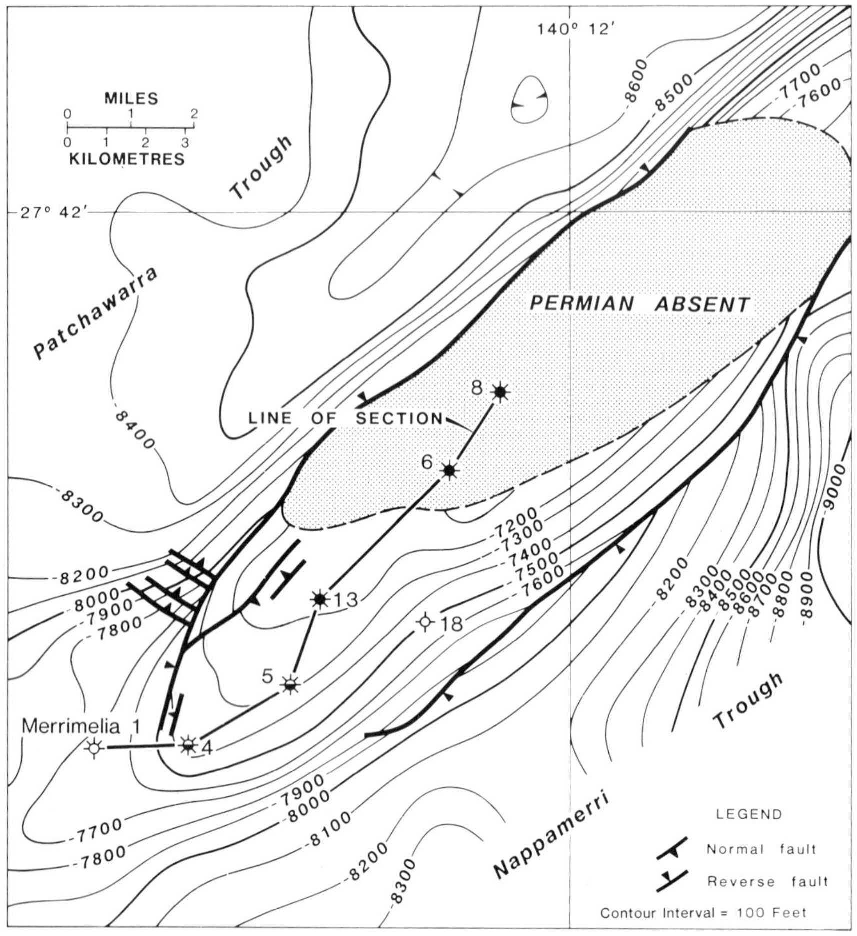

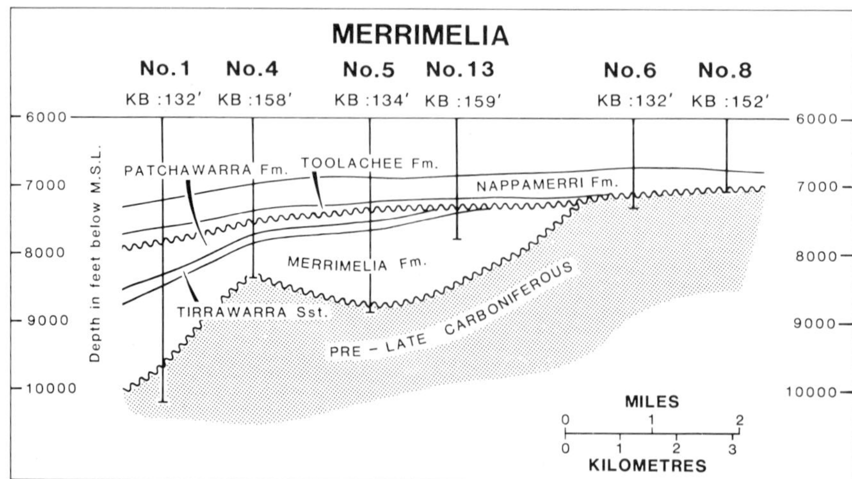

FIG. 3. Merrimelia Field. Above—Top Toolachee Formation contour map and structure inferred from P-wave seismic data. Below—structural cross-section with position of key wells.

thickness. It is coarser and more conglomeratic than the type section in the Tirrawarra Field, reflecting Early Permian syn-sedimentary tectonism.

Within the Merrimelia Formation of the Merrimelia Field the sequence exhibits much thickness variation from zero to greater than 363 m. The lithofacies present are also very variable encompassing a variety of glacigenic and paraglacial depositional settings. It is within this depositional array that the sandstones, interpreted as aeolian in origin, were discovered. The aeolian suite was first recognized in the core from Merrimelia 5 where it is located between 8603 and 8659 ft* drill depth. It was subsequently correlated with similar facies in Merrimelia 1 between 9649 and 9674 ft drill depth where contiguous lithofacies are found intimately associated with the aeolianites in the core. Analysis of wireline logs from uncored wells in the Merrimelia Field has shown that the aeolian sediments have also been penetrated in at least one other well, namely Merrimelia 13.

Aeolianite lithofacies in the Merrimelia Formation

The discovery of aeolianites adds to the kaleidoscope of facies patterns present within the Merrimelia Formation and brings new insight to the understanding of depositional processes that operated during the Gondwanaland glaciation in late Carboniferous–early Permian times (Sprigg 1978). For the first time core facies analysis has revealed sandstones of probable aeolian origin within the Gidgealpa Group of the Merrimelia Field.

Aeolian sandstones or aeolianites have been subjected to much discussion and evaluation in the last six years particularly in the light of their economic reservoir potential (see reviews of McKee 1979; and Ahlbrandt & Fryberger 1982). Most research on both modern (*ie* Quaternary, Holocene and Recent) and ancient (*ie* pre-Quaternary) aeolian sand systems has concentrated on deposits found in warm climatic regimes. However, a large number of dune fields are located in mid- to high-latitude areas and are subjected to cold-climate processes. Several recent papers (*eg* Ahlbrandt & Andrews 1978) focus attention on modern aeolian sand deposits that accumulated, or are accumulating, under cold-climate conditions. The present discovery of aeolianites in the Late Carboniferous–Early Permian Merrimelia Formation probably represents the first *full* description of cold-climate aeolian sandstones in the pre-Quaternary rock record (*cf* Clemmensen 1981). Cold-climate aeolian dunes may exhibit distinctive tensional, compressional and dissipation structures of sedimentary origin related to freezing, thawing and snowmelt processes in the cold-climate realm. Although aeolianites appear relatively homogeneous, they actually comprise a variety of deposits both texturally and depositionally, a variability that may be even further enhanced in the cold-climate setting.

Much discussion in recent years has been devoted to defining criteria by which aeolian sediments can be confidently identified. This debate arose largely from the discovery that several commonly invoked diagnostic features were equally applicable to sub-aqueous bedforms (see introduction in Kocurek & Dott 1981). These included the scale, dip, and style of cross-stratification, and the textural attributes of the sands. Little consideration was given to the internal composition of the cross-bedding. This has been rectified by Hunter (1976, 1977, 1981) who has shown that processes operative on subaerial dunes produce distinct types of internal stratification, some of which are uniquely aeolian. The presence of these laminations in conjunction with the more widely cited textural criteria, and other factors such as the occurrence of non-subaqueous features (*eg* raindrop imprints, granule lag monolayers, body and ichnofossils) is therefore powerful evidence for the aeolian origin of a sequence. In cores, where macroscale features are often difficult to determine, identification of these sedimentological features is of even greater significance, since their presence can confirm or deny an aeolian and, therefore, subaerial origin.

The aeolianites described from the Merrimelia Field consist almost exclusively of porous, poorly cemented medium- to fine-grained yellow, brown and cream-coloured sandstones which are dominated by low-angle parallel lamination with subsidiary foreset bedding (Figs 4 to 7). Sharp, planar truncation surfaces divide the suite into cross beds composed mainly of saltation deposits (climbing translatent wind-ripple lamination—hereafter termed wind-ripple lamination) with minor amounts of sandflow bedding. Primary depositional dip (post-compaction) observed in the cores varies between 0° and 25° suggesting preservation of both dune lee-slope and interdune/sand-sheet strata. Local deformation of avalanche deposits is also visible within the sequence and may be compared with similar phenomena described by McKee *et al.* (1971).

* Metric units are used where possible. As is the convention in most of the petroleum industry all drill depths are referred to in Imperial units.

MERRIMELIA 5

CORE 4

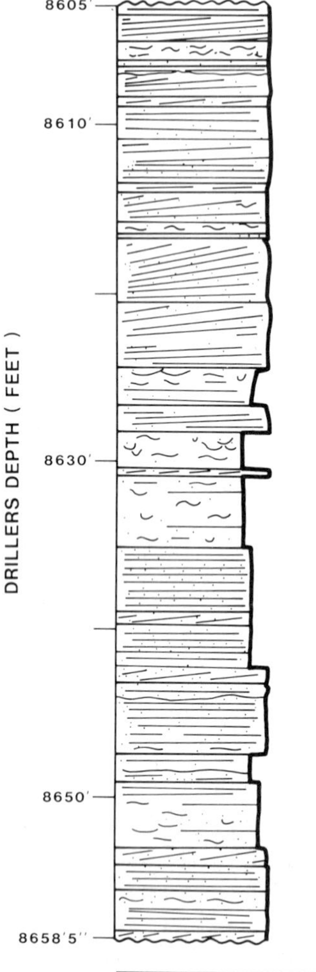

STACKED SEQUENCE OF MEDIUM GRAINED TO COARSE GRAINED SANDSTONES. LOW TO HIGH ANGLE FORESET BEDDING UBIQUITOUS WITH MARKED GRAIN SIZE SEGREGATION ON THE FORESET LAMINAE. LOCAL SOFT SEDIMENT DEFORMATION INCLUDING COMPRESSIONAL SLUMPING OF BEDS. EXCELLENT EXAMPLES OF TRUNCATION SURFACES PRESERVED IN THE CORE.

ENVIRONMENT

AEOLIANITE - DUNE LEE-SLOPE FACES DOMINATED BY WIND RIPPLE AND SAND FLOW LAMINATION, INTERDUNE AND SAND SHEET DEPOSITS MAY ALSO BE PRESENT.

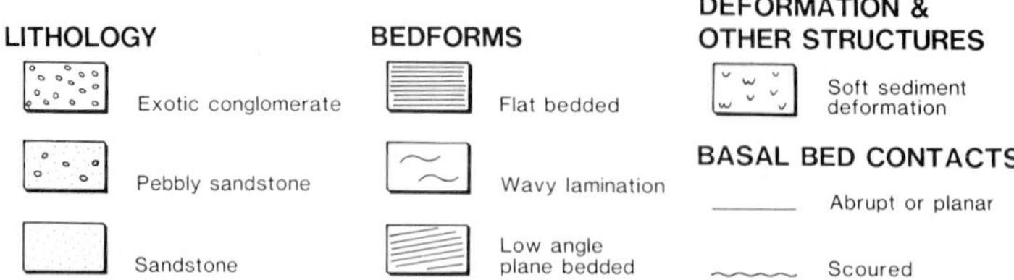

FIG. 4. Core facies log of aeolianite sequence within the glacigenic Merrimelia Formation, Merrimelia 5.

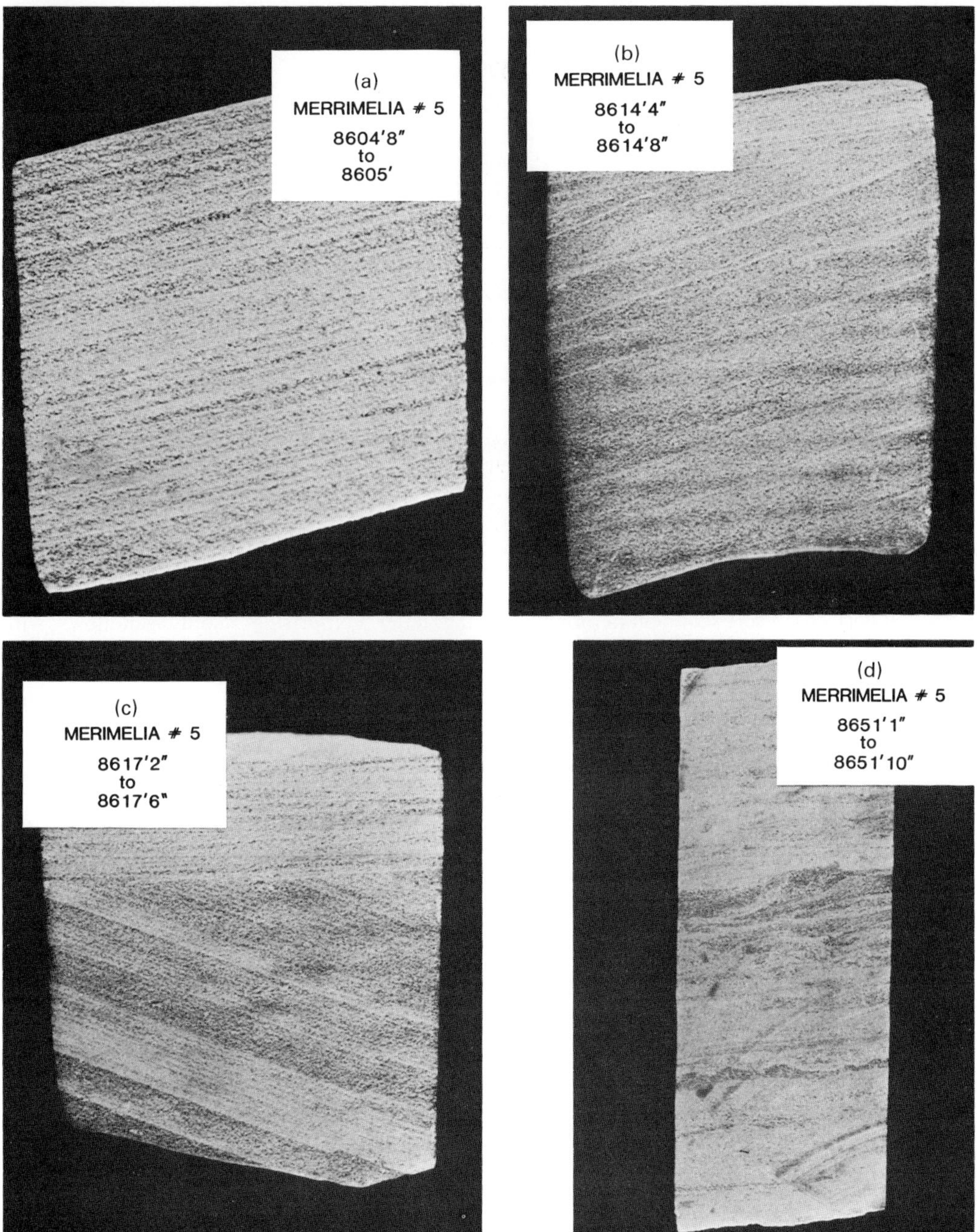

FIG. 5. Aeolianite bedforms in Core 4 of Merrimelia 5, Merrimelia Formation, Merrimelia Field. (*a*) Stacked, climbing translatent wind-ripple lamination. (*b*) Sandflow bedding with grainfall laminae overlain by bounding surface and climbing translatent wind-ripple lamination. (*c*) Sub-horizontal migration surface, truncating, and overlain by climbing translatent wind-ripple lamination. (*d*) Dewatered bedforms and adhesion ripples of wet interdune deposit.

Core 4 from Merrimelia 5 is entirely within the aeolian sandstones of the Merrimelia Formation (Fig. 4). Very little core loss is present within this section but the cored sequence is unfortunately bounded above and below by major core breaks (of 866 and 308 ft respectively). Core 3 was cut in the Tirrawarra Sandstone (*sensu stricto*) and Core 5 in 'basement' rocks (the so-called 'Innamincka Red Beds' of (?) Devonian age).

The aeolianite of Core 4 (Fig. 4) consists of a stacked sequence of cross beds of fine- to medium-grained sandstones separated by well-defined planar to concave-up bounding (truncation) surfaces. These are generally sub-horizontal and possess high angular discordances with the underlying set (Fig. 5*c*). This relationship suggests they are migration surfaces, although no bedform order, as proposed by Brookfield (1977), can be assigned to them.

Internally the cross beds are composed almost exclusively of parallel lamination and low- to moderate-angle foreset bedding. Marked grain size segregation is associated with the laminae and both normal and inverse grading is apparent. Localized horizons of soft-sediment deformation occur throughout the cored interval.

The aeolianite facies and transitions present within Core 4 are readily apparent in Figs 4 and 5*a–d*. The topmost 3.5 m of Core 4 is composed of three cross-bed sets, 0.69, 1.6 and 0.86 m thick, respectively. These are dominated by wind-ripple lamination which increases in dip from sub-horizontal at the cross-bed base to between 10–20° at its top. These saltation deposits comprise sub-critical climbing translatent wind-ripple lamination (Hunter 1981) and are on the scale of one grain to 1 cm thickness and produce a pattern commonly called 'pinstriping' or, erroneously, 'varving' in many aeolianites (Ahlbrandt & Fryberger 1982). These deposits are recognized by their low dip angles, their thinness, their similarity in thickness to adjacent laminae of the same type, their evenness, their sharp contacts, and their distinctness due in part to inverse size grading (Fig. 5*a*). Truncation surfaces with minor angular discordances are also present within these first three cross beds, and are interpreted as modifications of the surface brought about by cross winds on the dune lee-slope. Also included within this upper 3.5 m is one horizon of soft-sediment deformation possibly generated by post-depositional liquefaction (Fig. 4). The abundance of wind-ripple lamination and low angles of dip suggest preservation of only the base of the dune lee-slope, or intersection by the core of a cross-bed margin (*cf* Kocurek 1981).

Between 8614 and *c*. 8628 ft in Core 4 (Fig. 4), the aeolianite consists of three cross beds which range in thickness from 0.86 to 1.9 m. These show the expected increase upwards of depositional dip from parallel to the lower bounding surface to a consistent dip of 25°. They preserve, therefore, both basal apron and slip-face deposits, the latter comprising mainly sandflows with subsidiary wind-ripple lamination (Fig. 5*b*). The sandflows reveal inverse grading (Hunter 1977) and wedging, and, commonly, have abrupt textural contrasts between the coarser grains of the sandflow toes, and the finer-grained deposits produced by saltation processes and grainfall. The presence of both sandflows and wind-ripple lamination at high dip angles indicates that the dunes possessed active slip faces which were periodically reworked by winds of sufficient stress to generate wind ripples.

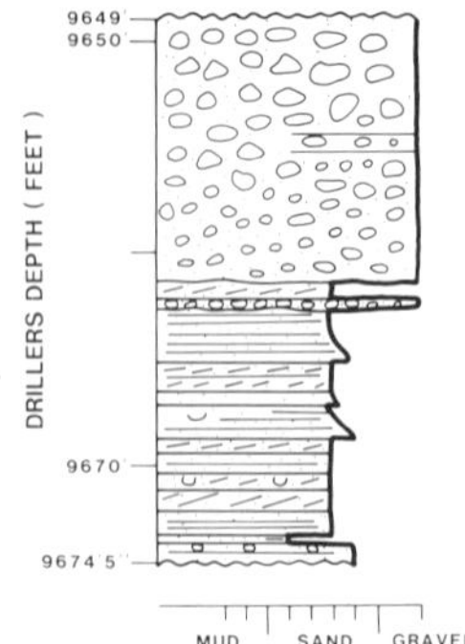

FIG. 6. Core facies log of interbedded fluvial conglomerates and aeolianite in the glacigenic Merrimelia Formation, Merrimelia 1.

Below 8628 ft, the core is composed of 4.75 m of sub-horizontal wind-ripple lamination, which above 8636 ft is almost entirely deformed by fluidization. Locally, high-angle foreset bedding is also present. The thickness and general low dip angles of this interval suggest deposition in an interdune area; the occurrence of dewatering phenomena indicates that there is comparability with the wet depositional interdune areas of Ahlbrandt & Fryberger (1981). The high-angle foreset bedding is therefore interpreted as representing the migration of a relatively small slip-faced bedform across the wet or at least damp interdune flat. Rare laminations with bimodal size distributions may be equivalent to the localized sand-sheet deposits described by Fryberger *et al.* (1979).

The lowermost 4.9 m of Core 4 (8643 to 8659 ft) displays a reversion to cross-bedded sandstone, with two sets, 3.2 and 1.5 m thick, respectively. Sharply defined migration surfaces bound the two cross beds, and are especially well seen at

FIG. 7. Interbedded conglomerate with aeolianite bedforms in Core 12 of Merrimelia 1, Merrimelia Formation, Merrimelia Field. (*a*) Framework-supported, petromict conglomerate with clasts exhibiting stylolitic contacts. (*b*) Wind-ripple lamination overlying a thin petromict conglomerate unit. (*c*) Sandflow bedding with diagenetic veins. (*d*) Compressional deformation (overthrusting) of slumped, thin sandflows on dune slip face.

8659 ft. Internally, both cross beds are dominated by wind-ripple lamination with occasional sandflows, although local deformation is still evident, testifying to the shallowness of the water table. At 8651 ft one deformed horizon is probably attributable to a dewatered, wet interdune deposit in which adhesion ripples may still be preserved (Fig. 5*d*). Wet interdune deposits (Ahlbrandt & Fryberger 1981) commonly exhibit horizontal (or $<10°$) stratification or structureless beds due to dewatering processes. They are also easily contorted by loading because of water saturation which renders them much less competent and also generates dissipation structures resulting after burial (Ahlbrandt & Fryberger 1982). Thus the sequence of facies preserved in Core 4 from the Merrimelia Formation of the Merrimelia Field point unequivocally to deposition by aeolian processes. These occurred on slip-faced dunes and in periodically wet interdune/(?)sand sheet areas.

The contiguous sedimentary facies of the aeolianite in Merrimelia 5 are not seen in the core unfortunately and thus the discovery of such sequences in Merrimelia 1 is enhanced in importance. The thickest proven section (363 m) of the

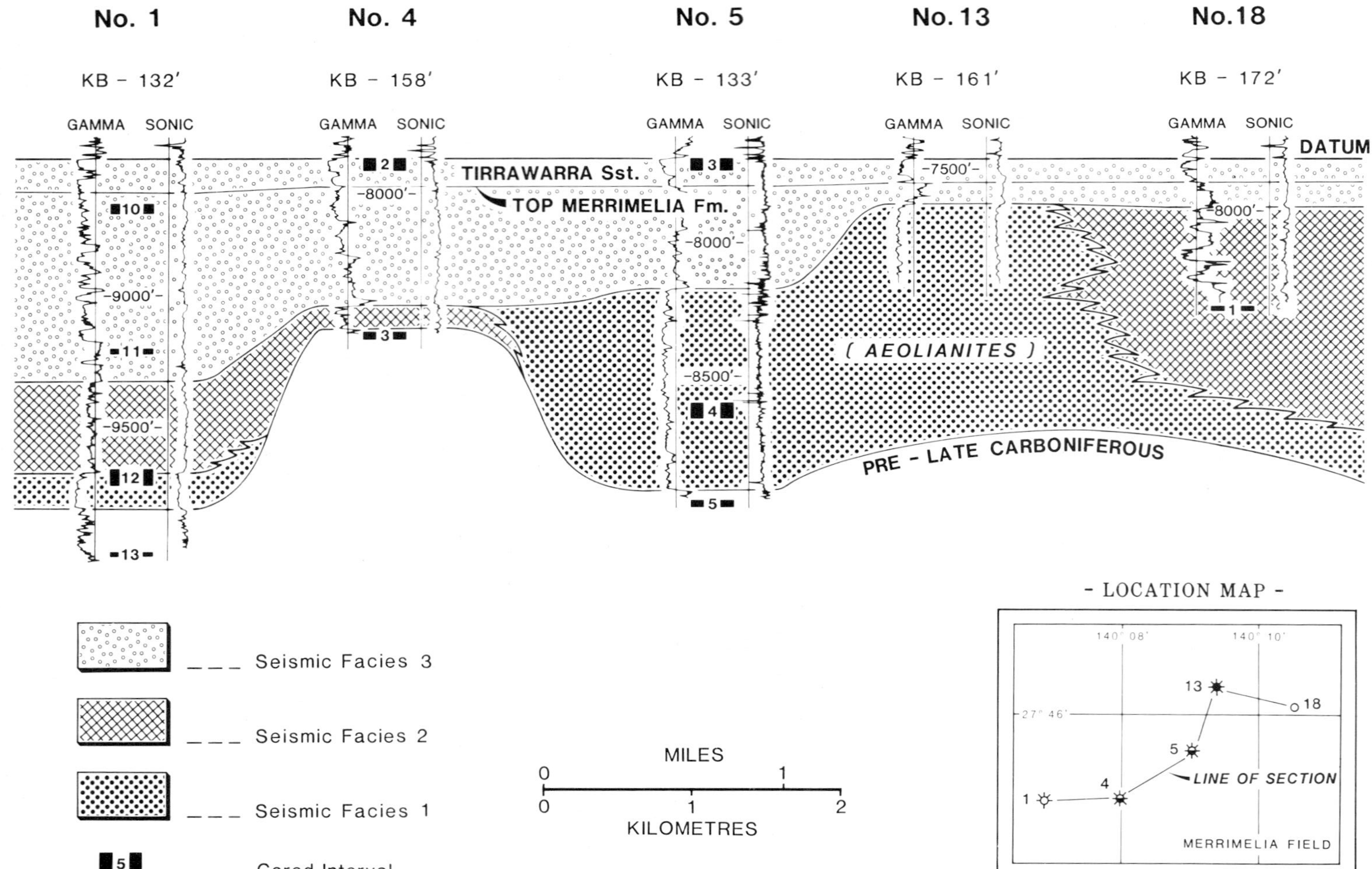

FIG. 8. Stratigraphic cross-section of the Tirrawarra Sandstone and Merrimelia Formation in the Merrimelia Field, showing location of cores and relationship of seismic facies to gamma-ray and sonic logs.

Merrimelia Formation yet encountered is present in Merrimelia 1. Cores 10 and 11 preserve lithofacies sequences attributable to glaciolacustrine and fluvioglacial deposition, but Core 12 contains an aeolianite sequence interbedded with, and abruptly overlain by, petromict conglomerates (Figs 6 and 7*a–d*). This core is bounded by two major core breaks of 131 m above and 184 m below. The lower core break encompasses the boundary between the Merrimelia Formation and the basement rocks of the Warburton Basin (Gatehouse 1986), since Core 13 is cut in the 'Innamincka Red Beds' of (?)Devonian age. Some core loss has occurred within Core 12 but, fortuitously, important bounding surfaces are still preserved.

Core 12 (9649–9684 ft) reveals the aeolianite sequence in its lower half where the cored section features wind-ripple lamination with minor amounts of sandflow bedding. The medium-grained sandstones are arranged in a stacked sequence of cross beds with pulses of coarser-grained units, locally crossed by diagenetic veins. Between 9665 and 9666 ft, wind-ripple lamination is the dominant structure where it is seen intimately associated with thin petromict conglomerate horizons (Fig. 7*b*). Above the main core loss at 9670 ft, the cross beds consist mainly of wind-ripple lamination with subordinate amounts of grainfall deposits and relatively thick sandflows (Fig. 7*c*). The core loss itself (9670–9673.5 ft) is most probably within a thick sandflow package which is usually more poorly cemented than the wind-ripple stratification. Between 9673.5 and 9675.5 ft wind-ripple lamination is again predominant with minor sandflow bedding. The lamination here is at varying angles, indicating the presence of several modified surfaces. From 9675.5 to 9680 ft the bedforms are wavy with few wind ripples, and may point to the preservation of a wet interdune horizon. Compressional deformation of sandflow foresets generating overthrust phenomena (Ahlbrandt & Fryberger 1982) is well displayed in the core at 9680.5 ft (Fig. 7*d*). The remainder of the core to the base at 9684 ft exhibits flat and wavy bedding, truncation due to both migration and modification of surfaces, and pervasive diagenetic siderite. This lower sequence is attributable to interdune or sand-sheet deposition with minor contributions from migrating dunes. The upper half of Core 12, where most of the core loss has been incurred, comprises massive to crudely horizontally stratified petromict, and framework-supported conglomerates (Figs 6 and 7*a*). The clasts are sub- to well-rounded and dominated by acid igneous debris (granites, porphyries, lavas) with subordinate fine-grained sandstones, cherts and phyllites.

In general there appears to be an upward coarsening in clast size throughout the unit together with an attendant coarsening of the matrix. The conglomerate is consistent with the deposits of sheet or longitudinal bars on a proximal braidplain (Bluck 1974), or with those of a low slope, wet, alluvial outwash fan (Boothroyd & Ashley 1975). Examples of this lithofacies are encountered elsewhere in the Merrimelia Formation (*eg* Lake Hope 1; Williams & Wild 1984*a*) but here it is recorded for the first time interbedded with, and overlying, sediments of aeolian origin. It is plausible that these gravel bars form part of the fluvioglacial extradune environment, the sand from which had been winnowed and reworked to form the Merrimelia dune field.

Geometry of the aeolianite sequence

The existence of aeolianite reservoirs in the Merrimelia Formation offers new opportunities in the development of hydrocarbon plays in the Cooper Basin. Mapping of the reservoir has been attempted to determine the geometry of the aeolianites and the presence of potential hydrocarbon traps. The method used integrated core data with gamma-ray and sonic (GR–S) well logs and seismic data.

The study of cores from the Merrimelia Field has identified aeolianites in two wells, Merrimelia 1 (Core 12) and Merrimelia 5 (Core 4). Correlation of these cored intervals with GR–S logs shows a characteristic response of low gamma-ray (45–50 API units) and low sonic (80–85 μs ft^{-1}) values, consistent with a mineralogically mature quartz arenite (Fig. 8).

In Merrimelia 1, 4 m of aeolianites were cored. Analysis of the GR–S logs (Fig. 8) shows that the total thickness is 41 m, and that the aeolianites lie directly on pre-late Carboniferous basement rocks of the Warburton Basin (Gatehouse 1986). The interbedding of the upper part of the aeolianites with outwash fan deposits, as observed in core and described above, is also visible from the GR–S log response (Fig. 8).

Merrimelia 5 contains a cored interval 18 m in thickness (Core 4), which consists only of deposits of aeolian, interdune and sand-sheet origin. The GR–S logs (Fig. 8) show that Core 4 is from approximately the middle of a thick 232 m sequence of sandstones and minor mudrocks. The total thickness is unlikely to be completely aeolianite and may therefore include the deposits of a sandy braidplain or outwash fan. Several high gamma-ray spikes, coincident with the occurrence of grey-green mudrocks in cuttings, probably represent wet interdune sediments.

GR-S logs indicate Merrimelia 13 is the only other well on the Merrimelia structure which penetrated aeolian facies. The interval was not cored (Fig. 8) and the well was completed in the aeolianite. However, its base has been mapped using its seismic character and this gives an estimated thickness of more than 82 m.

Synthetic normal polarity seismic sections constructed from the Merrimelia 1, 5 and 13 sonic logs show a characteristic response at the top of the aeolianites due to the low interval velocity of the porous sandstones. This response produces a mappable reflector on seismic sections marking the top of *Seismic Facies 1* (Fig. 8). Sonic log character suggests that two other intervals may produce characteristic seismic responses and these have been labelled *Seismic Facies 2* and *3* in Fig. 8.

Seismic Facies 2 corresponds to fluvioglacial outwash fan and glaciolacustrine deposition, interpreted from Merrimelia 18 core 1 and Merrimelia 1 core 12. This facies passes laterally into, and interdigitates with *Seismic Facies 1*. *Seismic Facies 3* which overlies *Seismic Facies 1* and *2*, contains sediments deposited within a wide variety of glacial and fluvioglacial settings, including fluvioglacial outwash, sub-aqueous flow and melt-out tills, lodgment tills and glaciolacustrine sediments. These represent the interfaced upper part of the Merrimelia Formation and the Tirrawarra Sandstone in the Merrimelia Field.

An isopach map derived from seismic data shows that the potential aeolianite reservoirs of *Seismic Facies 1* (Fig. 9) are confined to the southern part of the Merrimelia structural high. The geometry is of a 6 to 8 km wide zone trending NW–SE and bounded to the NE and SW by depositional edges where interdigitation with *Seismic Facies 2* occurs.

In the NE, pre-Toolachee Formation uplift

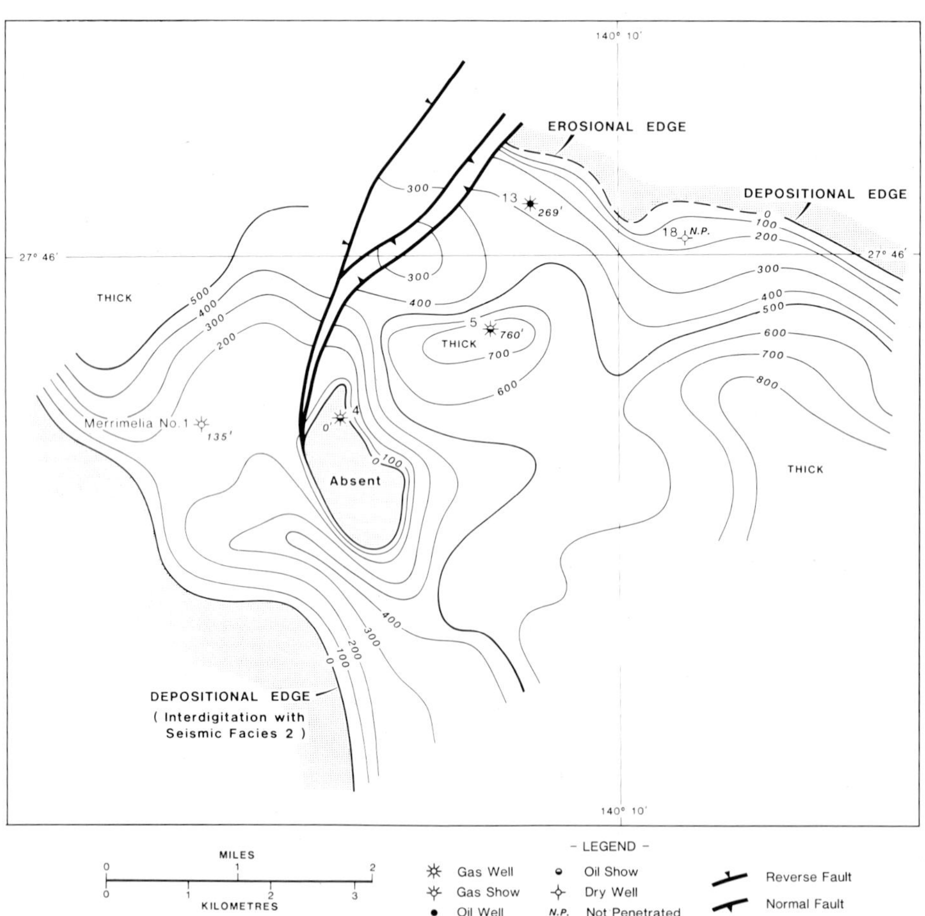

FIG. 9. Merrimelia Field, isopach map of the predominantly aeolian *Seismic Facies 1*.

gives an erosional edge to *Seismic Facies 1* at the crest of the structure (Fig. 9). Rapid thickening occurs from the depositional edges to the axis of the aeolianite zone along which it attains a thickness of over 213 m. Thickening is also evident away from the crest of the Merrimelia structural high, in the trough areas to the NE and SW.

A small (12 km^2) area of high basement occurs towards the western edge of the band of aeolianites. Merrimelia 4 was drilled on this basement high and encountered Merrimelia Formation sediments, but no aeolianites of *Seismic Facies 1*. Seismic interpretation indicates that the absence of this facies is due to non-deposition. The agreement between the two data types suggests that the facies mapping from seismic data is locally reliable .

The aeolianite sandstones have good log- and core-derived porosities of 12–18% and permeabilities of up to 75 md. These porosity and permeability values are abnormally high for the usually unprospective Merrimelia Formation. Petrographic studies confirm that large volumes of good-quality reservoir are preserved in the aeolianites as secondary and primary porosity. The porosity can be pervasive and may be controlled by selective dissolution of calcite cement. These observations support the potential development of good quality reservoirs in aeolianite sandstones at depth, perhaps beneath zones with little reservoir potential.

Merrimelia Formation aeolianites probably still remain undiscovered and untested in other areas of the Cooper Basin. Evidence of aeolian processes in other Merrimelia Formation core has already been tentatively identified in two additional wells, Minkie 1 and Tirrawarra 1, so aeolianites may yet be involved in future hydrocarbon plays in the Cooper Basin.

Aeolianite depositional setting

Models of glacigenic sedimentation are still relatively poorly known and those that have been documented relate mainly to Quaternary examples. However, in recent years more descriptions have become available for Permo–Carboniferous glacial sequences particularly from South Africa (Visser 1982), Oman (de la Grandville 1982), South America (Gravenor & Rocha-Campos 1983), and Australia (Alley & Bourman 1984; Harris 1981; Williams & Wild 1984*a*). The sedimentary facies represented in these sequences include tillites and other ice-contact deposits, glaciofluvial, glaciodeltaic, glaciolacustrine and glaciomarine sediments. Additionally here, for the first time, glacio-aeolian sediments are described from the Permo–Carboniferous.

From comparison with Recent and Quaternary examples of paraglacial environments (Ashley *et al.* 1985) the Merrimelia aeolianites are apparently related to broad outwash plains developed in front of a retreating ice sheet. These plains may comprise large braidplains (Bluck 1974) and/or extensive outwash fans (Boothroyd & Ashley 1975). The fluvioglacial environment would include a variety of bar forms (of sheet, longitudinal and linguoid type) developed in low-sinuosity, bedload-dominated channel systems. Such systems are known to have been present during Merrimelia/Tirrawarra times in the southern Cooper Basin (Williams & Wild 1984*a*, *b*) and may have terminated in lacustrine fan-delta and/or glaciolacustrine shoreline settings, for as yet no glaciomarine sediments have been positively identified in the Cooper Basin (Stuart 1976; Moore & Castro 1984). Proximity to an ice sheet is evidenced in these paraglacial environments by ice loading phenomena, by freeze/thaw structures, by dropstones in varvites, and by frozen clasts. Most of these features have been described from the Merrimelia Formation (Williams & Wild 1984*a*) as have the deposits of shallow and deep-water glacial lakes. Supraglacial and subglacial deposits are interbedded with, and are adjacent to, the paraglacial sediments, the vertical and spatial arrangement of which will depend on the number of ice-sheet advances and recessions (de la Grandville 1982).

Models for the development of cold-climate aeolianites in paraglacial settings are sparse. Diagnostic features of aeolian dune sedimentation and stratification in mid- to high latitudes have been proposed by Ahlbrandt & Andrews (1978) and Ahlbrandt & Fryberger (1982), and contrast with slip-face phenomena of warm-climate dune complexes (Ahlbrandt & Fryberger 1982). Several of these high-latitude features are present in the Merrimelia aeolianites.

When considering the generation of aeolian dune fields in paraglacial environments, the role of topography and climate (wind speed and rainfall) must be assessed. The Scott and Yana outwash fans of the NE Gulf of Alaska are located in front of the Malaspina Glacier. These extensive fluvioglacial systems exhibit many of the bedforms and channel forms found in the Merrimelia/Tirrawarra sequence (Williams & Wild 1984*a*). However, aeolian facies are lacking in this area due to the abundant vegetation cover of the inactive alluvial tracts.

In contrast, the paraglacial braidplains of Iceland do not support a strong vegetation cover. This is because precipitation in the Skeidararsan-

FIG. 10. Kobuk Dunes National Monument, Brooks Range, Alaska. A sand sea, 4 × 2 km, composed of barchanoid dunes migrating over vegetation. Sand transport from left to right. (Photo courtesy of Gail Ashley.)

dur area of southern Iceland is 1000 mm a^{-1} and the area is subjected to strong winds that are >10 m s^{-1} for at least 30% of the time. The result is that the Icelandic sandur plains support large areas of active aeolian dune fields in areas where fluvial activity is low (Boothroyd & Nummendal 1978; Figs 15 and 16). Many of the dunes in this setting exist as low mounds or small transverse dunes with or without slip faces. They range between 2–10 m in height with a spacing of 10–60 m between crest lines.

A similar possible modern analogue for the Merrimelia aeolianites is provided by a late Pleistocene–early Holocene aeolian deposit in northern Alaska (Ashley & Hamilton 1983). Here, the active dunes of the Kobuk Dunes National Park comprise a sand sea some 8 km^2 in area, composed of relatively low relief barchanoid dunes (Fig. 10). The two examples cited have been chosen for their high-latitude depositional setting rather than for the geometry and dimensions of the dunes they contain, since neither of these properties can be determined from core analysis.

Thus, the model envisaged for the Merrimelia aeolianites encompasses aspects of both the Alaskan and Icelandic depositional settings outlined above. There is no doubt that the Merrimelia aeolian dune field is associated with fluvioglacial sediments when the interbedded nature of the aeolian and alluvial sequence in Merrimelia 1 is considered. The Merrimelia aeolian dunes were probably of no great relief and might have possessed dimensions not unlike those of Skeidararsandur or Kobuk Dunes. The dune toes must have merged into extensive, ephemerally wet interdune areas and were locally adjacent to sand sheets. The Merrimelia dunes were not stabilized, as evidenced by the amount of avalanching of dune slip faces and the lack of root structures in the sediments.

The economic potential of aeolianite reservoirs in the Merrimelia Formation is of paramount importance. Aeolianites in general have proved to be complex, heterogeneous hydrocarbon reservoirs (Lupe & Ahlbrandt 1979). Problems associated with such reservoirs include the lateral discontinuity of reservoir zones; the variable permeability found in intercalated horizontal, low-angle plane- and cross-bedding; and the anisotropic permeability due to textural differences and differential cementation along and across laminae (Ahlbrandt & Fryberger 1982). Some of these problems have been documented for the Permo–Carboniferous Tensleep Sandstone of Wyoming and Colorado with respect for the exploration and development of hydrocarbon plays (Morgan *et al.* 1978). Performance studies of aeolian hydrocarbon reservoirs have recorded higher oil recovery with infill drilling and selective perforation. The possibility of hydrocarbon source beds within, or adjacent to, aeolian deposits should be carefully considered. In warm-climate aeolian settings evaporitic interdune deposits provide possible source beds, but obviously this model is not applicable to the Merrimelia Formation. In the present study glaciolacustrine mudrocks contiguous with aeolianites would be the most likely source rock and this is currently being investigated.

ACKNOWLEDGMENTS: The authors are extremely grateful to Dr Bevan Devine of South Australian Oil & Gas Corporation Party Limited for his enthusiastic encouragement throughout all stages of the study. Bill Kendall, Peter Moignard and the geological and geophysical staff of SAOGC are much thanked for their interest and input into the study and for their immense logistical support. The continuing core facies analysis programme could not have been undertaken efficiently if it were not for the enormous help of Brian Logan, John Curry and the staff at the Core Library Complex of the SA Department of Mines and Energy. Brenton Wells and his staff at SAOGC are thanked for their excellent drafting and Julie Batt and Joanne Woods for their flawless typing. The joint venture partners in the Cooper Basin of South Australia gave permission for the release of data which facilitated the publication of this paper; this is most gratefully acknowledged. Thanks are also extended to Lars Clemmensen and Jim Rose for their constructive comments on the manuscript and to Ian Reid for his excellent editorial work. Sabrina Sadri (University of Bristol) is thanked for the final typing of the manuscript.

References

AHLBRANDT, T. S. & ANDREWS, S. 1978. Distinctive sedimentary features of cold-climate eolian deposits, North Park, Colorado. *Palaeogeography, Palaeoclimatology, Palaeoecology* **25**, 327–351.

—— & FRYBERGER, S. G. 1981. Sedimentary features and significance of interdune deposits. *In*: ETHRIDGE, F. G. & FLORES, R. M. (eds) *Recent and ancient nonmarine depositional environments: models for exploration*. Society for economic Palaeontologists and Mineralogists, Special Publication **31**, 293–314.

—— & —— 1982. Introduction to eolian deposits. *In*: SCHOLLE, P. A. & SPEARING, D. (eds) *Sandstone Depositional Environments*. American Association of Petroleum Geologists Memoir **31**, 11–47.

ALLEY, N. F. & BOURMAN, R. P. 1984. Sedimentology

and origin of Late Palaeozoic glacigenic deposits at Cape Jervis , South Australia. *Transactions of the Royal Society of South Australia* **108**, 63–75.

ASHLEY, G. M. & HAMILTON, T. D. 1983. Epiguruk bluff—a late Quaternary environmental record from N.W. Alaska. *Geological Society of America Meeting Program with Abstracts* **15**, 191.

——, SHAW, J. & SMITH, N. D. 1985. *Glacial Sedimentary Environments*. Society for Economic Paleontologists and Mineralogists Short Course, 246 pp.

BATTERSBY, D. G. 1976. Cooper Basin oil and gas fields. *In*: LESLIE, R. B., EVANS, H. J. & KNIGHT, C. L. (eds) *Economic Geology of Australia and Papua New Guinea. 3, Petroleum*. Australian Institute of Mining and Metallurgy Monograph **7**, 321–370.

BLUCK, B. J. 1974. Structure and directional properties of some valley sandur deposits in southern Iceland. *Sedimentology* **21**, 533–554.

BOOTHROYD, J. C. & ASHLEY, G. M. 1975. Processes, bar morphology and sedimentary structures on braided outwash fans, north-eastern Gulf of Alaska. *In*: JOPLING, A. V. & MCDONALD, B. C. (eds) *Glaciofluvial and glaciolacustrine sedimentation*. Society for Economic Paleontologists and Mineralogists, Special Publication **23**, 193–222.

—— & NUMMEDAL, D. 1978. Proglacial braided-outwash: a model for humid alluvial-fan deposits. *In*: MIALL, A. D. (ed.) *Fluvial Sedimentology*. Canadian Society of Petroleum Geologists Memoir **5**, 641–668.

BROOKFIELD, M. E. 1977. The origin of bounding surfaces in ancient aeolian sandstones. *Sedimentology* **24**, 303–332.

CLEMMENSEN, L. B. 1981. Late Precambrian tillites of Peary Land, North Greenland. *In*: HAMBREY, M. J. & HARLAND, W. B. (eds) *Earth's pre-Pleistocene Glacial Record*. Cambridge University Press, 782–786.

DE LA GRANDVILLE, B. F. 1982. Appraisal and development of a structural and stratigraphic trap oil field with reservoirs in glacial to periglacial clastics. *In*: HALBOUTY, M. T. (ed.) *The deliberate search for the subtle trap*. American Association of Petroleum Geologists Memoir **32**, 267–286.

FRYBERGER, S. G., AHLBRANDT, T. S. & ANDREWS, S. 1979. Origin, sedimentary features and significance of low-angle eolian 'sand sheet' deposits, Great Sand Dunes National Monument and vicinity, Colorado. *Journal of sedimentary Petrology* **49**, 733–746.

GATEHOUSE, C. G. 1972. Formations of the Gidgealpa Group in the Cooper Basin. *Australian Oil and Gas Review* **18**, 10–15.

—— 1986. The geology of the Warburton Basin in South Australia. *Australian Journal of Earth Sciences* **33**, 161–180.

GRAVENOR, C. P. & ROCHA-CAMPOS, A. C. 1983. Patterns of Late Palaeozoic glacial sedimentation on the southeast side of the Parana Basin, Brazil. *Palaeogeography, Palaeoclimatology, Palaeoecology* **43**, 1–39.

GRUND, R. B. 1966. *The glacigenic sediments of the Coopers Creek Basin. BSc thesis*, Adelaide University, 127 pp.

HARRIS, W. K. 1981. Permian diamictites of South Australia. *In*: HAMBREY, M. J. & HARLAND, W. B. (eds) *Earth's pre-Pleistocene Glacial Record*. Cambridge University Press, 469–473.

HUNTER, R. E. 1976. Comparison of eolian and subaqueous sandflow cross-strata (Abs). *American Association of Petroleum Geologists Bulletin* **60**, 633–634.

—— 1977. Basic types of stratification in small eolian dunes. *Sedimentology* **24**, 361–387.

—— 1981. Stratification styles in eolian sandstones: some Pennsylvanian to Jurassic examples from the Western Interior U.S.A. *In*: ETHRIDGE, F. G. & FLORES, R. M. (eds) *Recent and ancient nonmarine depositional environments: models for exploration*. Society for economic Paleontologists and Mineralogists, Special Publication **31**, 315–329.

KAPEL, A. J. 1966. The Coopers Creek Basin. *Journal of the Australian Petroleum Exploration Association* **6**, 71–75.

—— 1972. The geology of the Patchawarra area, Cooper Basin. *Journal of the Australian Petroleum Exploration Association* **12**, 53–57.

KOCUREK, G. 1981. Erg reconstruction: the Entrada Sandstone (Jurassic) in northern Utah and Colorado. *Palaeogeography, Palaeoclimatology, Palaeoecology* **36**, 125–153.

—— & DOTT, R. H. 1981. Distinctions and uses of stratification types in the interpretation of eolian sand. *Journal of sedimentary Petrology* **51**, 579–595.

LUPE, R. & AHLBRANDT, T. S. 1979. Sediments of the ancient eolian environment—reservoir inhomogeneity. *In*: MCKEE, E. D. (ed.) *A Study of Global Sand Seas*. United States Geological Survey Professional Paper **1052**, 241–251.

MARTIN, C. A. 1967. The Gidgealpa and Merrimelia Formations in the Coopers Creek Basin. *Australian Oil and Gas Journal*, **14**, 29–35.

MCKEE E. D. (ed.) 1979. *A Study of Global Sand Seas*. United States Geological Survey Professional Paper **1052**, 429 pp.

——, DOUGLASS, J. R. & RITTENHOUSE, S. 1971. Deformation of lee-side laminae in eolian dunes. *Geological Society of America Bulletin* **82**, 359–378.

MOORE, P. S. & CASTRO, C. 1984. Petroleum exploration in fluvio-lacustrine sequences, with examples from the Cooper and Eromanga Basins, central Australia. *In*: JONES, B. G. & HUTTON, A. C. (eds) *Fluvio-deltaic systems: facies analysis in exploration*. Australasian Sedimentologists Specialists Group, Wollongong, 307–337.

MORGAN, J. T., CORDINER, F. S. & LIVINGSTON, A. R. 1978. Tensleep reservoir, Oregon Basin field, Wyoming. *American Association of Petroleum Geologists Bulletin* **62**, 609–632.

PATEN, R. J. 1969. Palynologic contributions to petroleum exploration in Permian formations of the Cooper Basin. *Journal of the Australian Petroleum Exploration Association* **9**, 79–87.

PRICE, P. L. 1973. *Cooper Basin Palynology*. Mines Administration Party Ltd, Palynological Report No. 13–95, unpublished, 13 pp.

SPRIGG, R. C. 1978. Proterozoic, Permo–Carboniferous and Pleistocene glacial cycles and cyclic sedimentation in relation to oil search. *Journal of the Australian Petroleum Exploration Association* **18**, 83–92.

STUART, W. J. 1976. The genesis of Permian and Lower Triassic reservoir sandstones during phases of southern Cooper Basin development. *Journal of the Australian Petroleum Exploration Association* **16**, 37–47.

THORNTON, R. C. N. 1979. Regional stratigraphic analysis of the Gidgealpa Group, southern Cooper Basin, Australia. *Geological Survey of South Australia Bulletin* **49**, 140 pp.

VISSER, J. N. J. 1982. Upper Carboniferous glacial sedimentation in the Karoo Basin near Prieska, South Africa. *Palaeogeography, Palaeoclimatology, Palaeoecology* **38**, 63–92.

WILD, E. K. & WILLIAMS, B. P. J. 1984. Fluvioglacial sandstone reservoirs and deposystem analysis in hydrocarbon exploration of the Permian Gidgealpa Group, southern Cooper Basin, South Australia (Abs). *American Association of Petroleum Geologists Bulletin* **68**, 539.

WILLIAMS, B. P. J. 1982. *Facies analysis of Gidgealpa Group reservoir rocks, southern Cooper Basin, South Australia*. Unpublished report, South Australian Oil & Gas Corporation Party Ltd, 247 pp.

—— 1984. *Facies analysis of Gidgealpa Group reservoir rocks from selected areas of the southern Cooper Basin, South Australia*. Unpublished report, South Australian Oil & Gas Corporation Party Ltd, 95 pp.

—— & WILD, E. K. 1984*a*. The Tirrawarra Sandstone and Merrimelia Formation of the southern Cooper Basin, South Australia—the sedimentation and evolution of a glaciofluvial system. *Journal of the Australian Petroleum Exploration Association* **24**, 377–392.

—— & —— 1984*b*. Late Carboniferous–Early Permian sandstone reservoir facies analysis in hydrocarbon exploration of the Gidgealpa Group, southern Cooper Basin, South Australia. *Geological Society of Australia, Abstracts 7th A.G.C. Sydney* **12**, 553.

B. P. J. WILLIAMS & E. K. WILD, Department of Geology, University of Bristol, Bristol BS8 1RJ, UK.

R. J. SUTTILL, South Australian Oil & Gas Corporation Party Ltd, Adelaide, SA 5000, Australia.

Aeolian sedimentation from the Galtymore Formation (Devonian), Ireland

R. A. Carruthers

SUMMARY: The Galtymore Erg (Devonian) of the Galtee Mountains, southern Ireland can be reconstructed by the coupling of field relationships with current hypotheses and observations from other ancient aeolian sequences and modern ergs. Stratigraphical relationships show that the Galtymore Erg formed at the northern, fault-controlled, margin of the Munster Basin. Northern sequences, close to the basin margin, are dominated by northerly derived, proximal braided stream deposits. Two types of aeolian-dominant sequences occur within the Galtymore Formation. A marginal erg facies, characterized by simple crescentic dunes showing S–N palaeocurrents is interbedded with fluvial deposits. To the S an erg facies is developed which is characterized by the vertical repetition of cosets of large-scale cross-strata and thin horizontal deposits. The cosets are interpreted as representing the downwind migration and climbing of large dunes and interdune areas, respectively. The geometrical relationship of the hierarchy of bounding surfaces suggests that the dunes of the erg facies were compound crescentic in form, with enclosed interdune areas. The distribution of stratification types and foreset dip dispersion similarly suggest crescentic dunes. Estimates of compound dune wavelength and some dune and interdune dimensions can be made. Crescentic dunes (simple and compound) were moved by south-southeasterly palaeowinds towards the foot of alluvial fans.

The Munster Basin is a large area of southern Ireland (Fig. 1) characterized by a relatively thick Old Red Sandstone succession (Naylor & Jones 1967). The major part of the basin-fill has been shown to be Upper Devonian in age, possibly extending down to the uppermost Middle Devonian (Clayton & Graham 1974; Higgs & Russell 1981; Russell 1978). Lower Devonian sediments are only suspected to exist in the non-marine Dingle Group (Holland 1981). There is everywhere an upward transition into rocks of Courceyan (Dinantian) age. The generalized isopachs for the basin (Fig. 1) are only minimum values away from the exposed basin margins. Graham (1983) suggests that these may demonstrate approximately the true thickness variations. This supports the Naylor & Jones (1967) construction of an asymmetrical basin profile, for which they propose a half-graben structure, fault bounded to the N and hinged to the S. Graham (1983) has proposed that the facies pattern of the basin is best explained by means of a major fluvial distributary system, and that although there was a local input of sediment from the basin margins, the main part of the fill was derived from a distant source, to the N.

The presence of aeolian sediments within the Munster Basin was first recorded by Horne (1971, 1975). He described the large-scale cross-stratified sandstones at Kilmurry Bay and Ballydavid Head, Dingle Peninsula, Co. Kerry (Fig. 1), situated palaeogeographically on the northern margin of the Basin. Along strike from Dingle, in the Galtee Mountains inlier (Counties Limerick and Tipperary, Fig. 1), Jackson (1972) noted that cross-stratification in the sandstones of the Galtymore Formation (Table 1), bears a close resemblance to that recorded from modern aeolian dunes.

The main purpose of this paper is to prove an aeolian origin for parts of the Galtymore Formation (Table 1), and to interpret these deposits in the light of recent advances in aeolian sedimentology.

Regional setting

The Old Red Sandstone facies of the Galtee Mountains outcrops as part of a large Variscan east-northeasterly trending anticlinal structure, fault-bounded on its northern limb. The Old Red Sandstone sequence overlies Silurian pelites and psammites with angular unconformity. It ranges in thickness from *c.* 1000 m on Slievenamuck (Fig. 2*B*) to over 2500 m on the southern flanks of the Galtees (Fig. 2*E*). Within this area, the presence of predominantly red coloration, mudcracks, pedogenic carbonates and rootlets all testify to the continental setting of these deposits. As in other parts of the Munster Basin, fluvial facies predominate.

The lowest unit in the area is the Pigeon Rock Formation, a locally derived lithic cobble–boulder breccia and conglomerate sequence (Table 1). It shows N–S palaeocurrents (Fig. 2*E*) and is interpreted as being of alluvial fan origin (Carruthers 1985). This unit is overstepped in the

From Frostick, L. & Reid, I. (eds), 1987, *Desert Sediments: Ancient and Modern*, Geological Society Special Publication No. 35, pp. 251–268.

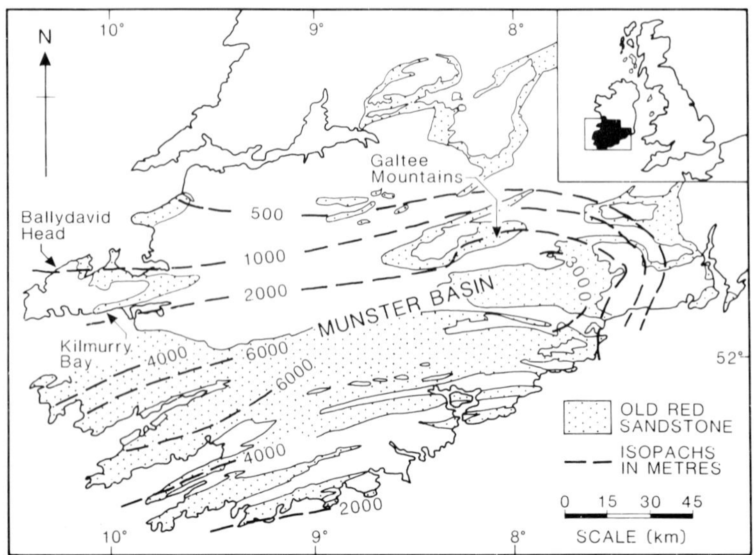

FIG. 1. Map of SW Ireland showing outcrops and isopachs of the Old Red Sandstone (isopachs modified after Graham 1983).

central part of the Galtee Mountains by a southerly derived sandstone sequence, the Galtymore Sandstone Formation (Jackson 1972; Fig. 2*C*; Table 1) which lies directly above the Silurian basement. To the N (Fig. 2*C*, *G*), these sandstones are replaced by a northerly-derived alluvial fan sequence. The Lough Muskry Formation, a mudrock-prone sequence of flood-plain origin, overlies the Galtymore Formation throughout the Galtee area and is in contact with the Silurian on Slievenamuck (Fig. 2*B*) where the Galtymore Formation is absent.

Good-quality exposure of the Galtymore Formation is present in numerous glacial corrie and crag sections surrounding the highest peak, Galtymore Mountain, 919.8 m (R 8781 2378). Elsewhere exposure is confined to the streams draining the slopes of the Galtee Mountains.

The Galtymore Formation—lithofacies

Six main lithofacies (two aeolian and four fluvial) are recognized in the Galtymore Formation; these are summarized in Table 2. The aeolian lithofacies are characterized by orange-red, pebble-free, very fine to coarse-grained sandstones (medium-grained predominating) showing S to N palaeocurrents. The fluvial lithofacies are typified by red-purple colours and contain variable amounts of lithic clasts (up to cobble-grade); these can be matched to the lithologies of the local Silurian basement. Southerly-directed palaeocurrents dominate in the fluvial lithofacies. Only the aeolian lithofacies are described in detail here. A fuller description of the fluvial lithofacies can be found in Carruthers (1985).

TABLE 1. *Lithostratigraphy of the Old Red Sandstone facies of the Galtee Mountains, Cos. Limerick and Tipperary, Ireland (Carruthers 1985)*

Formation	Thickness (m)	Depositional environment
Kiltorcan	250–300	Coastal plain (low–high sinuousity streams)
Ardane	70–200	Gravel braidplain (resistate detritus)
Ballydavid	120–480	Sandy braidplan
Slievenamuck	520–600	Gravel braidplain (resistate detritus)
Lough Muskry	40–330	Ephemeral streams and flood plain
Galtymore	0–200	Aeolian and alluvial fan
Pigeon Rock	0–200	Alluvial fan (detritus locally derived)

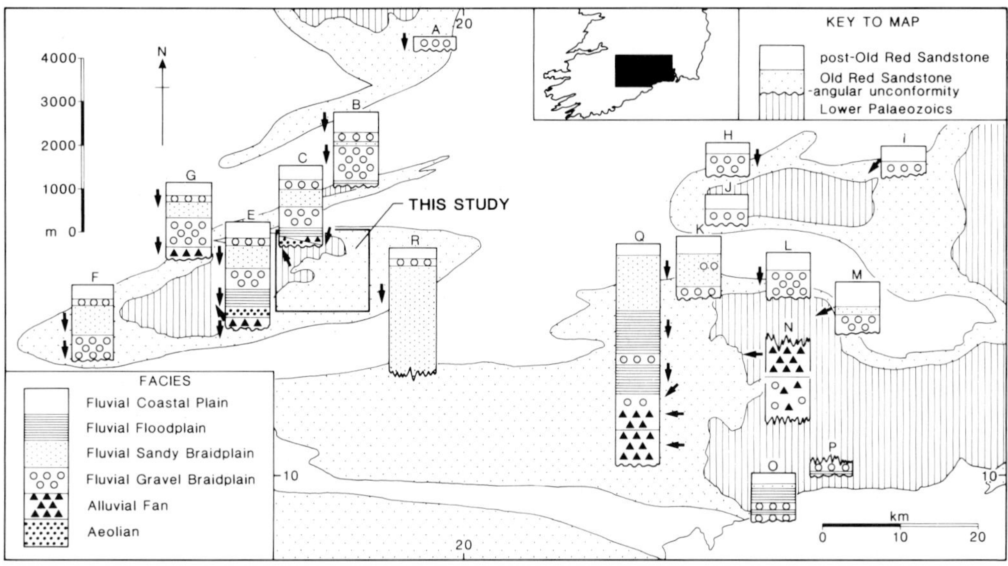

FIG. 2. Map of the eastern end of the Munster Basin showing the lateral and vertical distribution of facies and generalized palaeocurrents. Columns are composites with general location indicated by the base of the column. In addition to the author, data comes from Boland (1983), Boldy (1982), Colthurst (1978), Doran (1970) and Pilling (pers. comm.).

Large-scale cross-stratified sandstones

This lithofacies is composed of fine to medium-grained, orange-red pebble-free sandstones which display large-scale wedge-planar cross-sets (Fig. 3). The foresets have a concave upward asymptotic form and most show northerly-directed palaeocurrents. Set size varies from 0.3 m up to 15 m, although most are typically less than 5 m thick. Individual sets can be traced for up to 150 m parallel to, and over 80 m in sections perpendicular to the palaeoflow direction.

The stratification observed within the cross-sets is of two main types:

(1) Even, parallel and laterally extensive laminae (fine to medium sand size) varying in thickness from 1–30 mm (Fig. 4*a*). The thicker

TABLE 2. *Lithofacies of the Galtymore Formation*

Lithofacies	Description	Interpretation
(1) Large-scale planar cross-stratified sandstones	Single or composite cross-stratified (planar asymptotic) sets	Aeolian dunes
(2) Horizontally stratified to low-angle cross-stratified sandstones	Horizontally stratified to slightly inclined sandstone sheets (some showing wind-ripple lamination)	Aeolian interdune/dune bottom sets
(3) Horizontally stratified and massive lithic conglomerates	Horizontally stratified and massive lithic conglomerates with minor sand-wedges	Longitundinal gravel bars and channel lags
(4) Multistorey channel scour and fill	Stacked, cross-cutting channel forms up to 50 m wide and 2 m deep filled with trough cross-stratified sandstones/pebbly sandstones	Low-sinuousity multichannel streams
(5) Single channel-fill sandstones	Single channel forms up to 0.8 m deep and 20 m wide filled with massive or trough cross-stratified sandstones	Small distributary channels
(6) Horizontally stratified sheet sandstones	Horizontally stratified sandstones (0.2–0.7 m thick), commonly with mudcracked upper surfaces; width:height >100:1	Sandy sheet floods

FIG. 3. Large-scale cross-stratified sandstones (Lithofacies 1) from the W wall at Lough Curra (R 8650 2410). Note sweeping, concave-upward asymptotic form of foresets. Also note the major truncation surface (top right to centre left). This section is also illustrated in Fig. 9*a* and *b*. Hammer (bottom right) is 0.4 m in length.

units probably represent superimposed laminae of the same grain size. Individual laminae do not typically show any obvious grading; however, there are visible grain size changes between adjacent units. Inversely graded laminae up to 18 mm thick have only rarely been identified. These are composed of alternating units of very fine sand-grade material (1–2 mm thick) and thicker fine to coarse sand-grade units. Where foresets are exposed, ripple forms with a migration direction directly up or parallel to the strike of the sets are observed only rarely.

(2) Medium sand-size material forming ungraded lenticular laminae (5–80 mm thick). Laminae vary in lateral extent from 0.1 to over 60 m along foresets, both parallel and perpendicular to the palaeoflow direction. The base of individual laminae are often erosive and cut into underlying laminae (Fig. 4*b*).

Within each set, foresets are grouped into bundles by laterally impersistent erosional surfaces (Fig. 5*a*). Typically these cut into the underlying foresets but tend to disappear towards the toesets. Single planar sets often pass upward without any major erosional surface, from horizontally stratified, to low-angle cross-stratified sandstones of lithofacies 2 (Fig. 5*b*, *d*). This lithofacies is often intimately interbedded with sandstones interpreted as being of a sheet-flood origin (lithofacies 6; Table 2; Fig. 5*b*, *c*). In the central part of the Galtee Mountains cross-sets are grouped into cosets by laterally extensive truncation surfaces (Fig. 3) forming composite cross-stratified units.

Interpretation

The form, lamination and organization of this facies are all consistent with the interpretation that these deposits resulted from the migration of single aeolian dunes with distinct foreset-producing slip faces (*cf* McKee 1979*a*, *b*; Ahlbrandt & Fryberger 1982; Kocurek 1981*a*). The inversely-graded laminae identified in some of the cross-sets correspond to the subcritical climbing trans-

FIG. 4. Small-scale aeolian stratification. (*a*) Foresets of the upper part of a relatively thick dune unit displaying light grey wedges of grain-flow origin, Lough Diheen (R 8841 2401); (*b*) foreset of the upper part of a thick dune unit showing a light grey grain-flow unit with an erosive base, Lough Diheen (R 8840 2400). Clinometer is 150 mm in length.

latent wind ripples of Hunter (1977). The lenticular laminae with erosional bases (Fig. 4*b*) are ascribed to grain-flow lamination (Hunter 1977). The apparently ungraded laminae resemble grain-fall deposits, however Hunter (1977) considers that foreset lamination is rarely preserved or even developed in aeolian wind ripples and so these may also represent deposits from climbing translatent ripples. The inability to positively identify grain-fall strata in this study has made it impossible to map the distribution of the small-scale stratification types to any degree.

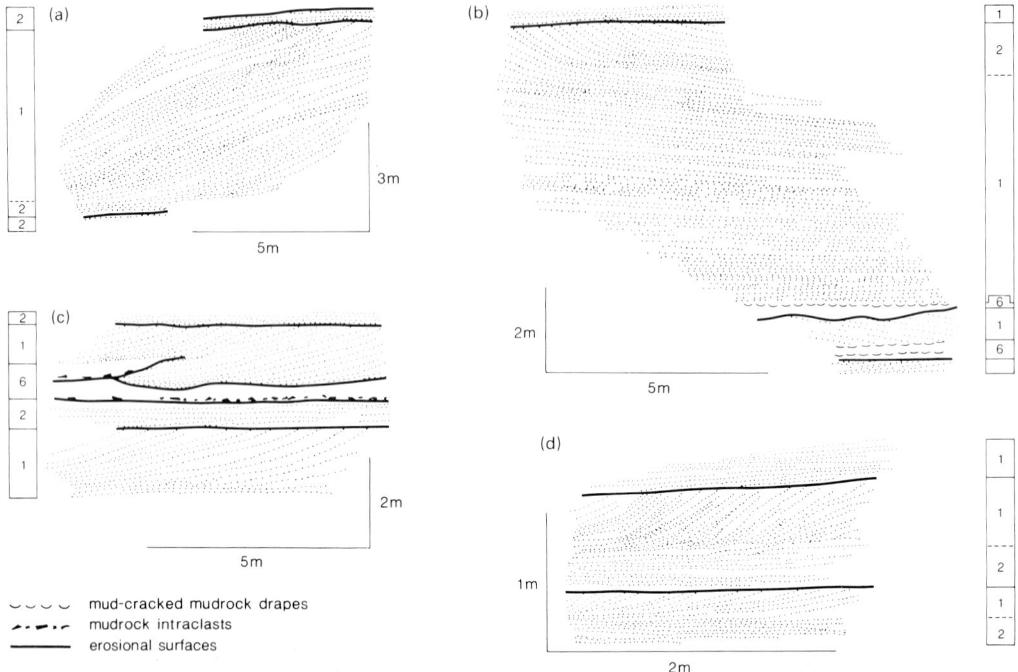

FIG. 5. Sketches (redrawn from photographs) to illustrate the aeolian lithofacies (interbedded fluvial facies are shown unornamented). Columns indicate vertical limits of facies. (*a*) Thick asymptotic dune set (lithofacies 1) showing impersistent (reactivation) surfaces, Galtymore Mountain (R 8810 2465); (*b*) horizontally stratified interdune (lithofacies 2) passing upward into planar asymptotic dune foresets (lithofacies 1), Lough Diheen (R 8841 2400); (*c*) interbedded dune (lithofacies 1) and fluvial sandstones (lithofacies 6), Lough Diheen (R 8842 2400); (*d*) small dunes (lithofacies 1) with interbedded interdune deposits (lithofacies 2); note possible grain-flow toe-pinchouts at base of upper set, Borheen Lough (R 8952 2440).

Horizontally stratified/low-angle cross-stratified sandstones

This lithofacies consists of orange-red coloured, horizontally stratified to low-angle cross-stratified sandstones, with units up to 3.0 m thick. These form laterally extensive sheets which can be traced for up to 800 m along strike. This lithofacies commonly passes upward, without intervening erosional surfaces, into the toesets of large-scale dune sandstones (lithofacies 1; Fig. 5*b*, *d*) and is intimately interbedded with fluvial sandstones (Figs 5*c* and 6).

Small-scale stratification is characterized by alternating laminae of fine sand-grade material (1 to 2 mm thick) and thicker laminae (up to 15 mm thick) of fine to coarse sand-grade material (Fig. 6) which commonly show distinct inverse grading. Individual laminae are laterally continuous and truncate underlying laminae. A second type is characterized by even, parallel and laterally continuous laminae which vary in thickness from 1 to 30 mm. Individual laminae do not show any obvious grading; however, there are common, visible grain-size changes between adjacent units.

Interpretation

These deposits show features consistent with an aeolian origin. The first type of lamination resembles closely the subcritical climbing translatent cross-strata described by Hunter (1977) and formed by the migration of wind ripples under conditions of net sedimentation. The origin of the second form of lamination is more problematical. It resembles the plane-bed lamination of Hunter (1977) which forms in aeolian sands when the wind velocity is high and ripple formation is suppressed.

It is probable that most of these sandstones represent the vertical aggradation of sand sheets by means of climbing wind ripples. The association of this lithofacies with the dune sandstones (lithofacies 1) is consistent with the interpretation that these were interdune areas. The lack of features indicating wet or damp surfaces, such as adhesion structures (Hunter 1973, 1980; Glennie

FIG. 6. Wind-ripple laminated (sub-critically climbing translatent cross-strata) interdune sandstones (lithofacies 2) overlying a fluvial sandstone (base of hammer). The unit is overlain erosively by trough cross-stratified sandstones (lithofacies 4), Lough Diheen (R 8841 2402). Hammer is 0.4 m in length.

1972; Kocurek & Fielder 1982) may be a function of the outcrop quality and the very subtle nature of such structures. The presence of interbedded fluvial facies (lithofacies 3–6) in some sections, is taken as evidence that some interdune areas were flooded episodically.

Lithofacies sequences

Two main types of aeolian sequences are recognized within the Galtymore Formation: (1) those composed almost entirely of dune and interdune sandstones (lithofacies 1 and 2, Fig. 7*a*), often forming composite cross-stratified units and containing little, if any fluvial deposits; and (2) those which contain abundant aeolian sediments, together with a significant fluvial component (lithofacies 1 and 2, 4, 5 and 6; Fig. 7*b*). The vertical and lateral distribution of these sequences and their relationship to fluvial-dominated sequences (Fig. 7*c–e*) is summarized in Fig. 8.

Erg sequences

Sequences ascribed to deposition within an erg or sand sea, consist of superimposed, large-scale, cross-stratified dune sandstone, with tabular to weakly trough-shaped foresets (lithofacies 1). Set size varies from 0.3 up to 15 m. These sequences contain laterally extensive, sub-horizontal erosional surfaces (major truncation surfaces) which group sets into cosets (Figs 3 and 9). These

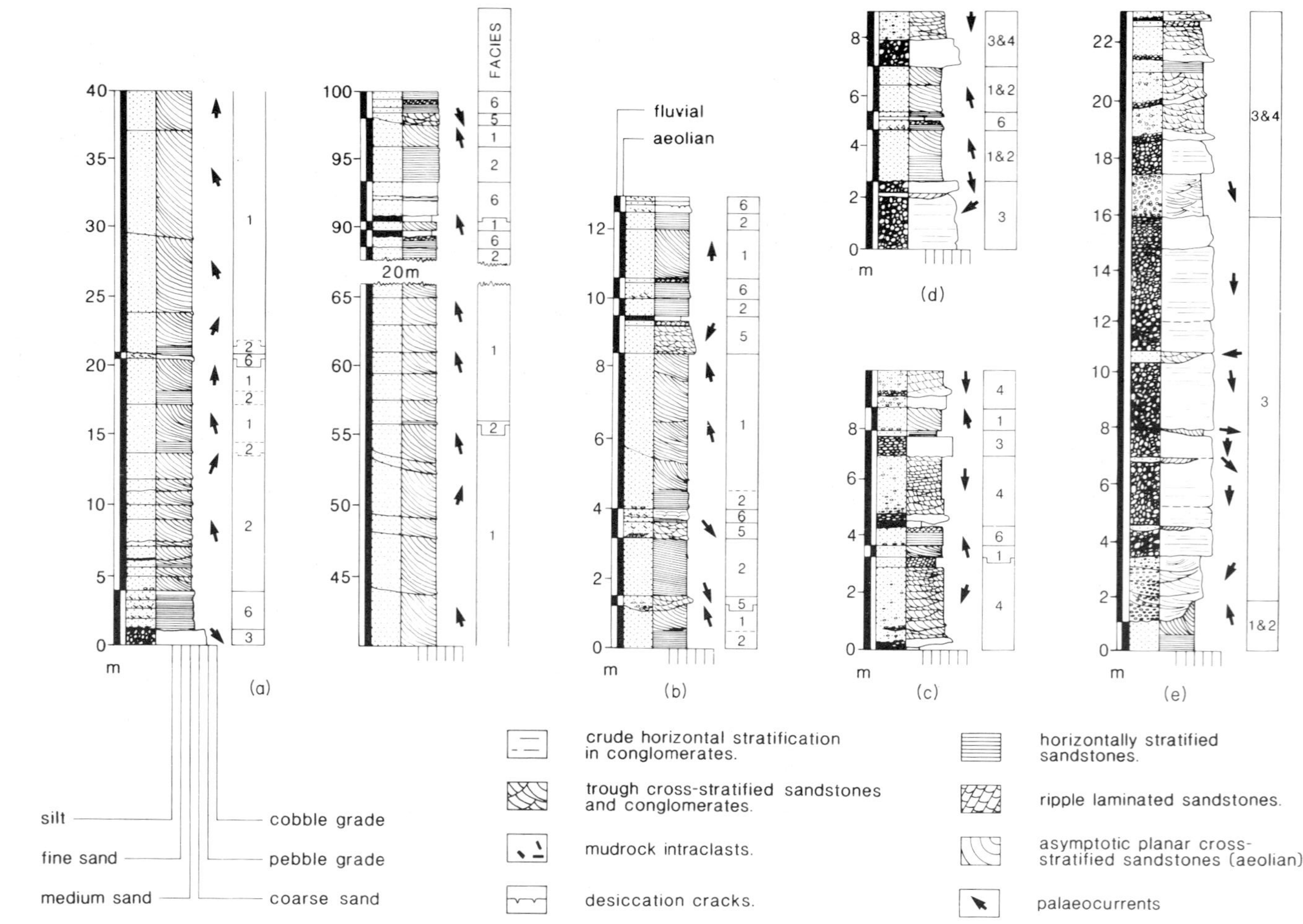

FIG. 7. Vertical profiles to illustrate the various lithofacies sequences present in the Galtymore Formation (note consistently opposing aeolian and fluvial palaeocurrent directions). (*a*) Lough Curra (R 8650 2410), erg sequence, note general absence of fluvial deposits; (*b*) Lough Diheen (R 8841 2400), marginal erg sequence; (*c*) Lough Curra (R 8645 2410), medial fluvial sequence with minor aeolian component; (*d*) Clydagh River (R 8746 2721), proximal fluvial sequence with minor aeolian component; (*e*) Lyraveg (R 8451 2315), proximal fluvial sequence.

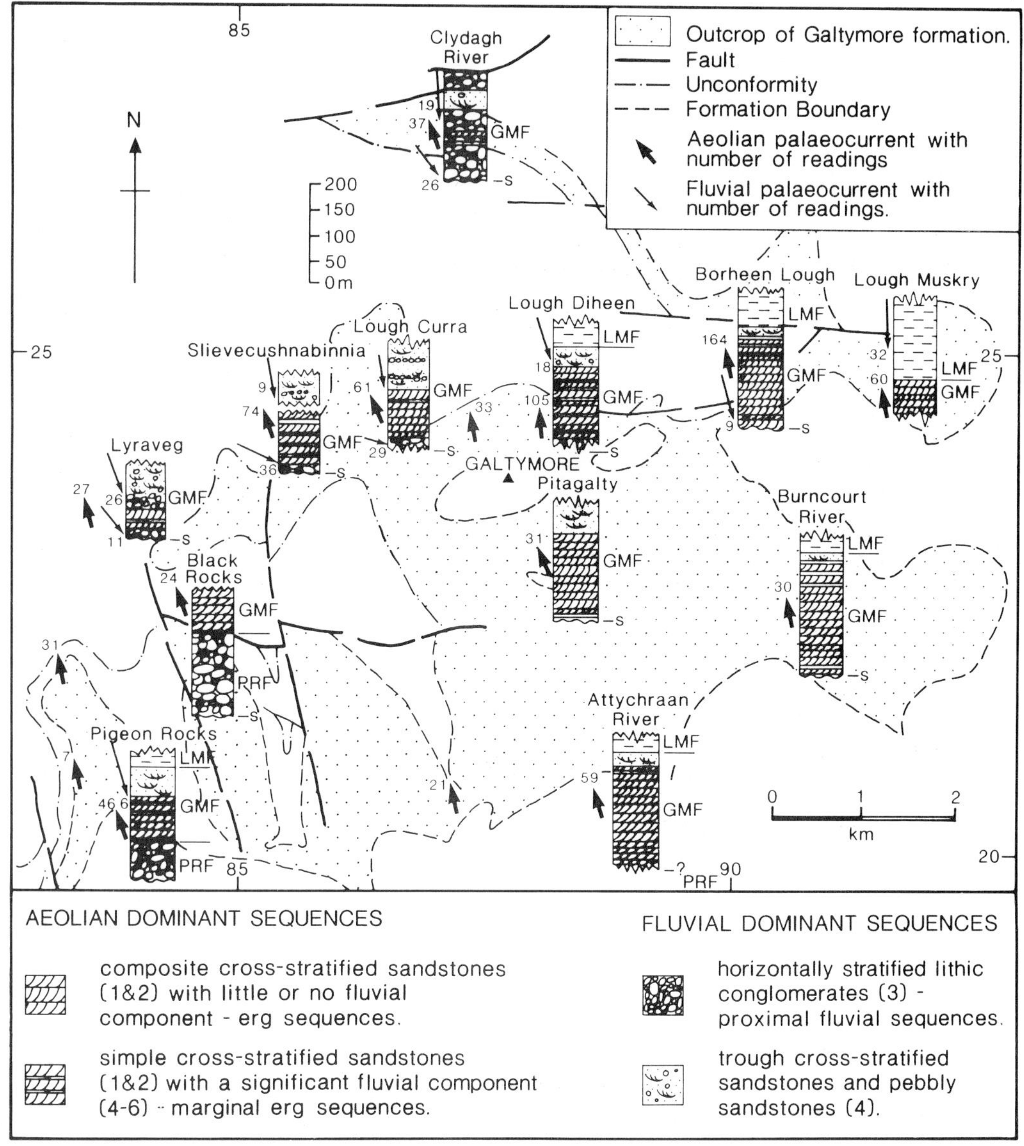

FIG. 8. Distribution of lithofacies sequences through the Galtymore Formation. Location of area is indicated on Fig. 2.

sequences can only be identified positively in the corrie sections. Vertical profiles show (Fig. 7*a*) that within this type of sequence, fluvial deposits are rare. Away from the corries the presence of these sequences can only be inferred using the absence of a significant fluvial component as a guide.

Bounding surfaces

The spacing of the major truncation surfaces varies between 2 and 20 m (mean 10 m). The surfaces may be overlain by the toe or bottomsets of a succeeding set, by interdune deposits, or, more rarely, by fluvial sandstones. Major truncation surfaces have been traced for over 125 m parallel to palaeoflow, exposure being the limiting factor. Sections perpendicular to the palaeoflow show surfaces that are traceable for a maximum of 800 m and show major truncation surfaces terminating laterally in surfaces bounding individual sets (Fig. 9*c*). Sets can be traced for 150 m parallel to the palaeoflow and for over 80 m

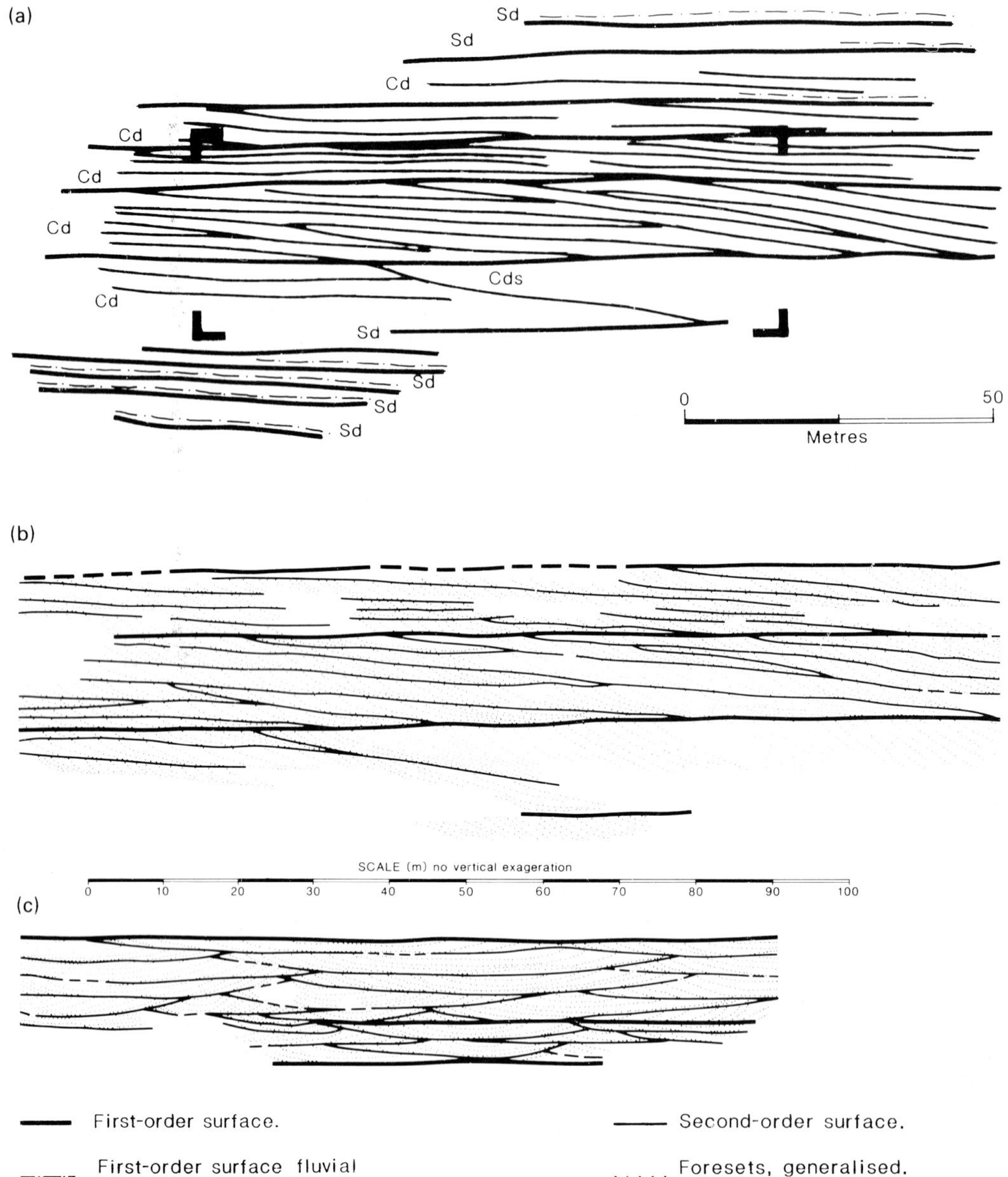

FIG. 9. Composite cross-stratification. (*a*) Plot of first- and second-order surfaces from the W wall of the corrie at Lough Curra (R 8650 2410), section broadly parallel to palaeoflow direction. Third-order surfaces not shown. Dune reconstruction: Sd, simple dune; Cd, compound dune without major slip face; Cds, compound dune with major slip face. (*b*) Detailed view of the composite cross-stratification from (*a*). Third-order surfaces not shown. (*c*) Lough Borheen (R 8952 2440) section broadly perpendicular to the palaeoflow direction. Note termination of first-order surface (centre) by second-order surface. Drawings made from photographs on which the structures were plotted in the field; parts of each sketch are therefore subject to distortion.

perpendicular to it (Fig. 9*a*, *b*). Impersistent erosional surfaces commonly bound groups of foresets within a single set. These surfaces are usually parallel to overlying foresets and dip at a steeper angle than underlying foresets.

The composite nature of the cosets bounded by the major truncation surfaces would indicate that at least two orders of aeolian bedforms are present within this type of sequence. Both the sets and the cosets were deposited by trains of bedforms climbing downcurrent (Allen 1970; Rubin & Hunter 1982).

Recent advances in the understanding of the migration of aeolian bedforms have led to the recognition of a hierarchical arrangement based on the extent of and truncating relationships between bounding surfaces (Brookfield 1977). Their origins have been attributed to the migration of a series of superimposed bedforms. Brookfield (1977) assumes that superimposed bedforms migrate together (see also Rubin & Hunter 1983), which conflicts with the suggestions of Wilson (1973) that only one bedform migrates at any one time. Rubin & McCulloch (1980) present evidence that bedforms of different sizes can be stable in the same steady flow. Large bedforms commonly support smaller superimposed bedforms because of fluctuating flow conditions (Allen 1973) and because large bedforms decelerate flow and produce small bedforms even under steady conditions (Rubin & McCulloch 1980).

The three-fold hierarchy of Brookfield (1977) has been shown to exist within the Irish erg sequences and a similar origin is therefore inferred. The most extensive major truncation surfaces—the first-order surfaces (terminology after Brookfield 1977)—are interpreted as resulting from the migration of interdune areas, and the horizontally-stratified sediments, where developed, are interpreted as interdune deposits (*cf* Kocurek 1981*a*, *b*). Considering the length of surfaces which bound individual cross-sets (second-order surfaces), relative to set thickness (Fig. 9*a*, *b*) they are unlikely to result from fluctuating flow conditions (*ie* they are not reactivation surfaces) (Rubin & Hunter 1983). They must, therefore, represent the migration of discrete dune foresets. The down-current dip of second-order surfaces suggests that the dunes migrated down the lee slopes of larger bedforms (the 'draas' of Wilson 1972 and 1973). The larger dune bodies may be 'compound' or 'complex' (dune bodies with smaller dunes of the same kind or different kind, superimposed respectively; terminology of Breed & Grow 1979). The low length-to-thickness ratio of foresets bounded by third-order surfaces is consistent with an origin in a fluctuating flow, the foresets marking the stages in the advance of a single dune (Brookfield 1977; Kocurek 1981*b*; Rubin & Hunter 1983).

Dune types

The foreset dip distribution of sets from the composite cross-stratified units is presented in Fig. 10. The unimodal distribution suggests that the majority of dune slip faces were orientated perpendicular to the wind direction and that the dunes were therefore of a crescentic type (*cf* Fryberger 1979). Crescentic dunes vary in form from straight-crested, transverse dunes to barchanoid forms (Breed & Grow 1979). A mean angular deviation ranging from 13° to 39° suggests dunes of transverse type. However, the undulose to gentle trough form of second-order surfaces (Fig. 9*b*) may indicate that these dunes had relatively sinuous crest lines.

The distribution and relative percentages of small-scale stratification types in ancient dune deposits have been utilized in interpreting dune type and shape (*eg* Kocurek & Dott 1981). Foreset lamination in the composite cross-stratified dunes of the Galtymore Formation is composed mainly of wind-ripple cross-strata. Grain-flow units are present as thin, typically less than 20 mm, slightly coarser grained units which can be traced along the strike of the foresets for over 60 m. These are most common in the upper parts of the sets and wedge out towards the toesets. This distribution is comparable to descriptions of other ancient dune sequences interpreted as being of crescentic form. As grain-flow laminae extend for greater distances along the strike of foresets than those investigated by Hunter (1977), there is a suggestion that the Galtymore dunes were considerably larger (*ie* greater than 3 m high).

Some approximation of the original height of the dunes (H) can be made using the Rubin & Hunter (1982) formula:

$$H \simeq \left(\frac{T \cdot D}{I} \right)^{0.5}$$

where: $I = 15$ (bedform spacing-to-height ratio, Wilson 1972); D = downcurrent depositional extent of cross-set; and T = set thickness. With T ranging from 0.5 to 15 m (typically less than 5.0 m) and D ranging from 80 m to over 150 m (exposure being the limiting factor), it is inferred that dune heights *c.* 7 m are feasible. Some indication of dune width is obtained from the extent of the second-order bounding surfaces viewed perpendicular to the palaeoflow direction (Fig. 9*c*). A minimum width of up to 80 m is estimated.

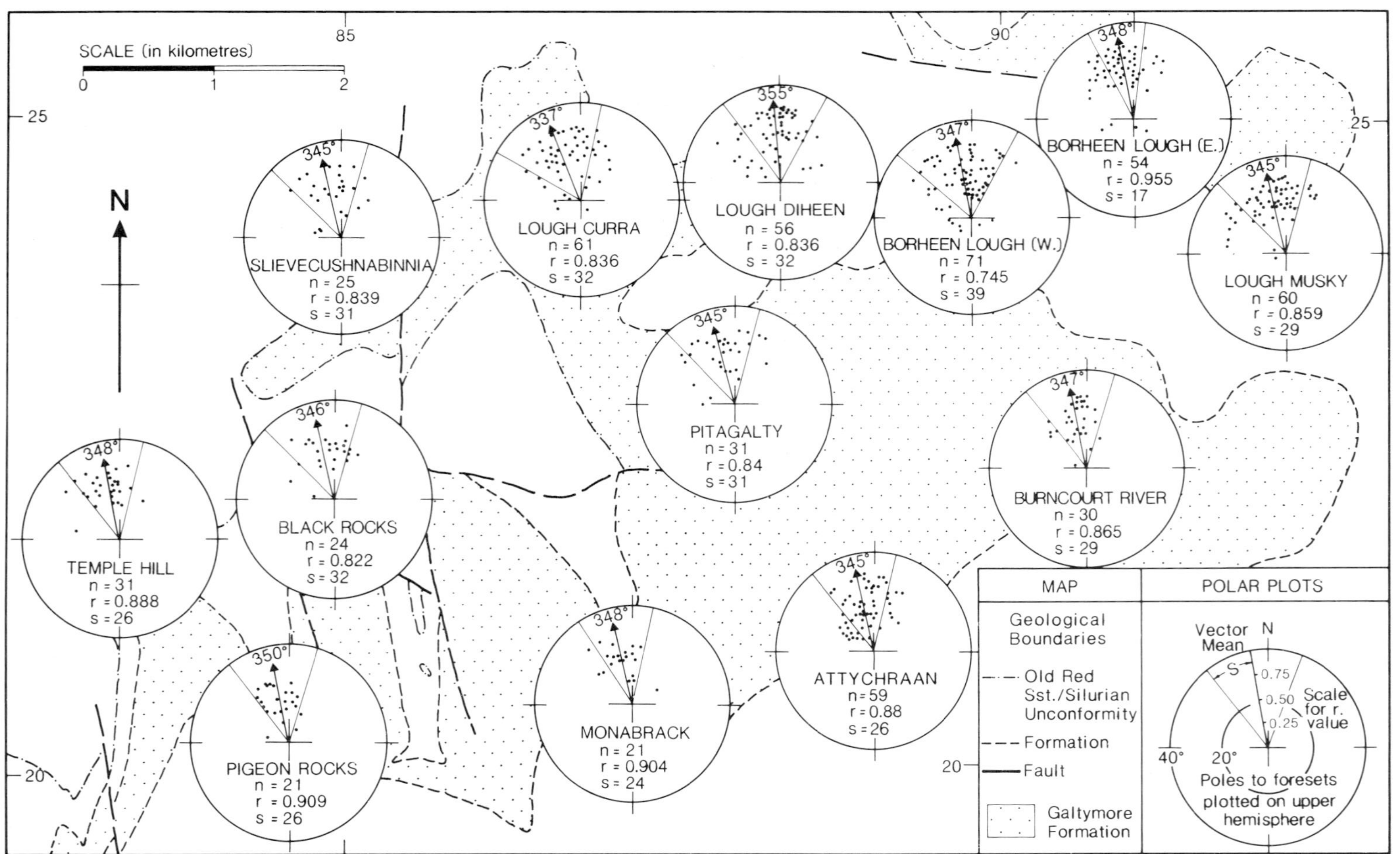

FIG. 10. Polar plots (upper hemisphere) of foreset dip dispersion of compound dunes, by station. Mean angular deviation (*s*) in degrees; estimates of spread of angular values around a unit circle (*r*). Statistical methods are described by Till (1974, pp. 38–43). Polar plots with points plotted on upper hemisphere. Location of area is indicated in Fig. 2.

Draas

Some specific constraints on the form and shape of the draas in the Galtymore Formation can be suggested. The geometry of first-order surfaces is illustrated in Fig. 9. Sections parallel to the palaeowind direction show second-order surfaces that dip consistently in a downwind direction (Figs 9*a*, 9*b*, 11*b*). The dune sets and their associated second-order surfaces are interpreted as having been formed by dunes migrating down the lee slopes of the draas. Sections orientated perpendicular to the palaeowind direction (Fig. 9*c*) show first-order surfaces that pass laterally into second-order surfaces. Associated with this, the overlying interdune deposits also wedge out. This fits well with observations made from other ancient aeolian sequences and has been used as evidence for crescentic draas (*cf* Kocurek 1981*a*, *b*). The combination of crescentic dunes and draas is comparable to the compound crescentic dunes described by Breed & Grow (1979).

The relatively low angular relationship between first-order and second-order surfaces, typically less than 15° (Fig. 11*b*), suggests that the compound dunes were comparatively small and of low relief. The general lack of very thick dune sets (most are between 1.5 and 4.5 m thick) would indicate that most of the compound dunes were without major slip faces. Larger slip faces only developed rarely, *eg* in a very thick set seen at Lough Curra (Fig. 9*a*, *b*).

Numerous workers (*eg* Hunter 1981) have shown that some compound dunes develop with crest lines aligned obliquely to regional wind currents. Some caution must therefore be exercised when interpreting the significance of palaeocurrent data. The downcurrent dip of second-order surfaces (Fig. 11*b*) is remarkably consistent in orientation over the area (showing a mean angular deviation of less than 10°). This suggests that the compound dunes were relatively straight crested. If sinuous crested forms developed, a greater mean angular deviation would be expected. The trend of cross-sets, expressing the migration direction of the dunes, is parallel to the downcurrent dip of second-order surfaces (compare Fig. 11*a* and *b*) indicating that the dunes migrated directly down the lee slope (Rubin & Hunter 1983).

Estimates of the dimensions of the compound dunes and interdunes within the erg sequences are made difficult by the lack of adequate exposure, even in the corrie sections. The most direct method, but probably yielding the least information, is the actual measurement of the preserved structures. Preservation of the coset thickness is primarily a function of net sedimentation (Rubin & Hunter 1982) and may bear little relation to the original compound dune height. The widths of these structures (in sections perpendicular to the main palaeoflow direction) may only be taken as estimates of original dune widths if one can be sure that the edges of the

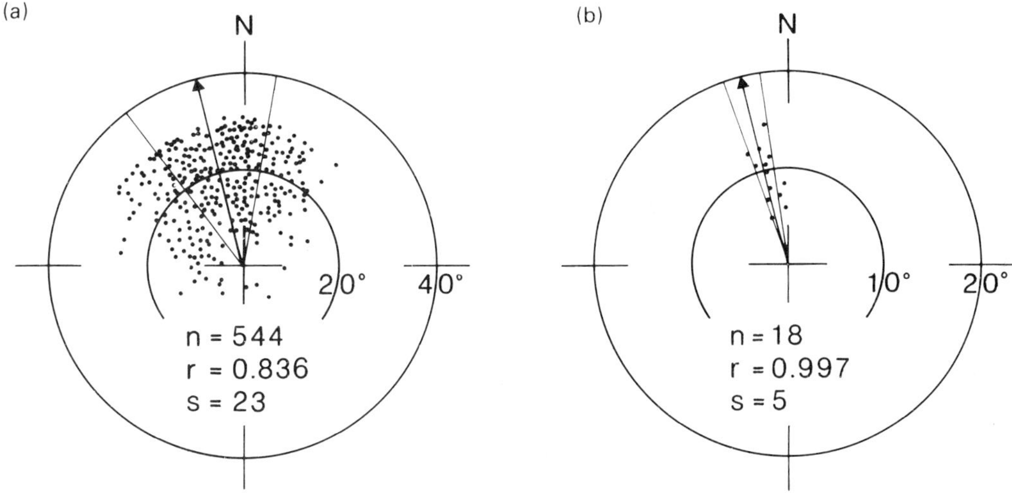

FIG. 11. Analysis of directional data from compound dunes. (*a*) Total data plot for foreset dip dispersion distribution. (*b*) Polar plot of dip of second-order surfaces (most readings come from the basal parts of compound dune units with first-order surfaces assumed to have been originally horizontal (Kocurek (1981*a*) suggests the depositional dip of first-order surfaces to be as low as 1.5°). Mean angular deviation (*s*) in degrees; estimates of spread of angular values around a unit circle (*r*). For statistical method see Fig. 10.

compound dunes are preserved and that they have not been truncated laterally. Further difficulties were encountered in this study arising from sections where significant amounts of interdune deposits do not overlie first-order surfaces. This hampers the delineation of first-order surfaces and, consequently, geometry and dimensions are often difficult to assess.

With the above reservations in mind some dimensions are inferred based on indirect evidence. One first-order surface was traced for *c.* 800 m perpendicular to the palaeoflow direction before it, and the associated interdune deposits, terminated. This gives some indication as to the minimum width of interdune areas. Observations from modern compound dunes suggest that the original compound dune width would have been in excess of this—*c.* 1200 m would be a reasonable estimate (*cf* Breed & Grow 1979). Using plots of dune wavelength versus width for crescentic compound dunes (Breed & Grow 1979, p. 274), dune lengths and wavelengths of *c.* 800 m and *c.* 1100 m, respectively, could be anticipated. The average downwind length of the interdune areas might then be of the order of *c.* 300 m. A correlation between modern draa height and wavelength has been made by Wilson (1972, 1973), who found it to be about 1:15. The compound dunes of the present study may therefore have achieved heights of over 70 m.

Aeolian/fluvial interactions

In the erg sequences, dune growth was only limited by the downwind migration of interdune areas creating first-order surfaces. In most cases the interdune deposits are entirely aeolian. Most interdune areas are therefore interpreted as having been dry. The presence of very rare fluvial deposits would indicate that some areas were flooded intermittently.

Marginal erg sequence

Sequences ascribed here to deposition on the margin of an erg or sand sea are characterized by the predominance of aeolian lithofacies (lithofacies 1 and 2), but also contain a significant fluvial component (lithofacies 3–6; Fig. 7*b*).

Dune and interdune facies

The aeolian dune lithofacies is typically composed of single dune sets, varying in thickness from 0.3 to 8 m thick. These are interbedded with the horizontally-bedded, or low-angle cross-stratified sandstones (lithofacies 2; Fig. 5). The dune sets contain localized erosional surfaces (third-order surfaces) which have been interpreted as reactivation surfaces. These cross-sets originated from the migration of simple dunes (*ie* single dune bodies without smaller dunes superimposed; terminology of Breed & Grow 1979), and their associated interdune deposits (*cf* Kocurek 1981*b*).

The unimodal foreset dip dispersion of simple dunes (Fig. 12) suggests that these simple dunes were of a crescentic type (Fryberger 1979). The mean angular deviation of foreset dips compares favourably with those of modern barchanoid dunes (*cf* Ahlbrandt & Fryberger 1980). The presence of interbedded fluvial deposits accords with this interpretation; barchanoid dunes tend to form on hard desert surfaces with sparse supplies of sand (Cooke & Warren 1973).

The preserved dune set heights range from 0.3 to over 8 m. The original dune height can only be estimated in a few small sets (less than 2 m thick) where the downcurrent depositional extent (D) is seen. Such sets have relatively steeply dipping foresets (up to 30°) composed predominantly of grain-flow laminae. These factors indicate that the original dunes were themselves small, rather than being larger dunes with only toesets preserved. Clearly then, these dunes are not representative of the full spectrum of bedforms. Observations indicate that in the smaller sets, T ranges from 0.3 to 2 m (average 1.2 m); D ranges from 10 m to 40 m and H therefore ranges from 1.09 m to over 1.78 m.

These sets can be traced perpendicular to the palaeoflow for distances of between 10 and 50 m (average of 20 m). This gives some indication as to the minimum width of the dunes.

The distribution of small-scale stratification types generally conforms well to those seen in aeolian dune deposits (*cf* Kocurek & Dott 1981), as the grain-flow laminae are thicker and most abundant in the upper parts of the larger sets. Grain-flow toe pinch-outs have been recognized tentatively in some of the smaller-scale sets (those less than 0.5 m; Fig. 5). Some of the thicker laminae observed in the larger sets extend for great distances along the strike of the foresets suggesting that such dunes were larger than those investigated by Hunter (1977), *ie* greater than 3 m high.

Aeolian/fluvial interactions

Verticial profiles show (Fig. 7*b*) that there is a complete mixture of facies (dune, interdune and fluvial) and that there is no apparent preferred sequence of facies. Dune deposits commonly overlie fluvial deposits without a significant interdune unit. Figure 5*c* shows a simple dune set overlying a sheet-flood sandstone unit. Development of the dunes was then interrupted by a second flood. This was in turn succeeded by the continued migration of the dune over the mud-

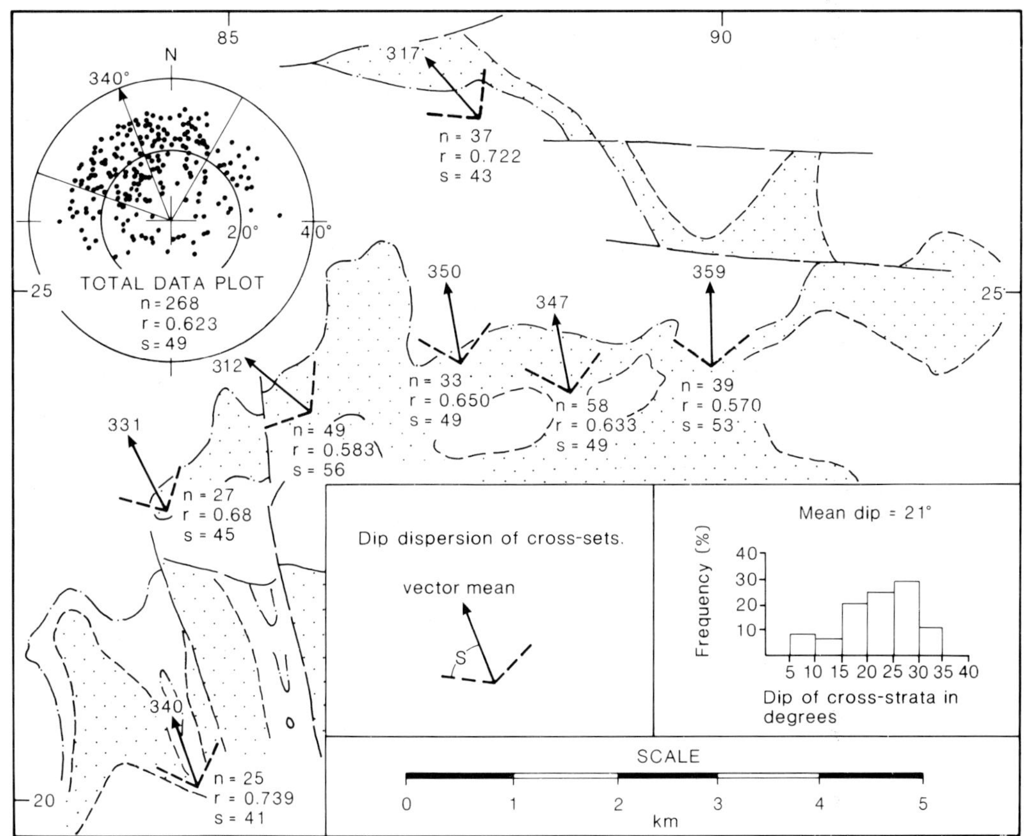

FIG. 12. An analysis of the foreset dip dispersion distribution of simple dunes. Mean angular deviation (*s*) in degrees; estimates of spread of angular values around a unit circle (*r*). For statistical method see Fig. 10. The location of the area is indicated in Fig. 2.

cracked siltstone drape of the flood deposit. The occasional presence of evenly spaced, small pebble to granule-grade, lithic clasts at the base of some simple dune units may represent deflation lags (Fig. 7*b*, at 4.0 m).

The overall dominance of aeolian units in this type of sequence suggests that aeolian deposition was interrupted only periodically by flooding. In some instances the development of dune and interdune sequences continued for a considerable period of time, enabling the growth of relatively large (greater than 8 m high) simple dunes.

Palaeogeography

Palaeoclimate and palaeowind

There is little evidence to suggest that the Galtymore Formation developed in a humid environment. Features which might indicate damp or wet conditions (bioturbation, adhesion structures, soft-sediment deformation structures) have not been observed in either the dune or the interdune facies. The close association of small dune and interdune deposits in the marginal erg sequences further points to relatively arid conditions. However, there are features within the fluvial facies (rare small vertical burrows 5–8 mm diameter, rootlet horizons, immature calcrete profiles), which suggest that conditions were not entirely dry and that the substrate was capable of supporting a limited biota.

Aeolian palaeocurrent observations show relatively consistent orientations over *c*. 200 m of section. There is no evidence to suggest subsidiary wind currents (*ie* dune modification structures are absent apart from reactivation surfaces). This implies a long-lived and relatively uniform, south-southeasterly palaeowind.

Fluvial palaeogeography

There is clear evidence of a NW to SE palaeoslope as indicated by (1) the decrease in maximum clast size; (2) overall palaeocurrent trend; and (3) the increasing southward influence of aeolian sedimentation. The distribution of facies (summarized in Fig. 8) is interpreted as representing a transition from an alluvial fan sequence in the N and NW, to an erg or sand sea developed to the SE. Downslope transitions from one sequence to another are not exposed. The consistent vertical and lateral distribution of facies throughout the outcrop of the Galtymore Formation suggests that these transitions exist over a maximum distance of 3–4 km (Figs 8 and 13).

The predominance of lithic clasts (mainly pelites and psammites from the local Silurian basement) in the proximal fluvial sequences is consistent with the idea that such sequences (Fig. 7*d*, *e*) represent small-scale alluvial fans with localized source areas.

Erg development

Ergs are created where there is downwind decrease in sand transport. This could be brought about by changing surface roughness, changes in climate, or a change in the regional sand drift direction (Wilson 1973). Ergs would usually form downwind of areas undergoing rapid deflation (Wilson 1972). The most extensive source areas in most modern desert basins are alluvial deposits since they contain large quantities of sand capable of being transported by the wind. The low degree of rounding of grains and petrological immaturity of the sandstones of the Galtymore Formation (Carruthers 1985) suggest local source areas. The most likely source area is contemporaneous unconsolidated fluvial deposits forming to the S within the Munster Basin. Where the erg sequences are well-developed, a complete sand cover is implied. Such situations require winds that are overloaded with sand. These could result in areas of wind deceleration or convergent flow (Wilson 1973). The local topography could result in a loss of transport capacity and hence sand accumulation.

Tectonic control

Contemporaneous faulting is thought to control the distribution of lithofacies. It is responsible for the direction of fluvial dispersal which in turn has had a marked control on the distribution of aeolian sequences. Tectonically, the Galtymore Formation has been interpreted as representing

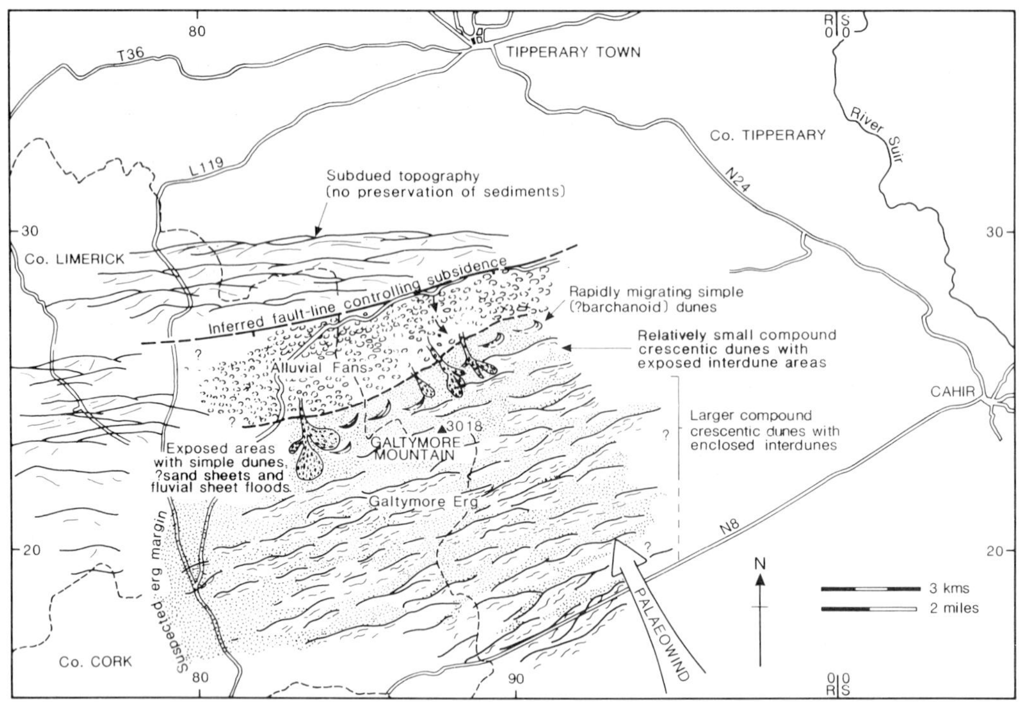

FIG. 13. Palaeogeography of the Galtymore Erg at its maximum extent.

a period of net deposition following faulting along the northern margin of the Munster Basin (Carruthers 1985). Movement along the fault which controlled the development of the underlying alluvial fan sequence (Pigeon Rock Formation) became less pronounced allowing the northward progradation of an erg. The northern margin of the erg was controlled by contemporaneous faulting, and facies variations suggest this was located immediately N of the Clydagh River section (Fig. 13). Fluvial input across this fault, whilst present within the northern part of the outcrop of the Galtymore Formation, did not extend sufficiently far S to prevent the development of thick aeolian deposits. Within the northern part of the area small simple crescentic dunes migrated rapidly northwards across the alluvial fan surface. Sand sheets may have developed in this area but their preservation potential is low. In more distal fluvial areas to the S (Fig. 13), where fluvial input was less, dune growth and migration was uninterrupted. Larger, simple, crescentic, and possibly small compound dunes and associated interdune areas migrated northwards. Although there is ample evidence of fluvial deposition within these sequences, aeolian deposits are only rarely eroded by water to any great extent. Both simple and compound dunes were eroded by the migration of interdune areas. It is only in the central and southern parts of the Galtee Mountains that fluvial input was reduced and allowed the larger-scale compound dunes to develop. Such bedforms require long periods of uninterrupted aeolian deposition in order to form. Wilson (1972) suggests at least 10 000 years. Erosion in these areas resulted from the migration of usually dry interdune areas.

Conclusions

A locally developed ancient aeolian dune–alluvial fan complex has been identified in the Old Red Sandstone of the Munster Basin (Upper Devonian), Ireland. Facies variations show that in the N of the area, proximal to medial braided streams flowed southwards on the surface of small alluvial fans. To the S, crescentic dunes (simple and compound) migrated NNW towards the alluvial fans. Fluvial input played an important role controlling the growth and distribution of dune types. In areas close to the basin margin, high fluvial input restricted the migration of aeolian bedforms to small simple crescentic dunes. Further S, where fluvial input was less, larger simple dunes and compound dunes were developed. The compound dunes were relatively small with a low relief, and for the most part without major slip faces. The tectonic setting is considered to be critical for the formation, development and preservation of this complex.

ACKNOWLEDGMENTS: Fieldwork carried out for this study forms part of a research studentship undertaken at Trinity College, Dublin and funded by the Geological Survey of Ireland. I would like to thank Dr Geoff Clayton for his patient supervision, and Dr John Graham who provided suggestions and criticism on an earlier version of this manuscript. I am indebted to Dr Chris Dodds (BP Exploration) who demonstrated to me the virtues of the Dingle aeolianites. I am grateful to many other friends and colleagues at Trinity College Dublin for their comments and assistance. Thanks are also due to Badley, Ashton & Associates Limited for technical support in the production of this paper, especially Sue Pittendreigh for drafting, and Diana Watts and Sylvia Toin for typing.

References

AHLBRANDT, T. S. & FRYBERGER, S. G. 1980. Eolian deposits in the Nebraska Sandhills. *United States Geological Survey, Professional Paper* **1120-A**.

—— & —— 1982. Introduction to Eolian deposits. *In*: SCHOLLE, P. A. & SPEARING, D. (eds) *Sandstone depositional Environments*. American Association of Petroleum Geologists Memoir **5**, 11–47.

ALLEN, J. R. L. 1970. A quantitative model of climbing ripples and their cross-laminated deposits. *Sedimentology* **14**, 5–26.

—— 1973. Phase differences between bed configuration and flow in natural environments, and their geological relevance. *Sedimentology* **20**, 323–329.

BOLAND, M. A. 1983. *The Geology of the Ballyvoyle–Kilmacthomas–Kilfarrasy area, County Waterford, with an account of the Lower Palaeozoic geology of County Waterford*. MSc thesis, University of Dublin.

BOLDY, S. A. R. 1982. *The Old Red Sandstone rocks of the eastern Knockmealdown Mountains and Monavullagh Mountains and adjacent areas, Counties Tipperary and Waterford*. PhD thesis, University of Dublin.

BREED, C. S. & GROW, T. 1979. Morphology and distribution of dunes in sand seas observed by remote sensing. *In*: MCKEE, E. D. (ed.) *A study of Global Sand Seas*. United States Geological Survey, Professional Paper **1052**, 253–302.

BROOKFIELD, M. E. 1977. The origin of bounding surfaces in ancient aeolian sands. *Sedimentology* **24**, 303–332.

CARRUTHERS, R. A. 1985. *The Upper Palaeozoic Geology of the Glen of Aherlow and Galtee Mountains, Counties Limerick and Tipperary*. PhD thesis, University of Dublin.

CLAYTON, G. & GRAHAM, J. R. 1974. Miospore assemblages from the Devonian Sherkin Formation of south-west County Cork. Republic of Ireland. *Pollen et Spores* **16**, 565–588.

COLTHURST, J. R. J. 1978. The Old Red Sandstone rocks surrounding the Slievenamon inlier, Counties Tipperary and Kilkenny. *Journal of Earth Sciences of the Royal Dublin Society* **1**, 77–103.

COOKE, R. U. & WARREN, A. 1973. *Geomorphology in Deserts*. Batsford, London.

DORAN, R. J. P. 1970. Palynological evidence for the age of the Old Red Sandstone near Cappagh White, Co. Tipperary. *Scientific Proceedings of the Royal Dublin Society* **A3**, 343–350.

FRYBERGER, S. G. 1979. Dune forms and wind regime. *In*: MCKEE, E. D. (ed.) *A study of Global Sand Seas*. United States Geological Survey, Professional Paper **1052**, 137–169.

GLENNIE, K. W. 1972. Permian Rotliegendes of Northwest Europe interpreted in light of modern desert sedimentation studies. *Bulletin of the American Association of Petroleum Geologists* **56**, 1048–1071.

GRAHAM, J. R. 1983. Analysis of the Upper Devonian Munster Basin, an example of a fluvial distributary system. *In*: COLLINSON, J. D. & LEWIN, J. (eds) *Modern and Ancient Fluvial Systems*. International Association of Sedimentologists, Special Publication **6**, 473–483.

HIGGS, K. & RUSSELL, K. J. 1981. Upper Devonian Microfloras Iveragh, County Kerry, Ireland. *Bulletin of the Geological Survey of Ireland* **3**, 17–50.

HOLLAND, C. H. 1981. Devonian. *In*: HOLLAND, C. H. (ed.) *A Geology of Ireland*. Scottish Academic Press, Edinburgh, 121–146.

HORNE, R. R. 1971. Aeolian cross-stratification in the Devonian of the Dingle Peninsula, County Kerry, Ireland. *Geological Magazine* **108**, 151–158.

—— 1975. The association of alluvial fan, aeolian and fluviatile facies in the Caherbla Group (Devonian), Dingle Peninsula, Ireland. *Journal of sedimentary Petrology* **45**, 535–540.

HUNTER, R. E. 1973. Pseudo-crosslamination formed by climbing adhesion structures. *Journal of sedimentary Petrology* **43**, 1125–1127.

—— 1977. Basic types of stratification in small eolian dunes. *Sedimentology* **24**, 361–388.

—— 1980. Quasi-planar adhesion stratification—an eolian structure formed in wet sand. *Journal of sedimentary Petrology* **59**, 203–216.

—— 1981. Stratification styles in some Pennsylvanian to Jurassic eolian sandstones of the Western interior, U.S.A. *In*: ETHERIDGE, F. G. & FLORES, R. M. (eds) *Recent and Ancient Non-marine Depositional Environments: Models for Exploration*. Society of Economic Palaeontologists and Mineralogists, Special Publication **31**, 315–329.

JACKSON, A. A. 1972. *The Silurian and Basal Old Red Sandstone of the Galty Mountain area*. PhD thesis, University of Dublin.

KOCUREK, G. 1981*a*. Erg reconstruction: Entrada Sandstone (Jurassic) of Northern Utah and Colorado. *Palaeogeography, Palaeoclimatology, Palaeoecology* **36**, 125–153.

—— 1981*b*. Significance of interdune deposits and bounding surfaces in aeolian dune sands. *Sedimentology* **28**, 753–780.

—— & DOTT, R. H., JR. 1981. Distinctions and uses of stratification types in the interpretation of eolian sand. *Journal of sedimentary Petrology* **51**, 579–595.

—— & FIELDER, G. 1982. Adhesion structures. *Journal of sedimentary Petrology* **52**, 1229–1241.

MCKEE, E. D. 1979*a*. Sedimentary structures in dunes with sections on the Lagoa dune field, Brazil. *In*: MCKEE, E. D. (ed.) *A study of Global Sand Seas*. United States Geological Survey, Professional Paper **1052**, 83–134.

—— 1979*b*. Ancient sandstones considered to be eolian. *In*: MCKEE E. D. (ed.) *A study of Global Sand Seas*. United States Geological Survey, Professional Paper **1052**, 187–238.

NAYLOR, D. & JONES, P. C. 1967. Sedimentation and tectonic setting of the Old Red Sandstone of Southwest Ireland. *In*: OSWALD D. H. (ed.) *International Symposium on the Devonian System*. Alberta Society of Petroleum Geologists **11**, 1089–1099.

RUBIN, D. M. & HUNTER, R. E. 1982. Bedform climbing in theory and in nature. *Sedimentology* **29**, 121–138.

—— & —— 1983. Reconstructing bedform assemblages from compound crossbedding. *In*: BROOKFIELD, M. E. & AHLBRANDT, T. S. (eds) *Eolian Sediments and Processes*. Developments in Sedimentology **38**, Elsevier, Amsterdam.

—— & MCCULLOCH, D. S. 1980. Single and superimposed bedforms: a synthesis of San Francisco Bay and flume observations. *Sedimentary Geology* **26**, 207–231.

RUSSELL, K. J. 1978. Vertebrate fossils from the Iveragh Peninsula and the age of the Old Red Sandstone. *Journal of Earth Sciences of the Royal Dublin Society* **1**, 151–162.

TILL, R. 1974. *Statistical Methods for the Earth Scientist: an introduction*. Macmillan, London.

WILSON, I. G. 1972. Aeolian bedforms. Their development and origins. *Sedimentology* **19**, 173–210.

—— 1973. Ergs. *Sedimentary Geology* **10**, 77–106.

R. A. CARRUTHERS. Department of Geology, Trinity College, Dublin 2, Eire. *Present address:* Badley, Ashton & Associates Ltd, Aveland House, Spilsby, Lincolnshire PE23 5JX, UK.

Grain size, process and dune environment

The statistical analysis of 'mixed' grain size distributions from aeolian sands in the Libyan Pre-Desert using log skew Laplace models

E. C. Flenley, N. R. J. Fieller & D. D. Gilbertson

SUMMARY: This paper describes the statistical analysis and modelling of simple and 'mixed' particle mass–size distributions. The particular sediments discussed derive from modern climbing, seif and barchan dunes. The statistical methods considered include calculation of sample moments (mean, sorting, skewness and kurtosis), in particular by semi-graphical techniques. Additionally, attention is given to the construction of log normal probability plots and their supposed interpretation in terms of distinct segments representing different components of a 'mixed' distribution. Whilst these techniques have been used with some effect in the case of analysing simple or pure sand samples, it is shown that their application to mixed size distributions can be misleading. The approach advocated as an alternative is to model the size distribution directly by one of a family of parametric statistical distributions and use the estimated parameters to characterize the individual samples. The particular family preferred is the three-parameter log skew Laplace, which can be extended readily to a seven-parameter family incorporating mixture densities. This family provides computational advantages over the log hyperbolic family, whilst retaining its geological interpretability and theoretical derivation.

The statistical analysis and modelling of mass–size distributions has received much attention in recent years. This paper examines the special problems that arise in the case of 'mixed' distributions. The particular sediments discussed here derive from modern climbing, seif and barchan dunes which were studied in connection with investigations of surficial geology, ancient floodwater farming and environmental change by the University of Manchester–University of Sheffield/UNESCO Libyan Valleys Project in the semi-arid desert area of Tripolitania known as the Libyan Pre-Desert (Fig. 1). For a description and background see Barker & Jones (1980, 1984), Gilbertson (1986), Gilbertson *et al.* (1984, in press), Gilbertson & Hunt (in press), Gale *et al.* (1986), Hunt *et al.* (1985) and Van der Veen (1985).

Statistical description and modelling of sand mass–size data

The study of the grain size distributions of sediments has been a major focus of sedimentological research. This is because these distributions reflect fundamental properties of the sediments, their origins and their resource potential. Two traditions in the associated statistical analyses can be distinguished, typified by whether the samples are merely described numerically, or whether some underlying model for the grain sizes is proposed and estimated from the data.

The most widely known and used approach for describing and analysing the particle mass–size distributions of aeolian and other types of sediments is that exemplified in the work of Friedman (1961, 1967, 1979*a* and *b*). These methods rely on the characterization of the size distributions by sample moments (mean, sorting, skewness and kurtosis). Often these are calculated semi-graphically from a log normal probability plot of the data, following the suggestions of Folk & Ward (1957) and Inman (1952). The calculated sample moments are used in further analysis; for example, plots of one against another may be a first step in distinguishing between sands from different environments. Whilst this approach has often been successful, there have been many recent expressions of doubt about the statistical assumptions the techniques employed (Bagnold 1979; Bagnold & Barndorff-Nielsen 1980; Fieller *et al.* 1984; Christiansen *et al.* 1984) and the 'geological' value of the whole approach (Ehrlich 1983).

A key criticism of the so-called Folk and Ward estimates, in particular, is that the numerical calculation of the measure of sorting by this semi-graphical procedure specifically requires that the size distribution is log normal, but the calculation of the sample skewness and kurtosis carries the implication that it is not so. Thus the calculated statistics cannot reflect the true properties of the sample, and must inevitably introduce unnecessary inaccuracy into the summary description of the data. Of course, with modern computing

From FROSTICK, L. & REID, I. (eds), 1987, *Desert Sediments: Ancient and Modern*, Geological Society Special Publication No. 35, pp. 271–280.

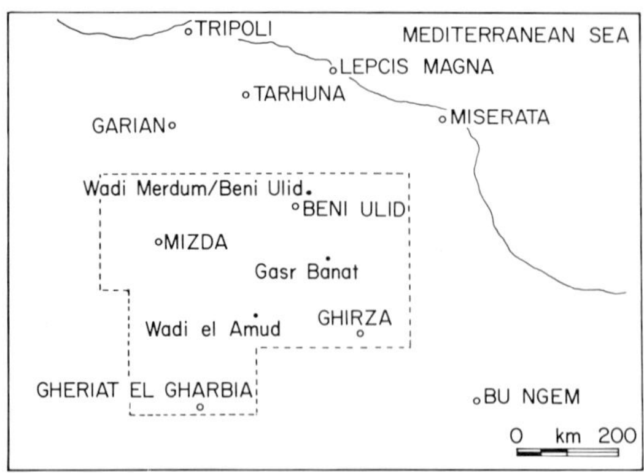

FIG. 1. Location of dune sites studied in Libyan Pre-Desert: climbing dunes at Wadi Merdum near Beni Ulid, barchan dunes near Gasr Banat, and seif dunes in the Wadi el Amud.

facilities it is easy to calculate sample moments directly without resort to dubious graphical aids. It is also straightforward to make some allowance for the grouping inherent in size data obtained by sieving, *eg* Sheppard's corrections (see Kendall & Stuart 1963). However, such techniques seem to be rarely used in routine practice.

An alternative approach is to postulate some underlying model for the distribution of particle sizes. Typically, this would be specified up to a few unknown parameters, these being estimated from the data to hand. This model might be that the distribution is log normal (*eg* Wyrwoll & Smyth 1985) with two parameters, or log hyperbolic (Barndorff-Nielsen 1977; Bagnold & Barndorff-Nielsen 1980) with four parameters or log skew Laplace (Olbricht 1982; Fieller *et al.* 1984; Fieller & Gilbertson 1985; Flenley 1985) with three parameters. The Laplace distribution is described in more detail below and its use is illustrated.

This alternative approach also aims to describe the size distributions of sand samples succinctly in terms of a few numerical quantities, just as is done by calculation of sample moment statistics. The distinction is that the modelling approach recognizes that the observed masses in the sequence of sieves are subject to sampling and measurement fluctuations and so provides a 'plausible' correction to allow for these. Naturally, calculation of a sufficient number of moments can characterize a specific sample more accurately than can any parsimonious distribution-based approach with its inherent smoothing. However, in general it is not the specific sample of sand that is of sole interest. Perhaps it is a sample from some larger body of material, or one of a sequence of samples taken along some direction which is known to exhibit progressive changes in features. In such cases, the specific data to hand are not of primary interest. They must be regarded as mere realizations of an underlying physical process incorporating a degree of randomness as well as being subject to sampling and other measurement errors. Under such conditions, a description incorporating some element of 'smoothing' (*ie* by statistical estimation) is the appropriate way to obtain information on the underlying quantities of interest from the (randomly contaminated) sample data. Christiansen (1984) compares analyses on the same sediments using both sample moments calculated from probability plots and the estimation of log hyperbolic parameters. He demonstrates the superiority of the latter technique.

The characterization of sand samples by a few estimated parameters (whether parameters of a proposed underlying distributional model or calculated sample moments) is usually just a first step in the analysis. For example, interest may centre on distinguishing depositional environments with a view to classifying further samples (*eg* Fieller *et al.* 1984). Alternatively, it may be directed towards understanding more subtle distinctions in the composition of the samples; perhaps they have been obtained from a sequence and the interest is in seeing how the structure of the samples changes with the sequence. This latter is the case with the present study, where samples have been taken along the surface of various dunes. In such situations, it might appear that the preliminary characterization of the individual samples can be by-passed and the end analysis be based directly on the raw sieve measurements. This is the approach advocated by Syvitski (1984) and Allen *et al.* (1972) who use factor analysis techniques. However, such methods not only ignore the ordering of the data imposed by the ordering of sieve sizes, but will be less statistically robust since they involve no preliminary 'cleaning' of the data. Further, they do not utilize the wealth of geological information available on the degradation and transportation processes involved which can be exploited to suggest appropriate families of mathematical models.

The arguments for believing the log hyperbolic family of distributions to be appropriate for modelling grain size distributions are presented in a sequence of papers by Barndorff-Nielsen (1977), Bagnold (1979), Bagnold & Barndorff-Nielsen (1980) and Barndorff-Nielsen *et al.* (1982). The principal advantage of the more parsimonious approach based on the log skew Laplace model (a limiting case of the log hyperbolic family) is that the estimation of the associated parameters is notably easier computationally, whilst the distribution retains the geological interpretability of the more complex model (Olbricht 1982 and Fieller *et al.* 1984). In particular, the arguments based on decomposition theory and random selection which lead to the proposal of the log hyperbolic family as suitable for modelling grain size data can be simplified so as to lead directly to the log skew Laplace family.

Much of the above discussion is directed particularly towards the analysis of 'single' distributions. Little attention has been given to the analysis of 'mixed' grain size distributions within a single sample. Several authors have indicated that such 'mixed' bimodal or trimodal distributions are likely to occur in various areas of aeolian, fluvial or lacustrine sedimentary environments as a result of the interplay of different sedimentary processes (*eg* Folk 1971; Warren 1972; Wood 1970). A commonly used approach, proposed by Visher (1969) as an *ad hoc* technique, is supposed to be based upon the identification of the components of the 'mixed' distribution from a log normal probability plot of the data. The technique is to identify distinct 'straight-line segments' in the mass–size data when plotted as a cumulative frequency curve on log normal probability paper. Each of these segments is then converted into a corresponding log normal density and considered to represent one of the components in a non-homogeneous mass–size population. The statistical reliability of this approach is questioned by Christiansen *et al.* (1984). The basis of the criticism is that data plotted in this way is being misinterpreted. In particular, Christiansen *et al.* demonstrated that a single log hyperbolic variate can generate plots on log probability paper which would be interpreted visually as consisting of three or more distinct line segments. Indeed, it is in general quite wrong to regard such 'segmentation' as evidence of the underlying distribution being a mixture of three log normal densities. The interpretation of 'segmented' probability plots is a potentially hazardous task. Not only can a 'single' distribution give rise to a multi-segmented probability plot, but 'mixture' distributions can give rise to probability plots with an arbitrary number of 'segments'. The hazards are well illustrated in Titterington *et al.* (1985). The only safe conclusion to be drawn from a log normal probability plot which is not reasonably straight (whether or not it consists of several apparently distinct straight portions) is that the underlying distribution is not log normal.

The use of log normal probability plots to determine the number of components in a mixture is a form of graphical analysis which is perhaps rather different from the two traditions outlined at the beginning of this section. However, it illustrates an unforeseen difficulty in using sample moments to characterize particle size distributions, particularly when these are calculated semi-graphically from log probability plots of the data. It is extremely difficult to tell from a cumulative frequency plot whether or not the distribution is truly a mixture of two or more sample densities, whether or not a probability scale is being used. Further, it is simplistic and often misleading to calculate moments such as the mean, skewness or kurtosis for anything other than unimodal distributions. Indeed, it is possible to construct examples of pairs of unimodal and bimodal distributions which have an arbitrary number of moments that are equal. That is, the values of the first few sample moments will not necessarily distinguish between unimodal and bimodal distributions, although bimodality may well be the most important feature, geologically, of the sample. Calculation of sample moments from mixture distributions will reveal little or nothing about the separate components of the mixture. Thus, routine use of sample moments for size data, whether approximated semi-graphically from log probability plots or whether calculated exactly, may well fail to reveal which samples are bimodal and which are not. It will not, in any case, reflect the properties of individual components in any mixture. Of course, in some clear-cut cases it is possible to separate the two components 'by eye' and calculate sample statistics for each separately. However, in the examples discussed below this would not be possible since the degree of overlap of the two components is substantial and arbitrary splitting would inevitably introduce inaccuracies.

The extension of the second tradition to 'mixture' distributions (that based on postulating an underlying statistical distribution for the sizes) has received little attention so far. It is clear that the computational difficulties inherent in estimating the parameters of even a single log hyperbolic model preclude the practical use of a model based on a mixture of two or more log hyperbolic densities. Flenley (1985) considers the modelling of some bimodal distributions by a mixture of two log skew Laplace densities. This is illustrated further below.

This brief review indicates that if size-related theories of sediment transport and deposition are to continue to be advanced and refined so as to apply to the present semi-arid aeolian or other sedimentary environments, then it is necessary to evaluate rigorously the objections to current procedures and to determine how to identify and model 'mixed' grain size distributions in the context of actual mass–size data. The implication of part of the above argument is that at least some inferred 'mixed' mass–size distributions are merely artefacts of the misunderstanding of the role of plotting conventions in the standard methods of graphical display described above. Therefore, the question arises of how frequently 'mixed' distributions actually occur in reality. The plots of mass–frequency data displayed below indicate that such mixed distributions do exist in aeolian sediments and re-emphasize the importance of plotting conventions in this context.

Illustrations of the methods used

We present here some further details of the various methods so far described, together with an account of their application to practical examples. The data relate to various sand samples obtained in an extensive study of the semi-arid desert area of Tripolitania known as the Libyan Pre-Desert. These data derive from standard dry sieving of the surface mineral sands (0–2 mm depth) for twelve minutes at quarter phi intervals. A complete statistical analysis of all the samples obtained is given in Flenley *et al.* (1986). Here, the account is limited to illustrations of the methodology.

Following a tradition established by Bagnold (1937), we present the mass–size distributions of various sand samples on 'log–log' plots. That is, the logarithm of the relative proportion of the mass in a particular class interval is plotted against the mid-point of the bounding log sizes of that class interval. Throughout, we have used natural rather than common logarithms, that is logarithms to base e ($=2.71828\ldots$) rather than to base 10. Of course the diagrams would not be materially altered if common logarithms were used, the change would be merely one of units. Usually, the class interval is determined by the two sieve sizes bounding it, but if the relative proportion of the mass in any particular sieve is very small, then there are statistical advantages to be obtained by combining the datum from that sieve with adjacent ones (see Flenley 1987).

On such a plot, it is typical that 'pure' samples taken from a simple depositional environment have a particularly simple form (Bagnold 1937, 1941). This form is well approximated as a pair of straight lines intersecting at the modal class. Barndorff-Nielsen (1977) and Bagnold & Barndorff-Nielsen (1980) took this as supporting evidence for proposing the family of log hyperbolic densities as models for simple particle size distributions. Additionally, they present mathematical reasons, based on theories of breakage and mixing, which also lead to the consideration of such a family of distributions. Log hyperbolic densities have the property that when plotted on log–log scales the resultant curve is a hyperbola with straight line asymptotes. On such a plot, a log normal density would appear as a parabola.

Estimation of the four parameters of a log hyperbolic density from sample data is a complex and numerically unstable procedure (Fieller *et al.* 1984). Instead, we have used a limiting case of the log hyperbolic density which we term the log skew Laplace distribution. On a log–log plot this density takes the simple form of a pair of straight lines, conforming with features widely observed in real examples. An alternative way of describing the density is that on a natural scale of relative frequency against log size the distribution appears as two 'back-to-back' exponential densities (*ie* an asymmetric double exponential density). The log skew Laplace distribution is cusped at the mode, which at first sight might be thought to be an unnatural feature. However, strictly, the distributions are used to model only the grouped data imposed by the sieve boundaries, so the idealized continuous density is averaged over each class interval in the estimation procedure and the cusp is thus not an intrinsic feature of the model.

The grain size analysis of the first of a sequence of samples (sample 276) taken along a transect across a barchan dune near Gasr Banat is given in Fig. 2. In Fig. 2*a* the data are plotted on log–log scales (*ie* as a 'log-histogram'); Fig. 2*b* shows a log normal probability plot of the same data. Also shown in Fig. 2*a* is the fitted log skew Laplace density. Fitting this density requires the estimation of three parameters. These are *alpha*, *beta* and *mu* which essentially reflect the slopes of the two lines and the abscissa of their point of intersection, and thus have the interpretation of expressing (respectively) the relative proportions of finer and coarser material in the sample, and the most common or, modal, particle size (on a logarithmic scale). The values of the estimates of these parameters are 0.170, 0.157 and -2.087, respectively, for this sample.

The fit of the estimated log skew Laplace density is extremely good, especially in the upper part of the distribution. The vertical logarithmic scaling exaggerates the apparent discrepancies in fit to the tails, but these last few points represent only a trivial fraction of the complete sample (about 0.05% by weight). The other 24 samples taken along the transect across the dune all exhibit equally good fits by log skew Laplace densities (Flenley *et al.* 1987) and examination of

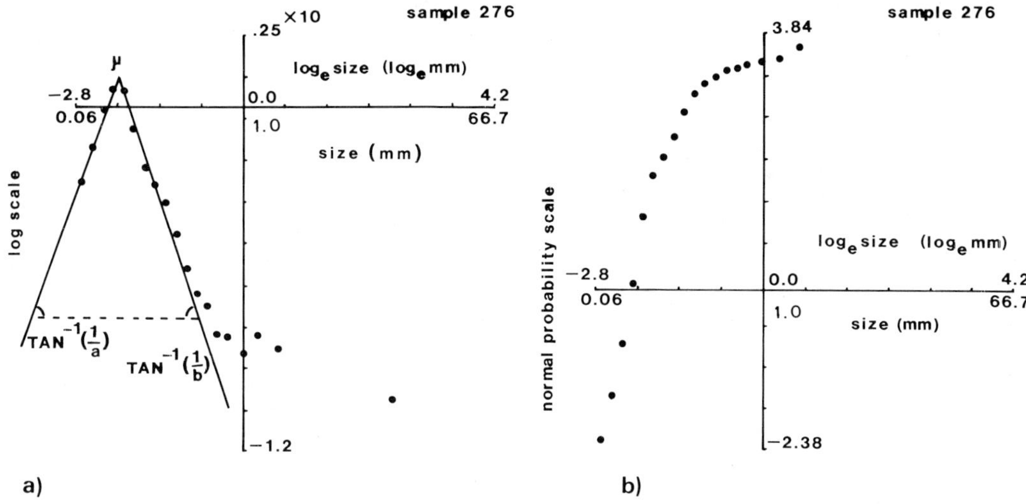

FIG. 2. Gasr Banat barchan dune sample 276 displayed on (*a*) log–log plot *ie* a histogram of log size plotted with a logarithmic vertical scale and (*b*) log probability plot, *ie* a cumulative plot of log size with a normal probability vertical scale. (*a*) shows the fitted log skew Laplace distribution and the definition of the three parameters *alpha* (denoted by a), *beta* (denoted by b) and *mu*.

the estimated parameters in sequence reveals the variations in grain size composition of the dune.

The log normal probability plot of the sample, Fig. 2*b*, shows at least two straight-line segments, even though the distribution is clearly not composed of a mixture. This highlights the potential hazards in identifying and interpreting separate segments of such a cumulative frequency curve. Of course, the value of any part of a cumulative curve will depend upon all the values before it, so it is dangerous to extract a 'segment' from the middle and attempt to give a separate explanation for that portion alone. The problem is exacerbated by using a normal probability scale on the vertical axis, since the curve then represents comparison with a cumulative normal distribution.

One of a sequence of samples (sample 550) taken along a seif dune in Wadi el Amud is shown in Fig. 3. In Fig. 3*a* is a log-histogram of the sample and Fig. 3*b* gives a log normal probability plot of the same data. Also presented in Fig. 3*a* is a fitted distribution composed of a mixture (or 'weighted average') of two separate log skew Laplace densities. This distribution is characterized by seven parameters, three from each of the two components (carrying interpretations for each component corresponding to those above) and the seventh expressing the relative proportions of the two components in the mixture. Estimation of the seven parameters from the sample data was performed essentially by maximum likelihood methods using specially written computer programmes (Flenley 1987). The values of the seven estimated parameters are 0.185, 0.273, −2.122, 0.315, 0.174, −1.196 and 0.716 respectively. Thus, the sample is modelled as a mixture, in the proportion of 71.6% to 28.4%, of two single log skew Laplace distributions, the first with estimated values of *alpha, beta* and *mu* of 0.185, 0.273 and −2.122, respectively, and the second with values of 0.315, 0.714 and −1.196.

The fit of the proposed distribution to the sample data is extremely close, apart from insignificant discrepancies in the tails which are exaggerated by the vertical logarithmic scaling. Again, these last few points represent less than 0.05% of the mass of the sample. This sample is typical of others taken from the seif dune (Flenley *et al.* 1986). Estimation of parameters of log skew Laplace mixtures for each sample provides a numerical description of them and permits interpretations to be attempted. Inevitably, this is more complex than with simple distributions that can be adequately characterized by three estimated parameters, but does provide essential insight.

The log normal probability plot in Fig. 3*b* illustrates further the difficulties in interpreting probability plots from non-normal data. At least four 'segments' could be identified, but these bear little relation to the more simple explanation that the sample is composed of a mixture of two simple constituents.

The natural histograms and fitted densities of the two samples (barchan dune sample 276 and

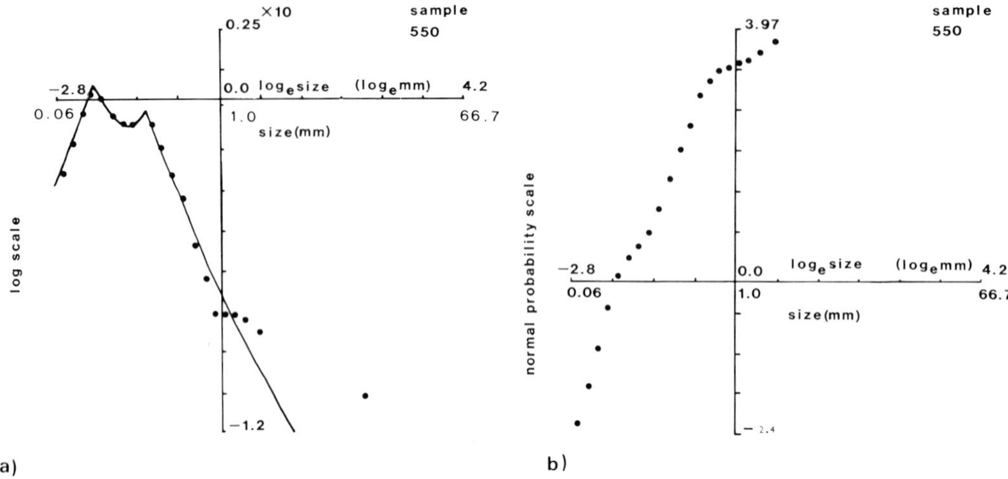

FIG. 3. Wadi el Amud seif dune sample 550 displayed on (*a*) log–log plot, *ie* a histogram of log size plotted with a logarithmic vertical scale and (*b*) log probability plot, *ie* a cumulative plot of log size with a normal probability vertical scale. (*a*) shows the fitted distribution composed of a mixture of two log skew Laplace densities.

seif dune sample 550) discussed and illustrated above are given in Fig. 4. For clarity, the vertical bars of the histogram have not been drawn, but symbols have been plotted at the top of the bars at the mid-points of the log class-intervals. These diagrams emphasize the extremely good fits to the data obtained by single and mixed log skew Laplace densities, and so demonstrate that use of these distributions captures all the information in the data. This scaling reaffirms how little importance should be attached to the slight discrepancies noticed in the tails on the log–log plots.

A situation rather more complex than the simple barchan and seif dunes discussed above is illustrated in Fig. 5. The diagram represents a

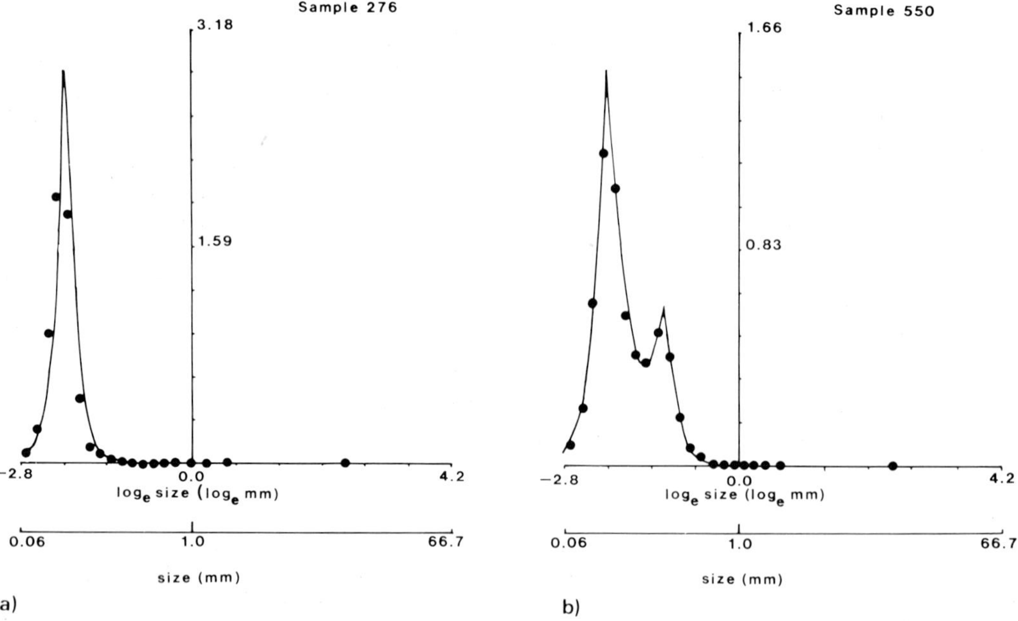

FIG. 4. Samples 276 and 550 displayed on natural versus log scales, together with their fitted densities: (*a*) sample 276 with fitted single log skew Laplace and (*b*) sample 550 with fitted mixture of two log skew Laplace densities.

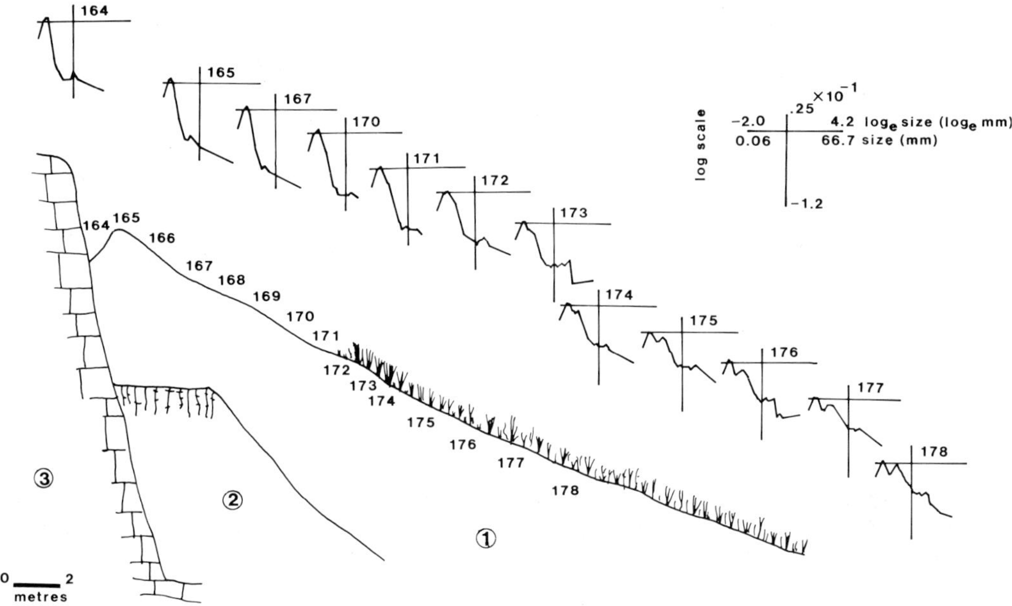

FIG. 5. Cross-section of a climbing dune in Wadi Merdum, near Beni Ulid, showing the location of vegetated area and of samples 164–178. The data from the samples is displayed on log–log plots (*ie* as log histograms) showing the transition from unimodality to bimodality. The fitted densities are not shown but their parameters are tabulated in Table 1. (1) is the modern climbing dune which overlies (2), an ancient climbing dune capped by a palaeosol. Both are banked up against (3), the limestone bedrock of the wadi edge.

cross-section of a climbing dune, the lower part of which is vegetated, which overlies ancient climbing dunes capped by a palaeosol. This dune is in Wadi Merdum near Beni Ulid (Fig. 1). Fifteen samples (164–178) were taken at 1 or 2 m intervals along a transect starting a short distance from the crest in a wind-scour hollow adjacent to the vertical cliffside of the wadi's edge, passing across the crest and descending vegetation-free sands before reaching dune grasses 0.2–0.4 m high. Also presented are log–log plots of the mass–size distributions from 12 of the samples (the omitted samples 166, 168 & 169 are intermediate in shape between their neighbours). These plots exhibit a clear progression from unimodality to bimodality along the transect. The apparent small peaks in the lower right-hand tails are again accentuated by the logarithmic scaling of the vertical axes and represent very small fractions (less than 0.05%) of the total. Careful inspection of the sequence of plots indicates the beginnings of the secondary peak in sample 171 (and possibly in 170) coinciding with the start of the sand-binding vegetation. Single and mixed log skew Laplace densities fit the earlier and later samples in the sequence, respectively, but are not shown on the diagram. Values of the estimated parameters of the distributions fitted to all 15 samples are given in Table 1.

Single log skew Laplace densities proved to be appropriate for samples 164–170. For samples 171–178 a mixture of two such densities provided a better model. Attempts to fit a mixture model to sample 170 (and earlier ones) resulted in a mixture of two densities, one with a mode corresponding to the apparent peak on the left (*ie* with *mu* estimated as about −2.0), the other corresponding to the very small peak on the extreme right (*ie* with *mu* estimated as about +2.2 which reflects a size of about 9 mm). The estimated value of p, the mixing proportion for this model, was extremely large, typically in the region of 0.9995 or more, indicating that this right-hand peak was insignificant and that the best 'mixture model' was in effect little different from a single log skew Laplace model.

Examination of the parameter estimates given in Table 1 shows how their values vary reasonably consistently with spatial sequence. Of particular interest is the growth of the secondary peak in the mixture as samples are taken progressively down the slope of the dune and across the vegetated sands. Starting with sample 171 (*ie* just before the start of the vegetation) this secondary

TABLE 1. *Parameter estimates of densities fitted to climbing dune samples 164–178. Samples 164–170 have a single log skew Laplace density and samples 171–178 have a mixture of two log skew Laplace densities. The mixing proportion* p *is the proportion of the left-hand component in the total sample*

Sample number	Left component			Right component			Mixing proportion (p)
	alpha	*beta*	*mu*	*alpha*	*beta*	*mu*	
164	0.189	0.077	−2.046	—	—	—	—
165	0.194	0.090	−2.120	—	—	—	—
166	0.173	0.066	−2.017	—	—	—	—
167	0.204	0.084	−1.971	—	—	—	—
168	0.290	0.126	−1.952	—	—	—	—
169	0.177	0.153	−2.075	—	—	—	—
170	0.225	0.140	−1.980	—	—	—	—
171	0.192	0.170	−2.049	0.286	0.088	−1.542	0.946
172	0.194	0.237	−2.080	0.255	0.144	−1.597	0.889
173	0.195	0.170	−2.083	0.004	0.349	−1.656	0.893
174	0.163	0.329	−2.184	0.135	0.061	−1.241	0.894
175	0.199	0.406	−2.136	0.153	0.108	−1.061	0.817
176	0.210	0.122	−2.143	0.397	0.308	−1.133	0.734
177	0.176	0.119	−2.210	0.437	0.411	−1.104	0.578
178	0.193	0.127	−2.147	0.288	0.316	−1.031	0.516

peak has a value of *mu* estimated as −1.542 (corresponding to a secondary modal peak size of 0.21 mm) and represents only about 5% of the sample (the value of the mixing parameter p is estimated as 0.946). This secondary modal size increases steadily down the dune, as does the relative proportion of the larger sized component in the sample, to reach a modal value of −1.031 (*ie* 0.36 mm) and a relative proportion of 48.5% at the base of the dune. Similar, though less consistent, progressive changes in the values of the other parameters are apparent. This relationship in parameter values between neighbouring samples can be exploited in the iterative estimation procedure by using estimates obtained from one sample as starting values for the iteration scheme for the next. This is particularly of use in the estimation of the mixing proportion parameter p.

The interpretation of these features in the parameter estimates is that sand samples from the vegetated area of the dune have an additional component of larger particles in their composition. Further, this component is represented in increasing proportions towards the base of the dune and is composed of progressively larger particles. Presumably, this is an effect of entrapment of the larger particles by the sand-binding vegetation. Additionally, it seems that the effect of the vegetation extends for some short distance beyond its boundary half-way up the slope of the dune.

Conclusions

The preceding discussion and examples demonstrate that, to make maximum use of particle mass–size data, it is inadequate to rely on log normal probability plots and the calculation of sample moments. Even though such methods have proved to be successful in many simple situations, they do not readily extend beyond these. In particular, probability plotting methods could lead to misleading interpretations of sample data by the unwary. Mixture samples do arise in practice, but perhaps less often than is suspected. When they do, it is of interest to unravel their constituents and this can only be done by proper modelling of the underlying distributions. The families of log skew Laplace densities and their mixtures are shown here to have sufficient flexibility to provide informative model for sand dune data of many different types.

References

ALLEN, G. P., CASTAING, P. & KLINGBIEL, A. 1972. Distinction of elementary sand populations in the Gironde Estuary (France) by R-mode factor analysis of grain-size data. *Sedimentology* **19**, 21–35.

BAGNOLD, R. A. 1937. The size-grading of sand by wind. *Proceedings of the Royal Society of London* **A163**, 250–264.

—— 1941. *The Physics of Blown Sand and Desert Dunes*. Methuen, London.

—— 1979. Acceptance of the Sorby Medal of the International Association of Sedimentologists. *Sedimentology* **26**, 159–160.

—— & BARNDORFF-NIELSEN, O. 1980. The pattern of natural size distributions. *Sedimentology* **27**, 199–207.

BARKER, G. W. W. & JONES, G. D. B. 1980. The UNESCO Libyan Valleys Survey 1980. *Libyan Studies* **12**, 9–48.

—— & —— 1984. The UNESCO Libyan Valleys Survey VI: Investigations of a Romano–Libyan Farm, Part 1. *Libyan Studies* **15**, 1–44.

BARNDORFF-NIELSEN, O. 1977. Exponentially decreasing distributions for the logarithm of particle size. *Proceedings of the Royal Society of London* **A353**, 401–419.

——, DALSGAARD, K., HALGREEN, C., KUHLMAN, H., MØLLER, J. T. & SCHOU, G. 1982. Variation in particle size distribution over a small dune. *Sedimentology* **29**, 53–65.

CHRISTIANSEN, C. 1984. *A comparison of sediment parameters from log-probability plots and log–log plots of the same sediments*. Geoskrifter Nr. 20, Department of Geology, University of Aarhus, Denmark.

——, BLAESILD, P. & DALSGAARD, K. 1984. Re-interpreting 'segmented' grain-size curves. *Geological Magazine* **121**, 47–51.

EHRLICH, R. 1983. Size analysis wears no clothes or have moments come and gone? *Journal of sedimentary Petrology* **53**, 1.

FIELLER, N. R. J. & GILBERTSON, D. D. 1985. Skew log Laplace distributions and the discrimination of shoreline environments at Dawlish Warren, Devon. *Proceedings of the Ussher Society* **6**, 271–272.

——, —— & OLBRICHT, W. 1984. A new method for the environmental analysis of particle size data from shoreline environments. *Nature* **311**, 648–651.

FLENLEY, E. C. 1985. Use of mixture distributions in the modelling of sand particle sizes. *In: Proceedings of the International Workshop on the Physics of Blown Sand*. Department of Theoretical Statistics, University of Aarhus, Denmark, 633–648.

—— 1986. *Statistical analysis of particle size data*. PhD thesis, University of Sheffield.

——, FIELLER, N. R. J. & GILBERTSON, D. D. 1986. *An analysis of particle size distributions of aeolian sands from the Libyan Pre-Desert using single and mixed log skew Laplace distributions*. Manchester–Sheffield School of Probability and Statistics: Research Report no. ECF86/2.

FOLK, R. L. 1971. Longitudinal dunes of the northwestern edge of the Simpson Desert, Northern Territory, Australia, 1. Geomorphology and grain size relationships. *Sedimentology* **16**, 5–54.

—— & WARD, W. C. 1957. Brazos River Bar: A study in the significance of grain size parameters. *Journal of sedimentary Petrology* **27**, 3–26.

FRIEDMAN, G. M. 1961. Distinction between dune, beach and river sands from their textural characteristics. *Journal of sedimentary Petrology* **31**, 514–529.

—— 1967. Dynamic processes and statistical parameters compared for size frequency distribution of beach and river sand. *Journal of sedimentary Petrology* **37**, 327–354.

—— 1979*a*. Address of the Retiring President of the International Association of Sedimentologists: Differences in size distributions of populations of particles among sands of various origins. *Sedimentology* **26**, 3–32.

—— 1979*b*. Differences in size distributions of populations of particles among sands of various origins: Addendum to IAS Presidential Address. *Sedimentology* **26**, 859–862.

GALE, S. J., HUNT, C. O. & GILBERTSON, D. D. 1986. The infill sequence and water carrying capacity of an ancient irrigation channel: Wadi Gobbean, Tripolitania. *Libyan Studies* **17**, 1–5.

GILBERTSON, D. D. (ed.) 1986. Runoff farming in rural arid lands. *Applied Geography Theme Volume 6 (1 and 2)*, 1–122.

—— & HUNT, C. O. in press. The Quaternary geology of the Wadi Merdum in the Tripolitanian Pre-Desert. *Libyan Studies* **17**.

——, HAYES, P. P., BARKER, G. W. W. & HUNT, C. O. 1984. The UNESCO Libyan Valleys Survey VII: an interim classification and functional analysis of ancient wall technology and land use. *Libyan Studies* **15**, 45–70.

——, HUNT, C. O., BRIGGS, D. J., COLES, G. M. & THEW, N. in press. The Quaternary geology of the Gasr Banat region of the Tripolitanian Pre-Desert. *Libyan Studies* **17**.

HUNT, C. O., GALE, S. S. & GILBERTSON, D. D. 1985. The UNESCO Libyan Valleys Survey IX: Anhydrite and limestone karst in the Tripolitanian Pre-Desert. *Libyan Studies* **16**, 1–13.

INMAN, D. L. 1952. Measures for describing the size distribution of sediments. *Journal of sedimentary Petrology* **22**, 125–145.

KENDALL, M. G. & STUART, A. 1963. *The advanced theory of statistics. Vol. 1 Distribution theory*. Second edn. Charles Griffin & Company Limited, London.

OLBRICHT, W. 1982. *Modern Statistical Analysis of Ancient Sand*. MSc thesis, Department of Probability & Statistics, University of Sheffield.

SYVITSKI, J. P. M. 1984. Q-mode factor analysis of grain size distributions. *Geological Survey of Canada: Open file report 965*.

TITTERINGTON, D. M., SMITH, A. F. M. & MAKOV, U. E. 1985. *Statistical analysis of finite mixture distributions*. John Wiley & Sons Ltd, Chichester.

VAN DER VEEN, M. 1985. The UNESCO Libyan Valleys Survey X: Botanical evidence for ancient farming in the Pre-Desert. *Libyan Studies* **16**, 15–28.

VISHER, G. S. 1969. Grain size distributions and depositional processes. *Journal of sedimentary Petrology* **39**, 1074–1106.

WARREN, A. 1972. Observations on dunes and bi-modal sands in the Ténéré Desert. *Sedimentology* **19**, 37–44.

WOOD, W. H. 1970. Rectification of wind-blown sand. *Journal of sedimentary Petrology* **40**, 29–37.

WYRWOLL, K.-H. & SMYTH, G. K. 1985. On using the log normal distribution to describe the textural characteristics of eolian sediments. *Journal of sedimentary Petrology* **55**, 471–478.

E. C. FLENLEY & N. R. J. FIELLER, Department of Probability and Statistics, University of Sheffield, Sheffield S3 7RH, UK.

D. D. GILBERTSON, Department of Archaeology and Prehistory, University of Sheffield, Sheffield S10 2TN, UK.

Grain-size variation on a 'complex' linear dune in the Namib Desert

I. Livingstone

SUMMARY: Against a background of increasing knowledge of regional grain-size variations, particularly in the Namib Desert, this paper reports an investigation of small-scale grain-size variations using 25 sample points on the cross-profile of a single linear dune. The results suggest that a discernible pattern of grain size across the dune does exist, and that changes are gradual rather than discrete. From an examination of the near neighbours of the main study dune there is also some evidence of measurable differences between dunes in the same area. The fact that grain sizes are not randomly distributed across the width of the dune indicates a response to spatial variations of process, but grain size does not seem to control dune form in the way that some earlier workers have envisaged. Furthermore, monthly sampling provides preliminary evidence that grain-size values on this dune respond to a seasonal wind regime, and that time of sampling may therefore be a crucial determinant of the results obtained.

For almost a century, grain-size analysis has been carried out on a large number of ancient and contemporary sediments in the belief that it is possible to define diagnostic properties of individual sedimentary regimes from the size frequency distribution of their deposits. As a result, there is now a considerable literature on beach/dune/river systems, but the number of detailed studies of the sands of linear dunes remains remarkably small.

In recent years, however, several workers have investigated regional patterns of grain size at sand-sea scale, notably Besler (1980) and Lancaster (1982*c*) for the Namib Desert, Lancaster (1986) for the SW Kalahari, and Warren *et al.* (1985) for the Wahiba Sands. Indeed, largely as a result of the endeavours of Besler and Lancaster there is now a fairly comprehensive picture of the regional variation of grain size throughout the Namib Desert. Little is known, though, about the changes that take place at a smaller scale, *eg* on a single linear dune. Data comparing inter-dune, plinth and crest sediments from several sand seas have been presented by various authors (Table 1), but until Watson's (1986) recent paper reporting data from 20 sample points on a linear dune in the Namib Desert, no study had exceeded seven samples from a single dune cross-profile. The object of the present paper is to intensify the investigation of small-scale grain-size changes by detailed sampling of an individual linear dune. The examination of grain-size variation is part of a wider study of the geomorphological dynamics of the dune (Livingstone 1985).

Study site

The main study dune lies at the northern edge of the Namib Sand Sea in Namibia, southern Africa, approximately 8 km SE of the Namib Desert Research Station at Gobabeb (23°34′S, 15°03′E) (Figs 1 and 2). The dune here is aligned roughly N–S, is approximately 350 m wide and 50 m high. It stretches some 30 km to the S of the study site and 3 km N to the Kuiseb River. Inter-dune corridors in the area are between 1.5 and 3.0 km wide. The study dune displays the morphometric asymmetry typical of dunes in this part of the Namib Sand Sea, with a relatively uniformly sloping west flank, but a series of secondary ridges and barchanoid features on the E flank. As a consequence of its size and its support of secondary dunes, this dune is termed 'complex' after McKee (1979) and Lancaster (1982*b*), although in Wilson's (1972*b*) classification it would be a 'draa'.

The central Namib Desert is subjected to a seasonal wind regime so that, broadly speaking, the dunes are affected by low- to moderate-force winds from the SW and NW in summer, and by high force but low frequency winds from the E in winter (Lancaster *et al.* 1984). Under the influence of this regime, the crest of the dune moves back and forth laterally by some 14 m each year (Livingstone 1985). Besides movement of the crest, there is a northward extension of the dune into the Kuiseb River valley of between 0 and 1.85 m a^{-1} (Ward 1984). Contrary to the belief of Rubin & Hunter (1985), there is no evidence that the imbalanced wind regime is leading to any lateral shift of the dune base.

The source of sand for the Namib dunes has been a matter of some debate. Beneath the present sand sea lie deposits of a former erg which have undoubtedly been partially reworked into the present dune system, but this is not the major source of sand. Besler (1980) regards the dunes as a Pleistocene reworking of fluvial deposits from a former alluvial plain system,

From FROSTICK, L. & REID, I. (eds), 1987, *Desert Sediments: Ancient and Modern*, Geological Society Special Publication No. 35, pp. 281–291.

TABLE 1. *Grain-size parameters of linear dunes (phi units)*

Dune Location and Sampling Site	Mean	Standard Deviation	Skewness	Kurtosis	Reference
Namib Desert, Namibia					
Crest	2.15	0.54	—	0.98	Besler (1980)
Base	1.98	0.85	—	1.03	
Crest	2.11	1.71	0.04	1.30	Goudie (1970)
Compound linear dunes					
Crest	2.25	0.39	0.19	0.50	Lancaster (1983*b*)*
Slip faces	2.33	0.41	0.04	0.47	
Plinths	2.01	0.86	0.22	0.46	
Interdunes	1.96	1.04	0.29	0.46	
Complex linear dunes					
Crest	2.49	0.36	0.13	0.51	Lancaster (1983*b*)*
Slip faces	2.51	0.37	0.03	0.50	
Upper W	2.40	0.51	0.09	0.49	
Plinths	2.07	0.76	0.32	0.48	
Interdunes	1.98	0.90	0.34	0.47	
E flank dune	2.30	0.44	0.17	0.52	
Crossing dunes	2.19	0.48	0.26	0.55	
1 (E base)	2.47	1.13	0.27	1.06	Watson (1986)
2	1.90	1.10	0.04	1.14	
3	2.02	0.88	0.29	0.87	
4	1.76	0.78	0.49	1.06	
5	1.88	0.88	0.55	0.91	
6	1.66	0.76	0.40	1.22	
7	1.96	0.81	0.41	0.89	
8	1.92	0.89	0.50	1.37	
9	2.12	0.80	0.30	0.80	
10	2.10	0.72	0.40	0.86	
11	1.95	0.67	0.26	0.88	
12	2.71	0.64	0.09	0.89	
13	1.58	0.65	0.46	0.92	
14	2.26	0.38	0.07	1.02	
15 (Crest)	2.56	0.30	0.01	0.98	
16	2.26	0.54	0.11	0.75	
17	2.30	0.46	0.26	1.03	
18	2.05	0.58	0.28	0.90	
19	1.91	0.56	0.39	1.04	
20 (W base)	1.82	0.85	0.25	0.77	
Kalahari Desert, southern Africa					
Crest	2.21	1.76	0.03	1.35	Goudie (1970)
Slope	2.51	0.88	−0.14	0.92	Lewis (1936)
Crest	2.37	0.57	0.07	0.86	
Street	2.49	0.89	−0.26	0.91	
Crest	2.16	0.49	0.14	0.52	Lancaster (1986)*
NE flank	2.21	0.62	0.05	0.52	
SW flank	2.26	0.59	0.07	0.53	
Interdune	2.12	0.90	0.02	0.52	
Negev Desert, Israel					
Base wind	1.41	0.70	0.88	3.79	Tsoar (1978)
Mid wind	1.70	0.47	0.31	7.32	
Crest	1.87	0.42	0.80	5.28	
Slip face	2.05	0.47	0.32	3.24	
Base lee	1.18	0.75	0.52	4.00	
Great Indian Sand Desert					
Crest	2.72	2.15	—	1.02	Goudie *et al.* (1973)
Simpson Desert, Australia					
Crest	2.53	0.43	0.11	0.52	Folk (1971)*
Flanks	2.75	0.57	—	—	
Reg	2.85	0.95	0.04	0.48	

* Lancaster (1983*b*, 1986) and Folk (1971) use transformed values of graphic kurtosis (KgI), such that $KgI = Kg/(Kg+1)$.

FIG. 1. A 'complex' linear dune in the northern Namib sand sea. Sample transects 1 and 2 lie across this dune.

while Lancaster & Ollier (1983), following Rogers (1977), provide substantial sedimentological evidence to support their belief that the majority of the Namib sand has been brought inland from the continental shelf under the influence of a predominantly southwesterly wind regime.

Techniques

Samples of surface sand were collected from 25 points across the width of the study dune at monthly intervals throughout the period April 1981 to April 1982, giving a total of 325 samples. Sample points were marked by steel posts. These were also used for measuring surface height changes (Livingstone 1985). The sampling sites were designated by the letters A to Y, from W to E across the dune, and were generally 20 m apart. However, near the crest and on the secondary dunes, samples were taken every 10 m. A sample was also taken from the dune crest.

In addition, and in order to examine the extent to which the main study dune could be considered representative of linear dunes in this part of the sand sea, samples were taken from eight cross-profiles on four other complex linear dunes: *ie* two dunes immediately to both the E and W of the study dune (Figs 2 and 3). All samples were collected in a two-day period to avoid the effect of changing wind regime. Samples were usually collected every 20 m across each dune profile, but where exceptionally long dune plinths were encountered this distance was extended to 40 m. A total of 180 samples was collected from these eight profiles. It should be noted that transects 1 and 2 of the present study are on the study dune of Watson (1986).

A portion of each sample was sieved at half-phi size intervals in the range +0.5 to +4.0 phi (0.707 to 0.063 mm); particles finer than +4.0 phi (0.063 mm) were collected in a receiver. The amount of sand retained in each sieve was weighed, and this information plotted as a cumulative frequency curve on arithmetic probability paper. From these graphical representations of the data, percentile values were extracted, and grain-size parameters (mean, standard deviation, skewness and kurtosis) were then calculated according to the formulae of Folk & Ward (1957).

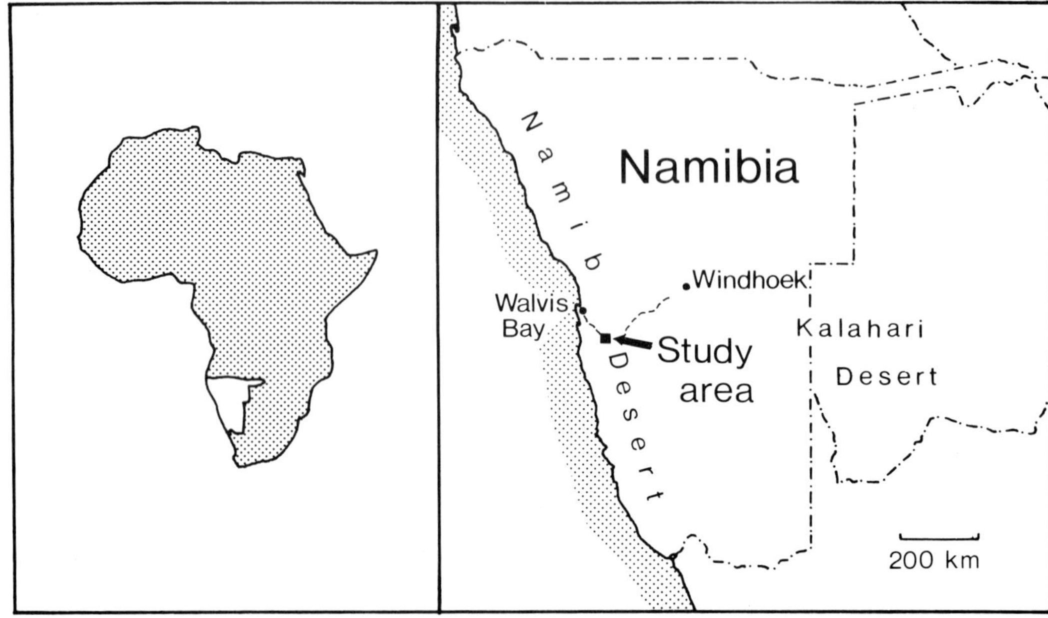

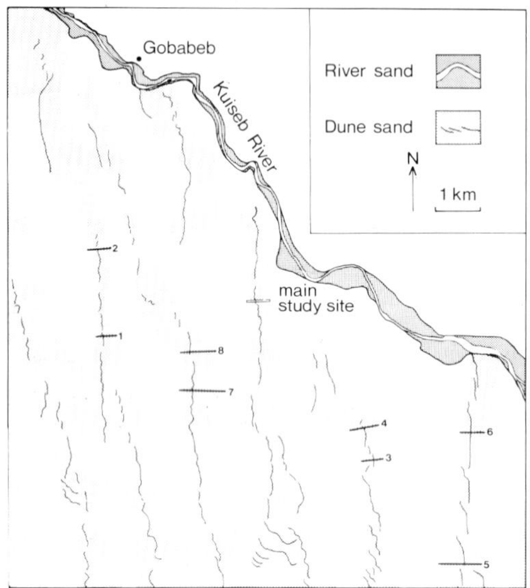

FIG. 2. Location maps of the main study dune and the eight supplementary dune transects.

A summary of the values from the main study dune is given in Table 2; the results are reported in full elsewhere (Livingstone 1985).

Discussion

The pattern of variation in grain size and sorting on a complex linear dune

The samples collected from the main study dune and from the other eight transects provide a very clear picture of the change in grain size and sorting across a linear dune profile. Figure 4 is a plot of the mean values of the four grain-size parameters against position across the main study dune, while Fig. 5 shows the changing pattern of mean grain size across the other eight transects. In Fig. 5 position in the cross-section is plotted as distance from the dune crest in an attempt to make the transects sit in comparable positions on the diagram; the values of mean grain size are five-point moving averages—aggregating the data overcomes the problems posed by occasional idiosyncratic values.

A feature of the data from all the dunes sampled is that the range in values of the size distribution parameters is not great. Thus, on the

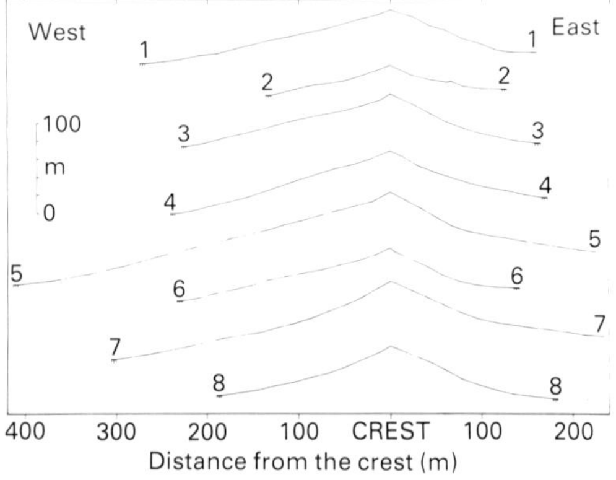

FIG. 3. Cross-sections of the linear dunes at the eight sample transects located in Fig. 2.

TABLE 2. *Average values and range of mean grain size and other size distribution parameters for the main study dune, northern Namib—derived from the 13 monthly samples*

Sample point*	Size distribution parameters					
	Mean		Standard deviation		Skewness average ϕ	Kurtosis average ϕ
	average ϕ	range ϕ	average ϕ	range ϕ		
A	1.86	1.72–1.98	0.50	0.36–0.61	0.36	1.31
B	1.89	1.76–2.22	0.44	0.28–0.57	0.32	1.41
C	1.98	1.78–2.08	0.48	0.35–0.55	0.33	1.10
D	2.11	1.93–2.56	0.47	0.37–0.52	0.32	1.00
E	2.19	2.04–2.59	0.43	0.37–0.50	0.31	1.02
F	2.21	1.95–2.62	0.37	0.29–0.45	0.25	1.07
G	2.30	2.08–2.64	0.38	0.32–0.46	0.18	1.00
H	2.34	2.03–2.63	0.36	0.32–0.41	0.13	1.02
I	2.40	2.15–2.66	0.37	0.27–0.43	0.12	0.99
J	2.47	2.27–2.73	0.32	0.25–0.36	0.03	1.02
K–L	2.52	2.32–2.67	0.33	0.28–0.40	0.05	1.01
M	2.50	2.16–2.63	0.32	0.27–0.38	0.03	1.03
N	2.44	2.25–2.56	0.30	0.25–0.36	0.01	0.98
O	2.08	1.88–2.44	0.38	0.25–0.49	0.16	0.98
P	2.40	2.10–2.58	0.33	0.24–0.40	−0.02	0.97
Q	2.49	2.41–2.63	0.32	0.23–0.42	−0.04	1.04
R	2.53	2.24–2.67	0.26	0.20–0.34	0.01	1.03
S	2.47	2.31–2.73	0.27	0.21–0.34	0.04	1.03
T	2.57	2.47–2.66	0.25	0.22–0.29	0.04	1.01
U	2.66	2.53–2.84	0.21	0.16–0.29	0.06	1.06
V	2.64	2.48–2.95	0.20	0.16–0.25	0.10	1.05
W	2.34	2.13–2.52	0.24	0.18–0.34	0.07	1.07
X	2.46	2.32–2.69	0.23	0.15–0.34	0.15	1.15
Y	2.15	1.96–2.29	0.38	0.27–0.49	0.02	1.08
Crest	2.37	2.16–2.71	0.30	0.23–0.35	0.07	0.99

* See Fig. 4 for location.

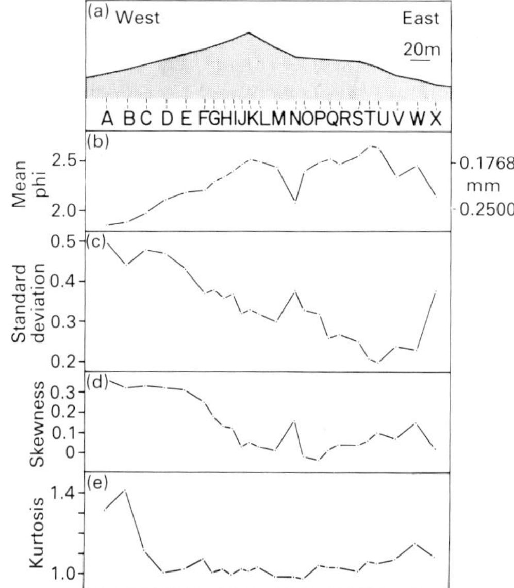

FIG. 4. Variation of the four grain-size distribution parameters across the main study dune.

main study dune, the range of values for mean grain size is from 1.72 to 2.84 phi (0.30 to 0.14 mm). This is directly comparable with the results from the other eight transects. It compares, too, with Lancaster's (1981) reported range of 1.80 to 2.55 phi (0.29 to 0.17 mm), and with Watson's (1986) values of between 1.58 and 2.71 phi (0.33 and 0.15 mm). The measures of grain-size sorting show equally small variations.

Despite the limited range of values, there exist very distinct patterns of variation in grain size and sorting over the dune cross-profiles. Results for mean grain size show that there is a progressive fining of sand from dune base to dune crest. This trend in mean grain size is matched by decreasing values of both standard deviation (*ie* sorting) and skewness from plinth to crest, indicating that the finer crestal sands are also better sorted and the size-distribution less skewed than the coarser plinth sands.

Kurtosis of the size frequency distribution does, however, not seem to be significant in discriminating geomorphic position on the main study dune, returning values close to 1.0 in all the

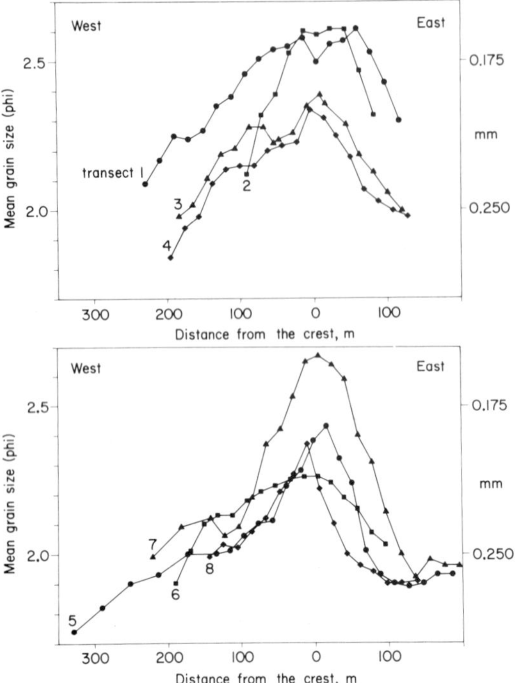

FIG. 5. Five-point moving averages of mean grain size across the eight dune transects located in Fig. 2.

samples except those close to the base of the W slope. Kurtosis is an indication of peakedness in the size distribution, and is therefore a valuable test of a normal distribution. The fact that all the values reported here are close to 1.0 suggests that grain-size populations in most of the samples tend towards a normal distribution. Whereas kurtosis has been used as a good discriminatory parameter in studies that compare widely different sedimentary environments, there is such a small range of values in the samples from the linear dunes in this investigation that it appears to be of little value as a discriminant in a single environment such as here.

In the past, several writers have attempted to show the existence of two or three distinct grain-size populations on linear dunes, and their sampling has often been based on identifiable morphological zones. Bagnold (1941), for instance, divides the dune cross-profile into three main components. First, there is an active crest moving back and forth in response to the seasonal wind regime. Here, finer grains move towards the crest. The crest surmounts a zone that has never been incorporated into a slip face; its sands are of virtually constant grain size. Below this are the coarser grains of the interdune corridor sands. In the Namib Desert, Besler (1983) has divided the dunes into zones of 'aeolian mobility' and 'aeolian stability', while Lancaster (1981) has distinguished between 'crest' and 'plinth' sands.

The present method of systematic sampling at 25 points on a dune cross-profile suggests that there are differences in grain-size distribution at a smaller scale than has been indicated by these previous studies. Indeed, as Watson (1986) has already noted, rather than demonstrating discrete populations, small-scale sampling indicates that there is a progressive change in the grain-size distribution parameters across the dune (Figs 4 and 5).

It is possible, however, to move beyond Watson's general observation of gradual change. Assuming that the values derived here for monthly samples at each point on the study dune are representative of the sand at that point, the 13 samples from one point can be compared with the 13 samples from another using a statistic such as the Mann–Whitney U. It is apparent, using this statistic for values of mean grain size, that there is a statistically significant change every 20 or 30 m along the dune surface, although the pattern becomes less clear in the upper part of the dune. This contrasts strongly with Warren's (1971) data that show uniform mean grain sizes over most of a seif dune E of Adrar Madet.

Given the progressive change in grain size across the dune, the implication is that it would be difficult to find a justifiably representative sampling point on morphological criteria alone. For example, samples from point C (Fig. 4) are no more or less representative of sand from the W plinth of the study dune than samples from point E, but they are statistically significantly different. When Folk (1971) observed that it was important to know how size parameters varied with the micro-morphology of the sampling site he was warning against one sample from a dune being taken as typical of the whole dune, but this warning can now be seen to be too conservative. In a system the size of a Namib linear dune, there are significant and progressive changes in grain-size distribution right across the dune profile.

Within the general pattern of grading from coarser, poorly sorted, and more skewed samples at the dune base to finer, better sorted and less skewed samples at the crest, some deviations do occur. On the main study dune, the progressive fining that is evident in traversing up the W slope continues beyond the crest to the zone of secondary dunes around points U and V. This pattern is broken only by the coarser sand found towards the base of the summer slip face around point O. This pattern is repeated on the only other transect with a secondary dune system (transect 1), and is also found in the results reported by Watson (1986). So, while Lancaster

(1983*b*) believes that the sands of E flank dunes tend to be slightly coarser than adjacent crest sands, the evidence from the present study suggests that this is not always the case. It would be difficult to generalize from the results of three dunes, but clearly Lancaster's assertion is not universally true.

Thus far, the discussion has centred on *intra*-dune variation of grain size and sorting parameters. The data from the main study dune along with the results from the other eight transects also provide an opportunity to investigate *inter*-dune grain-size and sorting variations on five neighbouring complex linear dunes. Once again, the Mann–Whitney U Test has been employed to examine the significance of any variation of mean grain size between dunes. It is apparent from the results that there is little basis for assuming homogeneity of the dunes in this part of the sand sea. While there appear to be no significant differences between any pair of transects from the same dune (*eg* transect 1 *cf* transect 2, *etc.*), there are statistically significant differences between dunes.

This has some implications for studies which seek to provide regional grain-size values. Studies such as those of Besler (1980) and Lancaster (1982*c*) for the Namib Desert, Folk (1971) for the Simpson Desert, and Warren *et al.* (1985) for the Wahiba Sands have based work on the premise that it is possible to calculate characteristic, regional values for grain-size values. Lancaster (1982*c*), for instance, has described a progressive fining to the N and W in the Namib Desert, reporting crestal sands with mean sizes coarser than 2.30 phi (0.2 mm) in the S and finer than 2.50 phi (0.18 mm) in the NW. The lack of any wide-ranging variation in the results of this study lends some support for this belief in a regional context. However, any assertions about transport processes and so on must be qualified by the fact that there may be considerable between-dune variance even though the range of values is relatively small.

A number of conclusions can be drawn from the spatial variation of grain sizes on the main study dune and its near neighbours. First, there are definite, discernible patterns of change in three of the grain-size distribution parameters—the mean, the standard deviation and the skew—but not in kurtosis. However, in general, the range of values for each parameter is small. The cross-profile trends demonstrated by the first three moments of the distribution are not dissimilar to those published by other workers, but greater detail highlights some important deviations from the general trends. Furthermore, the discovery of significant between-dune differences, when coupled with the fact that there are small-scale differences within a single dune, suggests that the exact location of the sampling point on the dune should be declared if comparisons are to be facilitated with other work and other places.

The relationship of grain size and sorting measures to process

In grain-size studies it is axiomatic that different processes provide different combinations of grain sizes, that certain modes of transport move grains of particular size preferentially, and, therefore, that cross-dune patterns describe dune dynamics in some way. Within a linear dune environment there are four discernible modes of aeolian transport: creep, saltation, suspension and the avalanching of sands on an active slip face. It would be of value to link these dynamic processes with grain-size measures, yet any attempt to discern process from grain size is fraught with difficulty, for it requires a high degree of inference, and there is always a danger that an argument will become circular. It is, none the less, possible to make some preliminary observations about the relationship between geomorphic process and sedimentary characteristics.

For the majority of samples, the cumulative grain size–frequency plots give more or less straight lines, indicating that the sand at each point comes from one population rather than from the mixture of two or more populations; this is confirmed by skewness values close to 0.0 and kurtosis values close to 1.0. The supposition might therefore be that these distributions, composed largely of sand within the size range from +2.0 phi to +2.5 phi (0.25 to 0.18 mm), indicate material which is predominantly moved by one process, presumably saltation. The apparent mixing of two populations at the base of the dune slopes, illustrated by Fig. 6, could then be the result of a combination of saltation and traction loads. The coarser fraction would represent material moved by creep along the interdune corridors and largely unable to ascend the dune slopes. This confinement would also explain the increased fining of sands towards the crest; a progressively smaller proportion of these coarse grains would be transported to the crest (Lancaster 1981).

Within this pattern of gradual fining towards the crest, some variation exists. There is some statistical similarly between dune base sand and sand at the base of the slip face, represented by samples from point O. As Bagnold (1941) noted, in an avalanching slip face, coarser grains tend

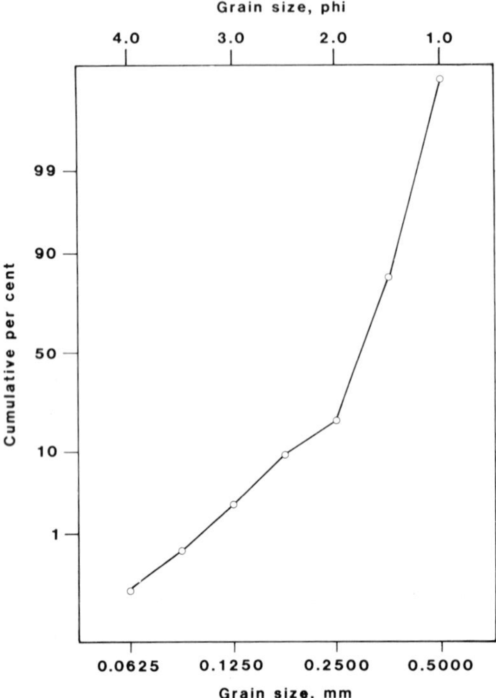

FIG. 6. Cumulative frequency grain size curves for a dune base sample. A probability scale is used for the ordinate.

to be pushed preferentially to the surface and thence to the base of the slope. Conversely, residual slip-face sand tends to be finer and skewed, the skewness representing a truncation of the coarse fraction.

It is not universally the case that dune crests are finer than plinths. In Australia, for instance, an explanation based on the nature of the source sand has been forwarded by Folk (1971) to explain the much finer reg deposits of the Simpson Desert linear dunes (Table 1). His suggestion is that the source of material for the dunes is fine-grained alluvial deposits, and that the fraction most mobile in air (around +2.5 phi, 0.18 mm) has been removed to build the dunes. This is reflected in average values for mean grain size on the crests of 2.53 phi (0.17 mm), and finer residual reg deposits with a mean grain size of 2.85 phi (0.14 mm). This reference to source material might also explain the results of Lewis (1936) and Lancaster (1986) for the Kalahari, although these sands show only small differences between crest and corridor samples (Table 1).

According to Buller & McManus (1972), the sand fraction which can be moved by the wind lies in the range 2 to 3 phi (0.25 to 0.13 mm). This would seem, however, to be only the fraction moved by saltation for coarser grains can be moved by creep, and Friedman (1961) suggests an absolute upper limit for sands to be moved by wind of +1.49 phi (0.36 mm). In fact, Tsoar's (1978) figures do not confirm that it is exclusively the 2 to 3 phi (0.25 to 0.13 mm) fraction from which dunes can be built. He reports sand at the crest of a linear dune in the Negev Desert, Israel, with a mean grain size of 1.87 phi (0.27 mm), compared with 1.41 phi (0.38 mm) and 1.18 phi (0.44 mm) at the base. Cooke & Warren (1973) report that most sand-sea deposits lie in the range −1.0 phi to +4.0 phi (2.00 to 0.06 mm), and this tallies with Ahlbrandt's (1979) reported range from −0.68 phi to 3.40 phi (1.60 to 0.09 mm) for 506 dune-sand samples. The evidence would suggest that, as long as sand (or indeed clay peds in lunettes) of a size which can be transported by the wind is available, factors other than grain size control dune form. Wherever possible, the processes of sand transport will move sand of around 2.5 phi (0.18 mm) to build dunes; but, as the results of Tsoar's (1978) study show, dunes can be built even when the source sand is outside this range.

In his discussion of bedform hierarchies, Wilson (1972*a*) proposed a causal link between grain size, represented by the size of the 20th percentile, and dune geometry (dune height and wavelength). Lancaster (1982*a*) has supported this argument, presenting a strong relationship between the size of the 5th percentile in a grain-size distribution and the spacing of transverse and barchanoid dunes of the Skeleton Coast in northern Namibia, though he has not been able to find such a clear relationship for Namib linear dunes (Lancaster 1983*a*). Watson (1986) finds a strong linear relationship between height above the interdune corridor and median grain size on his study dune. But while this same general trend towards finer sand at the crest has been found in this current study, there is no *general* relationship between dune height and sand size (Figs 4 and 5): taller dunes do not have finer crests.

The range of values for grain size from linear dunes in different sand seas suggests that grain size does not control dune form. This is supported by the work of Wasson & Hyde (1983*a*, *b*) in Australia, who also fail to find good relationships between grain size and dune geometry. They suggest that dune form is better understood by referring to the volume of sand supplied and to the wind regime.

Even though grain size may not control dune form, grain-size distributions are a response to process, though the nature of the relationship remains to be explained. Besler (1983) has used

Friedman's diagram (Friedman 1961) and plotted mean grain size against standard deviation for 393 samples from the Namib, the Kalahari, the Sahara and the Rub al Khali deserts. Besler relabels Friedman's zone of overlap between aeolian and fluvial deposits as a zone of 'aeolian stability', by which she mean deposits which are stable in an aeolian environment. Yet both Vincent (1985) and Livingstone (1987) have shown that active aeolian sediments may lie in Friedman's zone of 'overlap', and both argue that there is no justification for Besler's division of aeolian deposits. Livingstone demonstrates that there is a progressive rather than discrete change of surface activity across the dune profile, and that there is no more validity in dividing the sand samples on the basis of activity than there is on the basis of morphology.

It is apparent that the grain size of the source sand does not control aeolian dune form except in that material within the range moveable by the wind must be available. Grain-size distributions certainly do respond to process as shown by the fact that patterns across the Namib linear dunes are not random. Indeed, it is possible to make inferences about the relationships between grain size and process. However, it is not possible to divide the Namib dunes into discrete zones that carry connotations for grain size and process or grain size and morphology as Bagnold, Besler and Lancaster have done.

Variation of grain-size parameters with time

Given that grain-size parameters reflect process, even if not in the manner envisaged by some workers, it might be expected that there would be a temporal pattern in sympathy with changes in process. The seasonal, multi-directional wind regime found in the northern part of the Namib Sand Sea controls dune dynamics. Broadly speaking, in summer, there is erosion on the W flank of the dunes and deposition on the E flank, and in winter, the converse occurs (Livingstone 1985). Thus the slip face which develops in the lee of the crest is on the E side in summer and on the W side in winter. Consequently, we would expect to be able to recognize seasonal variations in surface grain size in response to these changing patterns of wind.

In order to test this, samples were taken from the main study dune every month for a year. The results for any one sampling point show no obvious pattern of change related to time. However, inability to find patterns at individual sampling points may reflect the fact that sand is so mobile that it is adjusted to highly localized factors. It is possible to smooth out the effects of some of these micro-scale, localized factors by aggregating data. Figure 7 shows the month-by-month variation of the average of two parameters (mean grain size and skewness) for the 25 sample points on the study dune. Difficult though it is to invoke a cyclic trend from the study of only one cycle, Fig. 7 shows that the dune sand becomes finer and its size distribution less skewed in the winter months of easterly winds. This pattern can be explained by the seasonal movement of finer sand from the E flank onto the crest and upper W slopes. In other words, the crest zone has a finer grained source area at this time of year. There is no seasonal trend in values of the standard deviation. However, a seasonal pattern of skewness values emerges. Negative (finer-grained) skewness in aeolian sediments is often associated with slip faces, and apart from a single value of −0.08 recorded at sample point H in January, the other 10 negative (fine) skews recorded on the W slope of the study dune all occurred in the winter months, five of these in September 1981 after particularly strong easterly winds.

It has been possible to demonstrate for the first time that, on one particular linear dune in the northern Namib Sand Sea, cyclical changes in the wind regime produce cyclical changes in

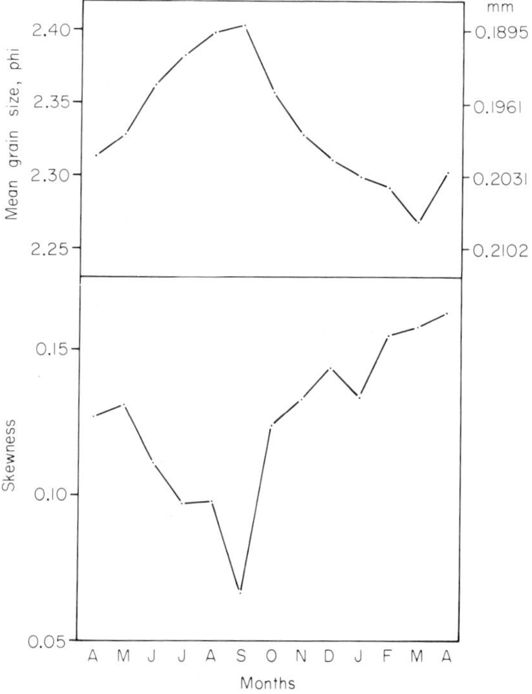

FIG. 7. Month-by-month change of mean values of: *Top*—mean grain size; *Bottom*—skewness.

sedimentary characteristics. This supports the belief that grain size at the surface of a dune responds dynamically to process. The implication of this finding is that it may not be possible to compare samples taken at different stages in the annual sedimentary cycle. It further implies that samples taken over long periods of time as part of studies of grain size over extensive geographical provinces might not be strictly comparable.

Conclusion

A number of points should be made factors that may affect grain-size parameters before moving to the specific conclusions of this study. Firstly, the sampling of more than a single lamina at the sand surface may well involve inclusion of a number of sedimentary events, although Bagnold believes that 'differences between such layers are usually of a very minor nature' (Bagnold 1941, p. 119). Secondly, in a zone of the dune where the surface is subject to erosion rather than deposition, the sampling of the surface may involve mixing the currently mobile load and the deposits of past events: a surface sand sample is not always the product of contemporary processes. Thirdly, micro-topographical features may induce very small changes in a sediment as mobile as dune sand.

This intensive study of grain-size variation on a single complex linear dune and its near neighbours in the Namib Desert has shown that a discernible pattern across the dune does exist, and that changes in the size distribution are gradual, not discrete. In general, crest samples are finer, better sorted and less skewed than samples from the dune plinths. There is also evidence that measurable differences occur between dunes in the same area, and that while they may not diverge far from regional averages, they may be significant. It is therefore important that studies reporting grain-size parameters should be more specific about the location of sample points, both within the sand sea and on each individual dune.

The fact that grain size is not random across the width of the dune indicates a sympathetic response to variations in process across the dune. However, grain size does not seem to control dune form in the way that some workers have envisaged.

Finally, there is some indication that grain size has a seasonal regime. As a whole, the dune sand becomes finer under the E winds of winter while the size distribution becomes negatively (fine) skewed. There is sufficient evidence to suggest a cyclical change in grain size.

ACKNOWLEDGMENTS: The funding for the project came from a NERC Research Studentship held at the School of Geography, University of Oxford. Permission to work in the Namib/Naukluft National Park was granted by the Department of Nature Conservation, Namibia, and facilities at the Namib Desert Research Station were made available through the Desert Ecological Research Unit. I am indebted to Jane Sinclair and Rose Whittington who helped with the laborious tasks of sample collection and analysis.

References

AHLBRANDT, T. S. 1979. Textural parameters of eolian deposits. *U.S. Geological Survey Professional Paper* **1052**, 21–51.

BAGNOLD, R. A. 1941. *The Physics of Blown Sand and Desert Dunes*. Chapman & Hall, London.

BESLER, H. 1980. Die Dunen—Namib: Entstehung und Dynamik eines Ergs. *Stuttgarter Geographische Studien* **96**.

—— 1983. The response diagram: distinction between aeolian mobility and stability of sands and aeolian residuals by grain size parameters. *Zeitschrift für Geomorphologie Supplementband* **45**, 287–301.

BULLER, A. T. & MCMANUS, J. 1972. Simple metric sedimentary statistics used to recognise different environments. *Sedimentology* **18**, 1–21.

COOKE, R. U. & WARREN, A. 1973. *Geomorphology in Deserts*. Batsford, London.

FOLK, R. L. 1971. Longitudinal dunes of the northwestern edge of the Simpson Desert, Northern Territory, Australia. 1. Geomorphology and grain size relationships. *Sedimentology* **16**, 5–54.

—— & WARD, W. C. 1957. Brazos River bar: a study in the significance of grain size parameters. *Journal of sedimentary Petrology* **27**, 3–26.

FRIEDMAN, G. M. 1961. Distinction between dune, beach and river sands from textural characteristics. *Journal of sedimentary Petrology* **31**, 514–529.

GOUDIE, A. S. 1970. Notes on some major dune types in southern Africa. *South African Geographical Journal* **52**, 93–101.

——, ALLCHIN, B. & HEGDE, K. T. M. 1973. The former extensions of the Great Indian Sand Desert. *Geographical Journal* **139**, 243–257.

LANCASTER, J., LANCASTER, N. & SEELY, M. K. 1984. Climate of the central Namib Desert. *Madoqua* **14**, 5–61.

LANCASTER, N. 1981. Grain size characteristics of Namib Desert linear dunes. *Sedimentology* **28**, 115–122.

—— 1982*a*. Dunes on the Skeleton Coast, Namibia (South West Africa): Geomorphology and grain size relationships. *Earth Surface Processes and Landforms* **7**, 575–587.

—— 1982*b*. Linear dunes. *Progress in Physical Geography* **6**, 475–504.

—— 1982*c*. Spatial variations in linear dune morphology and sediments in the Namib sand sea. *Palaeoecology of Africa* **15**, 173–182.

—— 1983*a*. Controls of dune morphology in the Namib sand sea. *In:* BROOKFIELD, M. E. & AHLBRANDT, T. S. (eds) *Eolian sediments and processes*. Developments in Sedimentology **38**, Elsevier, Amsterdam, 261–289.

—— 1983*b*. Linear dunes of the Namib sand sea. *Zeitschrift für Geomorphologie Supplementband* **45**, 27–49.

—— 1986. Grain-size characteristics of linear dunes in the southwestern Kalahari. *Journal of sedimentary Petrology* **56**, 395–400.

—— & OLLIER, C. D. 1983. Sources of sand for the Namib sand sea. *Zeitschrift für Geomorphologie Supplementband* **45**, 71–83.

LEWIS, A. D. 1936. Sand dunes of the Kahalari within the borders of the Union. *South African Geographical Journal* **19**, 22–32.

LIVINGSTONE, I. 1985. The dynamics of sand transport on a Namib linear dune. *Unpublished DPhil thesis, University of Oxford.*

—— 1987. Using the response diagram to recognise zones of aeolian activity: a note on evidence from a Namib dune. *Journal of Arid Environments*, in press.

MCKEE, E. D. (ed.) 1979. A study of global sand seas. *United States Geological Survey Professional Paper* **1052**.

ROGERS, J. 1977. Sedimentation on the continental margin off the Orange River and the Namib Desert. *Joint Geological Survey/University of Cape Town Marine Geoscience Group Bulletin* **7**.

RUBIN, D. M. & HUNTER, R. E. 1985. Why deposits of longitudinal dunes are rarely recognized in the geologic record. *Sedimentology* **32**, 147–157.

TSOAR, H. 1978. The dynamics of longitudinal dunes. *Final Technical Report, European Research Office, United States Army, London, DA-ERO 76-G-072.*

VINCENT, P. 1985. Some Saudi Arabian dune sands: a note on the use of the response diagram. *Zeitschrift für Geomorphologie* **29**, 117–122.

WARD, J. D. 1984. Aspects of the Cenozoic geology in the Kuiseb valley, central Namib Desert. PhD thesis, University of Natal.

WARREN, A. 1971. Dunes in the Ténéré Desert. *Geographical Journal* **137**, 458–461.

——, COOKE, R., GOUDIE, A. & JONES, D. K. C. 1985. Geomorphology Report. *In: Royal Geographical Society Oman Wahiba Sands Project, Mapping Phase Report 1985/86*, 47–137.

WASSON, R. J. & HYDE, R. 1983*a*. Factors determining desert dune type. *Nature* **304**, 337–339.

—— & —— 1983*b*. A test of granulometric control of desert dune geometry. *Earth Surface Processes and Landforms* **8**, 301–312.

WATSON, A. 1986. Grain-size variations on a longitudinal dune and a barchan dune. *Sedimentary Geology* **46**, 49–66.

WILSON, I. G. 1972*a*. Aeolian bedforms—their development and origins. *Sedimentology* **19**, 173–210.

—— 1972*b*. Universal discontinuities in bedforms produced by the wind. *Journal of sedimentary Petrology* **42**, 667–669.

I. LIVINGSTONE, School of Environmental Studies, Gloucestershire College of Arts and Technology, Oxstalls Lane, Gloucester GL2 9HW, UK. *Present address:* Department of Geography, Coventry (Lanchester) Polytechnic, Priory Street, Coventry CV1 5EB, UK.

Discrimination of depositional environments using sedimentary characteristics in the Mega Kalahari, central southern Africa

D. S. G. Thomas

SUMMARY: Unconsolidated Kalahari sand covers 2.5 million km^2 of central southern Africa, the 'Mega Kalahari', extending far beyond the Kalahari Desert of today. Relic aeolian, lacustrine and fluvial landforms indicate that in the Quaternary the Kalahari sand has been deposited and reworked by a variety of processes. Large areas of the sand have no identifiable landform associations, and because of this, sedimentological studies are necessary for a better interpretation of the palaeoenvironmental history of the Mega Kalahari. To this end, almost 200 Kalahari sand samples, in and not in identifiable landform associations, were collected and analysed in order to: investigate sedimentological characteristics; determine whether different processes have imparted distinct sedimentological characteristics to the sand; and thereby attempt to determine the depositional environments of samples without prior knowledge of the landform associations. Three methods have been used: (1) standard sedimentological techniques of grain size and shape determination; (2) multivariate discriminant analyses, to determine whether depositional environments could be better distinguished by considering grain size and shape parameters in combination; and (3) scanning electron microscope analyses of samples which lie centrally in any group identified by statistical analyses. Results indicate that aeolian processes have dominated the environmental history of the Kalahari sand, with only limited modification by subsequent processes. Multivariate discriminant analysis achieved only limited success in differentiating process environments. SEM investigations best identified the subtle modification of sand characteristics by non-aeolian processes, because individual grain textures are more readily adjusted than overall sample fabrics, and may therefore be the best technique for elucidating palaeoenvironments from sediment characteristics.

Unconsolidated Kalahari sand covers an area of 2.5 million km^2 in central southern Africa (Cooke 1964), extending from humid tropical Zaire to semi-arid Botswana (Fig. 1). This area has been termed the 'Mega Kalahari' (*eg* Thomas 1986*a* and *b*) in order to distinguish it from the smaller area known today as the Kalahari Desert (*eg* Cooke 1985; Jones 1982). Uncertainty surrounds the age and origin of the Kalahari sand. Maufe (1930), for example, considered it to be of Tertiary age, whilst Bond (1948) preferred the Pleistocene as the time of original deposition. Poldervaart (1957) and Bond (1948), utilized heavy mineral studies to infer the source areas and direction of transport of the surface sands of southeastern Botswana and Zimbabwe, respectively. However, in a systematic study in Botswana, Baillieul (1975) suggested that *in situ* subsurface bedrock weathering and bioturbation had made significant contributions to the sand. Boocock & Van Stratten (1962) also noted that even the youngest Kalahari sand may have been derived from underlying bedrock.

Despite these doubts about the provenance of the Kalahari sand, it is clear that it has been reworked on a number of occasions in the Quaternary period. The evidence for this was first recognized by Grove (1969) in the presence of a range of landforms produced under various palaeoenvironmental conditions. The most extensive are systems of fossilized sand dunes (*eg* Lancaster 1981; Thomas 1984), the northernmost of which is now covered by tropical rain forest in Zaire and Angola (Goudie 1983). Palaeolake beds and strandlines (*eg* Grey & Cooke 1977; Shaw 1985) and fossil drainage lines (Boocock & Van Stratten 1962) indicate more humid Quaternary conditions while pans, sometimes associated with lunette dunes (*eg* Lancaster 1978) testify to fluctuating hydrological conditions.

Palaeoenvironmental research in the Mega Kalahari has largely been concerned with evidence derived from landforms, although some investigations do include sediment descriptions (*eg* Lancaster 1978; Cooke 1980). Pettijohn *et al.* (1972) noted from the literature that over 80% of all sedimentological studies had been conducted to determine environments of deposition, however there has been limited application of grain size and shape studies to the sediments of the Mega Kalahari. The purpose of this paper is thus threefold: (1) to report the sedimentological characteristics of the Kalahari sand where found in known landform associations; (2) to see whether such sand has distinct characteristics imparted by the processes responsible for the

From Frostick, L. & Reid, I. (eds), 1987, *Desert Sediments: Ancient and Modern*, Geological Society Special Publication No. 35, pp. 293–306.

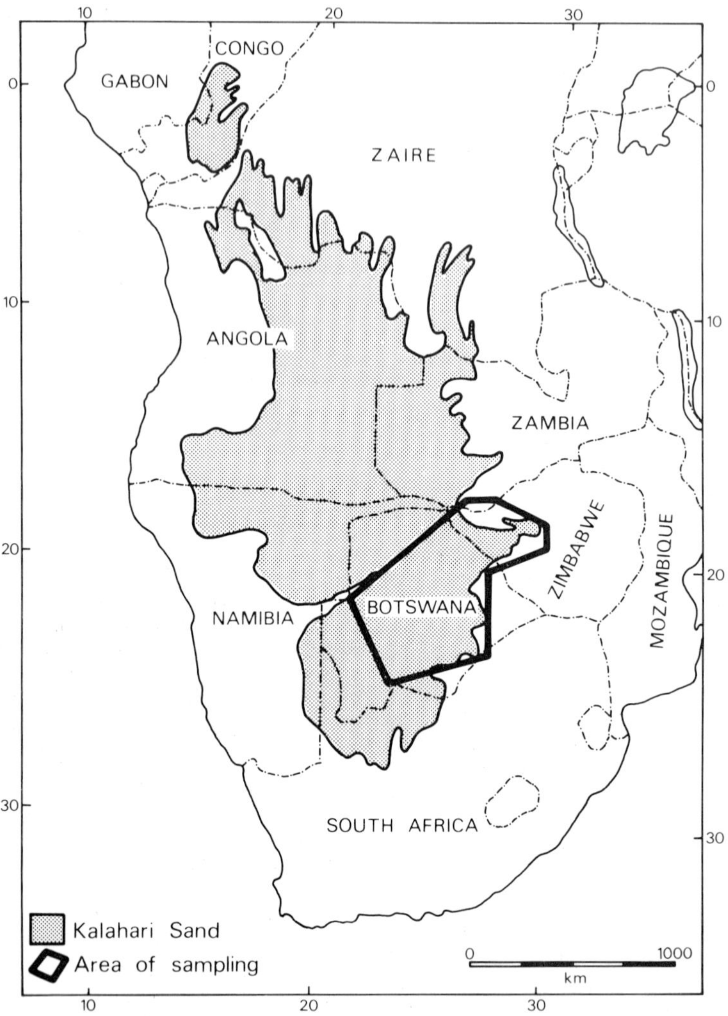

FIG. 1. Distribution of Kalahari sand in central southern Africa (after Cooke 1964). The area within which sampling was undertaken for this study is also shown.

formation of the relevant landforms; and (3) to examine whether palaeoenvironmental information may be gained from the sedimentological characteristics of the sand alone.

Sampling framework

The term 'Kalahari sand' is not well defined in the literature. It is used in a broad sense in this paper to include sediments which form the unconsolidated surface deposits in the Mega Kalahari, regardless of association with specific landforms. Various constraints limited sampling to only part of the Mega Kalahari (Fig. 1); nevertheless, some 250 000 km^2 were included in the study area. The full sediment sampling programme involved the collection of 254 samples each weighing approximately 300 g and taken using a 12 cm diameter auger from a depth of 30 cm to ensure consistency of sampling and to avoid the potentially disturbed surface layer. From the total sample collection, 198 were utilized in this study. Each was classified as coming from one of seven landform categories: relict linear dune ridge crest ($n=46$); interdune ridge trough ($n=25$); pan floor ($n=21$); pan surround ($n=12$); palaeolake floor ($n=18$); palaeodrainage floor ($n=21$) and no landform association ($n=55$). The first six represent the range of environments in which the Kalahari sand was deposited in the Quaternary period (*eg* Jones 1982; Thomas 1987).

The pans included in the study were all located in western Zimbabwe and northeastern Botswana. Unlike those in the southwestern Mega Kalahari, they lacked evidence, especially fringing lunette dunes, which would confirm their deflational origin (Goudie & Thomas 1985), and probably result from the interaction of both wind and zoological activity (Flint & Bond 1968). They are nevertheless a very important landscape component in the western Mega Kalahari, especially within the depressions between relict dune ridges. A pan surround, as the name implies, is an area separated from a pan floor and the surrounding environment by distinct breaks of slope and, unlike the pan floor, is not normally inundated seasonally with standing water.

The final category of sample classification was used for samples from areas which were devoid of recognizable landforms. These came from the Livingstone area of Zambia, and from E of the Gwayi river in Zimbabwe, but also from other locations throughout the study area.

Sediment analyses

Standard laboratory techniques were used to determine particle size and shape characteristics and the heavy mineral content of the samples. Following treatment to remove any organic material and to disaggregate the sediment, particle sizes in the sand fraction were determined by sieving at 0.25 phi intervals. The grain size distribution of the silt and clay fractions were investigated using a CILAS 715 laser granulometer (Cornillault 1972), which combines rapid determination with accuracy during repeated measurements. Descriptive sample statistics, or textural parameters, were then calculated from the grain size data.

Following separation of the quartz and heavy mineral components of the sand-size fractions of samples by the gravity method (Carver 1971), particle shape characteristics were determined for the quartz fractions. Shape and heavy mineral analyses only involved 48 samples because of the time-consuming nature of the methods. These samples were selected to retain the balance of both the spatial and landform characteristics in the overall sample framework.

Results

Textural parameters

Textural parameters are sample statistics used to describe particle size distributions within sediments, and those calculated according to the equations of Folk & Ward (1957) have been widely employed. Although their underlying assumption of a log–normal grain size distribution in sediments has been questioned recently (Barndorff-Nielsen *et al.* 1982; Christiansen *et al.* 1984) and the log–hyperbolic function favoured (Flenley *et al.* this vol.), the Folk & Ward (1957) parameters of mean grain size, sorting (standard deviation), skewness and kurtosis have been employed in this study. This may be justified by their common usage (*eg* Folk 1966; Besler 1983), and by the fact that Wyrwoll & Smyth (1985) found no apparent advantage in using the parameters of the log–hyperbolic as opposed to log–normal distributions in their study of dune sands.

Some studies utilizing Folk & Ward parameters have only investigated the grain size distribution of the sand fraction of a sample in detail. Friedman (1967), for example, treated the combined silt and clay fraction as one size class, characterizing it as either 4.25 or 6.00 phi, largely because of the added difficulties of measuring fine particles in the tail of the distribution. Yet Friedman (1967, 1979) has also noted that it is largely additions and subtractions to the fines which impart environmental significance to grain size distributions. The use of the laser granulometer allowed the particle size distribution in the fine fractions of samples in this study to be fully investigated and incorporated in the calculation of the textural parameters.

Table 1 shows the average values of grain size characteristics for samples grouped according to landform classification. General characteristics of the Kalahari sand emerge, such as the fact that the mean falls in the fine sand fraction, and that there is positive skewness and bimodality. These have been identified before and have been attributed to aeolian activity (Baillieul 1975; Cooke 1980; Thomas 1985). But such generalizations mask considerable variation between individual samples, and are of little value in attempting to determine depositional environments on sedimentological grounds. Alternatively, bivariate plots of different combinations of grain size parameters have been advocated by a number of authors as a method suitable to distinguish between modern sediments deposited by different processes. This method is based on the assumption that different processes of transportation result in variations in grain size distribution. These are reflected in the statistical parameters which, when plotted as scattergrams, produce bivariate clusters of samples affected by the same process, and separation of those samples

TABLE 1. *Graphical size distribution moments and percentage sand, silt and clay for Mega Kalahari sand samples from landform elements*

Landform element	*n*	sand particle grade %					graphical moment			
		coarse	med	fine	silt	clay	mean ϕ	sort. ϕ	skew.	kurt.
ridge crest	46	7.87	35.78	50.78	4.59	1.04	2.13	1.01	0.16	1.18
trough	25	9.81	30.64	45.08	11.42	3.52	2.43	1.31	0.15	1.53
pan	21	10.79	27.07	33.94	22.36	6.42	3.09	2.49	0.51	1.12
pan surround	12	10.61	26.31	43.60	10.32	5.59	2.88	2.07	0.41	1.76
palaeolake	18	13.19	25.45	32.59	19.60	10.40	3.32	2.46	0.48	1.16
fossil channel	21	7.60	30.90	43.63	14.71	2.24	2.58	1.72	0.44	1.40
unclassified	55	11.16	34.17	44.53	7.14	1.96	2.20	1.32	0.27	1.38

influenced by different processes. Figures 2 and 3 show plots of some combinations of parameters which have been attributed an ability to distinguish sands from different process environments. It is clear that the Kalahari sand displays considerable variation in the range of values of each parameter, and that such variation occurs both within and between landform categories.

Besler (1983) termed the plot of mean grain size versus sorting the 'response diagram' and felt sufficiently confident about its powers of discrimination to draw clear boundaries between plot clusters representing different desert process environments. Figure 2*a* shows that whilst there is some separation between samples from wind- and water-worked environments in the Mega Kalahari, considerable overlap also occurs. Other bivariate plots of Folk & Ward type parameters have been used successfully in the literature (*eg* Friedman 1961; Martins 1965; Moiola & Weiser 1968; Hails & Hoyt 1969). The percentage of silt and clay (<62 μm) was also successfully incorporated in bivariate plots by Friedman (1979), especially where skewness was involved.

In general, bivariate plots achieve some discrimination use in this investigation (Figs 2 and 3) but do not achieve a good and clear distinction of different environments, as has also been found by Schlee *et al.* (1964), Ahlbrandt (1979) and Vincent (1985) in other studies. This is because of the wide range of parameter values within each landform class. Ahlbrandt (1979) suggested that in the case of aeolian sands, ineffectiveness occurred because of variations in the grain size characteristics of the original sand sources. It is therefore not surprising that such variability in

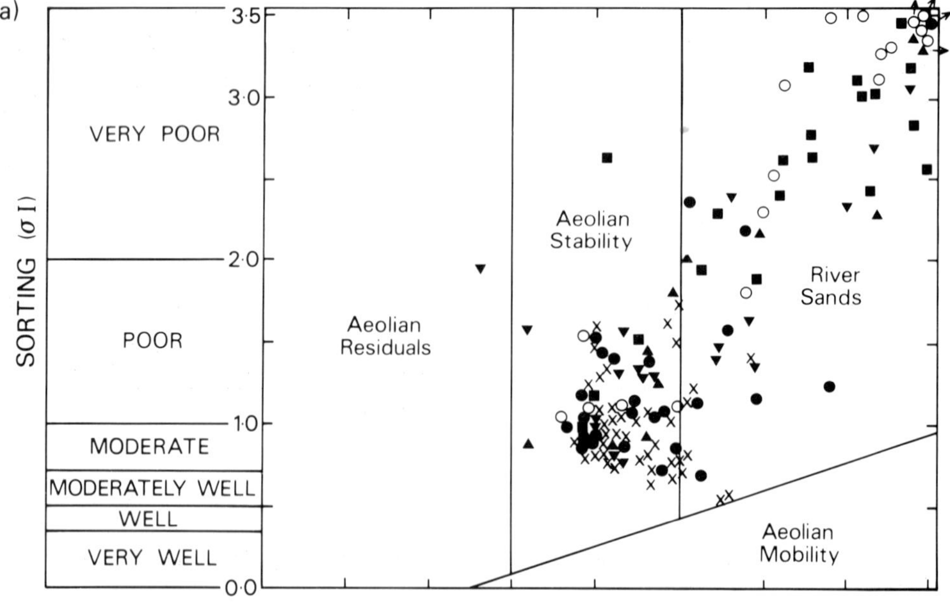

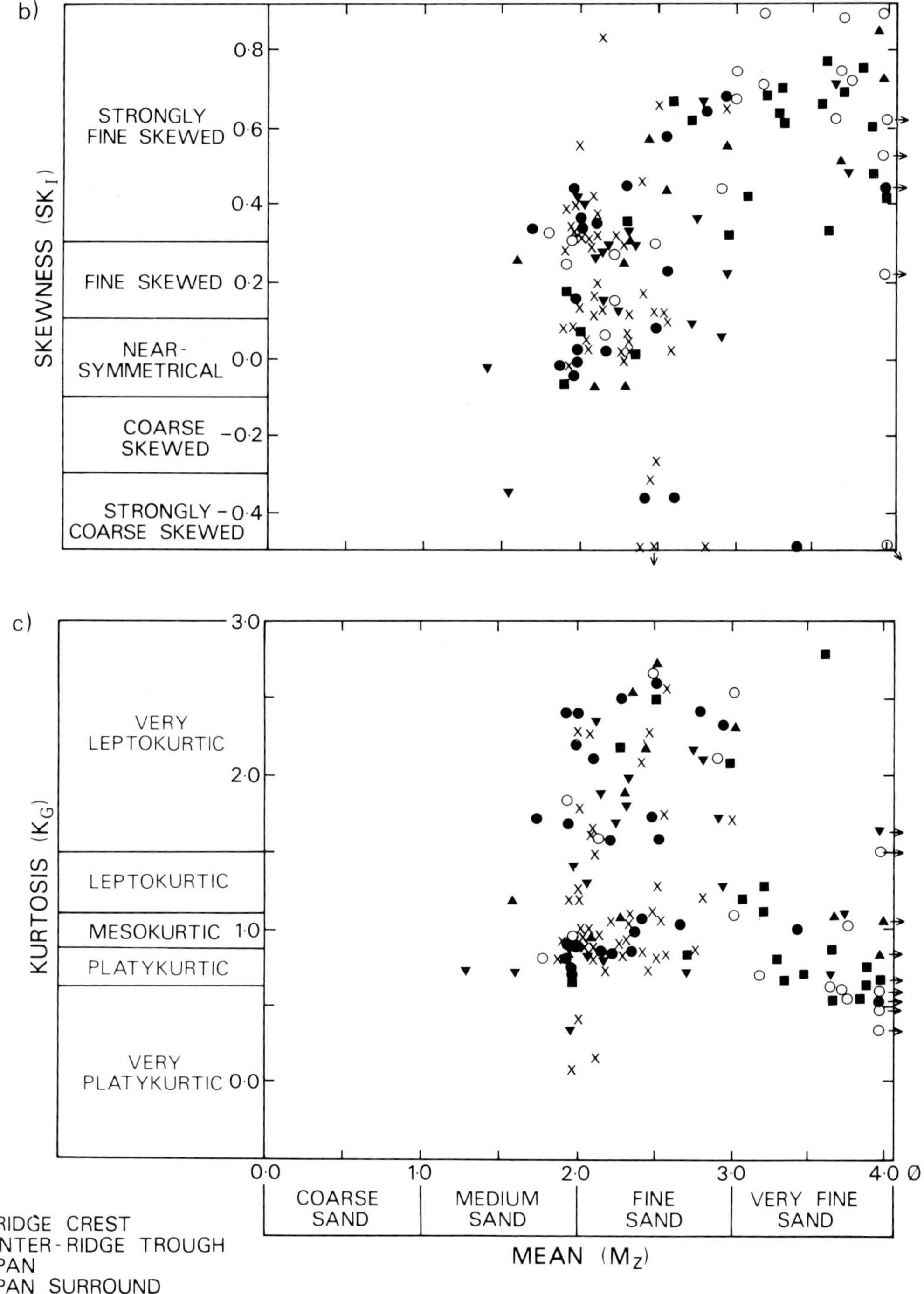

FIG. 2. Bivariate plots of (*a*) sorting, (*b*) skewness and (*c*) kurtosis against mean grain size, for samples classified according to their landform associations.

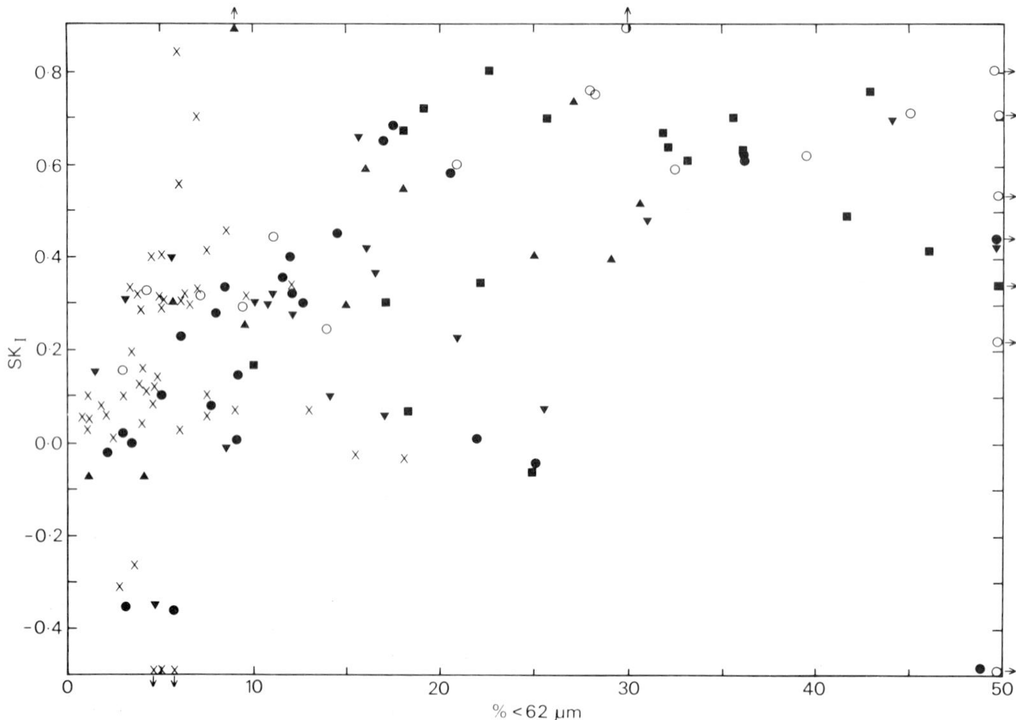

FIG. 3. Plot of skewness versus percentage finer than 62 μm (the silt and clay fraction). Symbols as in Fig. 2.

textural parameters occurs even within the sand from the relict dune forms of the Mega Kalahari, which are spatially extensive (Thomas 1984) and incorporate sand derived from distant provenances (Bond 1948; Poldervaart 1957) as well as from local sources (Baillieul 1975).

Shape

As Pettijohn *et al.* (1972) note, particle shape is very significant when investigating transport processes; but despite this, there have been relatively few studies which consider grain roundness and sphericity. Quartz grains in aeolian sands have often been attributed a high degree of roundness (Savory 1965; Ahlbrandt 1979). Goudie & Watson (1981) summarized the causes of the apparent roundness of desert sand grains as the superior abrasional potential of wind, aeolian shape sorting, chemical solution, and the roundness inherited from parent sediments. Variations in grain roundness characteristics between samples may therefore yield information imparted by different depositional processes in the Mega Kalahari.

Forty-eight samples were analysed, with quartz grains examined at 0.5 phi intervals in the sand fraction. One hundred grains were randomly chosen from each interval and, using a binocular microscope, were allocated to the appropriate roundness grade of Powers (1953). After examination, mean rho values (Folk 1966) were calculated for each size fraction and for each sample as a whole. Sphericity was similarly investigated used Rittenhouse's (1943) visual comparison chart and Waddell's (1933) numerical values.

Regardless of landform association, roundness increases with increasing grain size (Fig. 4*a*) but the degree of rounding is less than implied elsewhere in the literature (*eg* Savory 1965; Lockett 1979). Goudie & Watson (1981) and Folk (1978) noted similar trends in sands from other deserts, but it is also the case in the samples under consideration here that there is a reduction in roundness in the two coarsest size grades. Using the Wilcoxon matched-pairs signed ranks test, the data were analysed to test whether statistically significant differences in roundness existed between the samples from the different landform associations, but none emerged. Goudie & Watson (1981) found that within the Namib Desert, roundness did not vary across linear dune profiles. Thus from the current study and those conducted previously, grain roundness does not appear to be affected by sorting processes during aeolian transportation, or by other processes during reworking of the sand in the Mega Kalahari.

Figure 4*b* shows that, as with roundness, sphericity increases with grain size, but again no statistically significant distinction between sands from different landform associations exists. With reference to the particle form diagram of Dobkins & Folk (1970), it is probably most appropriate to describe sphericity values in terms of particle compactness and bladedness. None of the Kalahari sand samples, whatever the grain size, may be described as highly compact, but most mean sample values are compact (Fig. 4*b*).

Mean sample sphericity values occupy a narrow range of values. The degree of particle sphericity is important in the selective transportation of particles (Krumbein 1941; Mattox 1955; Morris 1957; Willetts *et al.* 1982). It may be especially important in the case of the larger grains mobilized by the wind as the creep load, since more compact grains are most likely to be mobilized. It may tentatively be suggested, therefore, that the degree of compactness of the Kalahari sand grains is an indication of the historical significance of sorting by wind transport. This is independent of the degree of rounding of the grains, which is a facet of edge abrasion (Barrett 1980) and is therefore an indication only of grain modification *during* transportation. The mutual independance of particle roundness and sphericity is attested to by the lack of a statistically significant correlation between the two shape parameters.

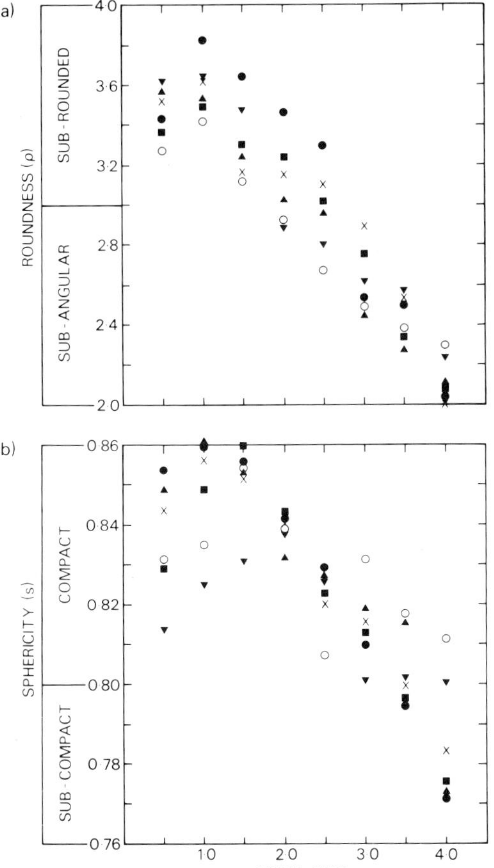

FIG. 4. Variations in (*a*) mean particle roundness and (*b*) mean particle sphericity with grain size, according to landform classification. Symbols as in Fig. 2.

Heavy minerals

Heavy mineral analyses of the samples were undertaken following standard procedures. The heavy mineral assemblages in sands can provide useful information about the character of the source rocks (Rubey 1933), and have also been used to differentiate sands deposited in different environments (*eg* Hand 1967). The main purpose of the heavy mineral studies of the Kalahari sand samples was to attempt to determine sub-regional variations in provenance following Bond (1948), Poldervaart (1957) and Baillieul (1975), and is therefore beyond the scope of this paper. Nevertheless, following the discriminatory powers which Hand (1967) and others found heavy minerals in sand to have, the percentages of zircon and staurolite (significant heavy minerals in the Kalahari sand) and total percentage of heavy minerals in each sample were included in the multivariate discriminant analyses.

Discriminant analysis

Discriminant analysis has been successfully used to differentiate modern sand bodies (Moiola *et al.* 1973, 1974; Moiola & Spencer 1979; Patro & Sahu 1977). It may therefore be an appropriate method of differentiating sands deposited in varied palaeoenvironments on the basis of their textural and shape characteristics (Pettijohn *et al.* 1972).

Multiple discriminant analysis is a statistical technique used to distinguish between more than two populations; the number of discriminating functions which may be produced is $g-1$, where g is the number of groups under consideration, or v, equal to the number of variables. The first function which is derived is the one which describes the maximum variance between groups and the minimum variance within groups. Accounts of the mathematical working of the technique are available in Klecka (1975). Various methods are available for the assessment of the individual contribution of each variable to the discriminating functions. Rao's method was the one which was selected for use in the analyses which follow. This method produces the greatest possible separation between groups of variables by choosing only the variables which contribute the largest increase in Rao's parameter, which is simply a measure of distance. Variables are therefore selected for inclusion in the discriminatory functions in decreasing order of importance. By so doing, it is then possible to distinguish variables which are important in group differentiation.

An attraction of this technique for the present research problem is that it is possible to classify cases of unknown origin within an already established framework. This is achieved by producing a classification function for each group, and then assigning individual cases to the groups on the basis of comparison of the group's classification functions and the individual case's discriminant functions. This can be used not only with cases whose group memberships are unknown, but also with cases where group affinity is known. By doing the latter, an assessment of the effectiveness of the discrimination is achieved and, in turn, it is possible to evaluate the ability of the variables to effect discrimination.

There are three ways in which the significance of the discriminant functions may be tested statistically. During the production of discriminant functions, eigenvalues are computed. These measure the relative importance of the function, when one function's eigenvalue is expressed as a percentage of the total eigenvalues of all functions. Wilks' lambda tests the significance of the discriminating information which is not accounted for in a function and all preceding functions. The larger lambda is, the less is the discriminating information remaining. Because it is possible to convert lambda into a chi-squared test, the statistical significance of functions can be determined. Finally, canonical correlations test the derived functions against predicted group memberships. The canonical correlation squared is the proportion of the variance in the discriminant function explained by the groups. Clearly, when the functions and groups have a high degree of correlation, the explanation provided by the functions is valid.

Multiple discriminant analysis (Nie *et al.* 1975) was used in this study to attempt to distinguish between the sediments sampled from different landform associations by considering variables in unison; it was also used to classify samples according to landform associations.

Data sets

Two data sets were used. Data set 1 consisted of all the samples in the study grouped according to landform associations and those without such associations. The six variables used to provide the discriminatory information were the four moments of the grain size distribution (mean, sorting, skewness and kurtosis), plus percentage silt and clay.

Data set 2 contained only the 48 samples which had been subject to shape and heavy mineral investigations. Therefore in addition to the six variables used with data set 1, five others were available for inclusion: mean roundness (rho), mean sphericity, heavy mineral percentage, percentage zircon and percentage staurolite.

Results

Using data set 1, five discriminating functions were produced, accounting for all the variance in

TABLE 2. *Measures for determining (a) the significance of and (b) contribution of variables to discriminant functions, data set 1*

	Function 1	Function 2	Function 3	Function 4	Function 5
(a) *Significance*					
Eigenvalue	1.910	0.171	0.117	0.018	0.003
% variance	86.07	7.71	5.27	0.83	0.12
Canonical correlation	0.810	0.382	0.324	0.135	0.053
Wilks' lambda	0.749	0.877	0.979	0.997	—
Chi squared	0.023	0.153	0.715	0.593	—
(b) *Variable*					
Mean	−0.113	−0.577	0.917	1.068	−1.005
Sorting	0.792	0.068	0.193	−1.184	0.541
Kurtosis	−0.194	0.869	0.355	0.521	−0.407
% Silt & clay	0.544	0.570	−1.049	0.351	0.427

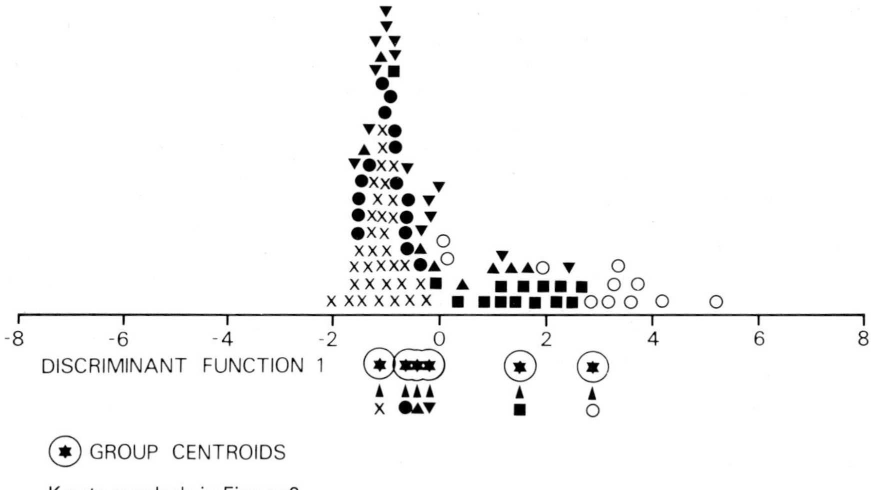

FIG. 5. Discriminant frequency plot of samples according to landform association for function 1, on the arbitrary scale −8 to +8.

the data set. Table 2a shows that most of the variance is accounted for by the first function (86.07%). Indeed, canonical correlation indicates that in this analysis only the first function is well correlated with the groups, and the chi-squared tests show that this function is the only one which is statistically significant. The variables of sorting, and silt and clay percentages contribute most to its discriminatory power (Table 2*b*).

Figure 5 demonstrates that even in the case of function 1, there is a considerable degree of group overlap and that group centroids are not well separated. The results of the classification stage of the analysis verify this, with only 49% of all samples being correctly classified into their appropriate landform association group. The greatest success rates are with the samples from dune ridges and not surprisingly, mudflats (77.1% and 72.7% correctly classified), which come, respectively, from the environments with the lowest and highest silt and clay contents. As under half of all samples were correctly classified, the variables included in data set 1 do not provide a reliable basis for attempting to establish the processes responsible for the deposition of Kalahari sand of unknown landform association except perhaps if classification was simplified and discrimination only on the basis of whether wind or water deposited.

When data set 2 was analysed using the technique, five discriminant functions were again produced, and again only function 1 was statistically significant, accounting for 84.79% of the variance between groups. The additional variables included did not enhance the discrimination between groups or contribute significantly to function 1.

Discussion

The lack of success in distinguishing depositional environments in the Mega Kalahari using bivariate plots of grain size distribution moments is unsurprising in view of the variable effectiveness of the method in separating modern sands. Discriminant analysis has, however, proved to be very effective (Moiola & Spencer 1979). Because of this it is necessary to explore the possible reasons for its relative ineffectiveness in *this* study. There are two major areas which need consideration: the nature of the variables used; and the depositional history of the Kalahari sand.

In studies of modern sands, the independent variables which formed the basis of the calculation of discriminant functions have been weight fractions in grain size classes or textural parameters (Moiola & Spencer 1979). The variables used in the present study correspond with those used previously. There is, therefore, no reason to suspect an incapacity of the technique to discriminate appropriately, though the validity of grain size parameters based on the assumption of a normal grain size distribution as used here has recently been subject to scrutiny.

Before the technique could be employed to classify samples of unknown origin into different groups, discriminant functions were established for samples from known landform elements. This assumes that samples taken from different landforms in the Mega Kalahari represent a sedimentary response to variable processes. The results of the discriminant analysis suggest that this assumption was incorrect insofar as the processes responsible for the different landforms either did not impart distinctive attributes on the Kalahari sand, or that any significant distinctions which did exist have been removed or blurred with the passage of time.

The extensive linear and other dunes testify to periods when aeolian processes have been effective in the Mega Kalahari during the Quaternary (Thomas 1984, 1986*a*). Provenance studies have also indicated the importance of aeolian processes (Baillieul 1975; Savory 1965; Binda & Hindred 1973). However, the origin of the sand is unclear: both Baillieul (1975) and Savory (1965) have suggested that in some areas rivers played an important role, whilst weathering of local sandstones has also made its contribution (Baillieul 1975). Despite this, Bond & Fernandes (1974), Binda & Hindred (1973) and Thomas (1985) have shown that the Kalahari sand displays sedimentological characteristics akin to those of modern wind-blown sands. It may therefore be argued that, regardless of the processes responsible for initial deposition, aeolian processes have played a significant role in the development of grain texture and shape, and that processes responsible for non-aeolian landforms have not caused significant modifications. Non-aeolian processes have merely re-worked the already wind-sorted sand. The action of water has made no mark that can be detected by discriminant analysis using coarse parameters such as mean grain size, percentage clay, *etc*.

SEM investigations

The discussion so far has suggested that the poor separation of samples from different landform associations in the Mega Kalahari is due to the overriding influence of aeolian processes. This has been checked by using another line of evidence. One sample from each of the landform elements that formed the basis of the sampling programme was inspected using a scanning electron microscope. The samples chosen fell closest to the centroids of groups identified by discriminate analysis function 1.

SEM investigations of sand-grain surface texture are now widely used to reconstruct the transportation history of sediments (Bull 1981). They are based upon the assumption that different combinations of grain-surface features are diagnostic of different processes (Krinsley & Doornkamp 1973).

Following usual practice, only quartz grains were investigated. The samples were pre-treated following standard procedures (Krinsley & Doornkamp 1973) and 20 or so grains of each selected from the 2.0–3.0 phi grain size range. The source of the samples was not identified during analysis in order to remove bias during observation.

Results

The characteristic grain-surface textures following aeolian transport are 'rolling topography', 'upturned plates' and 'dish-shaped concavities' (Krinsley & Doornkamp 1973), 'conchoidal fractures', 'star-cracks' and 'hertzian fractures' (Bull pers. comm.). These features were identified to varying degrees on grains in all six samples thus verifying the influence of aeolian processes regardless of present landform. Other textures were found to be superimposed upon this aeolian signature, a pattern also found on larger (0.0 phi) Kalahari sand grains by Bond & Fernandes (1974).

The textures identified on each sample are summarized below and in the micrographs of Fig. 6.

Dune ridge

This sample had a mixture of sub-rounded and sub-angular grains displaying surface textures produced by grain-upon-grain impacts. Small-scale mechanical impact pits were identified which coalesced to produce the upturned plates of Margolis & Krinsley (1971) and rolling topography (Krinsley & Doornkamp 1973).

Interdune trough

In terms of grain sphericity, grains of this sample were more compact than those from the ridge sample, and are perhaps good examples of 'millet seed' grains. The surface textures of grains in this sample were, in many respects, similar to the ridge sample, with a predominance of mechanical impact features including upturned plates.

Pan

As well as grains with concoidal fractures, dish-shaped concavities, smooth surfaces, and hertzian cracks—all indicative of aeolian transportation—there is also evidence of post-aeolian textural modification. This is in the form of a chemical surface modification to many grains

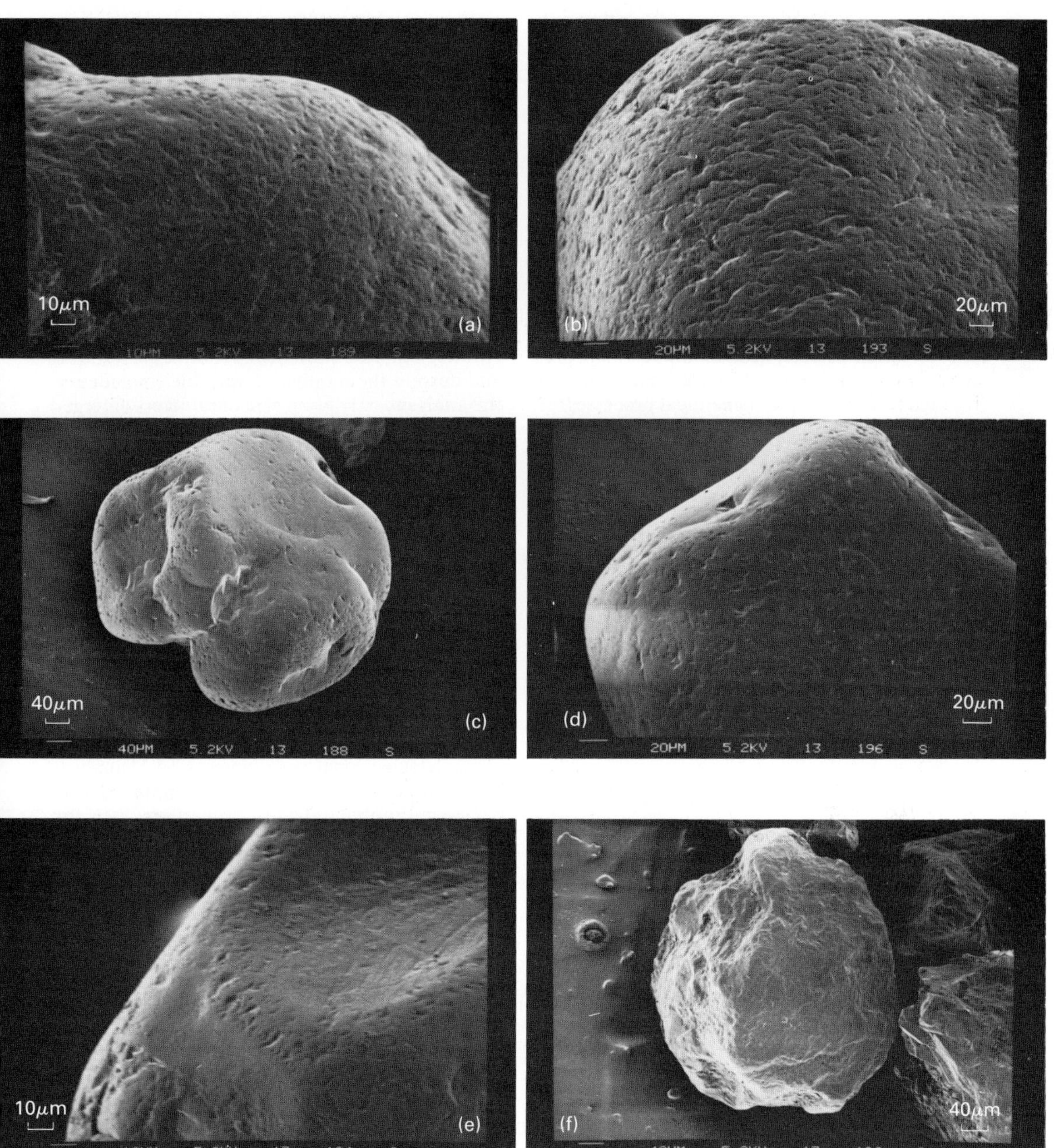

FIG. 6. (*a*)Detail of an aeolian quartz grain from the dune ridge crest sample, showing upturned plates. Scale bar 10 μm. (*b*) Grain from the inter-ridge trough sample. The whole grain surface is covered with upturned plates, resulting from the merging of small impact pits, each 20–40 μm long. The larger depression on the right is a dish-shaped concavity. Scale bar 20 μm. (*c*) Although from a pan, this grain displays typical aeolian textures, with rounded edges, dish-shaped concavities, impact pits (centre right) and hertzian fractures (centre). Scale bar 40 μm. (*d*) Detail of a grain from a pan surround: the hertzian fractures and edge rounding are due to aeolian action. The 'V' pits may be due to modification by chemical processes. Scale bar 20 μm. (*e*) Palaeolake sample, grain detail. The dish-shaped concavity has undergone sub-aqueous edge abrasion, giving rise to small, irregular impact pits. (*f*) Angular to sub-angular palaeochannel sand grains. Although upturned plates are identifiable on the central grain, the irregular surface is largely due to chemical weathering and to the precipitation of a varnish.

and is indicated by the presence of solutional 'V' pits, which Bull (1978) considered to be of sub-aqueous origin. The occurrence of many smoothed-grain surfaces, with a crypto-crystalline cover induced by solutional activity, could reflect changing grain-surface moisture experienced under fluctuating pan water-table levels.

Pan surround

This sample contained a mixture of sub-rounded and sub-angular grains whilst individual grains had much surface detail, dominated by hertzian fractures. Although these were undoubtedly produced by high-velocity aeolian impacts (Bull pers. comm.), the rounding of impact pit edges has resulted from subsequent chemical processes.

Palaeolake

The surface texture of some grains indicated aeolian activity. Others, though, showed modification by other processes, particularly chemical etching, and sub-aqueous edge abrasion. For example, Fig. 6*e* shows a grain where a dish-shaped concavity has been modified by edge abrasion as a result of water transport (*eg* see Goudie & Bull 1984), clearly showing that fluvial transport and lacustrine conditions have left their mark upon the aeolian sand grains.

Palaeochannel

The grains of this sample also showed textures resulting from a mixture of processes. Although aeolian textures were identified, there was a notable absence of high-energy impact features such as conchoidal fractures and star-cracks (Fig. 6*f*). Chemically weathered 'V'-shaped pits of sub-aqueous origin were also found, which bear witness to a fluvial environment.

In summary, the surface textures on individual grains from all of the landform elements indicate aeolian transport. However, these have been partially overprinted by non-aeolian processes in the pan, palaeolake and palaeochannel environments.

Conclusions

Not withstanding the doubts about Folk & Ward grain size distribution parameters, the homogeneous nature of Kalahari sand that is suggested by bivariate plots of these parameters is confirmed by discriminant analysis in which grain size, shape and mineralogical data are incorporated. Although this is a statistically complex technique, and despite the number of variables included in the analysis, only grain size parameters emerged as significant. The small separation of classified (*ie* discriminated) group centroids, and the large number of misclassifications of individual cases whose landform origin is known support the suggestion of homogeneity in the Kalahari sand. This homogeneity probably results from the overriding importance of aeolian activity in the environmental history of the Mega Kalahari.

The importance of aeolian processes in determining the sedimentological characteristics of the Kalahari sand is further demonstrated by SEM grain surface texture analysis. It has been possible, though, to distinguish secondary, non-aeolian textures on quartz grains from some present depositional environments. SEM analysis may, therefore, offer the best approach to palaeoenvironmental discrimination of the Kalahari sand, since individual grain surfaces appear to be more readily modified during reworking than is the overall grain size distribution or particle shape.

ACKNOWLEDGMENTS: The author would like to thank Dr Peter Bull for invaluable assistance in SEM analyses, and Liz Martin for the comments she made on a draft of the manuscript.

References

AHLBRANDT, T. S. 1979. Textural parameters of aeolian deposits. *In*: MCKEE, E. D. (ed.) *A Study of Global Sand Seas*. US Geological Survey Professional Paper **1052**, 21–53.

BAILLIEUL, T. A. 1975. A reconnaissance survey of the cover sands in the Republic of Botswana. *Journal of sedimentary Petrology* **45**(2), 494–503.

BARNDORFF-NIELSON, O., DALSGAARD, K., HALGREEN, C., KUHLMAN, H., MØLLER, J. T. & SCHON, G. 1982. Variations in particle size over a small dune. *Sedimentology* **29**, 53–65.

BARRETT, P. J. 1980. The shape of rock particles, a critical review. *Sedimentology* **27**, 291–303.

BESLER, H. 1983. The response diagram: distinction between aeolian mobility and stability of sands and aeolian residuals by grain size parameters. *Zeitschrift für geomorphologie Supplementbund* **45**, 287–302.

BINDA, P. L. & HINDRED, P. R. 1973. Bimodal grainsize distributions of some Kalahari type sand from Zambia. *Sedimentary Geology* **10**, 233–237.

BOND, G. 1948. The direction of origin of the Kalahari

sand of southern Rhodesia. *Geological Magazine* **85**, 305–313.

—— & FERNANDES, T. R. C. 1974. Scanning electron microscopy applied to quartz grains from Kalahari type sand. *Transactions of the Geological Society of South Africa* **77**, 191–199.

BOOCOCK, C & VAN STATEN, O. J. 1962. Notes on the geology and hydrogeology of the central Kalahari region, Bechuanaland Protectorate. *Transactions of the Geological Society of South Africa* **65**, 125–171.

BULL, P. A. 1978. A statistical approach to scanning electron microscope analysis of core sediments. *In*: WHALLEY, W. B. (ed.) *Scanning Electron Microscopy in the Study of Sediments*. Geo Abstracts, Norwich, 212–226.

—— 1981. Environmental reconstruction by electron microscopy. *Progress in Physical Geography* **5**, 368–397.

CARVER, R. E. (ed.) 1971. *Procedures in sedimentary petrology*. Wiley Interscience, New York, 653 pp.

CHRISTIANSEN, C., BLAESILD, P. & DALSGAARD, K. 1984. Reinterpreting 'segmented' grainsize curves. *Geological Magazine* **121**, 47–51.

COOKE, H. B. S. 1964. The Pleistocene environment in southern Africa. *In*: DAVIS, D. H. S. (ed.) *Ecological Studies in Southern Africa*. Junk, The Hague, 1–23.

COOKE, H. J. 1980. Landform evolution in the context of climatic change and neo-tectonism in the middle Kalahari of north central Botswana. *Institute of British Geographers, Transactions* **5**, 80–99.

—— 1985. The Kalahari today: a case of conflict over resource use. *Geographical Journal* **151**, 75–85.

CORNILLAULT, J. 1972. Particle size analyser. *Applied Optics* **11**, 265–268.

DOBKINS, J. E., JR & FOLK, R. L. 1970. Shape development on Tahiti-Nui. *Journal of sedimentary Petrology* **40**, 1167–1203.

FLINT, R. F. & BOND, G. 1968. Pleistocene sand ridges and pans in western Rhodesia. *Bulletin of the Geological Society of America* **79**, 299–314.

FOLK, R. L. 1966. A review of grain-size parameters. *Sedimentology* **6**, 73–93.

—— 1978. Angularity and silica coatings of Simpson Desert sand grains, Northern Territory, Australia. *Journal of sedimentary Petrology* **52**, 93–101.

—— & WARD, W. C. 1957. Brazos River bar—a study in the significance of grain size parameters. *Journal of sedimentary Petrology* **27**, 3–27.

FREIDMAN, G. M. 1961. Distinction between dune, beach and river sands from their textural characteristics. *Journal of sedimentary Petrology* **31**, 514–529.

—— 1967. Dynamic processes and statistical parameters compared for size frequency distribution of beach and river sand. *Journal of sedimentary Petrology* **37**, 327–354.

—— 1979. Differences in size distributions of populations of particles among sands at various origins. *Sedimentology* **26**, 3–32.

GOUDIE, A. S. 1983. The arid earth. *In*: GARDNER, R. A. M. & SCOGING, H. (eds) *Mega Geomorphology*. Oxford University Press, 152–171.

—— & BULL, P. A. 1984. Slope process change and colluvium depositions in Swaziland: an SEM analysis. *Earth Surface Processes and Landforms* **9**, 289–299.

—— & THOMAS, D. S. G. 1985. Pans in southern Africa with particular reference to South Africa and Zimbabwe. *Zeitschrift für Geomorphologie* **NF29**, 1–19.

—— & WATSON, A. 1981. The shape of desert sand grains. *Journal of Arid Environments* **4**, 185–190.

GREY, D. R. C. & COOKE, H. J. 1977. Some problems in the Quaternary evolution of the landforms of northern Botswana. *Catena* **4**, 123–133.

GROVE, A. T. 1969. Landforms and climatic change in the Kalahari and Ngamiland. *Geographical Journal* **135**, 191–212.

HAILS, J. R. & HOYT, J. H. 1969. The significance and limitations of statistical parameters for distinguishing ancient and modern sedimentary environments of the lower Georgia coastal plain. *Journal of sedimentary Petrology* **39**, 559–580.

HAND, B. H. 1967. Differentiation of beach and dune sand, using settling velocities of light and heavy minerals. *Journal of sedimentary Petrology* **37**, 514–520.

JONES, C. R. 1982. The Kalahari of Southern Africa. *Straiae* **17**, 20–34.

KLECKA, W. R. 1975. Discriminant analysis. *In:* NIE, N. H., HULL, C. H., JENKINS, J. G., STEINBRENNER, K. & BRENT, D. H. (eds) *Statistical Package for the Social Sciences*. Second edn. McGraw Hill, New York, 434–467.

KRINSLEY, D. H. & DOORNKAMP, J. C. 1973. *Atlas of Quartz sand surface textures*. Cambridge University Press, 91 pp.

KRUMBEIN, W. C. 1941. Measurement and geological significance of shape and roundness on sedimentary particles. *Journal of sedimentary Petrology* **11**, 64–72.

LANCASTER, I. N. 1978. Composition and formation of southern Kalahari pan margin dunes. *Zeitschrift für Geomorphologie NF 22*, 148–169.

—— 1981. Palaeoenvironmental implications of fixed dune systems in southern Africa. *Palaeogeography, Palaeoclimatology, Palaeoecology*, **33**, 327–346.

LOCKETT, N. H. 1979. The geology of the country around Dett. *Rhodesia Geological Survey Bulletin* **85**, 198 pp.

MARGOLIS, S. V. & KRINSLEY, D. H. 1971. Submicroscopic frosting on aeolian and subaqueous quartz grains. *Geological Society of America Bulletin* **82**, 3395–3406.

MARTINS, L. R. 1965. Significance of skewness and kurtosis in environmental interpretation. *Journal of sedimentary Petrology* **35**, 768–340.

MATTOX, R. B. 1955. Eolian shape sorting. *Journal of sedimentary Petrology* **25**, 111–114.

MAUFE, H. B. 1930. Changes of climate in Southern Rhodesia during late geological times. *South Africa Geographical Journal* **13**, 12–16.

MOIOLA, R. J. & SPENCER, A. B. 1979. Differentiation of eolian deposits by discriminant analysis. *In:* MCKEE, E. D. (ed.) *A Study of Global Sand Seas*. US Geological Survey Professional Paper **1052**, 53–61.

—— & WEISER, D. 1968. Textural parameters: an evaluation. *Journal of sedimentary Petrology* **38**, 45–53.

——, SPENCER, A. B. & MCKEE, E. D. 1973. Linear discriminant analysis—a technique for analyzing sand bodies. *Geological Society of America Abstracts with Programs 1973 Annual Meeting* **5**, 741–742.

——, —— & WEISER, D. 1974. Differentiation of modern sand bodies by linear discriminant analysis. *Gulf Coast Association of Geological Societies Transactions* **24**, 321–326.

MORRIS, W. J. 1957. Effects of sphericity, roundness and velocity on traction transportation of sand grains. *Journal of sedimentary Petrology* **27**, 27–31.

NIE, N. H., HULL, C. H., JENKINS, J. G., STEINBRENNER, K. & BRENT, D. H. 1975. *Statistical Package for the Social Sciences*. Second edn. McGraw-Hill, New York, 674 pp.

PATRO, B. C. & SAHU, B. K. 1977. Discriminant analysis of sphericity and roundness data of clastic quartz grains in rivers, beaches and dunes. *Sedimentary Geology* **19**, 301–311.

PETTIJOHN, F. J., POTTER, P. E. & SIEVER, R. 1972. *Sand and Sandstone*. Springer-Verlag, New York, 618 pp.

POLDERVAART, A. 1957. Kalahari sands. *In:* CLARK, J. D. (ed.) *Proceedings of the 3rd Pan African Congress on Prehistory. Livingstone, 1955*, 106–114.

POWERS, M. C. 1953. A new roundness scale for sedimentary particles. *Journal of sedimentary Petrology* **23**, 117–119.

RITTENHOUSE, G. 1943. A visual method of estimating two-dimensional sphericity. *Journal of sedimentary Petrology* **13**, 79–81.

RUBEY, W. W. 1933. The size distribution of heavy minerals within a water-laid sediment. *Journal of sedimentary Petrology* **3**, 3–11.

SAVORY, B. M. 1965. Sand of Kalahari type in Sesheke District, northern Rhodesia. *In:* SNOWBALL, G. J. (ed.) *Science and medicine in Central Africa*. Pergamon Press, Oxford, 189–200.

SCHLEE, J., UCHUPI, E. & TRUMBILL, J. V. A. 1964. Statistical parameters of Cape Cod beach and eolian sands. US Geological Survey Professional Paper **501-D**, D118–D122.

SHAW, P. 1985. Late Quaternary landforms and environmental change in northeastern Botswana: the evidence of Lake Ngami and the Mababe Depression. *Institute of British Geographers Transactions* **10**, 333–346.

THOMAS, D. S. G. 1982. Evidence of Quaternary palaeoclimates in western Zimbabwe. Chapter 1. *In:* WILLIAMS, G. J. & WOOD, A. (eds) *Geographical Perspectives on Development in Southern Africa*. Univ. of N. Queensland Press (in press).

—— 1984. Ancient ergs of the former arid zones of Zimbabwe, Zambia and Angola. *Institute of British Geographers, Transactions* **9**, 75–88.

—— 1985. Evidence of aeolian processes in the Zimbabwean landscape. *Transactions Zimbabwe Scientific Association* **62**, 45–55.

—— 1986*a*. Ancient deserts revealed. *Geographical Magazine*, 11–15.

—— 1986*b*. Dune pattern statistics applied to the Kalahari Dune Desert, southern Africa. *Zeitschrift für Geomorphologie* **30**, 231–242.

VINCENT, P. J. 1985. Some Saudi Arabian dune sands: a note on the use of the response diagram. *Zeitschrift für Geomorphologie* **29**, 117–122.

WADDELL, H. 1933. Sphericity and roundness of rock particles. *Journal of Geology* **40**, 443–451.

WILLETTS, B. B., RICE, M. A. & SWAINE, S. E. 1982. Shape effects in aeolian grain transport. *Sedimentology* **29**, 409–417.

WYRWOLL, K.-H. & SMYTH, G. K. 1985. On using the log hyperbolic distribution to describe the textural characteristics of eolian sediments. *Journal of sedimentary Petrology* **55**, 471–478.

D. S. G. THOMAS, Department of Geography, University of Sheffield, Sheffield S10 2TN, UK.

CHEMICAL SEDIMENTS

Silica and calcium carbonate replacement of plant roots in tropical dune sands, SE India

D. A. Hendry

SUMMARY: The Quaternary coastal sediments of SE India are composed of raised marine deposits capped by aeolianite and red weathered dune sands (Teri). The red weathering associated with these sands is irregular in depth and unusual in the fast rate at which it has developed. Associated at depth with this red weathering is precipitated silica and calcium carbonate which replace plant roots and act as a cement. In places the plant root cells are prefectly replaced to form rhizoliths which occur in the soil profile with and without surrounding cement. Ultimately, the silica source is silicic acid (H_4SiO_4) that is a product of the *in situ* weathering of unstable minerals in the overlying dune sands during the last 20 000 years. The calcium carbonate is thought to have had a much more localized source. Plants and the litter they produce play at least two possible roles in the silica cycle. First, silica–organic complexing occurs as the Teri sands are leached by organic acids derived from decomposing litter. Secondly, the plant roots create a favourable micro-environment for silica and calcium carbonate precipitation and replacement by altering partial gas pressures and hence the pH of soil solutions through their respiratory activity.

This paper presents the preliminary findings of silica and calcium carbonate replacement (petrifaction) of plant roots in the Quaternary coastal sediments of southeastern India (Fig. 1). These deposits are particularly interesting due to the highly detailed replacement of organic matter and their widespread occurrence along the coast. The time-scale of the diagenesis involved is short, geologically speaking, and the source of both the silica and calcium carbonate is thought to be local (Gardner 1981*a*).

Silica has been recognized as one of the principal agents of surface diagenesis, not only because of its ready availability, but also because of its tendency to precipitate at low temperatures (Krauskopf 1956). Chemical weathering of silicate minerals is the primary source of dissolved silica in most Earth surface conditions.

It is now well established that silica exists in natural waters predominantly in true solution as monosilicic acid (H_4SiO_4) rather than as a colloid. Up to about pH 9, silicic acid is largely undissociated, but above pH 9.5, $H_3SiO_4^-$ becomes the major species (Summerfield 1983). Wilding & Drees (1971) summarize the solubilities of silica types and as a result of this early work, it is now agreed that continental and subsurface waters are under-saturated with respect to quartz, but over-saturated with respect to amorphous silica.

With the exception of work on silcretes, detailed studies of silica within the near-surface weathering environment are few. The 'fossilized' roots seen in SE India occur as direct organic matter replacement with only limited inter-particle cement. They do not, therefore, fulfill the definition of silcrete, where the arbitrary lower limit of silica content is 85% weight SiO_2 (Summerfield 1983). It is possible that the features seen in SE India represent an initial stage of silcrete formation, although previous studies on silcretes have not found replaced root structures.

Organic matter that has been 'fossilized' by silica is a common feature in the literature. Almost all the published work is concerned with the process of permineralization, whereby silica is precipitated within voids, preserving the cell structure of the original organic matter in a silica matrix (Leo & Barghoorn 1976). Permineralized plant matter, composed largely of quartz, is found in geological formations ranging in age from Palaeozoic to Tertiary. These features are frequently thought to be the result of fluctuating groundwaters highly charged with silica, often derived from the leaching of volcanogenic sediments (Knoll 1985; Schopf 1971; Stein 1982). Stein summarizes the mechanism by which silica in solution is thought to become 'fixed' as a precipitate in the voids of the organic matter.

Studies of silica diagenesis in deposits of Quaternary age are rare. One exception is the work of Durga Prasada Rao *et al.* (1982) in eastern India. The sediments are apparently of similar age and lithology to those in SE India and the authors have recorded amorphous silica replacement of low-magnesium calcite columnar concretions, preserving algal structures in the process. The mechanism invoked is one of an external carbonate source that has reprecipitated in inter-granular spaces of the sediment by replacement of finer materials. Formation of columnar concretions is favoured by initial high

From FROSTICK, L. & REID, I. (eds), 1987, *Desert Sediments: Ancient and Modern*, Geological Society Special Publication No. 35, pp. 309–319.

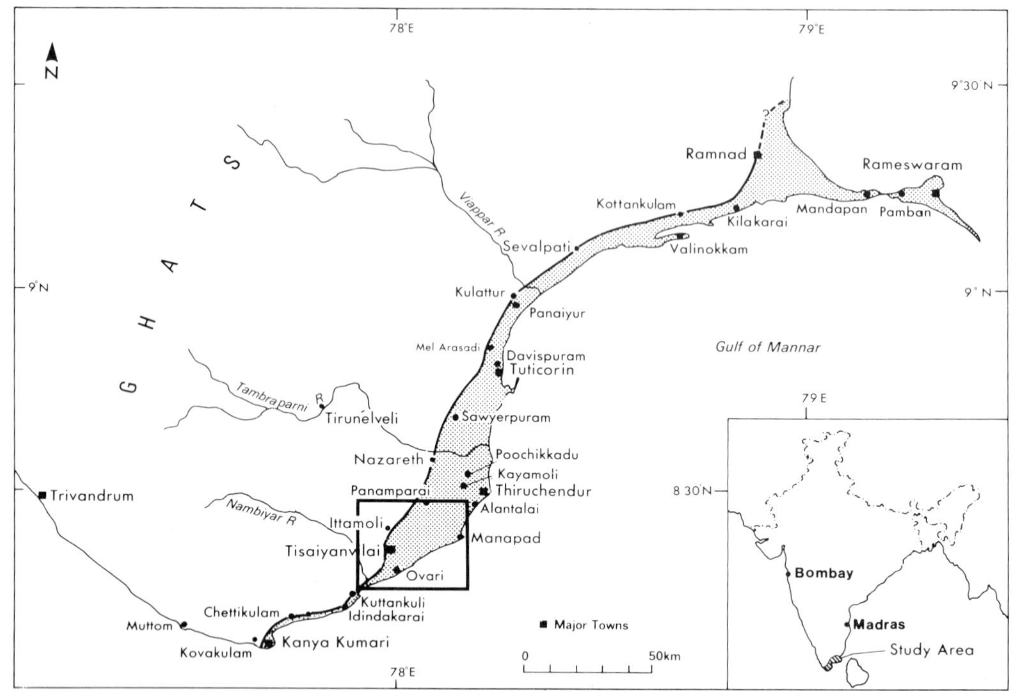

FIG. 1. Location of the Quaternary coastal deposits in SE India. Box inset indicates the location of the study area.

permeability of the sand deposits and intense evaporation in a freshwater environment during drier periods. The carbonate structures are then thought to have been replaced by amorphous silica, precipitated from saturated solutions existing in a protected environment associated with a post-Pleistocene rise in sea level.

In order to fully examine the process of silica replacement it is necessary to understand the release and movement of silica. Unfortunately, this is still not well understood. Experimental work (Morey *et al.* 1962) has shown that quartz has low solubility in water at pH 9 at 25°C. This solubility was seen to increase rapidly at higher pH values but, in nature, soil and weathering profiles usually have a pH value of less than 9 and most completely weathered modern soils are strongly acidic. Morey *et al.* (1962) reported that quartz continues to pass into solution indefinitely provided that the supply of water in weathered profiles result in silica concentrations in excess of 6 ppm as a result of weathering of more unstable alumino-silicates.

This fact does not explain field and laboratory evidence of weathered quartz sand grains. Features such as the rounding of protuberances and re-entrants within grain outlines, and the occurrence of skeletal forms have been noted, and several workers (Evans 1965; Carrozi 1967; Cleary & Conolly 1972) attribute them to dissolution, resulting from the presence of organic acids. It was found that the solubility of quartz was greatly enhanced by the presence of organic molecules. Cleary & Conolly (1972) found that the areas of greatest dissolution occurred within the root zone of the weathering environment, indicating that this zone produces organic complexes affecting dissolution. Laboratory studies (Evans 1965) took this a stage further by producing an artificial quartzite by suspending quartz sand in a variety of solutions of Na-ATP (sodium adenosine triphosphate).

Although organic complexing of silica appears to affect mobility, the mechanisms of the process are still not fully understood. For example, in the zone of greatest quartz dissolution it is common to find feldspars displaying various degrees of alteration, as well as a range of dissolution features on quartz. This suggests that there are much more complex chemical reactions occurring, such as those dependent on organic complexes, controlling micro-environments within the weathering profile. To further complicate the picture, it is now known that many metallic ions such as Al^{3+}, Fe^{2+}, Mg^{2+}, Ca^{2+}, Ag^{+}, Cu^{2+}, Pb^{4+} and Hg^{+}, when chemisorbed to silica surfaces, inhibit the dissolution of silica due to the formation of relatively insoluble silicate coatings (Krauskopf 1956; Lewin 1961; Jones & Handreck 1963).

The role of sesquioxides, and particularly aluminium, on silica dissolution has been widely studied (Okamoto *et al.* 1957; Iler 1973). The conclusions reached by these workers is that the sorption of iron and aluminium onto the surface of amorphous silica, opal, or quartz exhibiting a disrupted surface layer will decrease dissolution, whereas the sorption of monosilicic acid by sesquioxides may increase dissolution of amorphous silica on reactive uncoated surfaces.

Calcium carbonate associated with plant root 'fossilization' has also received much attention in the literature in varying amounts of detail (Glennie & Evamy 1968; Cohen 1982; Klappa 1979; Mount & Cohen 1984). One of the problems of this earlier work is the highly variable terminology. Klappa (1980) lists some of the previous terms and, in an attempt to reduce confusion, proposes a new terminology and classification scheme for the structures seen. Klappa has defined rhizoliths as 'organosedimentary structures resulting in the preservation of roots of higher plants, or remains thereof, in mineral matter'. This work is based on the abundant and characteristic features of Quaternary calcretes and aeolianites from coastal regions of the western Mediterranean. Field and laboratory observations led to the definition of five basic types of rhizoliths (Klappa 1980). Klappa's classification will be used in this paper and extended to lithologies present in SE India. All types of rhizolith identified by Klappa occur within the coastal sediments of SE India, but the most relevant to this paper is 'root petrifactions' which are defined as 'mineral impregnation or mineral replacement of organic matter which preserved anatomical features of roots partly or totally'.

Methods of study

All hand specimens and exposure descriptions were taken from river cliffs, quarries and recently excavated irrigation wells cut through or into the coastal sediments (Figs 2, 3, 4, 5*a* and *b*). Hand specimens were only taken from freshly exposed faces. The rhizoliths and associated sediments were examined in thin section after impregnation with resin.

Detailed observations of freshly broken rhizoliths were made using a scanning electron microscope (SEM). Elemental composition of the rhizoliths and variations within them were obtained from an SEM equipped with an energy-dispersive X-ray system (EDX). X-ray diffraction (Cu K_α) was used to determine crystallographic ordering within the silica. Carbonate composition was determined by staining techniques (Warnes 1962).

Regional geology and stratigraphy

Figure 1 depicts the location of the Quaternary coastal deposits in SE India. The outlined area indicates the location of the study area which is shown in greater detail in Fig. 3. These coastal deposits form a narrow strip that is continuous for a distance of approximately 350 km and includes a variety of sedimentary facies and weathering features (Fig. 2).

The red weathered sands of Fig. 2 (locally

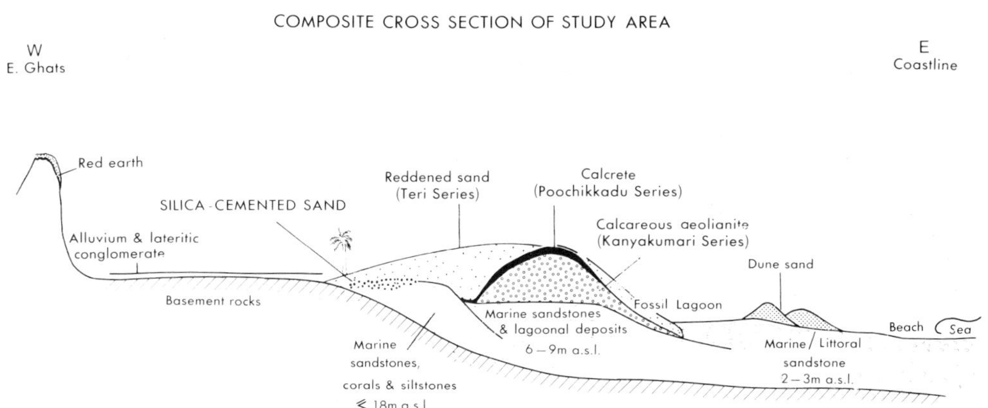

FIG. 2. Composite cross-section through the Quaternary coastal deposits. The location of the silica rhizoliths is indicated by 'SILICA CEMENTED SAND'. (Adapted from Gardner 1986.)

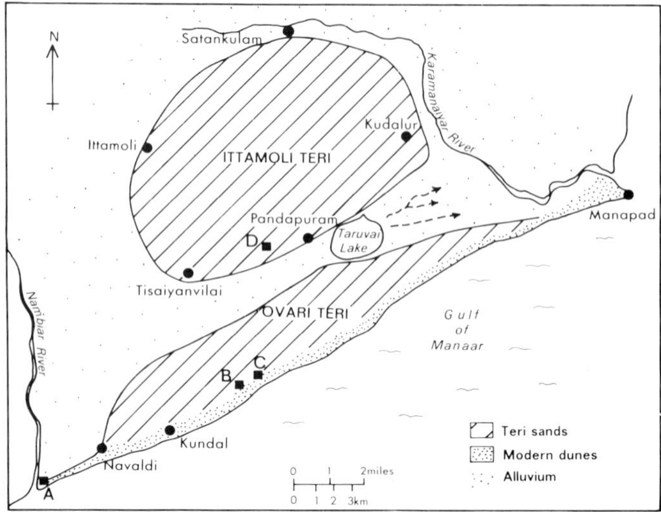

FIG. 3. Location map of exposures A–D within the study area. For regional setting see box in Fig. 1.

called Teri sands—colour 2.5YR 4/6 to 10R 4/8) form a cover deposit, and have been interpreted as coastal aeolian sands that have undergone *in situ* deep latosolic weathering (Gardner 1986). The weathering of garnet, feldspar, hornblende and opaque minerals has resulted in a loss of magnesium and calcium, and a partial loss of silica from the upper part of the profile. Kaolinite and illite have formed, and iron, released by weathering, has oxidized to haematite. Gardner (1981*b*) has shown that this aeolian deposit formed within the last 20 000 years at maximum, with probably the greatest intensity of weathering during the early Holocene.

Field observations

The replaced roots are found in many locations within the coastal sediments. They are, however, most abundant over the area seen in Fig. 3. Four exposures have been selected from this area to show the main forms of rhizolith occurrence and the stratigraphic relationships of the rhizoliths to the sedimentary units (Fig. 4).

Exposures A–D (Fig. 4) show a variety of common features:

(1) In all cases, the rhizoliths are *only* found in association with areas of red weathering.
(2) All rhizoliths are first found towards the base of the Teri sands, often continuing deep into the underlying sediments, which are typically calcareous sandstones.
(3) All rhizoliths have a circular/subcircular cross-section with an average diameter of 4–5 mm.
(4) All rhizoliths have a haematite/kaolinite cutan with a common thickness of 2 mm.
(5) All rhizoliths have a white, friable appearance.

However, major differences do occur between the carbonate and silica rhizoliths.

Carbonate rhizoliths

Calcium carbonate rhizoliths (low-magnesium calcite) are found in exposures A and B (Fig. 4). In exposure A, the rhizoliths occur as separate, distinct 'roots' within the basal Teri sands, and lower down along bedding planes composed of opaque and heavy mineral concentrations within a thin calcrete (maximum 2 mm thick). These bedding planes are connected by vertical or subvertical 'pipes' of red weathered sand, where the original low-magnesium calcite has been removed in solution, possibly being re-precipitated further down in the exposure. Often associated with these weathered bedding planes are present-day palm roots, frequently in a state of partial decomposition.

In exposure B however, the carbonate rhizoliths are only found within a sandy limestone. This limestone has an irregular morphology which, from field evidence, is interpreted as having undergone erosion before deposition of the overlying Teri sands and the subsequent formation of rhizoliths. Within depressions in the limestone, and coating its irregular topography, all the rhizoliths examined were composed of amorphous silica. Thus the *carbonate* rhizoliths are closely associated with calcareous sediments.

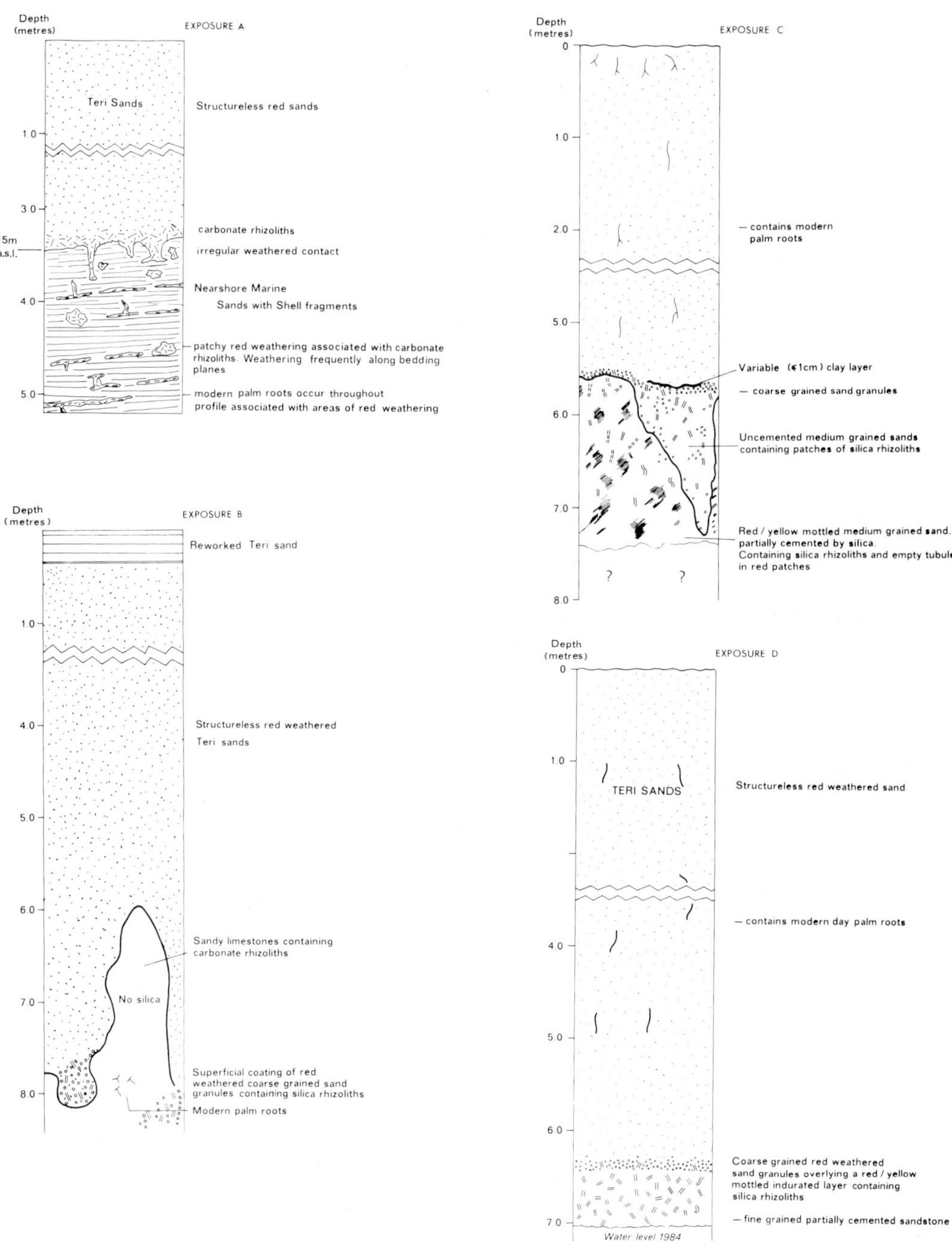

FIG. 4. Exposures A–D showing the variations in rhizolith occurrence and their relationship to the weathered Teri sands and underlying sediments. (*a*) Carbonate rhizoliths only. (*b*) Carbonate and silica rhizoliths. (*c*) and (*d*) Silica rhizoliths only.

Silica rhizoliths

Silica rhizoliths are more common and widespread than the carbonate variety. They occur in two distinct forms: as separate rhizoliths within unconsolidated reddened sands (Fig. 4, exposures B and C); or in greater concentrations within partially weathered and indurated sandstones lower in the stratigraphic sequence. This induration results from clay–silt and iron oxide coatings on the sand grains and an amorphous silica cement which is limited in extent and associated with the rhizoliths. Often, both forms of rhizolith are present within exposures (Fig. 4, exposures C and D).

The orientations of the rhizoliths (carbonate and silica) were found to be highly variable throughout the area, except where there is a local control such as along bedding planes as in exposure A. When comparing the four representative exposures, it can be seen that replacement occurs at variable intervals up and down the section. In the study area as a whole, rhizoliths were found to occur over a maximum depth interval of 14 m. The rhizoliths were never found to occupy only a specific horizon, even allowing for the fact that some rhizoliths have been completely weathered out. The evidence for this weathering is the presence of empty 'root tubules' (Klappa 1980) that are preserved by the incalcitrant nature of the cutans (Fig. 4, exposure D). From all this, it is clear that the rhizoliths were not associated with a specific horizon of lithification or diagenetic alteration.

Rhizolith analysis

The carbonate and silica rhizoliths appear very similar in hand specimen. However, on closer analysis there are fundamental differences.

Thin section

In both carbonate and silica rhizoliths the degree of replacement detail is variable, depending on the part of the root sectioned and its position in the soil profile. Both types show local areas in which detail is poorly preserved alongside areas of greater detail (Fig. 5*c*). In the more detailed areas the replacement/petrifaction is seen to affect only the cell walls leaving no other detail.

It is in thin section that the extent of the cement can be seen. This cement appears to be limited to within a distance of 1 cm around an individual rhizolith and its general pervasiveness in the sediments is therefore a function of the density of the rhizoliths. The cutans that are always present around the rhizoliths are revealed to be composed of micro-laminated haematite and kaolinite enclosing fine silt-sized clasts of quartz and opaque minerals. These enclosed grains tend to be aligned parallel to the outer surface of the rhizolith.

Scanning electron microscope

The SEM provides better differentiation of the silica and carbonate rhizoliths. The silica variety generally show greater preservation of anatomical detail compared to those of carbonate (Fig. 6*a*–*e*). The carbonate rhizoliths tend to show only vague cell outlines that are frequently masked by anhedral low-magnesium calcite overgrowths (Fig. 6*a*). On the other hand, various forms of silica preservation can be seen. The replaced cell walls appear to be made up of interconnected silica spheroids that occur in a variety of sizes up to a maximum of 2 μm (Fig. 6*d*). In some specimens these spheroids are not only restricted to the cell walls, but also occur as overgrowths, leading in some cases to a partial infilling of the cell interior (Fig. 6*e*).

Elemental analysis (EDX) of the silica rhizoliths reveals that they are almost pure silica. In the specimen shown in Fig. 6*d*, the silica content

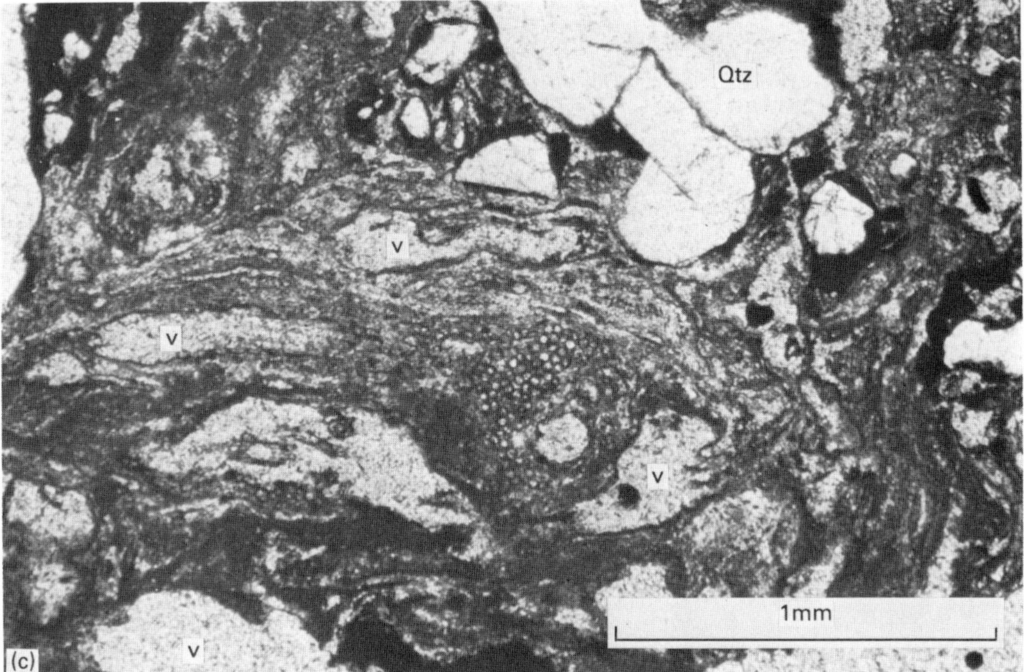

FIG. 5. (*a*) Irrigation well showing the irregular weathering front at the junction of the Teri sands and the underlying carbonate-cemented marine sands. For stratigraphic detail see Fig. 4*a*. Note the Palmyra palm in the distance with modern roots cutting through the reddened sands. (*b*) Sandy limestone containing carbonate rhizoliths enclosed within reddened Teri sands. (*c*) Thin section of a silica rhizolith. Note the variable detail preserved. V-voids in the section filled with the impregnating resin. Scale bar is 1 mm.

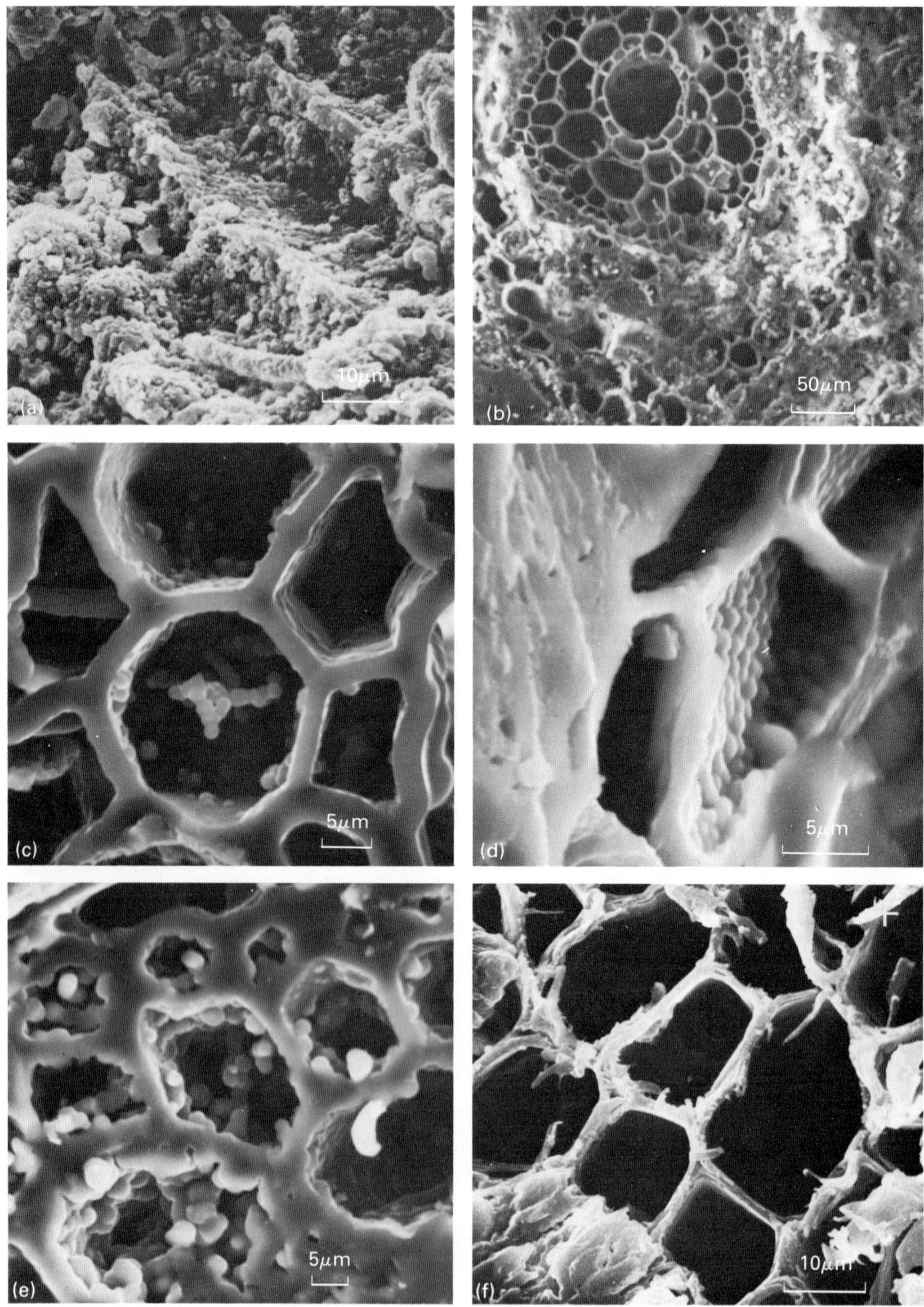

FIG. 6. (*a*) Carbonate rhizolith showing limited detail replacement masked by overgrowths of calcite. Scale bar is 10 μm. (*b*) Silica rhizolith showing variations in the preservation of anatomical detail. Scale bar is 50 μm. (*c*) Silica rhizolith showing cell walls completely replaced by amorphous silica. Scale bar is 5 μm. (*d*) Cell wall of a silica rhizolith composed of amorphous silica micro-spheroids. Scale bar is 5 μm. (*e*) Silica rhizolith showing possible later stage diagenetic overgrowths by amorphous silica micro-spheroids. Scale bar is 5 μm. (*f*) Present-day Palmyra palm root cells. Scale bar is 10 μm.

was found to have an average value of 97.5% (normalized to percentage oxides) with only trace amounts of aluminium, iron and calcium. A similar percentage value was found in other rhizoliths taken from a variety of locations within the study area. Initial results indicate that the spheroid silica overgrowths have an even higher silica content.

The carbonate rhizoliths occasionally show small enclosed areas of silica that are associated with preservation of cell detail. These enclosures have a silica content similar to the values obtained from the silica rhizoliths. Enclosed organic matter has never been found within any of the rhizoliths examined.

X-ray diffraction

The preserving silica is amorphous. No evidence of crystal structure was found.

Discussion

The fact that the rhizoliths are only found in association with areas of latosolic weathering indicates that their formation post-dates deposition of all the sedimentary facies seen in the area. The rhizoliths were never found to occupy a single definite horizon as would be expected if the roots had formed a mat immediately beneath a palaeosurface. Also, the rhizoliths are not associated with a specific horizon of lithification or diagenetic alteration as might occur in association with a former water table. This is further attested by the highly variable spread of rhizoliths in vertical exposures, again making groundwater an unlikely source.

Field evidence suggests that rhizolith formation is closely associated with a zone of replacement/petrifaction and cementation at the base of the reddened profile as weathering products move from the loose sands into the marine sandstones and limestones. A process of direct replacement of the original organic matter seems likely, rather than permineralization. This is because of the lack of enclosed organic matter within the rhizoliths, and also because permineralization is normally thought to be closely associated with groundwater fluctuations for which there is no evidence in this area.

Silica is present in the tissues of many plants, as is calcium carbonate, calcium pectate and calcium oxalate (Esau 1965). In contrast to animals, who only excrete excess inorganic materials, plants deposit such materials almost entirely within their tissues. Opaline silica, in an amorphous form, is deposited in cell walls and occasionally it forms bodies within the lumen of the plant cell. These bodies are called either phytoliths (Wilding & Drees 1971) or stegmata (Tomlinson 1961). Calcium carbonate rarely occurs as well-formed crystals but forms as overgrowths on cell walls (termed cystoliths) (Pireyre 1961).

These deposits in plants remain isolated from the weathering environment until the death of the plant, when they are returned to the soil. Opaline silica both from the cell walls and from the phytoliths then acts as a potential, relatively soluble, source of silica which is illuviated to lower horizons in the soil profile. A similar mechanism can be invoked for calcium carbonate.

Klappa (1980), in postulating a mechanism for carbonate petrifaction of plant roots, indicated that the plants may have a role to play. He states that the middle lamella between non-preserved cell interiors, originally composed of or containing calcium pectate, were preferential sites for calcification. This process could also be suggested in the case of the silica rhizolith of SE India, with precipitation of the silica preferentially on the existing silica deposits in the cells. Latosolic weathering provides a potential silica source while silica–organic complexing increases mobility. Both give rise to high silica levels in vadose water. Existing silica in the root cells would provide nuclei for precipitation, while high rates of evaporation would aid the process.

In the case of the silica rhizoliths, it can be seen that detailed replacement occurs in different areas within a single rhizolith cross-section. These are separated by areas of poorer preservation (Fig. 6*b*). It would appear that the process did not start at a single point and then move concentrically outward, preserving the original root, but rather that the silicification rapidly followed the cell pattern provided by the original cell walls. The overgrowths of silica spheroids may then indicate a later diagenetic stage. Areas of poorer preservation may be the result of partial root decomposition before the process was completed. Partly decomposed roots are occasionally found in the field associated with areas of weathering and rhizolith occurrence. Also, partly degraded organic matter is thought by some to play an important role in silica permineralization of wood (Leo & Barghoorn 1976). What is clear, however, is that the diagenetic process has occurred within the geologically short time-span of active latosolic weathering in the area, *ie* <20 000 years.

The differences between silica and carbonate rhizolith formation may be explained by the availability of the petrifying medium and the immediate geochemical environment. The carbonate rhizoliths only occur in areas with a comparatively shallow cover of red weathered sediments, thus possibly limiting the amount of available silica. Also, they occur in host sediments with a high carbonate content (Fig. 4, exposures A and B) and, with the exception of small, enclosed silica-rich areas showing detailed preservation, any preserved cell structures have been masked by carbonate overgrowths.

Field evidence of present-day roots frequently occurring in the side walls of water wells throughout the study area, has led to a comparison between the Palmyra palm (*Borassus fabellifer*) and the rhizoliths (Fig. 6*F*). Root morphology and size, and depth of penetration are similar and in many instances the modern roots are found growing in close proximity to the rhizoliths. Palmyra is known to contain silica within its roots, stems and leaves (Tomlinson 1961). In its roots, the stegmata or phytoliths occur next to cylindrical cortical fibrous strands and take the form of spherical or irregular shaped bodies, and are usually contained within silicified cells. However, comparison can only be tentative as these particular features are not obvious within the silica rhizoliths and no phytoliths have been identified within the surrounding sediment. Also, the state of root preservation at the time of petrifaction/replacement determines the morphology of the resulting rhizolith and in some cases it is probable that a limited amount of degradation has taken place.

Conclusion

This study has shown that replacement/petrifaction (not permineralization) of plant roots by amorphous silica and low-magnesium calcite has taken place within the last 20 000 years in SE India. This replacement is widespread throughout the Quaternary coastal deposits and postdates deposition of these sediments. The most obvious silica source is deep latosolic weathering of the overlying aeolian sands, with downward transfer of silica in solution as monosilicic acid. Replacement has resulted in the detailed preservation of cell structures by precipitation of silica under favourable micro-environmental conditions onto a template of biogenic silica that had formed within the root while the plant was alive. Field and laboratory evidence suggests that the plant involved in the replacement is the Palmyra palm (*Borassus flabellifer*).

The carbonate rhizoliths are less common and are only found in association with carbonate-rich host sediments. The low-magnesium calcite is thought to be derived locally from these sediments.

ACKNOWLEDGMENTS: I would like to thank Dr R. A. M. Gardner, for her helpful guidance during the course of this study, and Mr Harry Miller for his invaluable help in the early stages of the fieldwork. Thanks also go to the technical and secretarial staff in the Geography Department, King's College. This study was funded by a NERC Research Training Award, and forms part of a PhD thesis currently in preparation.

References

CARROZI, A. V. 1967. Recent calcite-cemented sandstone. *Journal of sedimentary Petrology* **37**, 597–600.

CLEARY, W. J. & CONOLLY, J. R. 1972. Embayed quartz grains in soils and their significance. *Journal of sedimentary Petrology* **42**, 899–904.

COHEN, A. S. 1982. Palaeoenvironments of root casts from Koobi Fora, Kenya. *Journal of sedimentary Petrology* **52**, 401–414.

CROOK, K. A. W. 1968. Weathering and roundness of quartz sand grains. *Sedimentology* **11**, 163–170.

DURGA PRASADA RAO, N. V. N., SHIHARI, Y. & BEHARY, A. K. A. 1982. Columnar concretions in the Visakhapatnam red sediments in the East coast of India. *Sedimentary Geology* **31**, 303–316.

ESAU, K. 1965. *Plant Anatomy*. 2nd edn. John Wiley & Sons.

EVANS, W. D. 1965. Facets of organic geochemistry. *In*: HALLSWORTH, E. G. & CRAWFORD, D. V. (eds) *Experimental Pedology*. Butterworths, London, 14–28.

GARDNER, R. A. M. 1981*a*. *Geomorphology and environmental change in southeastern India and Sri Lanka*. DPhil thesis, University of Oxford.

—— 1981*b*. Reddening of dune sands—evidence from SE India. *Earth Surface Processes and Landforms* **6**, 459–468.

—— 1986. The stratigraphy of Quaternary coastal sediments in SE India. *Man and Environment* **X**, 59–73.

GLENNIE, K. W. & EVAMY, B. B. 1968. Dikaka: plant and plant root structures associated with aeolian sands. *Palaeogeography, Palaeoclimatology, Palaeoecology* **4**, 78–87.

GOUDIE, A. S. 1983. Calcrete. *In*: GOUDIE, A. S. & PYE, K. (eds) *Chemical Sediments and Geomorphology: Precipitates and Residua in the Surface Environment*. Academic Press, London, 93–132.

ILER, R. K. 1973. Effect of adsorbed alumina in the solubility of amorphous silica. *Journal of Colloid Science* **43**, 399–408.

JONES, L. H. P. & HANDRECK, K. A. 1963. Effects of Fe and Al on silica in soils. *Nature* **198**, 852–853.

KLAPPA, C. F. 1979. Calcified filaments in Quaternary calcretes: organo-mineral interaction in the sub-aerial vadose environment. *Journal of sedimentary Petrology* **49**, 955–968.

—— 1980. Rhizoliths in Quaternary calcretes: classification recognition, genesis and significance. *Sedimentology* **27**, 613–629.

KNOLL, A. H. 1985. Exceptional preservation of photosynthetic organisms in silicified carbonates and peats. *Philosophical Transactions of the Royal Society* **B 311**, 111–122.

KRAUSKOPF, K. B. 1956. Dissolution and precipitation of silica at low temperatures. *Geochimica Cosmochimica Acta* **10**, 1–26.

LEO, R. F. & BARGHOORN, E. S. 1976. Silicification of wood. *Botanical Museum Leaflets Harvard University* **25**, 1–46.

LEWIN, J. C. 1961. The dissolution of silica from diatom walls. *Geochimica Cosmochimica Acta* **21**, 182–198.

MOREY, G. W., FOURNIER, R. O. & ROWE, J. J. 1962. The solubility of quartz in water in the temperature range 25°–300°C. *Geochimica Cosmochimica Acta* **26**, 1029–1043.

MOUNT, J. F. & COHEN, A. S. 1984. Petrology and geochemistry of rhizoliths from Plio–Pleistocene fluvial and marginal lacustrine deposits, East Lake Turkana, Kenya. *Journal of sedimentary Petrology* **54**, 263–275.

OKAMOTO, G. T., OKURA, T. & OCTO, T. 1957. Properties of silica in water. *Geochimica Cosmochimica Acta* **12**, 123–132.

PIREYRE, N. 1961. Contribution a l'etude morphologique, histologique et physiologique des cystoliths. *Revue Cytologie et Biologie Vegetation* **23**, 93–320.

SCHOPF, J. M. 1971. Notes on plant tissue preservation and mineralisation in a Permian deposit of peat from Antarctica. *American Journal of Science* **271**, 522–543.

STEIN, C. L. 1982. Silica recrystallization in petrified wood. *Journal of sedimentary Petrology* **52**, 1277–1282.

SUMMERFIELD, M. A. 1983. Silcrete. *In*: GOUDIE, A. S. & PYE, K. (eds) *Chemical Sediments and Geomorphology: Precipitates and Residua in the Surface Environment*. Academic Press, London, 59–91.

TOMLINSON, P. B. 1961. *Anatomy of the Mono-cotyledons 2. Palmae*. Clarendon Press, Oxford.

WARNE, S. ST. J. 1962. A quick field and laboratory staining scheme for the differentiation of the major carbonate minerals. *Journal of sedimentary Petrology* **32**, 29–38.

WILDING, L. P. & DREES, L. R. 1971. Biogenic opal in Ohio soils. *Proceedings of the Soil Science Society of America* **35**, 1004–1010.

D. A. HENDRY, Department of Geography, Kings College London, London WC2R 2LS, UK.

Spring mounds in southern Tunisia

C. R. Roberts & C. W. Mitchell

SUMMARY: Chotts Djerid and Fedjadj lie in the arid zone of southern Tunisia, and are chemically active, artesian environments which give rise not only to extensive emergent groundwater gypsum and salt deposits, but also to landforms of more complex architecture. In some geological situations belts of spring mounds up to 25 or 30 m high straddle former high-level late Pleistocene lake shorelines. Spring mounds are circular, gypsiferous or calcareous, volcano-like structures fed by point sources of artesian water rising from aquifers in the Continental Intercalaire and Complexe Terminal aquifer series. Similar mounds have been identified around playas in the USA, Australia, Egypt, Iraq and elsewhere. In some mounds a central well is occupied by a small pond with water conduits in the base, whilst others are dry and degraded. In these latter, water is only present at depth and wind erosion and redeposition are moulding the landforms. The morphology of spring mounds is determined by the balance between aquifer pressure head, output, chemical content, net evaporation, aeolian activity, and plant colonization. Their stratigraphy comprises alternations of crystalline materials, wind-blown gypsiferous sand, wash and splash deposits, and organic layers arranged in overlapping sequences which dip outwards from the water source. Such deposits offer a potentially sensitive index of environmental changes in the late Quaternary and Holocene.

Chemical conditions and the crystallization of salts within playa basins have attracted considerable interest in the last decade (Hardie *et al.* 1978), and a fuller explanation of brine evolution in these circumstances is emerging. By contrast, the landforms associated with saline precipitates, particularly those on the margins of artesian playas, have attracted relatively little attention. This paper examines the landforms known as spring mounds, and their associations with other sediments, within the context of Quaternary climatic and other changes.

On playa margins, springs can precipitate large amounts of gypsum and calcite which can be redissolved, recrystallized, eroded, redeposited and affected by diagenesis. Spring mounds are formed by the evaporation of mineralized water from artesian springs, sometimes with wind-borne additions, in the vicinity of playas in a number of arid areas. The main occurrences reported are in the Lake Eyre region of Australia (Madigan 1936; Taylor 1947; Fairbridge 1968; Twidale 1972; Watts 1975; Mabbutt 1977), southwestern USA (Blackwelder 1931; Stone 1956; Scholl 1960; Scholl & Taft 1964; Neal 1965; Neal & Motts 1967; Weaver 1977; Hardie *et al.* 1978), Djibouti (Fontes & Pouchan 1975), and Ethiopia (Englebert 1970). They have been identified on aerial photographs by the present authors in the Kharga Oasis, Egypt, northern Iraq, and near Mexico City.

Spring mounds are distinct morphologically from crusts of calcrete or gypcrete (Watson 1983*a*) formed at or near the ground surface, and from aeolian landforms such as dunes. They may be in the form of ridges, towers, pinnacles, 'mushrooms', 'tombstones', or conical hillocks with or without central craters. They vary from less than 1 m to over 30 m in height and may be up to 500 m in diameter. Most consist of particles of sand and silt cemented by varying amounts of tufa, travertine and gypsum. The formative spring may either form a pool at the top or issue from the side. The height cannot exceed that of the piezometric surface. Some mounds owe their development partly to algae and appear to have been formed underwater (Scholl 1960). Ridge features are formed from seepage along faults and fissures. The architecture of many show signs of changing environmental conditions during their formation.

Environment of the Tunisian spring mounds

The Fedjadj and Nefzaoua oases together cover an area of about 50 km N–S and about 30 km E–W centred on Kebili but separated by the Djebel Tebaga (see Figs 1 and 2). Their springs and spring mounds have provided irrigation water for centuries. Demands on the supply have recently led to a drop in artesian pressures and an increasing use of boreholes.

Geology and geomorphology

The Chott Fedjadj–Djerid basin forms the north-eastern and one of the lowest extremities of the immense, artesian, Bas Sahara basin (shown in

From FROSTICK, L. & REID, I. (eds), 1987, *Desert Sediments: Ancient and Modern*, Geological Society Special Publication No. 35, pp. 321–334.

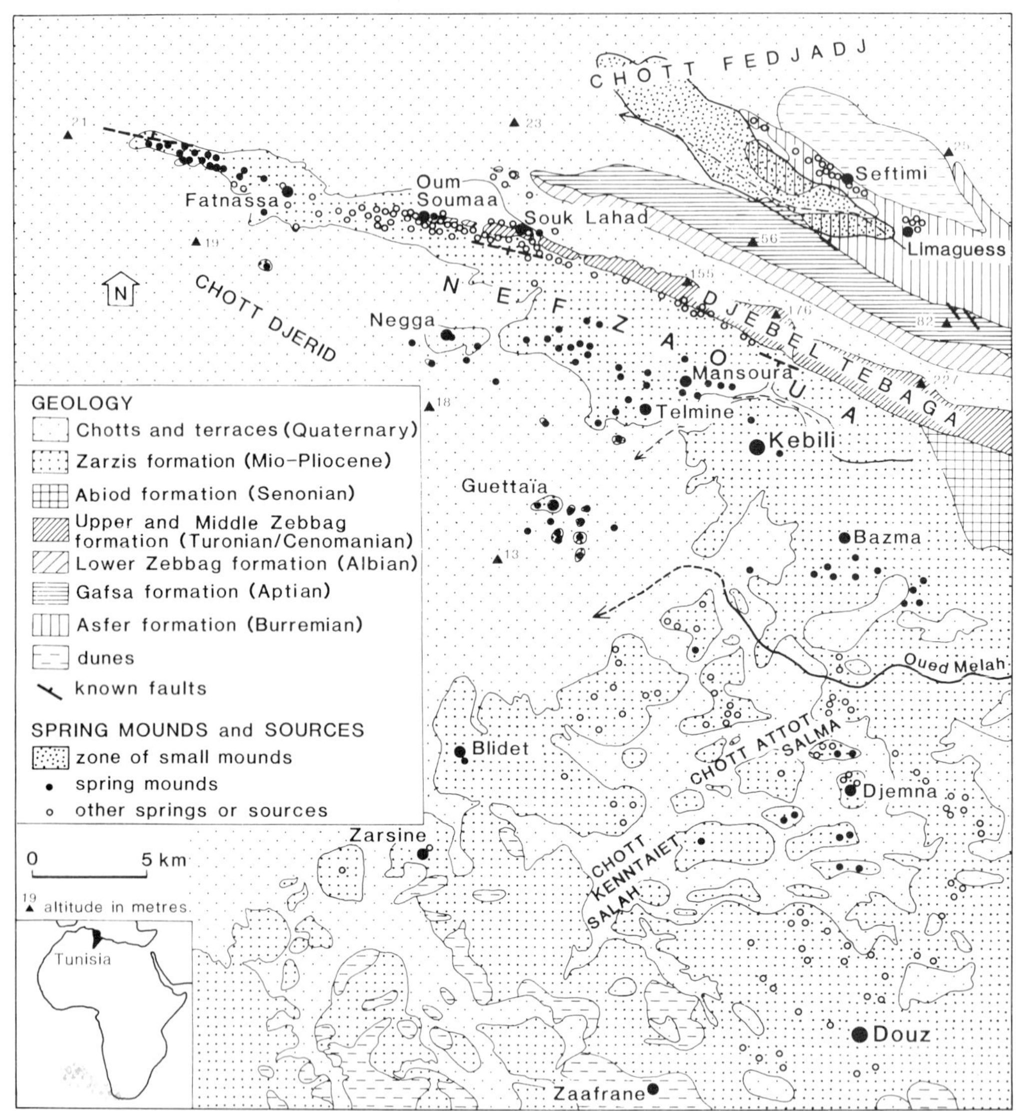

FIG. 1. Fedjadj and Nefzaoua Oases, southern Tunisia.

Fig. 3) which covers most of the Algerian and Tunisian Sahara, extends into Morocco and Libya, and encloses the whole of the Grand Erg Oriental (Drouhin 1958; Nesson 1978; Gischler 1979). The Bas Sahara is composed of a series of complex aquifers in the underlying and surrounding sediments, which emerge at the surface in and around the chotts. Chott Fedjadj receives discharge from the Cretaceous Continental Intercalaire via upward percolation through its base, supplemented by seepages from exposures on the escarpment and from spring mounds along the chott margins on the southern side of the enclosing anticline. Chott Djerid, in the overlying shallow syncline, is one of the major discharge areas from the Complexe Terminal, specifically the Turonian dolomites, Upper Senonian limestones, and in places the Mio–Pliocene Zarzis sands and conglomerates. The water emerges locally through an overlying clayey aquiclude. 'Aioun' (plural of Arabic 'ain' = spring) are artesian springs from the Terminal aquifer which emerge in the base of both chotts, but are most conspicuous in Chott Djerid. Their distribution appears to correlate with that of underlying Cretaceous limestones.

Geologically the chotts lie in peri-atlasic

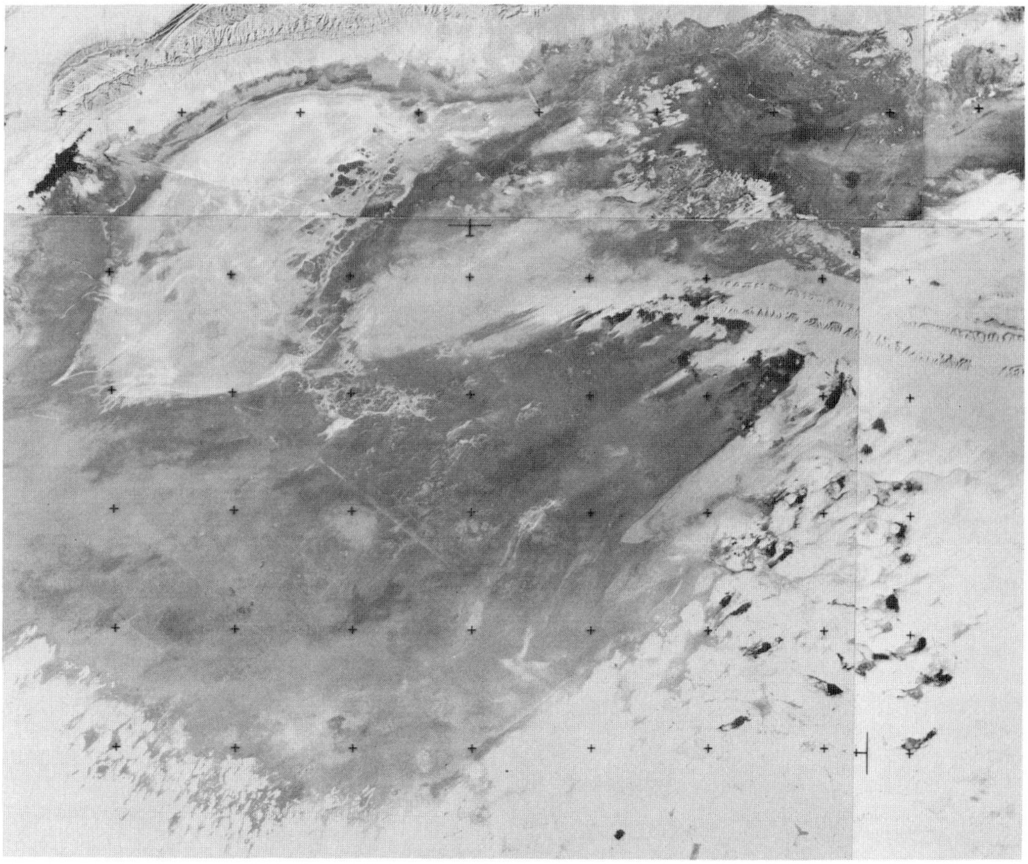

FIG. 2. *Landsat 3* mosaic of Djebel Tebaga–Chott Djerid area, taken November 1979 and April 1980. Note dark patches of oases, aeolian NE–SW streaks, Djebel Tebaga, and chott surface showing aioun area.

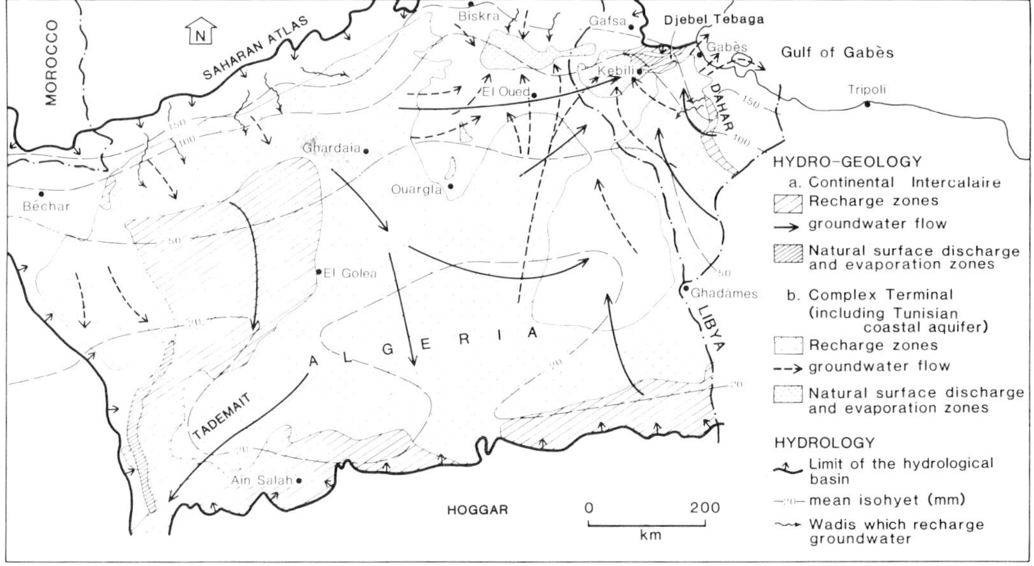

FIG. 3. Saharan artesian basin. (After Gonfiantini *et al.* 1974.)

Geologically the chotts lie in peri-atlasic cratonic basins between the Saharan shield to the S and the Kasserine island to the N (Burollet 1967). The area comprises Cretaceous and early Tertiary sediments, which were folded and uplifted in the Alpine 'atlasic' orogeny which in Tunisia began in the Oligocene, continued to the end of the Miocene, and experienced its most violent phase in the post-Villafranchian period (Bishop 1975). These movements created the chott depressions and the surrounding folded and faulted sedimentary cuesta-and-vale topography. Subsequently, the basins have experienced minor modifications from Quaternary bioclimatic changes (Coque & Jauzein 1967), and may also have been affected by tectonism both before and after a marine incursion (Richards & Vita-Finzi 1982).

Chott Fedjadj lies in a collapsed and faulted anticline bounded to the S by the Djebel Tebaga, a steep double cuesta capped with dolomite, which is in turn terminated to the S by a strike-fault (detectable on *Landsat* imagery) which downthrows to the S (see Fig. 4). Chott Djerid sits to the S of the fault and is floored and surrounded by almost horizontal Mio–Pliocene deposits up to 100 m thick. Superficial Quaternary deposits, some of which are gypcreted (Watson 1983), fine towards chott centres whose surfaces are crusted with halite and gypsum, halite increasing at the expense of gypsum towards the chott centres. The Quaternary sands and clays are usually less than 12 m deep. A notable feature is the remains of two marginal terraces, approximately 6 m and 2 m above the present local levels of the chotts. French workers call the lower and younger of these a 'niveau a Cardium' because it contains numerous detrital 'Cerastoderma [=Cardium] glaucum' shells. Radiocarbon ages range from 22 000 to 32 400 BP on shell fragments. These accord well with earlier estimates of its age (Richards & Vita-Finzi 1982), in spite of the high risk of fossil carbon contamination.

Climate and climatic change

The climate is pre-Saharan. At Kebili the mean monthly temperatures range from 9.4°C in January to 32°C in July with an annual mean of 20.9°C. Rainfall at the same station averages 89 mm, while in the region as a whole it decreases southwestwards from over 210 mm at Djerba and Matmata to about 100 mm over most of Chott Fedjadj and to below 70 mm at the southern margin of Chott Djerid. Evaporation measured by both Piche and 'bac' methods at Kebili gave values of 2520–2550 mm a^{-1}. Evaporation is at a maximum in the summer (Mamou 1976) and it exceeds rainfall by over 20 times.

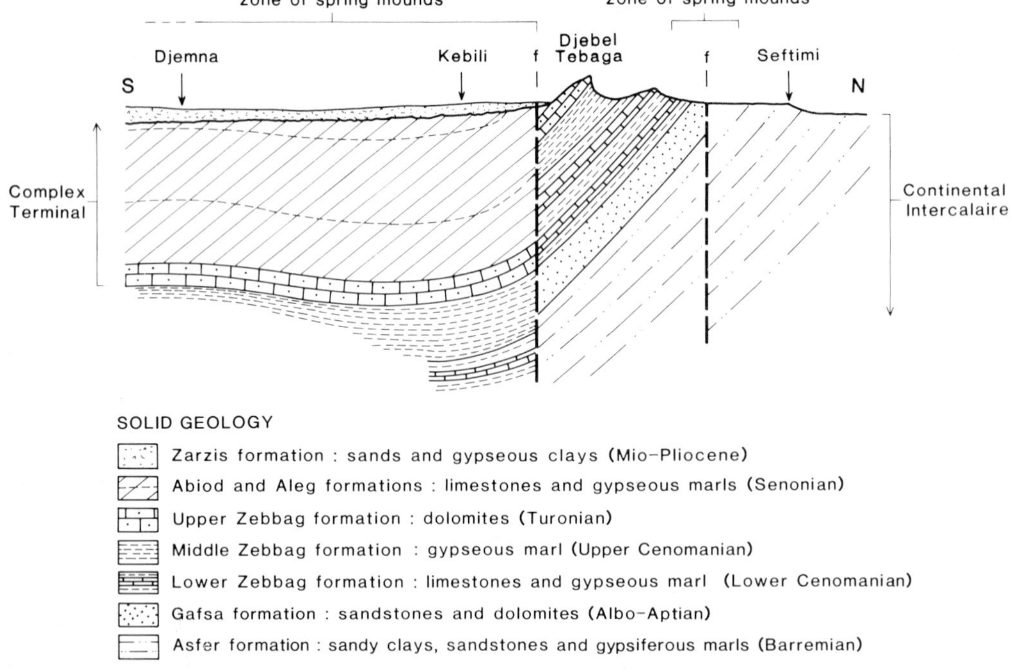

FIG. 4. Geological cross-section from Seftimi to Djemna. (After Mamou 1976.)

The current climate, however, can in no way be regarded as typical of the recent geological past. Alternations of wetter and drier periods throughout the Pleistocene and Holocene have been noted in numerous studies, particularly those of French geomorphologists. They cite the former high-level shorelines around the chott as evidence of 'pluvial' phases. There is some debate about the age of these phases, and of their association with the glaciations of northern latitudes. Rognon (1979) invokes the morphological evidence on slopes around the chotts to recognize two pluvial periods when lake levels were much higher, namely the Soltanien (Würm) 34 000–12 000 BP and the Holocene (Neolithic) 8000–5000 BP. He assigns the low-level 'niveau a Cardium' to about 10 000 BP in an interpluvial period. Richards and Vita-Finzi argue that the high lake levels during the earlier of these two periods were the result of a marine incursion 35 000–25 000 BP followed by an uplift of about 80 m. In historical time, dendrochronology in Algeria suggests a wetter period between 1600 and 1750 AD, whereas in central Tunisia similar conditions prevailed from 1700 to 1880, with a subsequent increase in aridity (Ginestous 1927; Nicholson 1980).

Types of spring mounds in Fedjadj and Nefzaoua

Fedjadj Oases

The spring mounds in the neighbourhood of the Fedjadj Oases lie not in the chott itself but in a depression, drained by a westward-flowing braided channel system, between the northern limb of the Djebel Tebaga (Lower Zebbag Formation) and the outcrop of the Asfer Formation on which Seftimi and Limaguess stand. Hundreds of spring mounds straddle the fossil shorelines found in this area, occurring in groups a few hundreds of metres apart. Within these groups they are spaced a few tens of metres apart, although sometimes they are sufficiently close to overlap and form 'multiples'. The spacing increases towards the W, and eastwards they disappear fairly abruptly at a NW–SE line halfway between Seftimi and Limaguess. Their form, when simple, is conical and unrilled with a central crater. In some of the craters there is a pond, in some a marsh, and in some shallow subsurface water. All tend to have more vegetation than the surrounding areas, and they frequently contain one or more date palms. The commonest situation is a sharply defined pool occupying a proportion of the central crater. Mounds are characteristically 1–4 m high, and surrounded by a gently sloping aureole 5–10 m wide. This aureole is covered with a fine-grained, rain-formed crust with a 'clay-pan' surface which merges outwards into the surrounding levels, sometimes passing over fluvial sediments in the chott. The rim, on whose inner side the stratigraphy is sometimes visible, is often broken through by one or more channels. Sometimes there is an inner rim covered by a saline crust currently being formed, indicating 'multiphase' development.

Towards the W the mounds become increasingly wind-modified. From evidence of sand streaks on *Landsat* imagery (Fig. 2) and the predominant 45°–55° bearing of aeolian features, it can be concluded that the dominant wind is northeasterly (McKee 1979). Wind action however, leads to somewhat complex geomorphological results (see Fig. 5). The amount of surface scour is indicated by the frequency of exhumed roots around exposed trees. This scouring attacks the windward end and sides of mounds, and fore, lag and lateral obstacle dunes are deposited. The foredunes are often nbak (singular: nebkha). The lag dunes are the largest, tending to be elongated and attached to the mounds. The lateral dunes are longitudinal to the wind and may either be attached to the flanks of the mounds or separated from them. Some modification of form is noticeable towards the W resulting from secondary southwesterly winds, which are increasingly prevalent away from the Gulf of Gabès.

A simple symmetrical type of spring mound is shown in Fig. 5 (SM2). It has five concentric zones: centre, inner rampart, rim, outer rampart and wash aureole. In the central well, decaying organic matter has produced organic-rich layers which are now buried by chemical precipitates and wind-blown sediments. The outer 15°–20° slopes have been affected by slumping and erosion in places, and the wash aureole has a gradient of less than 1°.

Figure 6 shows topographical relationships and water salinity in a group of small mounds a few kilometres S of Seftimi. Water tables are highest, groundwater freshest, and sediments least saline and least gypsiferous in the mound centres. Within the mounds themselves, pH values are at a minimum on the crests of the ramparts, which are grey and apparently gleyed.

Southern Nefzaoua Oases

The Nefzaoua Oases consist of two main groups of spring mounds related to the area of 'aioun' in Chott Djerid (Figs 1 and 2). The first group lies in the E and S and is the largest. It is distributed

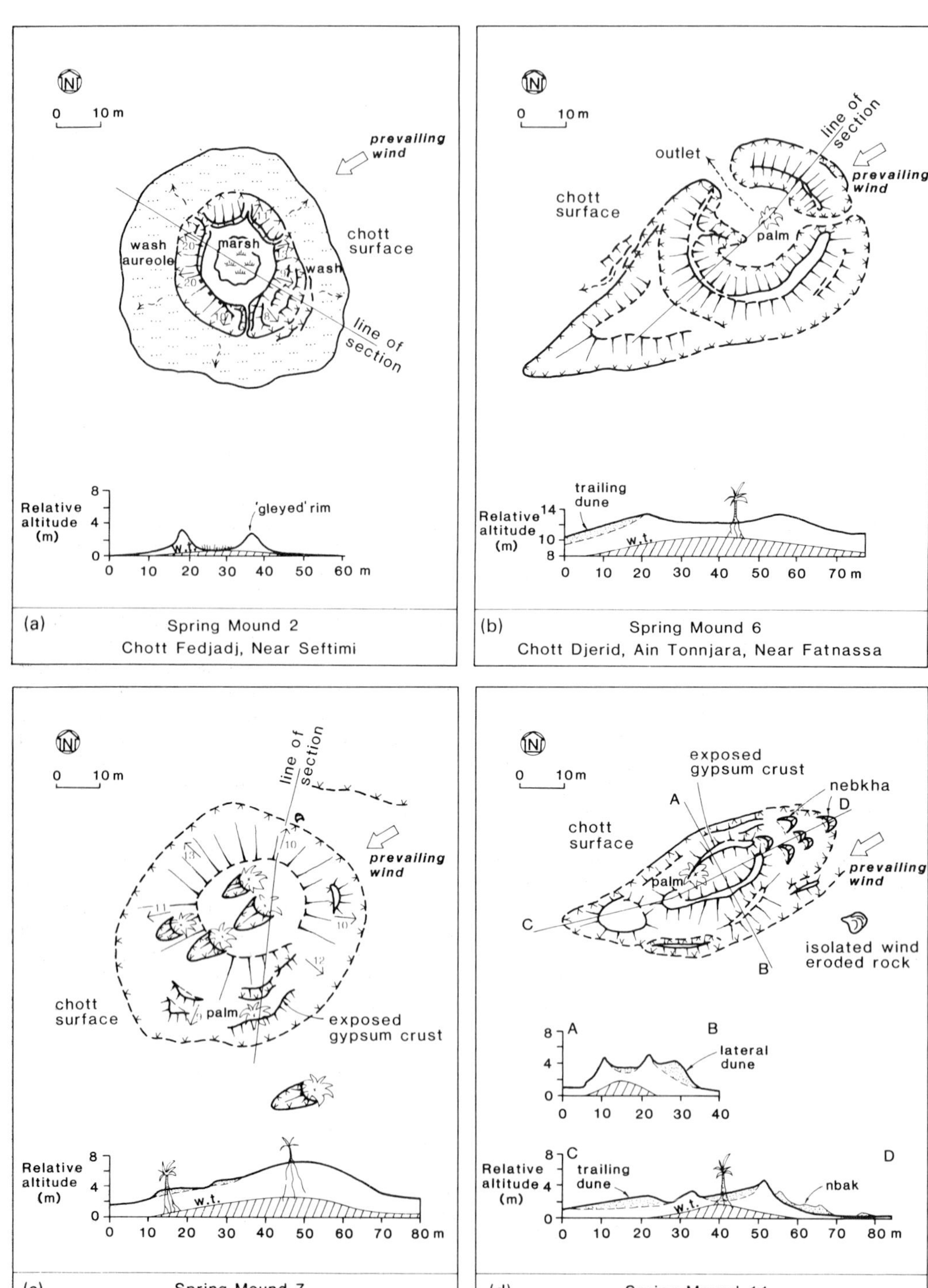

FIG. 5. Geomorphological maps and sections of selected spring mounds.

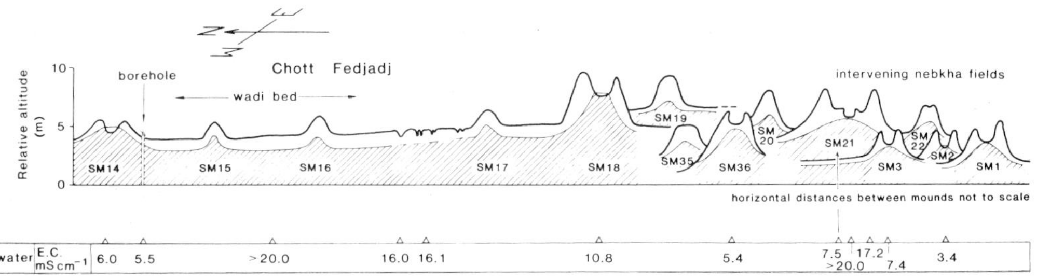

FIG. 6. Topographical relationships, water tables and water conductivities of a group of spring mounds in Chott Fedjadj, south of Seftimi. Total length of section is approximately 3 km.

over a wide area from Negga to Douz, decreasing in density southwards, but including an E–W trending mound-free zone between the lines Kebili–Telmine–Negga, and Guettaïa–Bazma. The occurrence of the mounds seems to correlate with outcrops of the Zarzis Formation which is often characterized by gypsiferous clays overlying conglomerates and sands (Mamou 1976). The spring mounds here are mainly single, isolated, and separated by several hundreds of metres to several kilometres, some forming 'islands' in the chott. They are relatively large, up to 30 m high and over 500 m in diameter, and characteristically have concave outer rim slopes of 12°–15° and inner rampart slopes of 20°–26°. Although some of the mounds contain ponds, these are often stagnant and almost none S of Djemna have surface water. Some are dissected and at least two (Guettaïa and Ain ech Cheikh, 1 km to the SW of Guettaïa) preserve an arch over a radial channel, indicating past tunnelling to draw water from the central crater. The artesian pressures have dropped so much in the past quarter century that mounds no longer contain usable springs and the villages on them must depend on deep boreholes for all their needs. Some mounds are so deflated that all that remains is a central platform lacking rim or dunes. Others are wind-scoured, leaving isolated sections of rims, making it possible to see the stratigraphy. Some are also partially overwhelmed by dunes.

At Negga the upper 3 m of the rim of a degraded spring mound reveals an alternation of fine cemented cross-stratified gypsiferous sand with some organic fragments and root pseudomorphs, gypsiferous rubble, thin cemented silt crusts showing occasional rainpits, and pure blown sand lenses (Fig. 7). 'Bulimus' (freshwater) shell fragments are found in some layers, and several sharp sub-horizontal erosional boundaries can be identified. Microscopic examination shows a substantial detrital quartz content in addition to twinned gypsum and low-Mg calcite. The well-rounded form of the quartz grains and the presence of garnet and tourmaline suggests lengthy aeolian transport, perhaps from the Atlas Mountains from which occasional winds blow (McKee 1979).

SM5 (Gezirat Eisa, 1.5 km SW of Telmine), shown on Fig. 8, is about 30 m high and is an example of the much larger spring mounds from the Complexe Terminal in the Kebili area. It is largely composed of hardened earthy material containing a mixture of calcium carbonate, gypsum and fine quartz particles, the last apparently of aeolian origin. The cross-section shows a

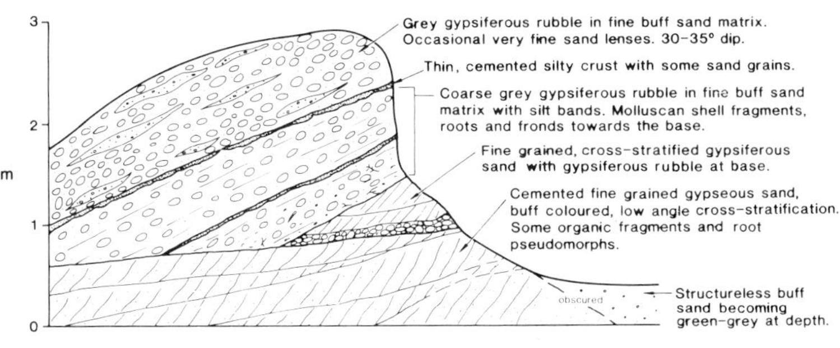

FIG. 7. Rampart section, spring mound 31, near Negga.

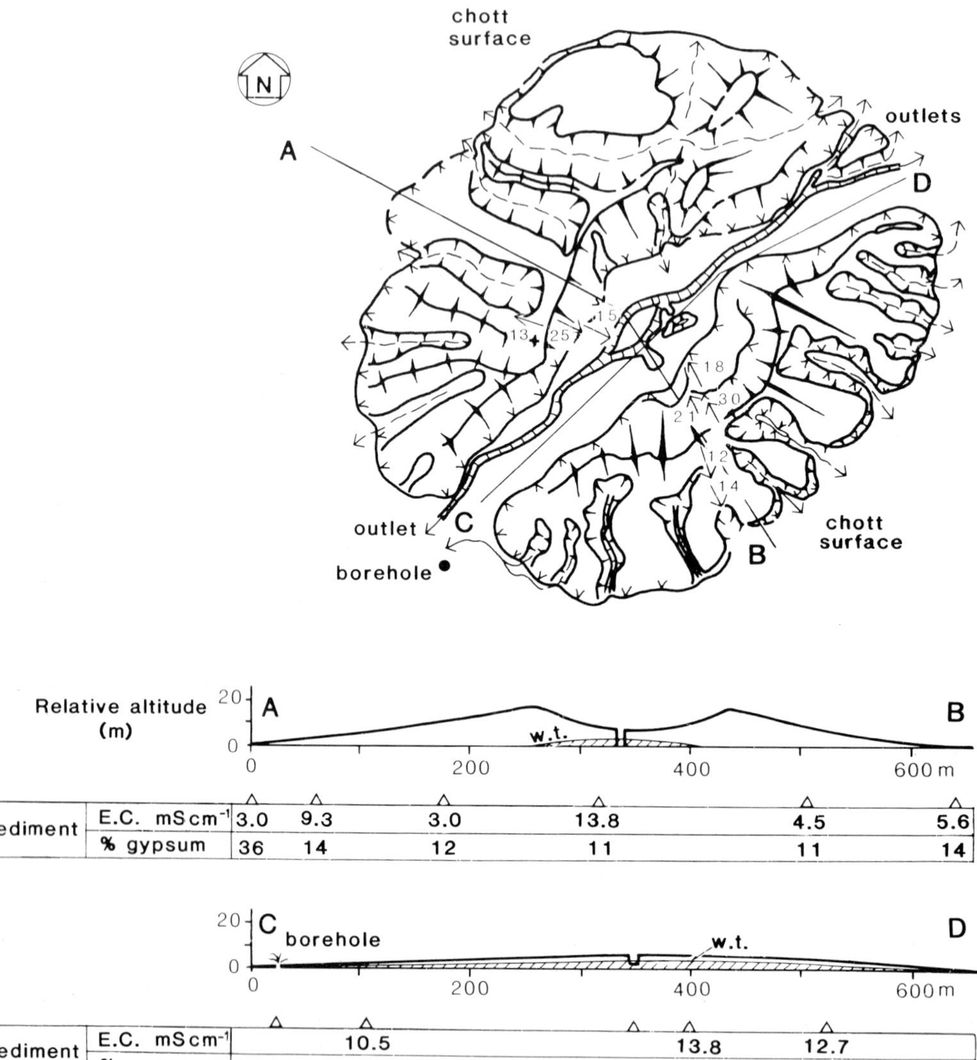

FIG. 8. Geomorphological map and sections of spring mound 5, Gezirat Eisa.

similar chemical profile to that found in the Fedjadj mounds with the salinity in the centre similar to an adjacent borehole and increasing in the direction of the overflow channel. No clear evidence was seen of the human excavation or artefacts associated with use of the spring for irrigation water.

The mound has clearly experienced at least two stages of development, a first stage in which it grew to its present height or higher, and a second in which water level has fallen and it has become degraded and trenched. Evidence of the former growth period can be seen in layering exposed in gullies on the mound flanks. These layers vary in colour from buff to grey or dark brown, and incorporate some organic matter.

Central Nefzaoua Oases

Along the foot of the southernmost dip-slope of Djebel Tebaga between Fatnassa and Kebili, spring mounds are absent. Instead, springs, ponds, and foggaras (irrigation tunnels) emerge along the edge of the aquiclude and, supplemented by boreholes, supply water to a line of villages, of which the chief are Souk Lahad and Oum Soumaa.

The Nefzaoua 'peninsula'

The second group of Nefzaoua mounds lies along Djebel Tebaga entirely W of Fatnassa and continues to below Chott Djerid. This covers a much smaller total area than the first group and the mounds are smaller and more closely spaced. They average perhaps 5–10 m in height and 50–200 m in diameter. Almost all are entirely fossil, extremely degraded, wind-eroded and partly overwhelmed by dunes. Although some support palms and have water tables within a metre or two of the surface in their centres, they have suffered from a drop in artesian pressures and are no longer used as a source of water.

The transect SM6–SM8 is shown on Fig. 9 and a plan and section of SM6 is given in Fig. 5. This group lies towards the eastern part of the peninsula. The land has a high, though variable, gypsum content due to the low elevation above the chott. SM6 represents a spring mound subject to falling water table but still with fresh groundwater below its centre. The mound has degraded, showing upwind scouring and a lee-dune. Vegetation is sparse on the upper, drier slopes except for the surviving palm, and only increases at some distance from the mound where the lower elevation permits halophytes to flourish within reach of the water table around the edges of the chott. SM7 exhibits further degradation with the rim virtually absent. Subsidence cavities up to 1 m deep are occasionally found in the surrounding area, covered by a thin cemented-gravel carapace.

Further W along the peninsula, the spring mounds are even more degraded and wind-modified, although isolated stunted date palms on the tops of a few show that groundwater is still accessible. SM11 (Fig. 5) is an example of this type. It is so degraded that the cratered conical form is only just discernible. The surface is in places wind-scoured, to expose gypsiferous gravel, and elsewhere raised into nbak of gypsiferous material behind shrubs, and lateral and trailing dunes. Vegetation is very sparse. There are traces of buildings, brick fragments, and pottery on the top of a nearby mound 10 m higher.

The aioun

Aioun are visible only when the surface of the chott is dry. They appear as dark patches of damper clay surrounded by a circular or slightly oval annulus of lighter coloured salt efflorescence. The overall diameter of aioun varies between about 20 and 50 m, and their relief is imperceptible, the centres being perhaps 5–6 cm higher than the surrounding chott surface. Vertical sections show a complex stratigraphy of algal and saline material (particularly halite), and water tables are never more than 30–40 cm from the surface. Fig. 10 shows a detailed transect through two aioun and the salinity of the groundwater sampled at shallow depth under crests and aioun centres; centres are characterized by lower salinity than the surrounds.

Hydrological relations

Storages

The Continental Intercalaire is a complex aquifer approximately 1000 m thick, covering about 600 000 km². It consists, in the Kebili area, of aquifers at several levels linked by percolation along beds and faults. Faults also provide limited connection with the overlying Terminal through a marly aquiclude. Radiocarbon analyses, quoted by Mamou (1976) give an age for Chott Fedjadj water in excess of 20 000 years and, for that in the Intercalaire upstream of the Nefzaoua, 27 000

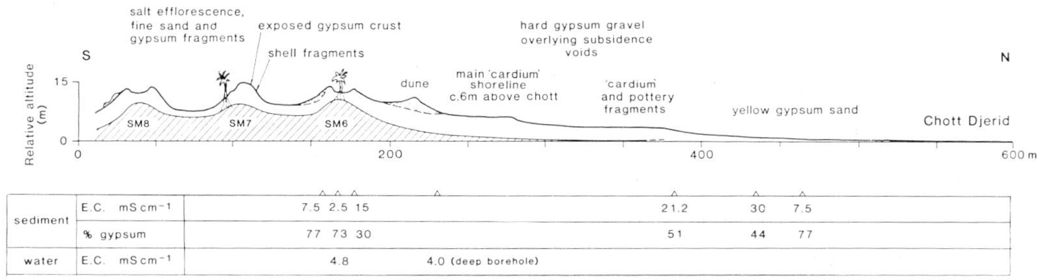

sediment	E.C. mS cm⁻¹	7.5 2.5 15		21.2	30	7.5
	% gypsum	77 73 30		51	44	77
water	E.C. mS cm⁻¹	4.8	4.0 (deep borehole)			

FIG. 9. Transect between spring mounds 6 and 8, Nefzaoua peninsula area.

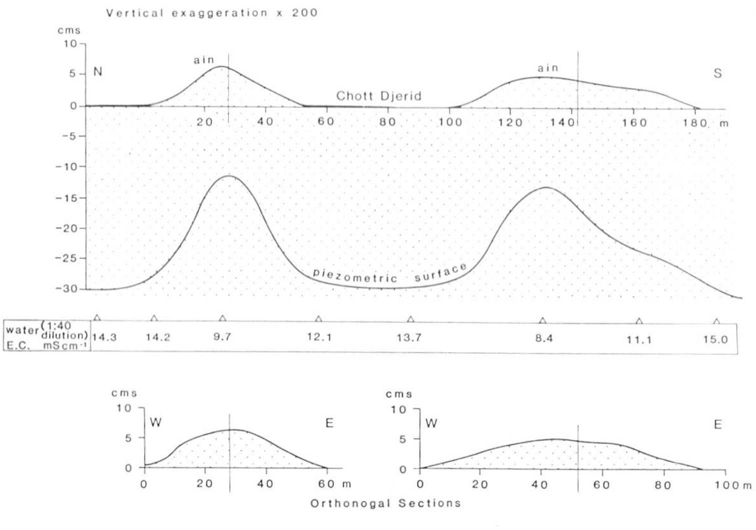

FIG. 10. Detailed topographic survey of aioun surfaces and water tables, Chott Djerid.

years (Ksar Rhilane) and 36 000 years (Bir Zobbeus). These ages imply that the aquifer was mainly filled during pluvial periods of the Quaternary and that the current recharge from sporadic rainfall must be relatively small. Outflow is mainly via vertical leakage into Chott Fedjadj, although there is some percolation into a coastal aquifer via the El Hamma fault zone (Pallas 1972; Mamou 1976).

The Complexe Terminal is some 350 000 km^2 in extent and up to 400 m thick. It consists of five major beds ranging in age from the dolomitic limestones and dolomites of the Cenomano–Turonian to Mio–Pliocene sands, the most important bed being the intermediate Upper Senonian limestones (Fig. 3). Recharge comes from rainfall on the exposed and accessible beds of the Complexe Terminal, but the distribution of piezometric pressures in Nefzaoua oases suggests that much of the recharge for the spring mounds must come from the Dahar supplemented by some localized additions from Djebel Tebaga. Subterranean flow occurs along the strata and through the extensive fissures which connect the aquifers. Radiocarbon analyses give dates for Terminal waters ranging from 30 000 years (Zarsine) to 3500 years (Douz), and give an average speed of percolation between Douz and Djemna of about 1 m a^{-1}.

Losses: past and present

In 1976 the Continental Intercalaire piezometric surface ranged from over 130 m in the E, to 64 m near Souk Lahad and 55 m at Oum El Ferth, so that water movement was from E to W. Boreholes in the Fedjadj Oases, mostly installed since 1974, now account for almost all of the output from the aquifer in this area, and yield a total flow of about 300 l s^{-1}. Some loss of piezometric pressure in springs is evidenced from the reports of local farmers that it has declined in their lifetimes.

Groundwater movement in the Complexe Terminal is from SE to NW. In 1975 the piezometric level was about 70 m at Douz and about 35 m at Fatnassa, an overall gradient of about 0.45%. The estimated yield from natural springs in the Nefzaoua Oases, 80% of which comes from the 'peninsula', was 660 l s^{-1} in 1900. The first artificial borehole, yielding 604 l s^{-1}, was installed in 1907. Further boreholes reduced the yield of the springs to 560 l s^{-1} by 1950, and to 250 l s^{-1} when abstractions achieved 2570 l s^{-1} in 1976. This caused some springs to dry up completely. For the same year, Mamou estimated the losses from lateral and downward percolation as 1570 l s^{-1}, and evaporation from the chott as up to 9700 l s^{-1}. The presence of abandoned spring mounds in topographic positions more than 10 m above the present piezometric levels also reflects a decline pre-dating borehole exploitation (Roberts 1986).

Aquifer hydrochemistry

The strong mineralization of groundwaters in southern Tunisia is the result of the intensity of evaporation from influent subterranean water percolating through carbonate and gypsiferous

TABLE 1. *Representative analyses of groundwaters*

Aquifer	Location	Ionic concentrations (mg l^{-1})							Conductivity mS cm^{-1} at 25°C
		Ca	Mg	Na	K	HCO_3	SO_4	Cl	
Intercalaire									
	Seftimi hot bore	200	66	380	20	70	817	773	3.5
	Seftimi cold bore	295	128	750	55	49	756	1610	5.7
	Spring Mound 3	485	160	990	85	131	1210	2200	7.5
	Spring Mound 24, pool	255	260	2340	90	258	1830	4260	12.0
	Spring Mound 24, rim	370	350	3400	135	172	2305	5500	16.0
	Chott adjacent to SM 24	410	410	4600	90	92	2650	7480	20.0
	Chott adjacent to SM 3	1200	1050	14400	350	143	4915	21330	60.0
Terminal									
	Faouar bore	178	92	250	15	102	393	550	2.4
	Djedida lower spring	180	144	550	20	115	899	616	4.1
	Spring Mound 11	240	175	520	10	131	1445	616	5.0
	Spring Mound 6	430	450	2100	60	98	2675	2480	11.0
	Spring Mound 39, Blidet	300	270	2400	125	110	1950	3770	12.0
	Aioun, Chott Djerid	1150	3400	100000	3800	57	15010	163900	380.0

bedrocks. Allochthonous waters, such as additions from wadi flow, are negligible in their effects. Salinity generally increases upwards from the aquifers to the chott surfaces, down the piezometric gradient towards the chott centres, towards areas where borehole exploitation has been especially intense, and in association with a fault along the alignment of the Wadi el Melah N of Kebili which facilitates the percolation of more saline water into the Upper Senonian aquifer.

The background salinity of the Intercalaire aquifer lies mainly between 0.3% and 0.35%. That in both springs and deep boreholes in the Complexe Terminal (143 analyses from Mamou (1976), 48 from the present authors) ranges between extremes of about 0.1% and 1%, but averages 0.2–0.3%. The dominant cations are everywhere Na, Ca and Mg in that order of concentration (Table 1). Anions are dominated by sulphate and chloride in almost equal quantities with bicarbonate between about 1/5 and 1/10 of their concentrations (Fig. 11). There is little gross difference between the two aquifers, except for the higher salinities in the Intercalaire and a possible indication that in the Complexe Terminal, levels of bicarbonate may be high relative to the other anions.

Water conductivities range from less than 2 mS cm^{-1} away from the chotts up to several hundred mS cm^{-1} in their centres. pH is usually in the range 7–8. Complexe Terminal water conductivities in boreholes range from 1.6 to 2.6 mS cm^{-1}, generally increasing northwards from Faouar towards the main populated areas of the Nefzaoua and towards the centre of Chott Djerid. Within spring mound pools it ranges from 2 to 4 and below dry mounds from 4 to 12 mS cm^{-1}. In the Intercalaire, similar salinity relationships pertain although levels are generally higher. Powerful springs and boreholes show conductivities of between 3.5 and 6 mS cm^{-1} spring mound pools up to 12 mS cm^{-1} and near-surface water adjacent to mounds up to 20 mS cm^{-1}. Mounds relatively close together may show considerable differences in conductivity depending on the size of the water discharge. For comparison, rainfall locally yields only 0.05 mS cm^{-1}, while ephemeral channel flows have a relatively high value of 2.4–3.8 mS cm^{-1}.

The ratios of major ions show trends consistent with the precipitation of first Ca, then Mg, and then Na salts. The two aquifers are distinguishable by their Mg/Ca molar ratios. Intercalaire waters increase from 0.3–0.4 to around 0.6 towards springs, reaching 1.0 in and around spring mounds. Terminal waters show higher initial ratios of 0.45–0.8 in borehole samples, increasing to 0.75–1.5 in spring mounds. Ratios in the aioun range from 2.5 to 3.0. Na/Mg ratios also increase with evaporation as Mg is lost by precipitation. The ratios range from 3.0 for better quality springs and bores to 8.0 or 9.0 for brines with conductivities in excess of 10 mS cm^{-1}. Generally Na/Mg ratios for Intercalaire samples are slightly higher than the Terminal ones of the same conductivity, again suggesting relatively less Mg in solution before evaporation begins. The anion ratios show sequential loss of first carbonate then sulphate and chloride by precipitation, but there is no clear distinction between the two aquifers.

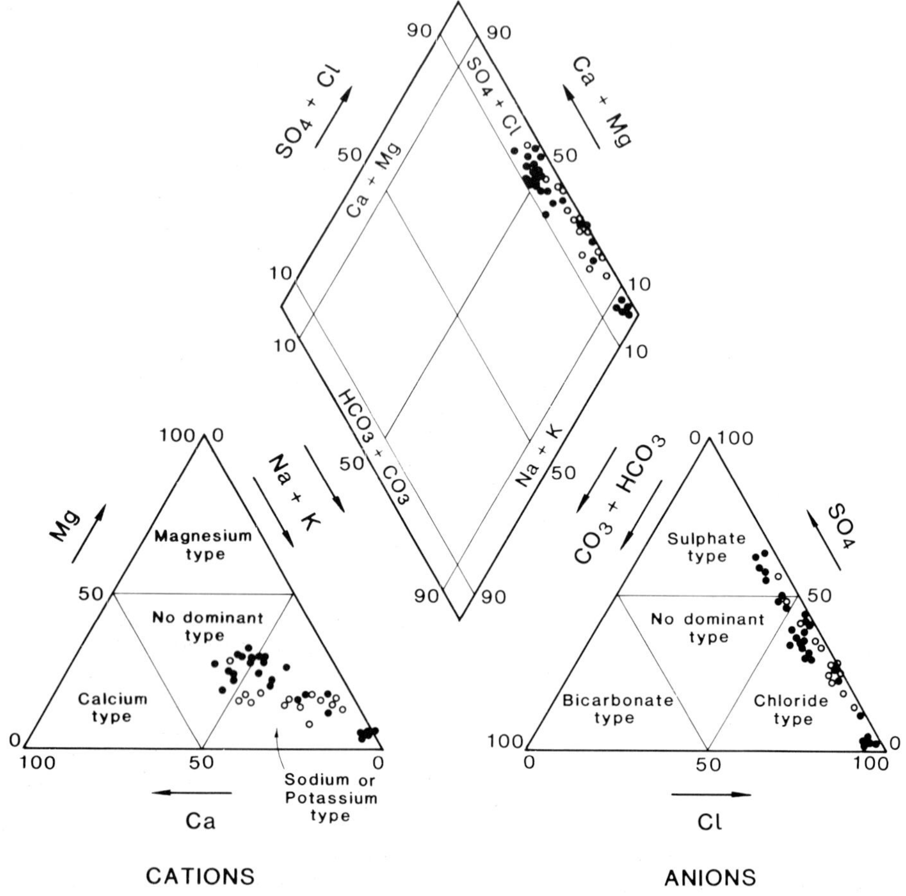

FIG. 11. Hydrochemistry of springs and bores.

The development of the spring mounds

The Tunisian spring mounds must have developed with the landscape in its present form since there is no morphological or fossil evidence that they have been affected by the lake level changes during the late Quaternary. Some of the largest, such as Zarsine and Telmine, appear to rise straight from the chott surface and thus postdate the last of the 'niveau a Cardium'. They must be regarded as potentially ephemeral landforms since their gypsum content makes them vulnerable to solution by rainwater, although rapid burial by aeolian materials might allow preservation into the geological record.

The modern spring mounds are predominantly inorganically cemented with no evidence of stromatolitic or algal structures. They appear to have been built mainly by groundwater from the two aquifers rising to the surface up chimneys or fissures, and seeping through the overlying clayey aquicludes at sharply defined points which may be related to fault zones or karstification in the underlying limestones. Although the water in the two aquifers has been somewhat salinized before it moves upwards, and becomes more so before it emerges, it is still relatively fresh in comparison with its surroundings. As it spreads from its point of emergence and spills over the top or seeps through the rim of the mound, evaporation makes it more saline and the precipitation of Ca and Mg compounds increases the dominance of the Na ion. Where the point of emergence is in the regularly flooded parts of the chott, only aioun can develop because the eminences are destroyed seasonally. Elsewhere, spring mounds grow upwards incrementally into a conical cratered form containing a high proportion of calcite, gypsum, and soluble salts, alternating locally with additions from human excavation. Aeolian additions to the damp surface during phases of growth vary. In the Fedjadj Oases these processes led to the formation of small spring mounds. In Nefzaoua, where artesian pressures were greater, the process continued until it had formed the large spring mounds up to 30 m or more in height, which were used for irrigation and excavated in historic times.

Even when, as here, the discharge and mineral content of the springs and the size and chemical composition of the mounds are known approximately, it remains very difficult to make accurate calculations about the time taken for their formation, especially in view of the difficulties of obtaining reliable radiocarbon dates. The most obvious model assumes that the artesian pressures were sufficient for water to reach the mounds' highest points continuously. It would then spill out radially over the surface in a film, or migrate to the surface by capillarity until evaporation initiates mineral precipitation. Under these conditions periods as short as a few hundred years would be sufficient for the construction of even the largest mounds.

Growth must however have alternated with substantial periods of decay when the moisture balance was less favourable. This is evidenced from some mounds by alternations, in central wells, of peat, wind-blown sand and clastic debris, and in the rims, of 'wet phase' debris and vegetation fragments, and 'dry phase' wind-borne deposits cemented during a later return to moister conditions. The evidence of rainsplash on some buried stratigraphic surfaces suggests that there have been rain-induced erosional phases as well as periods of intense wind erosion leaving truncated stratigraphic boundaries. Consequently, spring-mound formation is more likely to have occurred intermittently over thousands, rather than hundreds, of years.

After the formation of the mounds in the Nefzaoua Oases, and to a lesser extent in Fedjadj, there followed a period of relative inactivity. In order to maintain the flow of irrigation waters, increasingly deep channels and tunnels were cut through the sides of the mounds to reach freshwater in the central wells. It is difficult to determine when this first occurred, but the decline in piezometric pressure has been exacerbated since about 1950 by the proliferation of artificial boreholes to lower and lower aquifers. This has maintained and even enhanced the flow of irrigation water, but at the expense of depleting the spring mounds so seriously that degradation by wind and rain has been severe in the majority of Complexe Terminal-fed mounds. The Continental Intercalaire mounds, at present only slightly affected by rilling, wind scour and burial, provide modern analogues of the original methods of construction of the degraded Terminal mounds, and also highlight the necessity of incorporating considerable spatial variation in morphology into any evolutionary model.

References

BISHOP, W. 1975. Geology of Tunisia and adjacent parts of Algeria and Libya. *In*: MARTIN, L. (ed.) *Guidebook to the Geology and History of Tunisia*. Petroleum Exploration Society of Libya, Ninth Annual Field Conference, Tripoli, Libya, 51–58.

BLACKWELDER, E. 1931. The lowering of playas by deflation. *American Journal of Science* **221**, 10–144.

BUROLLET, P. F. 1967. General geology of Tunisia. *In*: MARTIN, L. (ed.) *Guidebook to the Geology and History of Tunisia*. Petroleum Exploration Society of Libya, Ninth Annual Field Conference, Tripoli, Libya, 51–58.

COQUE, R. & JAUZEIN, A. 1967. The geomorphology and Quaternary geology of Tunisia. *In*: MARTIN, L. (ed.) *Guidebook to the Geology and History of Tunisia*. Petroleum Exploration Society of Libya, Ninth Annual Field Conference, Tripoli, Libya, 227–257.

DROUHIN, G. 1953. The problem of water resources in northwest Africa. *In*: *Reviews of Research on Arid Zone Hydrology*. UNESCO, Paris, 9–41.

ENGLEBERT, V. 1970. The Danakil, nomads of Ethiopia's wasteland. *National Geographic Magazine* **137**(2), 186–211.

FAIRBRIDGE, R. W. 1968. Mound springs. *In*: *Encyclopedia of Geomorphology*. Reinhold Book Corporation, New York.

FONTES, J.-C. & POUCHAN, P. 1975. Les cheminees du lac Abbe (TFAI): stations hydroclimatiques de l'Holocene. *Comptes Rendus de l'Academie des Sciences*, E280, Series D, Paris, 383–386.

GINESTOUS, G. 1927. Le chêne zeen D'Ain Draham. *Bulletin Direction Générale de l'Agriculture, Commerce, et Colonisation* **31**(130), 329–338.

GISCHLER, C. E. 1979. *Water Resources in the Arab Middle East and North Africa*. Menas, Cambridge.

GONFIANTINI, R. *et al.* 1974. Etude isotopique de la nappe du Continental Intercalaire et de ses relations avec les autres nappes du Sahara septentrional. *In*: *Isotope Techniques in Groundwater Hydrology*. Proceedings of the Symposium of the International Atomic Energy Authority.

HARDIE, L. A., SMOOT, J. P. & EUGSTER, H. P. 1978. Saline lakes and their deposits: a sedimentological approach. *In*: MATTER, A. & TUCKER, M. E. (eds) *Modern and Ancient Lake Sediments*. Special Publication No. 2, International Association of Sedimentologists, Blackwells, Oxford, 7–42.

MABBUTT, J. A. 1977. *Desert Landforms*, MIT Press, Cambridge, Massachussetts.

MADIGAN, C. T. 1936. *Central Australia*, Oxford University Press.

MAMOU, A. 1976. *Contribution à l'Etude Hydrogéologique (de la Presqu'ile de Kebili)*. PhD thesis, Université Pierre et Marie Curie, Paris.

MCKEE, E. D. (ed.) 1979. A study of global sand seas. *United States Geological Survey Professional Paper 1052*.

NEAL, J. T. 1965. Environmental setting and general surface characteristics of playas. *In*: NEAL, J. T. (ed.) *Geology, Mineralogy, and Hydrology of US Playas*. Environmental Research Paper 96, Air Force Cambridge Research Laboratories, Massachussetts, 1–29.

—— & MOTTS, W. S. 1967. Recent geomorphic changes in playas of western United States. *Journal of Geology*, **75**(5), 511–525.

NESSON, C. 1978. L'évolution des ressources hydrauliques dans les oasis du Bas-Sahara Algérien. *In*: *Recherches sur l'Algerie*. Centre National de Recherche Scientifique, Mémoires et Documents **17**, 10–91.

NICHOLSON, M. E. 1980. Saharan climates in historic times. *In*: WILLIAMS, M. A. J. & FAURE, H. (eds) *The Sahara and the Nile*. Balkema.

PALLAS, PH. 1972. Water resources in the northern Sahara. *Nature and Resources* **8**(3), 9–17.

RICHARDS, G. W. & VITA-FINZI, C. 1982. Marine deposits 35,000–25,000 years old in the Chott el Djerid, southern Tunisia. *Nature* **295**, 54–55.

ROBERTS, C. R. 1986. Forecasting long term water needs on the Saharan margins of southern Tunisia. *In*: COALBECK, B., GARDINER, V. & HERRINGTON, P. (eds) *Water Demand Forecasting*. Elsevier, Amsterdam.

ROGNON, P. 1979. Evolution du relief et palaeoclimats depuis 40,000 ans sur la bordure nord du Sahara. *Bulletin de l'Association Géographique de France*, Paris, 462.

SCHOLL, D. W. 1960. Pleistocene algal pinnacles at Searles Lake, California. *Journal of sedimentary Petrology* **30**(3), 414–431.

—— & TAFT, W. H. 1964. Algae, contributors to the formation of calcareous tufa, Mono Lake, California. *Journal of sedimentary Petrology* **34**(2), 309–319.

STONE, R. O. 1956. *A geologic investigation of playa lakes*. PhD thesis, University of California.

TAYLOR, T. G. 1947. *Australia: a Study of Warm Environments and their Effect on British Settlement*. Methuen, London.

TWIDALE, C. R. 1972. Landform development in the Lake Eyre region, Australia. *Geographical Review* **62**, 40–70.

WATSON, A. 1983*a*. Gypsum crusts. *In*: GOUDIE, A. S. & PYE, K. (eds) *Chemical Sediments and Geomorphology*. Academic Press, London and New York.

—— 1983*b*. Evaporite sedimentation in non-marine environments. *In*: GOUDIE, A. S. & PYE, K. (eds) *Chemical Sediments and Geomorphology*. Academic Press, London and New York.

WATTS, S. H. 1975. Mound springs. *Australian Geographer* **13**, 52–53.

WEAVER, K. F. 1977. Geothermal energy: the power of letting off steam. *National Geographic Magazine* **152**(4), 566–579.

C. R. ROBERTS, School of Geography and Geology, College of St Paul and St Mary, Cheltenham, UK.

C. W. MITCHELL, Department of Geography, University of Reading, Reading RG6 2AB, UK.

REMOTE SENSING OF DESERT SEDIMENTS

Sand sheets of the eastern Sahara and ripple blankets on Mars

C. S. Breed, J. F. McCauley & P. A. Davis

SUMMARY: *Landsat* images of the eastern Sahara (North Africa) show widespread sandy plains covered by broad, gentle undulations (not dunes) that are interpreted as giant aeolian ripples. This regionally extensive bedform pattern embodies the entire upper unit of the Selima Sand Sheet. This unit, which we interpret as modern and presently active, contains very coarse sand, granules, and small pebbles, medium to very fine sand, and silt in horizontally laminated pairs that form a tabular deposit 1 cm to 10 m thick. It is commonly underlain by a stabilized sand sheet unit, and one or both units are disconformable on an erosion surface of truncated alluvium, mixed alluvial–colluvial–aeolian deposits, or bedrock. Sediment in the sand sheet is derived from alluvium, eroded mostly from sandstones in the Nubia Formation and widely distributed in dry, relict valleys. The upper surface of the sand sheet is a layer of pebbles and granules identical to coarse particles in the internal layers and in the truncated alluvial valley fill. Elongate, thin ripples composed of small pebbles are driven across this surface by prevailing north winds. Fine to medium sand is widely dispersed across the sand sheet from isolated dune fields. Silt particles are added seasonally by dusty winds (Khamsins) from the S. The Selima Sand Sheet may provide an analogue for some thin, horizontally-bedded aeolian sandstones. It offers an explanation for ripple blankets on Mars, where locally-derived coarse particles transported by surface creep and saltation are mixed with globally-derived fine particles deposited from dust storms.

Nearly flat or gently undulating sandy plains cover extensive areas in southwestern Egypt and northwestern Sudan (Figs 1 and 2). The plains are blanketed by horizontally-laminated deposits of sand, silt, granules, and pebbles, which are typically armoured by a surface layer of the coarsest particles. These sand-sheet deposits have accumulated independently of, and are overlain by, the occasional dune trains that migrate across their surfaces (Fig. 3). The Saharan sand sheets appear to be different in character (and perhaps in origin) from deposits in less arid deserts, where sand sheets are described (Ahlbrandt & Fryberger 1981; Kocurek & Nielson 1986) as the products of interdune or near-extradune environments. One large area of virtually unbroken sandy plains in Egypt and Sudan (at least 60 000 km^2) has been named the Selima Sand Sheet (Bagnold 1931, 1933; Haynes 1982*b*), but other names (Sand Plains of Misaha, Atmur el Kibeish) have been applied in various localities (Fig. 1). Altogether, the sand-sheet deposits cover at least 100 000 km^2; in this report all are included informally under the title Selima Sand Sheet.

Evidence from more than 150 hand-dug pits and backhoe trenches excavated in the Selima Sand Sheet from the border of Egypt with Sudan to the Dakhla Depression, and from the Gilf Kebir Plateau to Bir Kiseiba (Fig. 1) shows that the sand sheet is typically only a few centimetres to metres thick, and that it disconformably overlies either wind-truncated sediments of alluvial, colluvial, and mixed origin, or shallow bedrock (McCauley *et al.* 1986). Some of the underlying alluvial deposits contain artefacts of the Acheulian culture, provisionally dated as at least 250 000 years old, whereas the sand-sheet units contain artefacts of much younger cultures (Late Palaeolithic and Neolithic; McHugh *et al.* in press). Previous workers (Said 1980; Haynes 1982*a*; Haynes & Johnson 1984) included some of the alluvial deposits in the Selima Sand Sheet, but we suggest that the term 'sand sheet' should be restricted to the sediments of aeolian origin. The regional extent of the aeolian disconformity marks a change from dominantly fluvial to almost wholly aeolian processes of sedimentation during the late Quaternary in the eastern Sahara.

Today most of the eastern Sahara lacks surface water and, as a consequence, vegetation is absent except in isolated, scattered groundwater oases. This region is the hyperarid core of North Africa, where average annual rainfall is less than 5 mm (Henning & Flohn 1977; Nicholson & Flohn 1980). Its extremely flat, virtually featureless surface (Fig. 2) has been described as 'broad, flat peneplains mantled with small pebbles, all of about the same size and equidistant from each other, a result of countless aeons of wind abrasion, deflation, and sorting' (Wendorf & Schild 1980). New evidence of regional erosion and aggradation by ancient (Tertiary?) streams has been detected by spaceborne imaging radar (SIR), whose signal penetrated the sand sheets and revealed a subjacent fluvial topography (Breed *et al.* 1983; McCauley *et al.* 1982, 1986; Schaber *et al.* 1986). This evidence corroborates the earlier view of Peel (1939), Caton-Thompson & Gardner

From Frostick, L. & Reid, I. (eds), 1987, *Desert Sediments: Ancient and Modern*, Geological Society Special Publication No. 35, pp. 337–359.

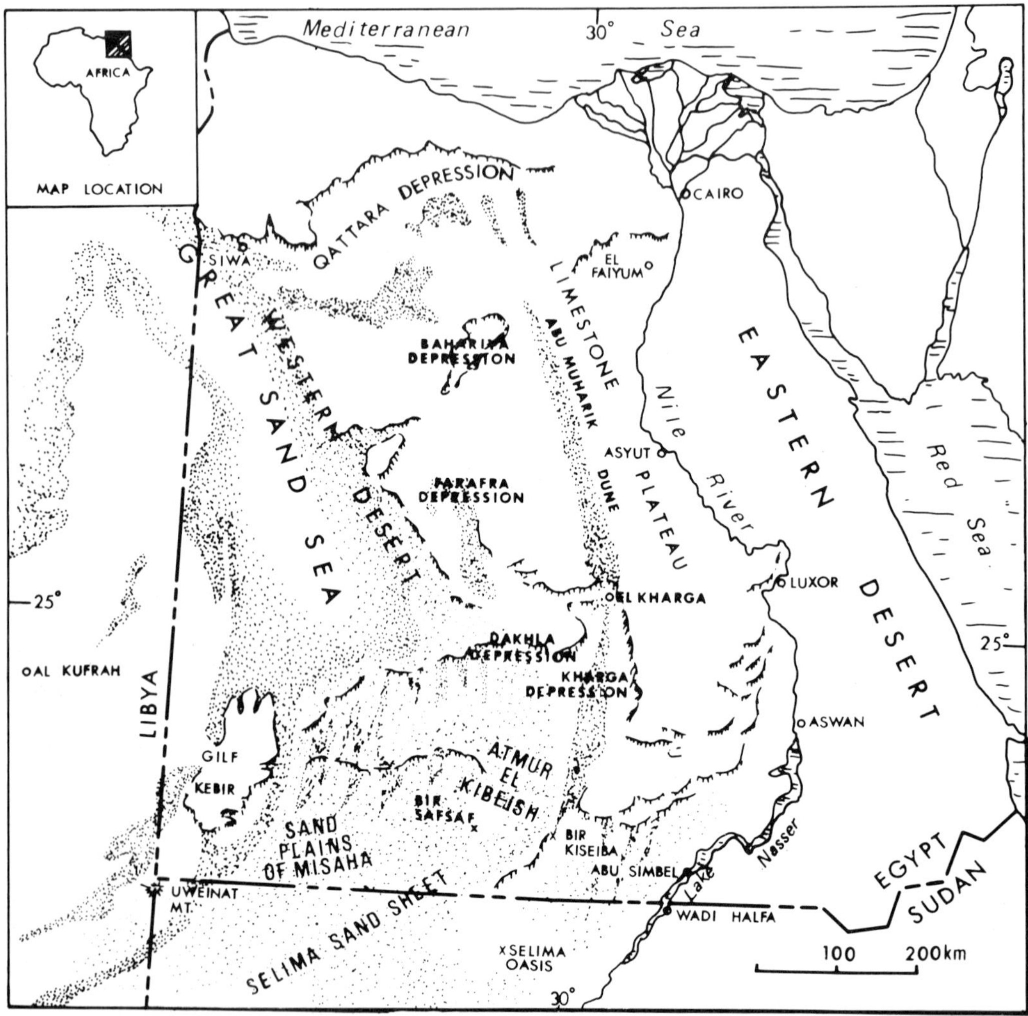

FIG. 1. Index map of the eastern Sahara shows regional extent of desert floor at least partly mantled by sand-sheet deposits. The Selima Sand Sheet of Bagnold (1931, 1933) is the central area along the border of Egypt with Sudan where these deposits are virtually unbroken.

(1932), and others that the physiographic 'framework of the eastern Sahara is primarily the legacy of pre-Quaternary fluvial erosion. Much later, during the Quaternary pluvials, this region was probably a savannah (as are areas farther S today; Wendorf *et al.* 1976), and occasional flash flooding occurred in the relic valleys. The sand sheet, although the dominant landform today, is an aeolian overprint on a much earlier fluvial landscape. Now only the artefacts of Stone Age people, preserved in and under the pervasive sand sheets, suggest the former availability of surface water and vegetation.

Field and remote-sensing observations of the Selima Sand Sheet

Variations in surface morphology

On *Landsat* Multispectral Scanner (MSS) and Thematic Mapper (TM) images of the Selima Sand Sheet, several subtle but distinct physiographic features can be discriminated. Each feature has a different surface expression (Fig. 3), ranging from bland, virtually flat plains to giant

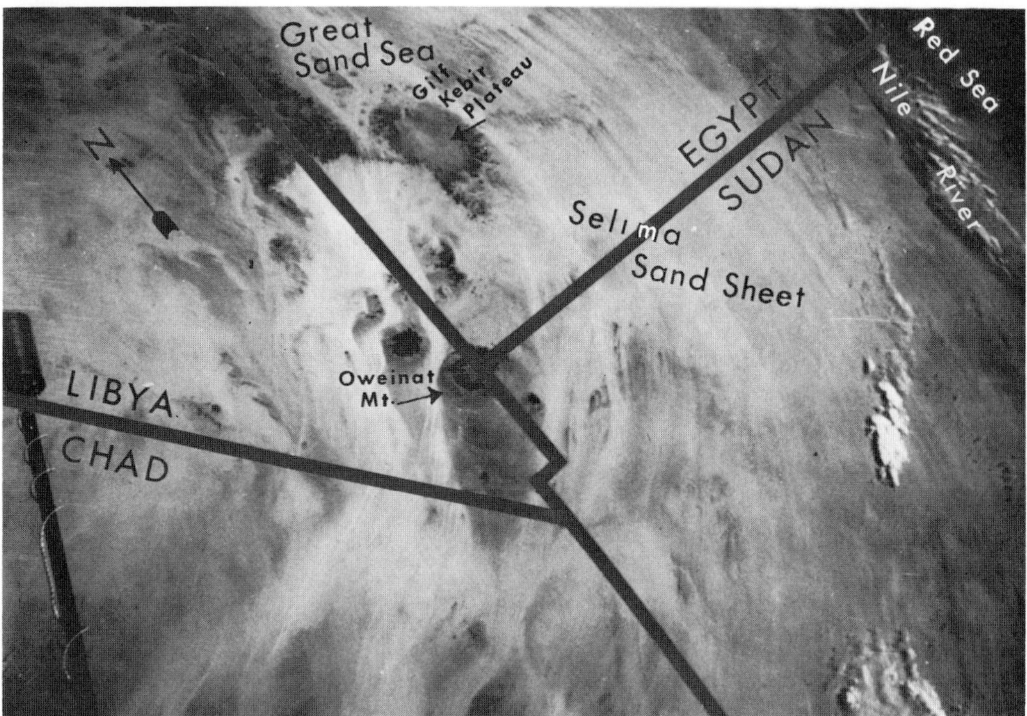

FIG. 2. Photograph taken during the *Gemini XI* space mission (1966) shows widespread aeolian sand cover and absence of surface water and vegetation. *Gemini* photograph S66-54529.

ripples; also present are local chevron patterns of dark material interrupted by traces of large, bright, parabolic features. Although each of these patterns is distinct on *Landsat* images, ground studies show little or no corresponding topographic relief, except in areas of the gentle undulations that we call giant ripples. Rather, the striking variations shown on the *Landsat* images of the sand sheet are due to differences in spectral reflectance of the surface sediments, and signify variations either in their grain size, in the extent of iron oxide or carbonate coatings on the grains, in their mineral composition or in a combination of these factors (Fig. 4).

The patterns seen both on the *Landsat* images and on the ground indicate that the entire veneer of loose particles, including concentrations of the coarsest grains, is mobile. On the ground in any given place, the surface particles on the sand sheet have a characteristic maximum size distribution that suggests an equilibrium response to the strongest winds at that locality. During a 10-year period of remote-sensing observations, *Landsat* images show that some of the patterns have changed. On the ground in some areas shown as elongate, dark chevrons on *Landsat* images, we found thin spreads of pebble gravel; some of the pebbles are ventifacted and most are coated with desert varnish (iron and manganese oxides). These pebbles, generally 1 or 2 cm in diameter, are commonly organized in very low ripples (a few centimetres high) on top of the sand sheet (Fig. 5*a*, *b*). These pebble-ripple spreads are invariably elongate in a southward direction, indicating that even the coarsest grains on the sand sheet are being driven across its surface by the strong N winds.

We have not yet located the bright parabolic features on the ground because they are transient—their positions as marked on older *Landsat* images used for field navigation in 1982–84 (Fig. 3) were probably unreliable. The high reflectivity of these features suggests, however, that they are composed of fine particles—sand or possibly granules—and some are reported by Haynes (1985) to be composed of laminated sand. Ordinary 'normal' sand ripples a few centimetres high (Sharp 1963) are seen on the sand sheet only as thin patches near sand dunes (Fig. 5*c*). However, peculiar 'truncated' ripples, which have unusually flat tops that intersect steep lee slopes (as if their upper parts have been sheared off) are common in many localities (Fig. 5*d*). At the flat surface of the truncated ripples is a layer,

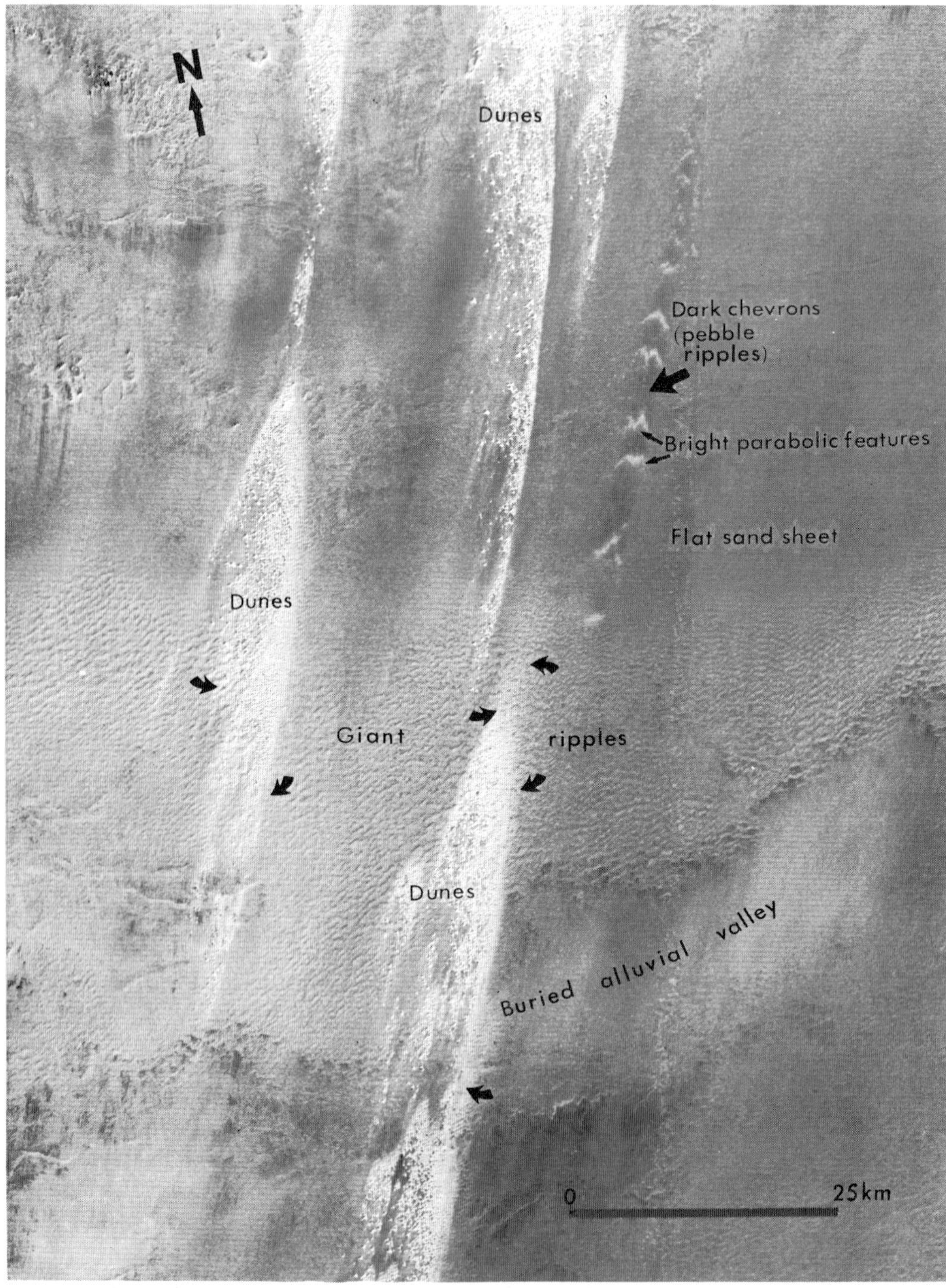

FIG. 3. The surface of the Selima Sand Sheet is marked in many areas by giant ripples and overridden in a few places by fields and trains of barchan dunes. Small arrows mark aprons of dune fields where the giant ripple pattern on the sand sheet can be seen undisturbed beneath a halo of sand dispersed from the dunes. The pattern of giant ripples is shown in more detail in Fig. 7. Dark chevron patterns (large arrow) are spreads probably of pebble ripples (Fig. 5*a*, *b*). *Landsat* image 1112-08085-7 (11/12/72).

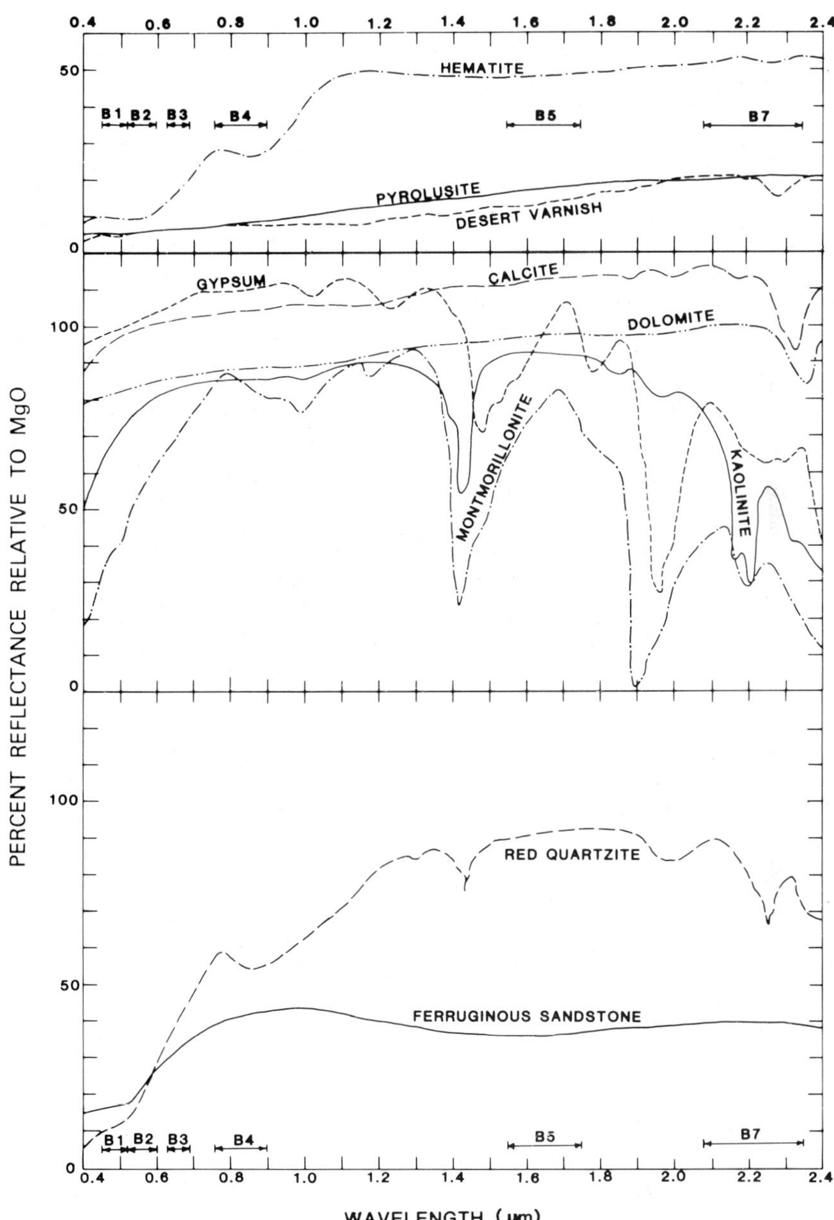

FIG. 4. Differences in spectral reflectance of several materials commonly found on desert surfaces (including the eastern Sahara) and observed on *Landsat* Thematic Mapper (TM) images. The TM sensors record these differences in several visible and infrared (IR) wavelength bands: B1, blue; B2, green; B3, red; B4, near-IR; B5, mid-IR; B7, mid-IR; (B6, thermal infrared, 10.3–12.5 μm, is not shown).

one grain thick, of very coarse sand. High-relief granule mega-ripples like those described by Bagnold (1954) and Sharp (1963) do not occur on the broad, open plains but they are common near barriers to wind flow such as bedrock outliers and dune fields (Fig. 6).

In areas marked by the giant ripple pattern on *Landsat* images, the entire surface of the sand sheet is gently undulatory. The relative relief of the ripple swells and hollows is as much as 10 m (Bagnold 1931, 1933; Haynes 1982*a*), and the average spacing of the swells, measured on

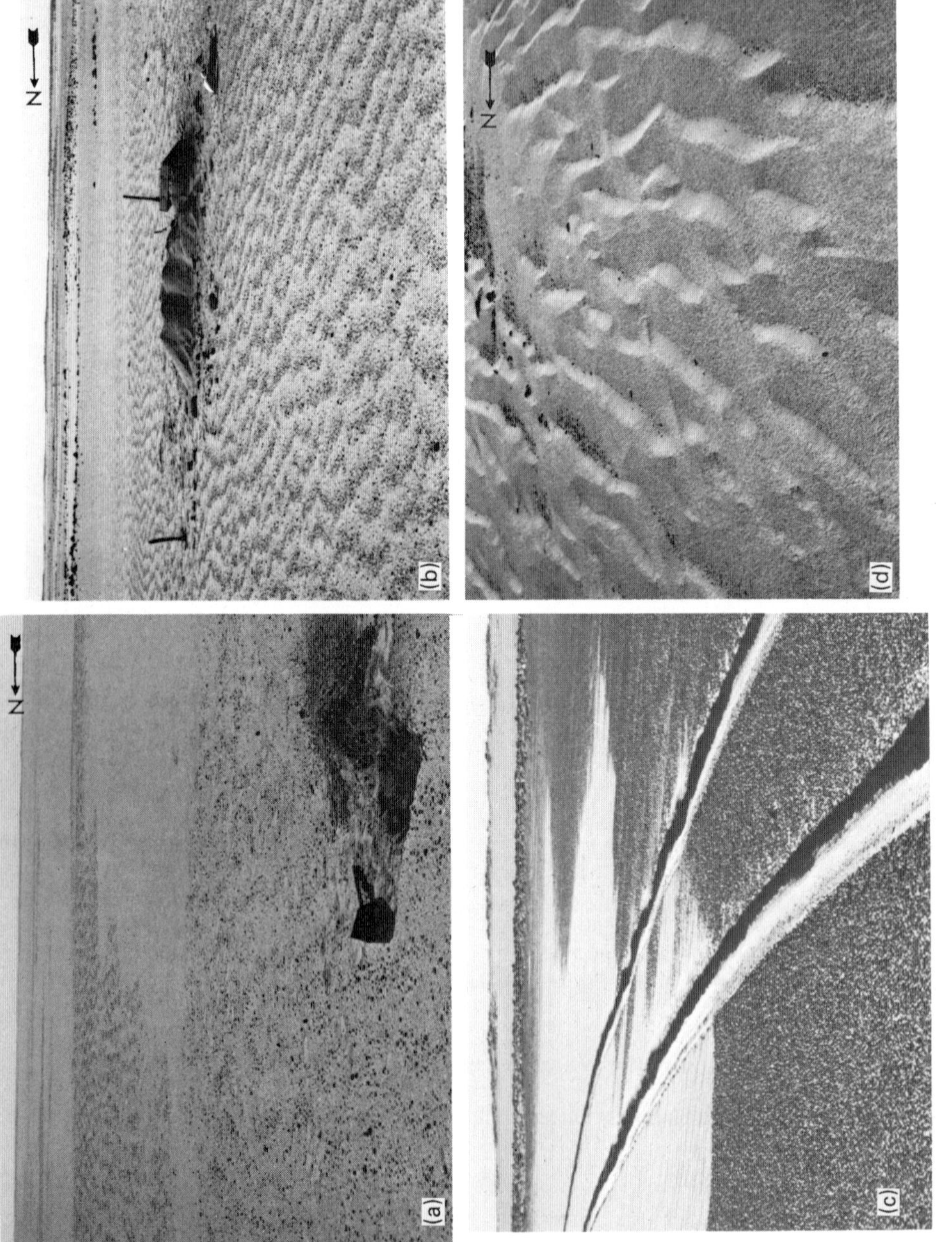

FIG. 5. (*a*, *b*) Pebble ripples migrating southward in two different localities on the Selima Sand Sheet consist of very coarse sand, granules, and small pebbles identical with those in alluvium exposed in shallow pits and trenches. (*c*) Thin layer of sand capable of saltation has accumulated on top of coarse lag particles on sand sheet about 20 km downwind of dune field. Sand is migrating in low ripples, which add a deposit one ripple thick to the surface of the sand sheet (Fig. 3). (*d*) 'Truncated' ripples typically occur near topographic barriers such as dunes and plateau outliers, where winds are funnelled. Lee slopes of ripples are about 10 cm high. Photograph (*a*) by C. V. Haynes.

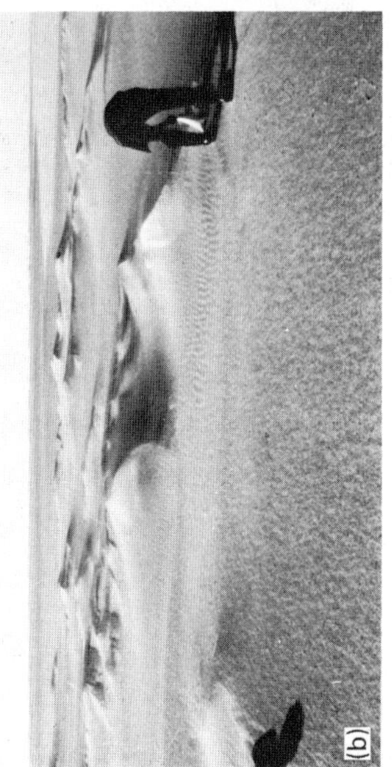

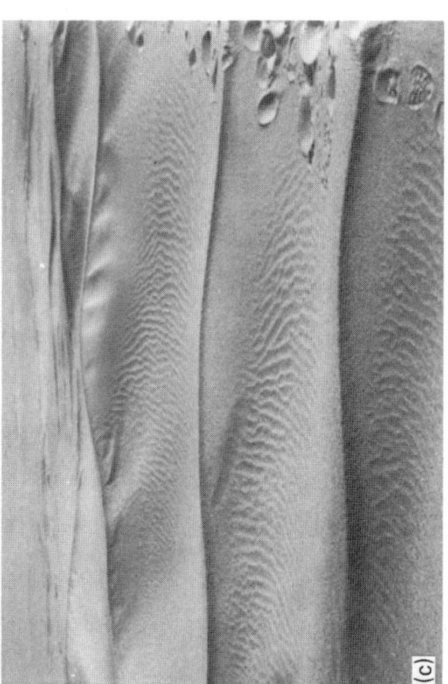

FIG. 6. (*a*) Granule ripples with 0.5 m high relief; the mega-ripples (large arrow) occur in Wadi Bakht, a sand-choked wadi of the Gilf Kebir Plateau (see Fig. 1 for location), where wind flow is constricted and accelerated. The size of 'normal' sand ripples on the dune in the foreground (with footprints) contrasts with the size of the granule ripples (figures for scale indicated by small arrow). (*b*) Granule ripples also occur near sand dunes and other obstacles to wind flow across the sand sheet. (*c*) Close-up view of ripples in (*b*). Photograph (*c*) by Maurice Grolier.

Landsat images, is just under 0.5 km (Fig. 7). The spacing/height ratio of the giant ripples is therefore unusually high, of the order of 50:1. The giant ripples are oriented generally transverse to the prevailing north wind, but their orientation varies locally in response to very subtle differences in topography that correspond to the locations of subjacent, relic alluvial valleys (Fig. 3). The sand sheet is thickest in the swells of the giant ripples and thinnest in the flat-floored hollows between. In the places where truncated bedrock, calichified alluvium, colluvium, or an older, pedogenically altered sand-sheet deposit is exposed in the hollows between giant-ripple swells, the contrast of surface materials is sufficient to make the rhythmic pattern obvious (Fig. 8). Elsewhere on the ground the pattern is barely noticeable, but it is pronounced on *Landsat* images throughout the region (Figs 3 and 7). The spectral contrast between the giant ripples and hollows on *Landsat* images appears similar to the contrast between dunes and inter-dunes that produces 'sand sea' patterns (Breed & Grow 1979). However, in the case of sand seas, the contrast can be explained by obvious differences in grain size, composition, and other characteristics, but the variations in spectral response within the giant ripple pattern are not so obviously explained and need further study.

Morphologically, the giant ripples do not resemble dunes either in their external geometry or their internal structure and grain-size distribution. Unlike most active dunes, they have no slip faces and do not avalanche; they therefore lack grainflow cross-stratification. Neither do they resemble the much smaller ripples and megaripples of the aeolian regime (Sharp 1963; summarized by Reineck & Singh 1980), most of which are the products of local wind activity. They may be extremely large examples of *zibars*, described by Holm (1960) in the Arabian desert as 'rolling transverse ridge[s] without slipface[s] . . . which [are] present in many interdune areas and which in the central Rub' al Khali [are] a dominant form.' However, few features as large and as regionally extensive as the giant ripples of the Selima Sand Sheet have been termed *zibars*. In the Qoz area of central Sudan, similar but stabilized undulations that are spaced as much as 400 m apart have been considered to be probable degraded dunes (Warren 1970). Undulations on the sandy plains in the Ténéré Desert of Niger have been termed *zibars* (Warren 1972; Fig. 2) but these have spacings five times less than those of the giant ripples in the Selima Sand Sheet. Nielson & Kocurek (1986) described still smaller *zibars* in the Algodones Dunes of southern California as 'coarse-grained, low-relief, slipface-less aeolian bedforms that occur on sand sheets and within interdune corridors of many aeolian sand seas'. However, unlike the giant ripples (which are overridden in only a few places by dunes) the *zibars* recognized by these and other authors typically occur in interdune corridors between rows of dunes, and they appear to feed into the dunes (Tsoar & Yaalon 1983). Some of the differences between the Saharan giant ripples and the *zibars* in the American deserts,

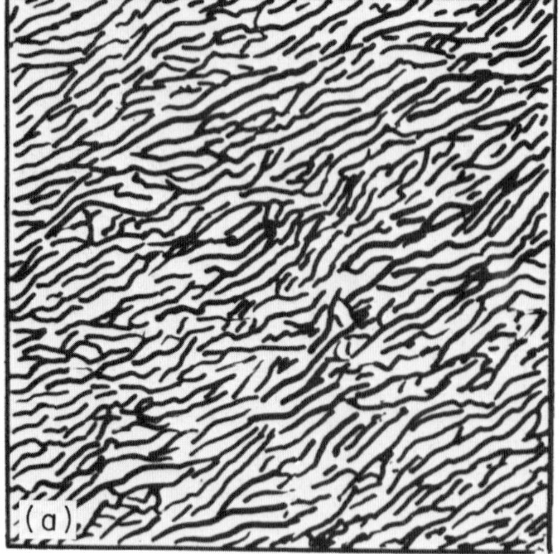

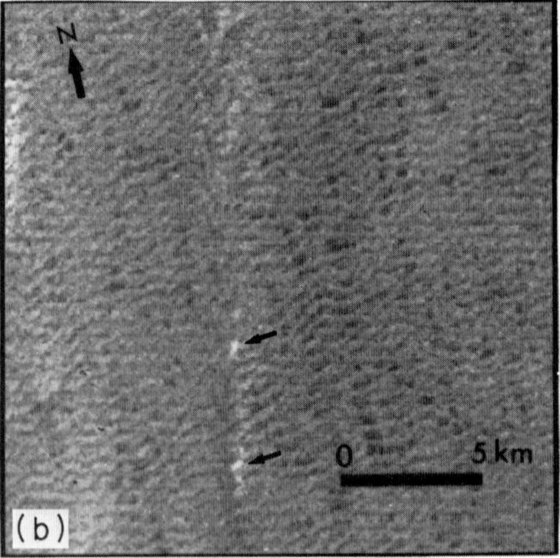

FIG. 7. (*a*) Sketch shows pattern of giant ripples drawn from enlarged *Landsat* image of the Selima Sand Sheet (*b*). The bright spots on the *Landsat* image (arrowed) are trains of isolated barchan dunes. Compare the pattern on *Landsat* imagery with ground view of giant ripple swells and hollows in Fig. 8.

FIG. 8. (*a*) Giant ripples on the Selima Sand Sheet are not readily apparent on the ground except where the inter-ripple hollows expose dark soil, calichified alluvium, or shallow bedrock. Vehicles in hollows for scale. (*b*) Flat-bedded laminae exposed in shallow pit dug in surface of the giant ripple in the foreground of (*a*). The coarse grain size mode here is 1.2 mm. The surfaces of the giant ripples are made of very coarse sand and lack sand ripples.

which we think can be explained by differences between regional and local environments of deposition, are discussed later.

Relation of the Selima Sand Sheet to dunes

Dunes in southwestern Egypt and northwestern Sudan are mostly barchans, and their distribution is limited to trains and minor fields migrating southward across the sand sheet (Fig. 3). The sand in the dunes has come probably from alluvial sources several hundred kilometres to the N, such as the ancient watercourses that have been inferred for the Dakhla region (Brookes 1983) and north-central Egypt (Salem 1976); but some may have been blown from dunes in the Great Sand Sea of northwestern Egypt. The isolated, throughgoing pattern of the dunes suggests little local contribution from the pre-existing sand sheet deposits over which the dunes are migrating. Passage of the dunes leaves a local diffuse halo of sand on the sand sheet, but does not otherwise change its surface (Figs 3, 6*b*, 7*b*). The localized concentration of fine to medium sand in the dunes, as opposed to the regional dispersion of the same size fraction in the Selima Sand Sheet, makes it unlikely that the sand sheet represents the 'trailing margins' of the present dune fields, or even the relict, degraded remnants of dunes that passed by long ago.

Composition and internal structure of the Selima Sand Sheet

Excavations show that the Selima Sand Sheet is built of unconsolidated to very weakly consolidated wind-blown sand, silt and small pebbles. These sediments are highly mature (Folk 1968), consisting of more than 99% quartz. Pairs of cyclic, concordant laminae within the sand sheet are commonly less than a centimetre thick. Each pair consists of a lower layer of coarse silt and very fine to medium sand and an upper layer, typically one grain thick, of very coarse sand, granules, or small pebbles (Fig. 9). All the internal structures that have been seen in more than 150 excavations are virtually horizontal and parallel. No clear evidence has been seen of cross-laminae or climbing-ripple foresets. Stringers of small pebbles, one grain thick, commonly occur on the truncated alluvial surface at the base of the sand sheet and are repeated through the aeolian deposit at several levels. These pebbles are identical to those on the present surface, where they are commonly organized into low ripples spaced a few centimetres from crest to crest (Fig. 5*a* and *b*). Vegetation is absent; not a blade of grass can be seen for hundreds of kilometres in any direction, and the sand-sheet surface is broken only occasionally by low outcrops of mostly wind-roughened Nubia Sandstone of Cretaceous age that are commonly darkened by desert varnish.

An upper and a lower sand-sheet unit can be distinguished in many localities on the basis of slight differences in colour and degree of consolidation. We interpret the upper light-brownish-yellow unit (7.5YR 7/4, Munsell Color Co., 1954) as the modern, presently active aeolian unit; this is the unit that forms the giant ripples. Locally, and particularly in the hollows between giant ripple crests, the modern sand unit consists of only a single depositional pair of fine and coarse laminae and is only a centimetre or two thick (Fig. 10). Elsewhere, the modern unit consists of several concordant pairs of laminae and is commonly 5 to 30 cm thick; but in some giant ripples, it is as much as 10 m thick (Fig. 8).

In many places, the modern unit overlies an older, redder sand sheet (7.5YR 6/6, typically) usually only a few tens of centimetres thick, but in some places (Haynes 1982*a*) as much as a few metres thick. This older unit is slightly more consolidated, such that it forms benches in the walls of excavations (Fig. 9*a*); it tends to be more massive and to lack the distinct coarse–fine laminae of the active unit. Where the older deposit is absent, the active sand sheet lies directly upon wind-truncated alluvium, colluvium, deposits of mixed origin, or upon shallow bedrock. Desiccation cracks commonly mark the upper surface of colluvial deposits containing very red soils (Fig. 10). These soils and their palaeoclimatic implications are currently under investigation (Haynes & Johnson 1984; Haynes 1985). Granular material in what is a deflated valley fill is identical with the reworked material of the sand sheets, and although the sediment in the sand sheet is typically bimodal or polymodal in size as a result of aeolian reworking, the range of grain sizes in the sand sheet closely resembles that of the regionally extensive alluvial deposits (Fig. 11). The primary source for the alluvium is the Nubia Formation of Cretaceous age, which forms the bedrock floor in most of this part of the eastern Sahara (Issawi 1981).

Grain sizes in both the younger and older sand sheets have a bimodal distribution and are poorly sorted (Fig. 11). Modes are typically in the very fine to fine sand classes (about 0.125 mm) and in the very coarse sand class (about 1.5 mm). The grains are typically well rounded and commonly are frosted. A puzzling characteristic was noted earlier by Maxwell (1982) was also recognized by Warren (1972) in sand sheets of the Ténéré Desert (Niger): this is that a considerable proportion of these deposits consists of medium to fine sand, *ie*

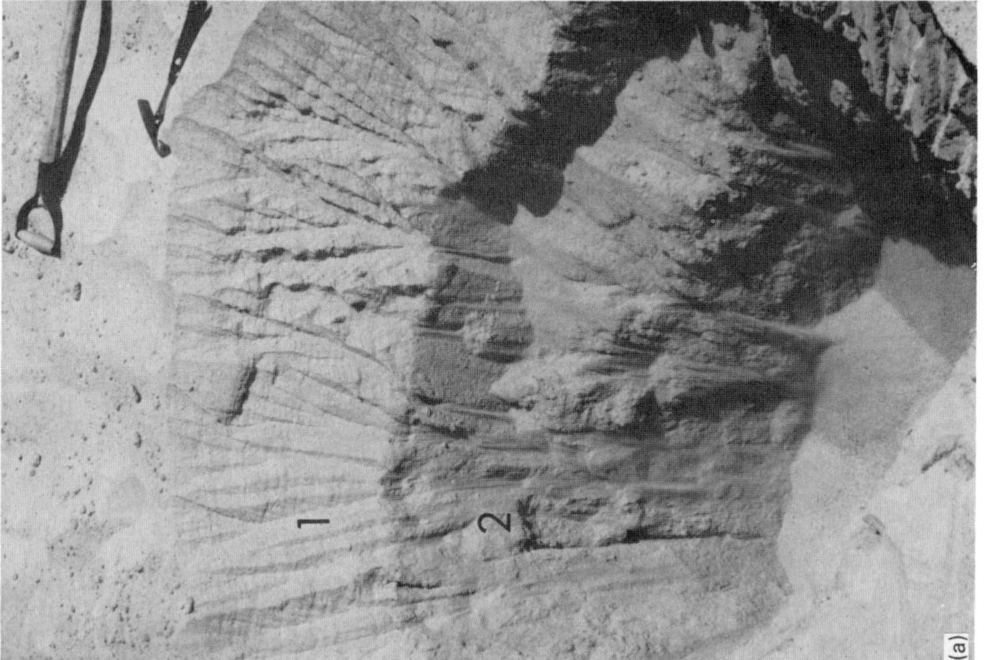

FIG. 9. Pits in the Selima Sand Sheet at different localities show typical development of near-horizontal laminae. (*a*) 30 cm of active, modern sand sheet (1) lies over older sand sheet with weakly developed soil structure (2); depth of pit is about 1 m (pit did not reach the bottom of sand sheet). (*b*) Modern sand sheet only above disconformity (arrow) that marks wind-truncated surface of alluvial deposits.

FIG. 10. Pit in hollow between giant ripples of the Selima Sand Sheet exposes an older, reddened colluvial deposit beneath the modern, active sand sheet which is only a few centimetres deep. Desiccation cracks indicate pedogenesis in the older deposits during episodes of climatic conditions less arid than those at present. Photograph by G. G. Schaber.

grains capable of saltating and forming sand dunes. The Selima Sand Sheet, unlike the 'bimodal supermature' deposits described from other deserts (Folk 1968), is not 'impoverished': it does not consist entirely of coarse and very fine grains left behind by the concentration of medium to fine sand in the dunes. Warren (1972) and Peel *et al.* (1974) suggest that much of the finer size fraction of the saltation load has been trapped in the sand sheets in voids between larger particles, leaving the coarser size fraction to be collected in the dunes. A gap does exist in the distribution of grain sizes, but the missing fraction is coarse sand, and this coarse sand is not stored in the dunes (Fig. 11). Coarse sand *is* present in the alluvium that is widely distributed throughout the region, and although the alluvium is the putative source material for both the sand sheets and the dunes, coarse sand is deficient in these aeolian derivatives. Perhaps the coarser sand grains are broken along microfractures by impact during saltation, thus effectively disappearing from the size distribution while contributing to the abundant smaller size fractions.

One explanation for the growth and migration of sand dunes in spite of the retention of abundant fine-to-medium sand in the sand sheet is that the dunes are not necessarily built of sand winnowed directly from the sand sheet. These dunes may have migrated into the region from points of origin hundreds of kilometres upwind of their present positions. Present-day annual migration rates for barchan dunes on the Selima Sand Sheet in Sudan range from 6 to 7 m (Haynes 1985). Distant relations of dunes to source areas is common in many inland desert regions, especially where the migration of dunes has not been impeded by topographic barriers (Breed *et al.* 1979; Fryberger & Ahlbrandt 1979). In this scenario, the dune fields are displaced southward by the repetitive removal of sand from their upwind to their downwind margins; they are not replenished by sand removed locally from the sand sheet. Indirect support for this concept can be seen in the patterns typical of the dune fields in this part of the Sahara: each field or train extends southward from a narrow point source that appears unrelated to any supply of sand from the sand sheet (Fig. 3).

Field observations also indicate that the supply of new sand to the dunes is minor. The transport of loose sand in dune fields is primarily accom-

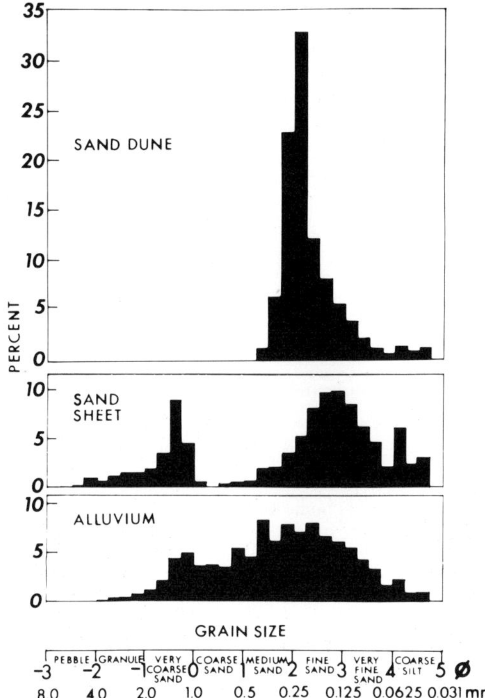

FIG. 11. Histograms illustrating the different grain-size distributions typical of the three types of Quaternary sediments that mantle much of the eastern Sahara. The alluvium, found throughout the region in aggraded valleys and truncated by deflation, is considered the parent material for both the sand sheets and the scattered, isolated dunes.

plished by ordinary sand ripples (Hunter 1977), but loose sand has been observed only locally distributed in rippled patches on the sand sheet during the autumn, winter, or early spring field seasons, when north winds dominate (Fig. 5*c*). During six field seasons, the air was generally clear when north winds of sand-moving strength were blowing, perhaps because winds from this direction blow mostly over surfaces that are now stable, *ie* desert pavement, bedrock, alluvium deflated to a $CaCO_3$-cemented and resistant horizon, and granule-armoured sand-sheet deposits that do not yield fine particles unless disturbed (Breed *et al.* 1980). Only where north winds blow over the isolated dune fields is any sediment visibly in motion, and there, saltating sand is abundant but limited to the vicinity of the migrating dunes.

Sand ripples probably develop widely in the region—that is, dispersed on the sand sheet beyond the aprons of the dune fields—only during the late spring and summer Khamsin season, when winds reverse direction and blow from the S. These wind storms are reportedly severe enough to knock down tents and redistribute sand on the plains of SW Egypt (Wendorf & Schild 1980). The Khamsins blow into the region from the southern margin of the Sahara, where fine particles, initially blown southward by the prevailing winds, have accumulated. These particles are recycled into the region of the Selima Sand Sheet as dust. During and after the Khamsins, both the sand that is dispersed in haloes around the dunes and the silt particles deposited from suspension undoubtedly infiltrate the coarser particles of the sand sheet in the same way that aeolian silt and sand infiltrate desert pavements in parts of the Mohave Desert (Wells *et al.* 1985). With the seasonal shift of winds to the N, loose sand on the surface would again be swept southward; sand grains carried across the sand sheet to the dune fields would be trapped and incorporated there, and the dunes would resume their march S. This interpretation provides a mechanism for the widespread dispersion of sand in the sand sheet concurrent with its local concentration in the dunes, rather than the growth of the latter at the expense of the former.

Problems of origin of the Selima Sand Sheet

Comparison with sand sheets *(zibars)* in N American deserts

Kocurek & Nielson (1986) summarized the environmental factors that promote accumulation of sand-sheet deposits in the American deserts. Differences between the external and internal characteristics of the Saharan giant ripples and the American *zibars* are summarized in Table 1. Vegetation is one key factor that encourages accretion (by promoting the accumulation of flat layers of sediment in grassy areas and of variably inclined layers in hummocky areas). Other factors are high water tables and the presence of dissolved salts, which promote cementation of sand-ripple laminae. The resulting crusts prevent further sediment transport and lead to the accumulation of thick sand-sheet deposits in interdune basins and extra-erg sabkhas. On the trailing edges of dune fields, the coarse residuum of this dune-building process is organized into *zibars* (Fig. 12). Coarse materials are periodically introduced into the aeolian system by rare flood events on nearby alluvial fans, from which particles are removed by deflation and redistributed by extremely high velocity wind storms. Successive migrations of sand ripples across the surface rework the coarse

TABLE 1. *Characteristics of bedforms on sand sheets:* zibars *in the Algodones Dunes, California (after Nielson & Kocurek 1986) and giant ripples in the eastern Sahara. Characteristics common to both are listed centrally between columns*

Zibars, Algodones	Giant ripples, Sahara
Striped or wavy pattern in plan views	
Inter-*zibar* and inter-ripple slope nearly horizontal	
No slip face	
($\delta_I = 0.7$ to 1.1) Poorly sorted ($\delta_I = 1.7$)	
Orientation transverse or oblique to resultant direction of transport	
Amplitude <3 m Well-developed stoss slopes <5° Lee slopes <15°	Amplitude <10 m Stoss and lee slopes extremely low angle <5°
Spacing averages 60 m	Spacing averages nearly 0.5 km
Ubiquitous sand ripples on surface	Patches of granule and small-pebble ripples; sand ripples rare
Grass and shrub vegetation (coppice dunes)	No vegetation except in groundwater oases
Stacked packages or cross-cut bundles of wind-ripple laminae dipping <15°, bounded by truncation surfaces dipping <10°	Concordant depositional pairs of coarse and fine laminae in sets bounded by horizontal truncation surfaces
Preserved ripples common	Preserved ripples not seen
Root mottling, bioturbation	No observed roots or bioturbation
Grains coarser than dune sand (0.1 to 1.0 mm size range)	Contains large proportion of sand equivalent to dune sand (medium to very fine size range)
Bimodal size distribution (modes at 0.1–0.3 mm and 0.3–0.65 mm)	Bimodal size distribution (modes at 1.2 mm and 0.125 mm)
Local deposits marginal to dunes	Regional deposits disconformable beneath *superimposed* dunes
Contain two types of depositional sequences; low and high angles of climb	Contain planebed(?) and probable grainfall laminae; climbing ripple structures not seen
High potential for preservation	Potential for preservation as thin beds on first-order bounding surfaces
Form in ergs in locations unsuitable for dunes but marginal to dune fields	Form on desert plains downwind from sources of alluvial sediment and distant from dune fields
Semi-arid to arid regions	Hyperarid to arid regions

and fine sediments into *zibars* that have characteristic internal structures (low-angle laminae, multiple truncation surfaces).

The giant ripples and the *zibars* may represent end members of a class of bedforms produced by aeolian processes that operate somewhat differently where climatic conditions are different and where topographic influences on wind flow directions are regional rather than local in scale. The environment of the sand sheets in the eastern Sahara differs in almost all respects from that of the sand sheets in N America. Only one of the five factors considered by Kocurek & Nielson (1986) to be favourable for sand-sheet development is common to both regions: this is a significant coarse-grained sediment fraction. The other factors (vegetation, high water table, surface cementation or binding, and periodic flooding) do not apply to the active Selima Sand Sheet. A further key difference is that the sand sheets described in American deserts are all elements or byproducts of the dune-building process; they occur on the lateral fringes, the progradational edges, or the trailing edges of active dune fields. These localities are generally characterized by gentle slopes on which climbing sand ripples are common and secondary wind directions are obvious (Fig. 12). In contrast, the environment of deposition in the eastern Sahara is characterized by almost imperceptible slopes and by a regional wind fetch that is virtually unbroken by topographical barriers.

Most vertical accretion of coarse and fine sediments in low-angle laminae in American

FIG. 12. Low, rolling ridges without slipfaces lie in the foreground of the field of star dunes adjacent to inselbergs in Gran Desierto, Sonora, Mexico. These sand ridges are *zibars* that form on the aprons or trailing margin of dune fields. Photograph by Peter Kresan.

deserts (type 'a' laminae of Fryberger *et al.* 1979; and *zibars* of Nielson & Kocurek 1986; and of Kocurek & Nielson 1986) is attributed to deposition by climbing sand ripples that migrate across the surfaces, most commonly along the aprons of dune fields. The only localities where climbing sand ripples could build up the modern Selima Sand Sheet to a thickness of more than a centimetre or so are on the very gentle lee slopes of the giant ripple swells (Hunter 1977, and pers. comm. 1986). There, removal of grains from the upwind slopes of the giant ripples and their deposition on the lee slopes by saltation and creep presumably causes migration of the entire giant ripple in the direction of the strong, prevailing north wind. The coarsest particles, left behind as a lag in the hollows between giant ripples, are buried by the oncoming swells and incorporated as the pebble stringers in the sand sheet.

The small-scale cut-and-fill structures, the low-angle ripple structures, the multiple, gently dipping truncation surfaces, and the bioturbation traces common to sand-sheet deposits formed in the American deserts are virtually absent from the active sand sheet in southwestern Egypt. Perhaps the extreme horizontality of structures within the Selima Sand Sheet is a type of planar lamination, which occurs where particles are driven by traction (saltation and surface creep) across a smooth surface under high-speed winds (Sharp 1963; Hunter 1977; type 'b' laminae of Fryberger *et al.* 1979). Field observations during and after severe storms are obviously needed to resolve problems of the origin of the near-horizontal laminae and the mechanics of entrapment of fine particles in the sand sheet. So far these observations have not been possible because the extremely high temperatures (average summer maximum 42°C; Wendorf & Schild 1980) have made logistical support unobtainable.

Climatic significance

The geological, climatic and archaeological history of the latest Pleistocene and the Holocene in southwestern Egypt and northwestern Sudan is represented by a few centimetres to metres of aeolian sediment preserved in the Selima Sand Sheet. The shallow stratigraphy observed in our pits and trenches, including the artefacts of cultural groups widely separated in their times of occupation (McCauley *et al.* 1986; McHugh *et al.* in press) suggests that the materials in the sand sheet have been reworked by wind during several long intervals when hyperarid conditions prevailed. Palaeoclimatic and geoarchaeological studies by other workers indicate that the aridity, which set in at about the Pliocene–Pleistocene boundary, recurred episodically, most recently in the middle Holocene (Williams 1975; Haynes 1980, 1982*b*; Nicholson & Flohn 1980). The modern unit with its giant ripple bedforms appears to post-date the onset of the present hyperaridity. This unit buries many grinding stones (Fig. 13) and other implements left by Neolithic occupants who left the region about 4000 years ago. Some suggestion of the time required to accumulate the active sand sheet is given by the buried or partially buried positions of these Late Stone Age implements and of undisturbed objects such as small food cans and bottles left behind by World War II military convoys (Fig. 14).

The active sand sheet is generally much thinner than the underlying older sand sheet (Fig. 9*a*); the older unit probably accumulated throughout longer intervals of hyperaridity that prevailed between successive human occupations (Wendorf *et al.* 1976). During the late Pleistocene and early to middle Holocene pluvials that punctuated the general aridity, a widespread grass cover is believed to have existed (Haynes 1982*a*). Repeated growth of vegetation would have helped

FIG. 13. (*a*) Grinding stones abandoned by Neolithic occupants of southwestern Egypt suggest the former presence of vegetation and a shallow water table in a relic alluvial valley near Bir Safsaf (see Fig. 1 for location), now covered by the active unit of the Selima Sand Sheet. Implements are shown partly buried in the thin sand sheet. (*b*) The artefacts after excavation. The region was abandoned in the middle Holocene, after the onset of the most recent episode of hyperaridity.

FIG. 14. Position of bean can abandoned about 40 years ago suggests length of time required to accumulate uppermost, undisturbed laminae in the active unit of the Selima Sand Sheet. (The laminae beneath the can have apparently been disturbed by it, but those above are horizontal and undisturbed.) The thickness of the undisturbed laminae—less than 2 cm—represents about 40 years' accumulation under present wind conditions. The stringers of small pebbles are left by pebble ripples like those shown in Fig. 5*a*, *b*. Layers of fine sediment are sand dispersed from dune fields (as in Fig. 5*c*) and silt deposited from seasonal dust storms (Khamsins).

to stabilize the older sand-sheet deposits and could account for their more massive, less distinctively layered, bioturbated and pedogenically altered character (Haynes 1982*a*, 1985).

Applications to the geological record

The differences in internal structure and morphology between sand sheets that have accumulated in hyperarid and semi-arid environments (Table 1) suggest some applications in interpreting the geological record. Aeolian sediments that accumulated on plains during geological intervals before the evolutionary appearance of sand-trapping plants, especially grasses, or in hyperarid depositional environments where plants were scarce, may be represented in the geological record by relatively thin but widespread, horizontally-laminated deposits like the Selima Sand Sheet. Such deposits may resemble the very thin, flat-bedded layers in the Eramprunya Sandstone of early Triassic age described by Marzo (pers. comm.), or parts of the Navajo Sandstone of Triassic(?) and Jurassic age illustrated by Middleton & Blakey (1983; Facies D, Figs 5 and 9).

Ripple blankets on Mars

The recognition of giant ripples as diagnostic features of the widespread sand-sheet deposits in the eastern Sahara has prompted another look at *Viking* high-resolution orbital images, to see whether similar bedforms might be present on Mars. Many other desert features in the eastern Sahara are morphologically similar to features on Mars (McCauley *et al.* 1979, 1981; El-Baz *et al.* 1979; Breed *et al.* 1979*b*, 1982) suggesting that aeolian processes have operated in mechanically similar ways in both places (Fig. 15). Most attention to deposits of aeolian origin on Mars has focused on either sand dunes or 'dust blankets', and on the related problems of sand sources, the physics of saltation in the thin Martian atmosphere (more than 100 times less dense than Earth's), the geomorphic characteris-

FIG. 15. Surface of Mars at *Viking Lander* site shows the characteristics of the northern mid-latitude plains of Mars, including aeolian ripples and drifts (Sharp & Malin 1984), and pitted rocks that resemble the wind-pitted rocks of the eastern Sahara (McCauley *et al.* 1979). Deflation of the drifts indicates that post-depositional winds at this locality have been more effective than the winds that deposited the drifts. Large rock ('Big Joe') is about 2 m wide and 1 m high. *Viking* image 11B139.

tics of Martian dunes, the variable nature and distribution of streaks, and the sedimentation of particles from global dust storms. The 'debris blankets' widely distributed on Mars have generally been interpreted as dust deposits (Christensen 1986). Maxwell (1982), who has also studied the Selima Sand Sheet, is one of the few to have suggested the widespread occurrence of sand sheets on Mars. The variations in Martian surface mechanical properties and chemical signatures that would be implied by the presence of regional sand-sheet deposits have not been widely considered.

Martian giant ripples

Figure 16 is a *Viking* high-resolution image of Martian debris-mantled plains in the Tharsis volcanic province about 500 km NW of the great volcano, Olympus Mons. The location suggests that the surficial materials should include an abundant supply of volcanic particles. A gently rolling topography, on which a distinct ripple-like bedform pattern is developed, extends over hundreds of square kilometres in the Arcadia Planitia and Diacria Patera quadrangles. The spacing of these ripple-like bedforms is remarkably uniform and averages about 200 m. The geological unit is mapped as smooth plains materials, and has been interpreted as a thin mantle of wind-blown material partly stripped by wind (McCauley 1973; Soderblom *et al.* 1973; Zimbelman & Kieffer 1979; Scott *et al.* 1981). The Martian undulations, like the giant ripples of the Sahara, appear to be an integral part of a widely distributed debris blanket.

These Martian bedforms are interpreted here as giant ripples rather than as dust blankets or dunes. It is unlikely that dust alone can build such bedforms, for ripple development requires the repeated migration of particles across the surface in traction mode. These bedforms lack the crisp definition and sharp breaks of slope that mark the slip faces of dunes, which, in any case, are concentrated in the north polar erg and intracrater dune fields (Breed *et al.* 1979*b*; Tsoar *et al.* 1979). Unlike most dunes (whether on Mars or on Earth) the ripple-like features shown in Fig. 16 are neither deflected by nor concentrated by topographic irregularities; rather, their pattern is

FIG. 16. Ripple-like bedforms on the surface of Mars at lat. 40°N, long. 147°, suggest that the region is mantled with sand-sheet deposits similar in character to Selima Sand Sheet in the eastern Sahara. Compare the rhythmic pattern with the Saharan giant-ripple pattern (Fig. 7). *Viking* image 112A22.

virtually uniform throughout the region, even around craters that would be expected to deflect dunes. The prominent crater shown in Fig. 16 appears to have been exhumed from beneath a surficial sedimentary unit that contains the giant ripples. The irregular margins of the rippled unit, and the apparent absence of ripples in topographically and stratigraphically lower units, suggests that this area was mantled by a rippled unit that has been partly stripped by wind erosion, exposing an unrippled substrate. No dune field on Earth known to us has this appearance.

If these bedforms are indeed giant ripples, their regional development on the mantled plains has several implications for the nature of the Martian surface. First, it implies that the Martian giant ripples are the topographic expression of widespread, tabular, sand-sheet deposits, probably only a few metres thick, which consist of thin layers of very coarse particles (deposited from surface creep), interspersed with layers of fine particles (deposited from suspension in dust storms) plus an increment of sand (deposited from saltation). Such a sequence of laminated coarse and fine sediment could be expected to have physical and mechanical properties much like those of the Selima Sand Sheet. It should include the entire size range of aeolian particles that have been active in the Martian environment, including granules (Sharp & Malin 1984). The stripping back of such a layered sequence provides a stratigraphic explanation for the peculiar appearance of some streaks on Mars, and for the grain-size distributions that have been inferred from orbital measurements of thermal inertia and colour relations (Thomas 1982; Thomas *et al.* 1986).

Second, the formation of giant ripples signifies that effective (sand-moving) winds have been active much further S on Mars than the boundaries of the circumpolar sand sea would indicate. The region shown in Fig. 16 is probably affected by the wind regime associated with Olympus Mons, where strong slope winds have been inferred from streak directions (Thomas & Veverka 1979). The orientation of the ripples is transverse to the resultant wind direction inferred from the streaks in this vicinity. The relatively coarse-grained particle size distribution implied by the presence of the giant-ripple bedforms requires surface winds at least as strong as those required to move dune sand in saltation; surface creep has probably been a major process for the redistribution of the coarse fraction in the Martian ripples. This interpretation agrees with the conclusions of Sharp & Malin (1984) regarding the capabilities of wind to move sand and granules across plain surfaces at the *Viking Lander* sites (Fig. 15).

The third implication of giant-ripple development is that the chemical signatures of Martian ripple blankets should reflect both the variety of sources for the sediments in the sand sheets and their rhythmic patterns of spatial distribution. The locally-derived coarse grains may have a chemical composition (as well as a range of grain sizes and sorting characteristics) entirely different from those of the finer grain-fall deposits that are globally distributed in dust storms and that have built up the 'universal regolith' of Mars (Clark 1982). The coarsest grains probably form a lag surface no more than one grain thick, but a layer that is repeated at near-horizontal angles through the deposit. On the surface the coarse grains may be organized in smaller-scale granule or pebble ripples superimposed on the giant ripple pattern. The grainfall deposits from global dust storms on Mars (Christensen 1986) are likely to infiltrate and be stacked between the layers of relatively coarse-grained particles that have been transported by surface creep and saltation.

Fourthly, the presence of sand sheets can explain some inconsistencies in the correlation between the thermal inertia of the Martian surface, measured from *Viking Orbiter 2*, and the mapped *bedrock* geological units (Zimbelman & Kieffer 1979). These authors showed thermal inertia values to be strongly controlled by aeolian debris. They suggested that some factor has stabilized the surface of an aeolian debris mantle no more than about 10 m thick, and that this prevents the suspension of fine particles by saltation. Christensen (1986) estimated the thickness of the low-inertia deposits to be between 0.1 and about 2 m, and suggested that variations in their average thermal inertia are controlled not so much by differences in the relative abundance of rock type but by variations in particle size and degree of consolidation. Although these differences in thermal inertia on Mars have been attributed to the burial of sand by dust blankets, we suggest that sand sheets stabilized by a surface layer of coarse particles offer a more plausible geological explanation.

Conclusions

The Selima Sand Sheet in the eastern Sahara is an aeolian deposit of regional extent, covering more than 100 000 km^2 in southwestern Egypt and northwestern Sudan. It can be recognized on *Landsat* images by its characteristic surface morphology, dominated by rolling plains of giant ripples. The giant ripples may be an extreme form of *zibar* but they differ in morphology, internal structure, and their relation to dunes from *zibars* described in the American Southwest which are localized on the aprons of dune fields. The Selima Sand Sheet, in contrast, has dune fields on its *surface* but in only a few places.

The Saharan giant ripples incorporate the active upper unit of the Selima Sand Sheet, which ranges in thickness from a centimetre or less in some hollows between giant ripples to about 10 m in the largest ripple swells. This active unit is built of pairs of virtually horizontal laminae, each pair generally less than a centimetre thick, and consisting of sand and silt alternating with granules and very coarse sand. The surface is a layer of the coarsest particles, identical with those in repeated thin stringers of small pebbles within the sand sheets and with coarse particles in the underlying truncated alluvial deposits. In some places these coarse surface particles form a smooth lag pavement, but in many localities they are organized in pebble or granule ripples.

An alluvial origin for the sediments that have been redeposited by wind as sand sheets explains their composition, their broad areal extent, their great volume of material, their sorting characteristics, and their range of grain sizes. The giant ripples and their superimposed small-pebble and granule ripple bedforms have formed during the transport by strong north winds of material derived mostly from deflated alluvial deposits exposed in aggraded, relic river valleys. Most of the silt in the sand sheet has probably been recycled back into the region in suspension during late spring and summer Khamsins, which carry fine sediment from areas of accumulation in the S. Many sand grains trapped in the sand sheet are of a size capable of saltating and forming dunes. Some have probably been contributed to the sand sheet by the dispersion, during the Khamsins, of sand ripples from the broad, thin aprons of the dune fields.

High-resolution *Viking* images of the debris-mantled northern plains of Mars, centred at about lat. 70°N, long. 147°, show blankets of ripple-like aeolian bedforms that are morphologically similar to the giant ripples of the Selima Sand Sheet. We interpret the Martian giant ripples as intrinsic features of regional sand-sheet deposits, rather than as dunes or simple dust blankets. The Martian deposits are likely to consist of thin layers of relatively coarse grains deposited by surface creep and saltation, alternating with layers of fine sediment deposited seasonally from global dust storms. This interpretation has implications for Mars' surface chemistry and mechanical properties. The locally-derived coarse grains almost surely differ in mineral composition from the fine, globally-derived sediment. Near-horizontal, alternating coarse and fine layers are to be expected rather than a well-sorted grain population distributed uniformly throughout the deposit. Hollows between successive ripple swells may contain only a very thin veneer of sediment, or may even expose a more consolidated substrate.

This interpretation of the Martian giant ripples as bedforms diagnostic of regional sand-sheet deposits is in agreement with observational evidence based on other studies (*ie* colour, thermal inertia) of Martian surface characteristics. It offers a better geological explanation for the surface characteristics in certain parts of Mars than do simple dust blankets or sand dunes. When coupled with the earlier recognition of similar aeolian features on the two planets, this interpretation implies that a full suite of aeolian processes has operated widely on Mars, and that these processes have resulted in considerable remodelling of the surficial sediment. This has consequent implications for Martian small-scale topography, shallow stratigraphy, sedimentology, and surface geochemistry.

ACKNOWLEDGMENTS: This work has been a cooperative effort resulting from six field seasons in the eastern Sahara, initially supported by the Smithsonian Institution (1978–80) and later supported in part by the National Aeronautics and Space Administration Office of Space Science and Applications, the US Agency for International Development (USAID–Cairo), and the US Army Research Office (ARO Contract DRXRO-GS-18405). We thank Charles Elachi of the Jet Propulsion Laboratory, Pasadena, California, for his invitation to the senior author to participate as a Guest Investigator in the initial Shuttle Imaging Radar (SIR-A) experiment, and for his continuing support of our efforts as members of the SIR-B team led by his Co-Investigator, Gerald Schaber of the US Geological Survey in Flagstaff, Arizona. We thank the Egyptian Geological Survey and Mining Authority (EGSMA), and the General Petroleum Company (GPC) for their logistical support, without which we could not have made this effort. This report was improved by comments from Ralph Hunter and David Rubin of the US Geological Survey, Menlo Park, California.

References

AHLBRANDT, T. S. & FRYBERGER, S. G. 1981. Sedimentary features and significance of interdune deposits. *In*: ETHRIDGE, F. G. & FLORES, R. M. (eds) *Recent and ancient non-marine depositional environments: models for exploration*. Society of Economic Palaeontologists & Mineralogists, Special Publication **31**, 293–314.

BAGNOLD, R. A. 1931. Journeys in the Libyan Desert 1929–1930. *Geographical Journal* **78**, 13–39.

—— 1933. A further journey through the Libyan Desert. *Geographical Journal* **82**, 103–129.

—— 1954. *The physics of blown sand and desert dunes*. Chapman and Hall, London, 265 pp.

BREED, C. S. & GROW, T. 1979. Morphology and distribution of dunes in sand seas observed by remote sensing. *In*: MCKEE, E. D. (ed.) *A study of global sand seas*. US Geological Survey Professional Paper **1052**, 253–301.

——, EMBABI, N. S., EL-ETR, H. A. & GROLIER, M. J. 1980. Wind deposits in the Western Desert. *Geographical Journal* **146**, 88–90.

——, FRYBERGER, S. G., ANDREWS, S., MCCAULEY, C. K., LENNARTZ, F., GEBEL, D. & HORSTMAN, K. 1979*a*. Regional studies of sand seas using Landsat (ERTS) imagery. *In*: MCKEE, E. D. (ed.) *A study of global sand seas*. US Geological Survey Professional Paper **1052**, 305–397.

——, GROLIER, M. J. & MCCAULEY, J. F. 1979*b*. Morphology and distribution of common 'sand' dunes on Mars: Comparison with the Earth. *Journal of Geophysical Research* **84**, 8183–8204.

——, MCCAULEY, J. F. & GROLIER, M. J. 1982. Relic drainages, conical hills, and the eolian veneer in southwest Egypt—Applications to Mars. *Journal of Geophysical Research* **87**, 9929–9950.

——, SCHABER, G. G., MCCAULEY, J. F. & GROLIER, M. J. 1983. Subsurface geology of the Western Desert in Egypt and Sudan revealed by Shuttle Imaging Radar (SIR-A). *In*: *Spaceborne Imaging Radar Symposium, January 17–20, 1983, Pasadena, California*. National Aeronautics and Space Administration, Jet Propulsion Laboratory Publication **83-11**, 10–11.

BROOKES, I. A. 1983. Dakhleh Oasis—A geoarchaeological reconnaissance. *Journal of Society for Studies of Egypt Antiquities* **13**, 168–186.

CATON-THOMPSON, G. & GARDNER, E. W. 1932. The prehistoric geography of Kharga Oasis. *Geographical Journal* **80**, 369–409.

CHRISTENSEN, P. R. 1986. Regional dust deposits on Mars: Physical properties, age, and history. *Journal of Geophysical Research* **91**, 3533–3545.

CLARK, B. C. 1982. Chemical composition of Martian fines. *Journal of Geophysical Research* **87**, 10059–10067.

EL-BAZ, FAROUK, BREED, C. S., GROLIER, M. J. & MCCAULEY, J. F. 1979. Eolian features in the Western Desert of Egypt and some applications to Mars. *Journal of Geophysical Research* **84**, 8205–8221.

FOLK, R. L. 1968. Bimodal supermature sandstones: Product of the desert floor. *International Geological Congress, 23rd Proceedings, Prague, 1968 section 8*, 9–32.

FRYBERGER, S. G. & AHLBRANDT, T. S. 1979. Mechanisms for the formation of eolian sand seas. *Zeitschrift für Geomorphologie* **23**, 440–460.

——, —— & ANDREWS, S. 1979. Origin, sedimentary features, and significance of low-angle eolian "sand sheet" deposits, Great Sand Dunes National Monument and vicinity, Colorado. *Journal of sedimentary Petrology* **49**, 733–746.

HAYNES, C. V. 1980. Geological evidence of pluvial climates in the Nabta area of the Western Desert, Egypt. *In*: WENDORF, F. & SCHILD, R. (eds) *Prehistory of the Eastern Sahara*. Academic Press, New York, 353–371.

—— 1982*a*. Great Sand Sea and Selima Sand Sheet, Eastern Sahara: Geochronology of desertification. *Science* **217**, 626–633.

—— 1982*b*. Quarternary geochronology of the Western Desert. *First Thematic Conference: Remote Sensing of Arid and Semi-arid Lands. ERIM, Cairo, Egypt, Proceedings*, 297–311.

—— 1985. Quaternary studies, Western Desert, Egypt and Sudan—1979–1983 field seasons. *National Geographic Society Research Reports* **19**, 269–341.

—— & JOHNSON, D. L. 1984. A provisional soil chronosequence for the Quaternary of the eastern Sahara. *Geological Society of America Abstract with Programs*. Annual Meeting, Reno, Nevada, 534–535.

HENNING, D. & FLOHN, H. 1977. Climate aridity index map. *Abstracts, United Nations Conference on Desertification*. UN Environment Programme, Nairobi, Kenya.

HOLM, D. A. 1960. Desert geomorphology in the Arabian Peninsula. *Science* **132**, 1369–1379.

HUNTER, R. E. 1977. Terminology of cross-stratified sedimentary layers and climbing-ripple structures. *Journal of sedimentary Petrology* **47**, 697–706.

ISSAWI, BAHAY 1981. Geology of the southwestern desert of Egypt. *In*: ISSAWI, B. (ed.) *Annals of the Geological Survey of Egypt* **11**, 57–66.

KOCUREK, G. & NIELSON, J. 1986. Conditions favourable for the formation of warm-climate eolian sand sheets. *Sedimentology* **33**, 795–816.

MARZO, M. 1986. Fluvio-aeolian sequences in the lower Triassic Eramprunya Sandstone (Catalonian Coastal Range, Barcelona, Spain). *Abstract Desert Sediments, Ancient and Modern*. Geological Society of London, Special Scientific Meeting, May 20–21, 1986.

MAXWELL, T. A. 1982. Sand sheet and lag deposits in the southwestern desert. *In*: EL-BAZ, F. & MAXWELL, T. A. (eds) *Desert landforms of southwest Egypt, a basis for comparison with Mars*. National Aeronautics and Space Administration Contractors Report CR-3611, 157–174.

MCCAULEY, J. F. 1973. Mariner 9 evidence for wind erosion in equatorial and mid-latitude regions of Mars. *Journal of Geophysical Research* **78**, 4123–4138.

——, BREED, C. S., EL-BAZ, F., WHITNEY, M. I., GROLIER, M. J. & WARD, A. W. 1979. Pitted and fluted rocks in the Western Desert of Egypt: Viking comparisons. *Journal of Geophysical Research* **84**, 8222–8232.

——, —— & GROLIER, M. J. 1981. The interplay of fluvial, mass-wasting, and eolian processes in the eastern Gilf Kebir region. *In*: ISSAWI, B. (ed.) *Annals of the Geological Survey of Egypt* **11**, 207–239.

——, ——, SCHABER, G. G., MCHUGH, W. P., ISSAWI, BAHAY, HAYNES, C. V., GROLIER, M. J. & EL KILANI, ALI 1986. Paleodrainages of the eastern Sahara—The radar rivers revisited (SIR-A/B implications for a mid-Tertiary trans-African drainage system). *Institute of Electrical and Electronics Engineers Special Volume* **GE-24**, 624–648.

——, SCHABER, G. G., BREED, C. S., GROLIER, M. J., HAYNES, C. V., ISSAWI, BAHAY, ELACHI, C. & BLOM, R. 1982. Subsurface valleys and geoarchaeology of the eastern Sahara revealed by Shuttle Radar. *Science* **218**, 1004–1020.

MCHUGH, W. P., MCCAULEY, J. F., BREED, C. S., SCHABER, G. G. & HAYNES, C. V., in press. Radar, paleorivers, and geoarchaeology in the southern Egyptian Sahara. *Geoarchaeology*.

MIDDLETON, L. T. & BLAKEY, R. C. 1983. Processes and controls on the intertonguing of the Kayenta and Navajo Formations, northern Arizona: Eolian-fluvial interactions. *In*: BROOKFIELD, M. E. & AHLBRANDT, T. S. (eds) *Eolian sediments and processes*. Developments in Sedimentology **38**, Elsevier, Amsterdam, 613–634.

MUNSELL COLOR CO. 1954. *Munsell soil color charts*. Baltimore, Maryland.

NICHOLSON, S. E. & FLOHN, H. 1980. African environmental and climatic changes and the general atmospheric circulation in late Pleistocene and Holocene. *Climatic Change* **2**, 313–348.

NIELSON, J. & KOCUREK, G. 1986. Climbing *zibars* of the Algodones. *Sedimentary Geology* **48**, 1–15.

PEEL, R. F., COOKE, R. U. & WARREN, A. 1974. The study of desert geomorphology. *Geography* **59**, 121–138.

PEEL, R. F. 1939. The Gilf Kebir. *In:* BAGNOLD, R. A. *et al.* An expedition to the Gilf Kebir and Uweinat, 1938. *Geographical Journal* **93**, 295–307.

POLLACK, J. B., COLBURN, D. G., FLASAR, F. M., KAHN, R., CARLSTON, C. E. & PIDEK, D. 1979.

Properties and effects of dust particles suspended in the martian atmosphere. *Journal of Geophysical Research* **84**, 2929–2945.

REINECK, H. E. & SINGH, I. B. 1980. *Depositional sedimentary environments*. Springer-Verlag, New York, 549 pp.

SAID, RUSHDI 1980. Quaternary sediments of southern Western Desert of Egypt, an overview. *In*: WENDORF, F. & SCHILD, R. (eds) *Prehistory of the eastern Sahara*. Academic Press, New York, 281–289.

SALEM, RAFIK 1976. Evolution of Eocene–Miocene sedimentation patterns in parts of northern Egypt. *American Association of Petroleum Geologists Bulletin* **60**, 34–64.

SCHABER, G. G., MCCAULEY, J. F., BREED, C. S. & OHLHOEFT, G. 1986. Space Shuttle Imaging Radar: Physical controls on signal penetration and subsurface scattering in the eastern Sahara. *Institute of Electrical and Electronics Engineers Special Volume* **GE-24**, 603–623.

SCOTT, D. H., TANAKA, K. L. & SCHABER, G. G. 1981. Map showing lava flows in the southeast part of the Diacria Quadrangle of Mars. *US Geological Survey Atlas of Mars, 1:2,000,000 geologic series* I-1276 (MC-2 SE).

SHARP, R. P. 1963. Wind ripples. *Journal of Geology* **71**, 617–636.

—— & MALIN, M. C. 1984. Surface geology from Viking landers on Mars: A second look. *Geological Society of America Bulletin* **95**, 1398–1412.

SODERBLOM, L. A., MALIN, M. C., CUTTS, J. A. & MURRAY, B. C. 1973. Mariner 9 observations of the surface of Mars in the north polar region. *Journal of Geophysical Research* **78**, 4197–4210.

THOMAS, P. 1982. Present wind activity on Mars: Relation to large latitudinally zoned sediment deposits. *Journal of Geophysical Research* **87**, 9999–10 008.

—— & VEVERKA, J. 1979. Seasonal and secular variation of wind streaks on Mars: An analysis of Mariner 9 and Viking data. *Journal of Geophysical Research* **84**, 8131–8146.

——, ——, GINERIS, D. & WONG, L. 1986. 'Dust' streaks on Mars. *Icarus* **60**, 161–179.

TSOAR, H. & YAALON, D. H. 1983. Deflection of sand movement on a sinuous longitudinal (seif) dune: Use of flourescent dye as a tracer. *Sedimentary Geology* **36**, 25–39.

——, GREELEY, R. & PETERFREUND, A. R. 1979. Mars: The north polar sand sea and relation to wind patterns. *Journal of Geophysical Research* **84**, 8167–8180.

WARREN, A. 1970. Dune trends and their implications in the central Sudan. *Zeitschrift für Geomorphologie Supplement Band* **10**, 154–180.

—— 1972. Observations of dunes and bi-modal sands in the Ténéré Desert. *Sedimentology* **19**, 37–44.

WELLS, S. G., DOHRENWEND, J. C., MCFADDEN, L. D., TURRIN, B. D. & MAHRER, K. D. 1985. Late Cenozoic landscape evolution on lava flow surfaces of the Cima volcanic field, Mohave Desert, California. *Geological Society of America Bulletin* **96**, 1518–1529.

WENDORF, F. & SCHILD, R. 1980. *Prehistory of the eastern Sahara*. Academic Press, New York, 409 pp.

——, ——, SAID, R., HAYNES, C. V., GAUTIER, A. & KOBUSIENWIEZ, M. 1976. The prehistory of the Egyptian Sahara. *Science* **193**, 103–114.

WILLIAMS, M. A. J. 1975. Late Pleistocene tropical aridity synchronous in both hemispheres. *Nature* **253**, 617–618.

ZIMBELMAN, J. R. & KIEFFER, H. H. 1979. Thermal mapping of the northern equatorial and temperate latitudes of Mars. *Journal of Geophysical Research* **84**, 8239–8251.

CAROL S. BREED, JOHN F. MCCAULEY & PHILIP A. DAVIS, US Geological Survey, 2255 North Gemini Drive, Flagstaff, Arizona 86001, USA.

Surficial deposits of Qatar Peninsula

M. M. Ashour

SUMMARY: The present work is the first step of a geomorphological mapping of the surface of Qatar. In this work a regional map of the surficial deposits at a scale of 1:200 000 was prepared on the basis of *Landsat* imagery of aerial photographs and topographic maps. The Qatar Peninsula has been exposed since the Miocene and consequently a variety of erosional and depositional landforms have been formed. Features of depositional origin (surficial deposits) can be classified into seven main groups according to their occurrence and textural properties, namely hamadas, sand accumulations, saline flats, coastal sediments, depression sediments, fluvial sediments and duricrusts. The deposits may be of a local or exogenous origin.

Physiography of Qatar

The Qatar Peninsula is located in the Arabian Gulf at the northeastern margin of the Saudi Arabian land mass. It lies between latitudes 24°40′ N and 26°10′ N and longitudes 50°45′ E and 51°40′ E and has an area of about 11 000 km^2 (Fig. 1).

Qatar is a low-lying country, with its highest point 103 m above sea level and its lowest point less than 3 m below sea level. The surface is primarily composed of lower and middle Eocene limestones and gypsiferous rocks, overlain unconformably by Miocene marls, shales and limestones towards the S and SW. Two main folds affect the relief; the Dukhan anticline in the W and a broad N–S trending anticline in central Qatar. Extensive Recent deposits cover most of the area.

Being a part of the sub-tropical arid region, a hot desert climate dominates with mild winters and very hot summers.

Classification and occurrence of surficial deposits

Several classifications of surficial deposits in Qatar have been suggested (Cavelier 1970; Seltrust 1980). The classification adopted in the present work is based on the mode of occurrence and textural properties of the sediments. The following sediment types have been delineated in this study (Fig. 2):

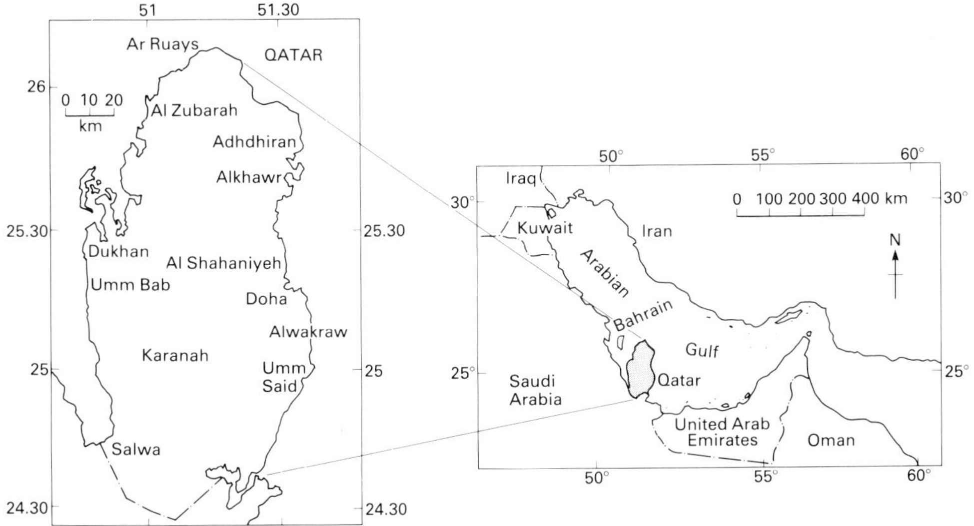

FIG. 1. Location maps of study area.

From FROSTICK, L. & REID, I. (eds), 1987, *Desert Sediments: Ancient and Modern,* Geological Society Special Publication No. 35, pp. 361–367.

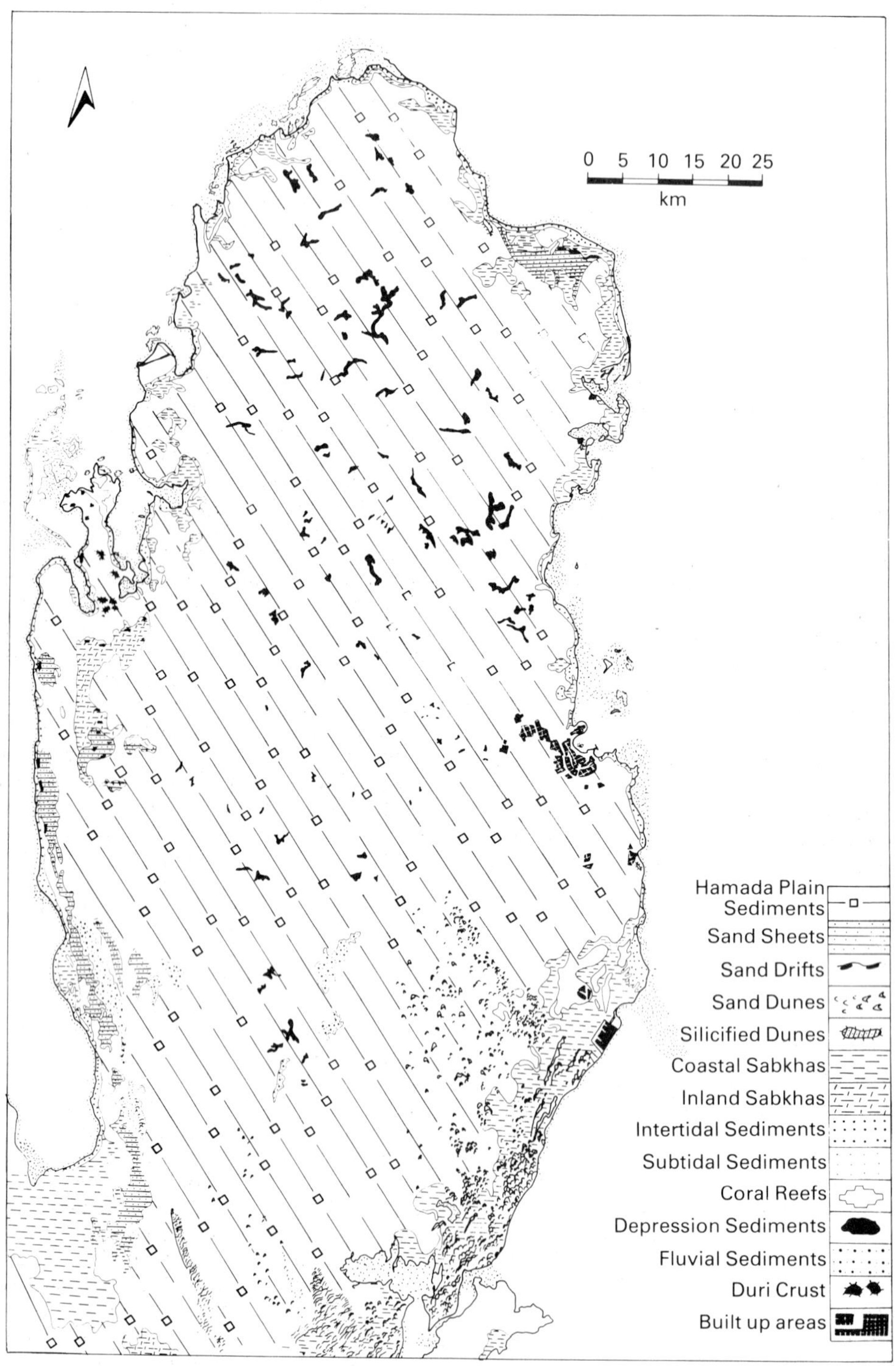

FIG. 2. Surficial deposits of Qatar Peninsula.

(1) Hamada plain sediments
(2) Sand accumulations
(3) Saline flats
(4) Coastal sediments
(5) Depression sediments
(6) Fluvial sediments
(7) Duricrusts

Hamada plain sediments

Sediments of hamada are rather coarse-grained and angular with little evidence of prolonged transport. They are mostly the product of *in situ* disintegration. The coarse particles can either be primary or secondary. Primary particles are similar in almost all respects to the coarse particles found in the underlying material. Secondary particles are derived from primary particles by processes of disintegration. Blocky disintegration and exfoliation could be responsible for most of the rock weathering on the hamada rocky surface in Qatar, forming the stone pavements.

The pavement materials and rock fragments are normally identical to the bedrock beneath the mantle, suggesting direct derivation from the underlying rocks. The mantles normally range between 25 and 50 cm in thickness and form flat or gently sloping surfaces. Some of the pavements are underlain by soils.

The coarse particles are usually veneered with desert varnish. The varnish is a brownish dull dark rind on the exposed surface of the fragments, on the buried side it is usually an orange-brown stain. The surface stain arises from migration of iron and manganese to the rock surface.

In some areas, an indurated strong carapace is evident. For instance, at the Dahl El-Misfer and in the surrounding area (Dahl is a local Arabic term denoting a small cave), there is a 2 m thick carapace. They are formed by infiltration of matrix fines into the interstices between coarse particles (Fig. 3*a*).

Sand accumulations

Wind-blown sand accumulations cover about 15% of the surface of Qatar. These were most evident in spectral band 7 and in the false-colour composite. It is believed that the sands in Qatar were derived from the Arabian landmass during the Quaternary when the Gulf of Bahrain and Salwa Bay were dry land (Seltrust 1980). Sand is still being deposited mainly from large-scale recirculating winds in the Gulf (Embabi & Ashour 1985).

Based on the published work of Embabi & Ashour (1983, 1985), sand accumulations in Qatar can be classified into four subgroups according to their mode of occurrence.

Sand sheets are recent sand deposits covering flat-lying areas. They are characterized by relatively flat surfaces and are associated with scattered sand drifts and patches of residual gravels. They are well developed in two localities. In the NE there is a 200 km^2 area of calcareous sands similar in composition to the marine calcareous sands of the shoreline. A second sand sheet occurs to the E of Dukhan hills in western Qatar. These deposits are of siliciclastic and calcareous sands and are believed to be the main source of the sand dunes developed in southern Qatar.

Nebkhas are commonly observed as accumulations of sands behind or around vegetation in the path of sand-laden wind. On air photographs they are seen to be well developed in the Zikrit area and on the margins of some sabkhas. They rarely exceed 50 cm in height and are composed of moderately well-sorted fine quartz sand ($MZ = 2.13\ \phi$; Ashour 1985).

Sand dunes occur as individual and compound barchans and as dune fields containing both longitudinal and transverse dunes (Fig. 3*b*). Barchans are mainly concentrated in southern Qatar in two localities, an eastern and a western field. The eastern field covers about 10% of the peninsula. The dunes are essentially composed of silica sands with small quantities of calcareous fragments, argillaceous dust and trace amounts of heavy minerals. Qatar dunes are built of sand grains which fall into the fine and medium size ranges ($MZ = 1.93\ \phi$). Scanning electron microscopy reveals that the most common surface features of the grains are upturned plates, equidimensional or elongate depressions, arcuate, circular and polygonal cracking and V-shaped notches or pits. Heavy minerals are of two types, a group of local origin and an exogenous group.

Using aerial photographs of three different dates (1963, 1971 and 1976) it has been found that the barchans migrate southwestwards in Qatar. The general mean annual rate of movement for all dunes is 7.9 m in the eastern field (Hunting Surveys Ltd 1977), and 20.7 m in the western field (Embabi & Ashour 1985).

Silicified dunes are well developed along the eastern coast at Jassasyah, Wakrah, Fuwairet and Al-Wussail, and in embayments in the escarpment on the eastern side of the Ras-Abrouk. They consist essentially of rounded carbonate grains cemented to form calcarenites. Houboult (1957) described these deposits as rounded calcarenites while Cavelier (1970) classified them as pseudo-oolitic limestones. These silicified dunes probably accumulated during a period when sea level was lower than at present.

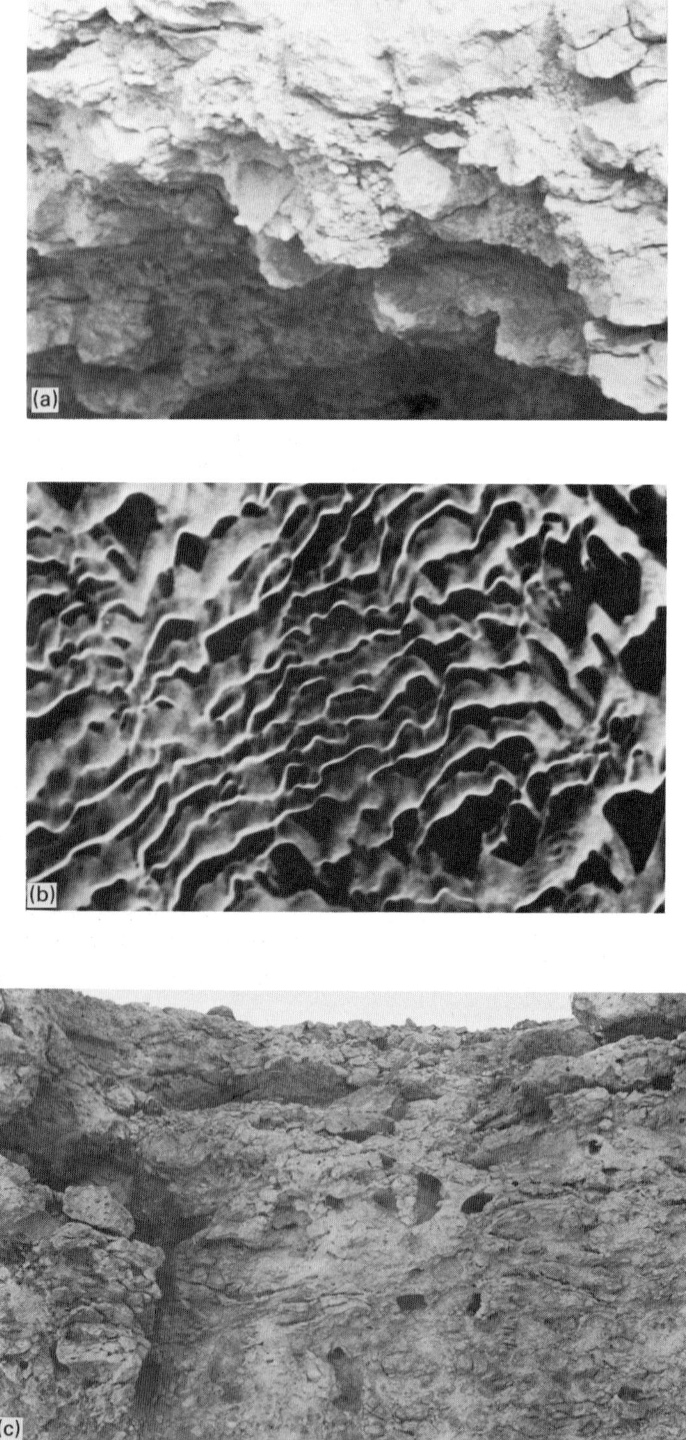

FIG. 3. (*a*) Indurated carapace forming the roof of Dahl El-Misfer and surrounds. (*b*) Aerial photograph of dune field S of Umm Said. (*c*) Hard cap-rock (duricrust) forming the top of Zikrit hills.

Saline flats (sabkhas)

Sabkha is an Arabic term applied to an inland or coastal saline flat (Evans *et al.* 1973; Perthuisot 1977). Two kinds of sabkha deposits can be recognized in Qatar. Firstly, coastal sabkhas (supratidal flats) bordering the peninsula, for example the Umm Said sabkha which occupies the southeastern corner of the peninsula covering an area of about 500 km^2. Secondly, inland or continental sabkhas not connected with the sea at the present time, for example the Dukhan and Souda-Nethil sabkhas. Sabkhas were delineated accurately using the *Landsat* image spectral band 4.

These very flat areas characteristically contain a shallow (0.2 to 2.0 m depth) water table, with highly saline groundwater. The groundwater may be either directly connected with the sea or with continental groundwater, but surface water from tidal flooding or run-off also partially recharges the system.

Umm Said sabkha: a 30 m thick sequence of sabkha deposits in Umm Said sabkha was described by Shinn (1973*a* and *b*). The sequence is composed mainly of medium to fine siliciclastic sand containing 5%–15% sand-sized fragments of Eocene dolomite and minor traces of other non-dolomitic carbonates. These sands are well rounded and sorting ranges from moderate to good. According to Shinn (1973*a* and *b*), the sabkha sands are assumed to have been deposited during two separate phases of regressive sedimentation.

Dukhan sabkha: occurs some 6–8 km from the coast, in a 130 km^2 depression ranging between 1.8 m above sea level and 3 m below sea level. It is not directly connected with the sea and no recent marine sediments have been observed in and around the depression. Perthuisot (1977) claimed that it may be a part of a structural low occupying the area between Dukhan anticline in the W and Qatar anticline in the E. Highly saline loamy soil is found in the sabkha. It varies in depth from a few centimetres to more than 3 m. The salt content is generally very high throughout the profile and in some places a surficial halite crust is well developed. These salts cement the surface material and prevent their erosion by wind action. Gypsum and anhydrite occur at or near the surface of the sabkha deposits as hard layers 5–10 cm thick. Greyish clays are found mixed with fine sands in the sabkha area, sometimes with additional calcareous material (mainly calcite).

Coastal sediments

The coastline of Qatar is emergent and irregular in outline with numerous capes, bays and inlets. The beaches are partly constructional, but erosional features are found near Doha and Salwa. Coastal deposits can be classified into five main types, namely, supratidal flats, intertidal flats, subtidal flats, carbonate 'chenier' beaches and coral reefs.

Supratidal flats: these flats are dealt with in the previous section under coastal sabkhas.

Intertidal flats: these occur in a belt from a few metres to more than a kilometre in width, depending on tidal range and local relief (Embabi 1984). Within the northern embayments intertidal sediments are predominantly carbonate muds and silts with varying amounts of skeletal material. Along the SE coast they are mainly sandy.

Subtidal flats: these are formed in the open sea or coastal inlets and are permanently below low tide. In NE Qatar subtidal deposits are entirely a mixture of clay and silt-sized carbonates, nevertheless the sediments contain varying amounts of skeletal grains. Shinn (1973*a*) concluded that these sediments were partly derived from inland sources by sporadic rain and strong winds.

Carbonate 'chenier' beaches: narrow cheniers rest on intertidal, supratidal and even channel deposits. Cross-bedding in the intertidal and supratidal part of the straight segment of chenier beaches is distinct from that of the hook-shaped spits at the southern prograding end of chenier beaches.

Coral reefs are found at many places along the coast as well as around many small islands. In water shallower than 16 m several local 'highs' occur, one of them being Al-Oudaied Shool. These are built almost entirely of coral limestone.

Depression deposits

In the surface of Qatar a large number of depressions are carved. They lie below the surrounding land surface at a depth ranging from a few metres to 20 m. They vary greatly in shape and size. Numerous depressions are filled with Recent silts and muds (30 cm to 3 m thick) which are deposited both from surface runoff carrying weathering products from surface rocks of neighbouring areas and from the wind. This association of depression sediments is known locally as the Roda soil.

Deposits are characteristically fine grained, including clays, silts and sand-sized granular particles. Salts are precipitated *in situ* from evaporating saline water. As a group these sediments normally show clear evidence of grading towards the centre of the basins.

Fluvial sediments

These can be differentiated on the basis of age

into pre-Pleistocene deposits and Recent wadi sediments. The former consist of alluvial gravels sands and silts. They occur in the SW of Qatar capping middle Miocene rocks. The most prominent area lies S of Umm Bab along the ridge from Kharaij to Nakhash, at Tus-Al-Karanah and Huriyah, and also to the N and S of Karanah. These sediments are of exogenous origin. They were brought to Qatar by a major river system flowing E and NE across Saudi Arabia spreading out to form a vast alluvial fan. At Al-Hofuf in Saudi Arabia the deposits reach 95 m in thickness, but in Qatar only some 12 m has survived erosion (Seltrust 1980). Re-worked gravel deposits frequently occur around Miocene hillocks. The characteristics of the pebbles suggest they were derived from the Hofuf Formation (Cavelier 1970). Absence of these deposits in wadis and depressions is an indication that there has not been significant fluvial activity since the deposition of the Hofuf Formation.

Duricrust

Calcareous duricrusts are extensively developed in Qatar. They form the surface of the hamada plain in parts of central and northern Qatar. They are very well consolidated and are similar in appearance to older sedimentary rock strata, covering a rugged topography developed on Eocene and Miocene Formations. These crusts (calcretes) are widespread in arid and semi-arid areas (Goudie 1973). They are recorded at Al-Hasa in Saudi Arabia (Chapman 1974), in Bahrain (Doornkamp *et al.* 1980) and in Kuwait (Khalaf *et al.* 1984). A grey buff calcareous duricrust is a conspicuous feature of the Zikrit Hills. Although partly eroded, it still caps some hills (Fig. 3*c*) and geomorphological processes dissect the exposed surface. It also caps isolated rock pillars and pedestals, indicating that these forms are younger than the crust. Its present thickness is generally about 0.5 m, but originally it must have been thicker and perhaps more evenly distributed.

During the formation of this calcareous duricrust, the climate prevailing was wetter than it is now. It was formed during periods of semi-aridity during the Pleistocene, when rainy seasons alternated with dry ones.

Conclusion

From the foregoing discussion, it is obvious that the origin and evolution of the surficial deposits in Qatar have been greatly influenced by geological setting, climatic changes and sea-level fluctuations.

The surface of Qatar is composed essentially of limestone, gypsum and marls with uncomplicated structure. This has led to the subdued topography of the country, the existence of karstic landforms and the characteristic surficial deposits. Karstic features, duricrusts and wadi sediments suggest wetter conditions during their formation. The silicified dunes are evidence of a period of lowered sea level 20 000 to 30 000 years BP. Sea-level began to rise about 17 000 to 20 000 BP, before reaching its present level some 5000 years ago. During the period of lower sea level the dunes were able to cross the Gulf of Bahrain and Salwa Bay.

References

ASHOUR, M. M. 1985. Textural Properties of Qatar Dune Sands. *Journal of Arid Environments* **8**, 1–14.

CAVELIER, C. 1970. *Geological Description of the Qatar Peninsula*. Department of Petroleum Affairs, Government of Qatar, Bureau de Recherches Geologiques et Minieres, Paris, 39 pp.

CHAPMAN, R. W. 1974. 'Calcareous Duricrust in Al-Hasa, Saudi-Arabia'. *Geological Society of America, Bulletin* **85**, 119–130.

DOORNKAMP, J. C., BRUNSDEN, D. & JONES, D. K. C. 1980. *Geology, Geomorphology and Pedology of Bahrain*. Geo Abstracts Ltd, University of East Anglia.

EMBABI, N. S. 1984. *Sea Intrusions in the Coasts of Qatar* (in Arabic). Bulletin **70**, University of Kuwait, 54 pp.

—— & ASHOUR, M. M. 1983. *Sand Dunes in Qatar Peninsula*. University of Qatar (in Arabic) **I**, 224 pp.

—— & —— 1985. *Sand Dunes in Qatar Peninsula*. University of Qatar (in Arabic) **II**, 237 pp.

EVANS, G., MURRY, J. W., BIGGS, H. E. J., BATE, R. & BUSH, P. R. 1973. The Oceanography, Ecology, Sedimentology and Geomorphology of Parts of the Trucial Coast Barrier Island Complex, Persian Gulf. *In*: PURSER, B. H. (ed.) *The Persian Gulf*. Springer-Verlag, New York, 233–277.

GOUDIE, A. 1973. *Duricrusts in tropical and subtropical landscapes*. Clarendon Press, Oxford.

HOUBOULT, J. J. H. C. 1957. *Surface Sediments of the Persian Gulf Near the Qatar Peninsula*. Mouton, Den Haag.

HUNTING SURVEYS LTD 1977. *Technical Report, Sand Dune Movement Study, South of Umm Said 1963–1976*. Ministry of Public Works, Doha, Qatar, 7 pp.

KHALAF, F. O., GHARIB, I. M. & AL-HASHASH 1984. Types and Characteristics of the Recent Surface

Deposits of Kuwait, Arabian Gulf. *Journal of Arid Environments* **7**, 9–33.

PERTHUISOT, J. PIERRE 1977. 'La Sebkha de Doukhane (Qatar) et La Transformation: gype anhydrite + eau'. *Bulletin de Societé Geologiques France* (7) **XIX**, No. 5, 1145–1977.

SELTRUST ENGINEERING LTD 1980. *Qatar Geological Map Explanatory Booklet*. I.D.T.C., Doha, Qatar, 20 pp.

SHINN, E. A. 1973*a*. Carbonate coastal accretion in an area of longshore transport, NE Qatar, Persian Gulf. *In*: PURSER, B. H. (ed.) *The Persian Gulf*. Springer–Verlag, New York, 179–191.

—— 1973*b*. Sedimentary accretion along the leeward, SE coast of Qatar Peninsula, Persian Gulf. *In*: PURSER, B. H. (ed.) *The Persian Gulf*. Springer-Verlag, New York.

M. M. ASHOUR, Department of Geography, University of Qatar.

Remote sensing of sediment transfer processes in playa basins

A. C. Millington, A. R. Jones, N. Quarmby & J. R. G. Townshend

SUMMARY: *Landsat* Thematic Mapper and Multispectral Scanner data have been used to detect changes in the spectral reflectance and absorption of three playas in south-central Tunisia. Change detection images have been used to analyse changes between wet and dry seasons and between different years. Changes in geomorphological phenomena are interpreted from changes in soil and foliar moisture levels, differences in reflectance between different salts and sediments, and the spatial expression of geomorphological features. The intra-annual changes that can be detected from multidate imagery include surface moisture, texture and chemical composition, vegetation cover and the extent of aeolian activity. Detected inter-annual changes are divisible into those restricted to playa margins—sedimentation from sheetwash and alluvial fans, erosion from surface runoff, and cliff retreat—and those found at central playa locations which are related to the redistribution of water, salt and sediment.

Sediment transfer processes in deserts are characterized by: (1) high-magnitude and low-frequency events; (2) a strong seasonal pattern (which is often so marked that the dominant sediment transfer process can change from fluvial to aeolian as the (year progresses); (3) events which are spatially discrete and therefore uncorrelated even in small areas (Brunsden & Thornes 1979; Dott 1983; Schumm 1979). Consequently, geomorphological processes are difficult to monitor using conventional ground-based instrumentation because there is a very low probability of being on-site at the time of an event such as a flash flood. However, the probability is increased if either the temporal or the spatial dimensions are increased. One method of achieving this is to utilize the synoptic capability provided by data from satellite sensors.

This approach is currently being evaluated in an area of south-central Tunisia (Fig. 1) using data from the *Landsat* Thematic Mapper and Multispectral Scanner for four major process domains—hillslopes, alluvial fans, braided rivers and playas. A preliminary visual study of multidate satellite imagery from the area indicates that geomorphological change can be detected readily. However, to use the digital data provided by satellite imaging systems effectively, computer-assisted image processing techniques need to be utilized.

This paper is therefore concerned with evaluating the use of suitable image processing algorithms for the detection and monitoring of geomorphological change in playa basins of south-central Tunisia. Data from the *Landsat* Thematic Mapper and Multispectral Scanner sensors, as well as the SPOT High Resolution Visible sensor are utilized.

The study area

Geomorphological processes

Water plays a central role in the transfer of sediments in south-central Tunisia. Rivers head in actively eroding montane drainage basins, and discharge to closed depressions containing playas (known locally as chotts). It is clear that the transfer of sediment between process-domains is far from continuous, and episodic erosion and deposition prevail, with long intervening periods during which material is stored on hillslopes, in channels, or on the playa margin.

Aeolian activity is important, especially in the dry season from May to September when material is deflated from the playa surfaces and redistributed on adjacent mountain slopes, in nebkhas, in depressions, and in river channels. Aeolian deposits in channels are then flushed back to the playas by floods. Fine-grained material is also blown into the area from the S by dust storms that occur late in the wet season.

Geology

The area includes both the highly folded Atlas Mountains to the N and the relatively undeformed Saharan Shield to the S. Detailed descriptions of the geology can be found in Burollet (1972). These two areas are separated by the Saharan Boundary Fault; the northern area is still tectonically active (Coque & Jauzien 1967).

Climate

The mean annual rainfall is 80–157 mm and is mainly restricted to the period between September and May. Temperatures and evaporation

From FROSTICK, L. & REID, I. (eds), 1987, *Desert Sediments: Ancient and Modern*, Geological Society Special Publication No. 35, pp. 369–381.

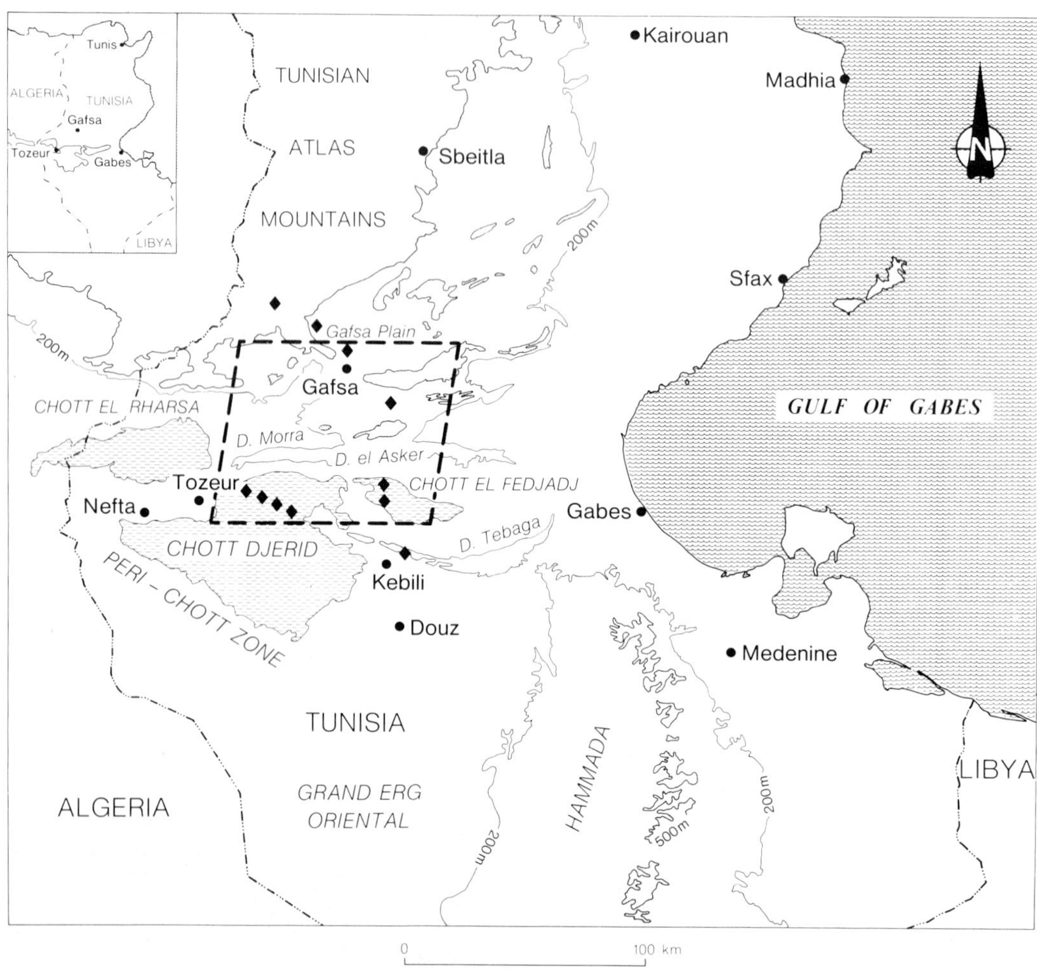

FIG. 1. Location map. The Thematic Mapper quarter scene used in this analysis is indicated by a dashed line; the control sites are indicated by ♦.

rates increase markedly in May, remaining high until late September. Runoff events are therefore restricted to the wet season and streams are dry throughout the summer since there is no base flow and soil moisture deficits are high. The climate is more arid in the S where runoff events are even more rare.

Remote sensing and desert geomorphology

Remotely sensed data have been used for geomorphological investigations in arid and semi-arid areas by several workers. Most applications have involved landform mapping (*eg* Mitchell *et al.* 1982; Sunha & Venkatachalam 1982) or surficial material surveys (Ashour 1987, this volume; Asem *et al.* 1982; Bird *et al.* 1982; Davies *et al.* 1982; Townshend & Hancock 1981). With the exception of Graetz & Pech (1982) and Klemas & Abdel-Kader (1982), few workers have attempted to monitor geomorphological change in these environments. Three problems are apparent from these studies:

(1) The Multispectral Scanner sensor has relatively coarse spatial resolution (79 m). This situation has been improved with the higher resolutions of *Landsat* Thematic Mapper (30 m) and the SPOT High Resolution Visible sensors (20 m and 10 m) (Millington & Townshend 1986).

(2) The Multispectral Scanner has restricted spectral resolution. The inclusion of middle infrared bands (1.55–1.75 μm and 2.08–2.85 μm)

on the Thematic Mapper has greatly enhanced the possibility of discriminating surficial materials (Bodechetel 1983; Gladwell 1982; Hunt 1980; Kahle 1984). These bands are crucial to any interpretation of sediment dynamics as they provide a potential method of identifying river and fan gravel sources, duricrusted and varnished surfaces on fans, and the accumulation of aeolian sand in channels and on playas.

(3) Geomorphological monitoring in this environment using *Landsat* data has been severely restricted by the availability of imagery. This is because of the problems of obtaining large quantities of imagery due to the costs involved, because of ground station receiving policies, and because of unfavourable atmospheric conditions during sequential overpasses of the sensor.

Change detection

Jones (1986*a*, *b*) and Munday (1985) have both shown the value of digitally processed Thematic Mapper and Multispectral Scanner imagery for geomorphological mapping in southern Tunisia. However, any change detection study involves scene-to-scene image registration in order to ensure that pixels in each image correspond to the same ground location. To accomplish this, six ground control points were used to co-register the Thematic Mapper quarter scenes used in this study (Fig. 2); an average root mean square error of ± 0.37 pixels (equivalent to 11 m on the ground) was achieved. This can be compared with the 27 ground control points required to register 512×512 pixel Multispectral Scanner images to an average error of ± 0.45 pixels (equivalent to 35 m on the ground) to indicate the advantage of Thematic Mapper over the Multispectral Scanner in change detection studies. A nearest-neighbour resampling procedure was used for the image registration in order to preserve the spectral values of the raw image data. The ground control points chosen in the co-registration procedure were permanent features such as road junctions, road–rail crossings, corners of well-defined plantations and sharply defined points along mountain fronts.

Slight misregistration of the imagery will produce errors in the change detection output image, since boundary pixels corresponding to one ground surface type in one image may be compared with the boundary pixels of an adjacent type of surface in subsequent images. This inevitably results in spurious change detection.

'Difference images' were produced by subtracting the filtered image of the first date from that of the second, and after adding a constant to ensure that all output values were positive. 'Ratio images' were produced by dividing the image of the first date by that of the second. For subsequent principal component analyses, the two images being compared were treated as one data set; Multispectral Scanner Bands 1 (0.5–0.6 μm), 2 (0.6–0.7 μm) and 4 (0.8–1.1 μm) were used because of their widespread acceptance as a standard in false colour composite production. However, a limitation of principal components analysis is that the change which is detected cannot be attributed directly to changes in reflectance for any particular spectral band (Howarth & Boasson 1983). Nevertheless, previous work on the use of principal component analysis for change detection indicates that gross differences due to overall radiation and atmospheric changes are contained in component one, and that the statistically minor changes that are associated with changes at the land surface are incorporated in higher-order components (Byrne *et al.* 1980; Lodwick 1979; Richardson & Milne 1983). Consequently, components two and three were used to detect geomorphological change in this study.

Various change detection algorithms have been applied to Multispectral Scanner data (see Nelson 1983), but not in semi-arid or arid areas. Difference images, ratio images and images arising from principal components analysis were found

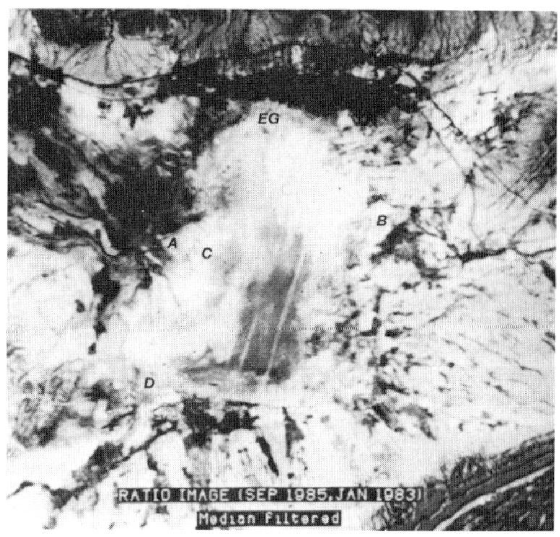

FIG. 2. Change detection image of Chott el Guettar produced by ratioing Thematic Mapper 2.08–2.25 μm data of January 1983 with that of September 1985; El Guettar oasis is indicated (EG). The areas of highest surface moisture change can be readily identified—the spring pots and necks on the fan delta (A); Guettaria (B); the area of high summer moisture levels on the playa relate to groundwater discharge from the Gafsa Fan (C). The gullied areas in the SW can also be seen (D).

to be the most useful for the detection of change in the study area. Vegetation indices (*ie* near IR: red; Howarth & Boasson 1983; Singh 1984) were unsuccessful in detecting change in this environment, but they were initially considered because of their successful deployment in Temperate and humid Tropical environments. Their failure here is in part due to the inherent noise in each data set—a problem which is compounded when different data sets are ratioed. This problem is particularly evident with Multispectral Scanner data, because it has a lower signal-to-noise ratio than that of the Thematic Mapper. However, the spectral bands providing some useful information were the near-infrared band (0.8–1.1 μm) of the Multispectral Scanner, and the red (0.63–0.69 μm) and middle-infrared (2.08–2.25 μm) bands of the Thematic Mapper. These bands provide information about vegetation and soil moisture.

In the preliminary stages of the investigation, no attempt was made to convert 'digital numbers' (*ie* raw data) to absolute radiance values, or to correct for variations in sun angle and atmospheric scattering and absorption. Consequently, there may well be changes detected from one date to the next which are attributable to these factors rather than to real changes. However, these complications have been minimized by setting thresholds for the acceptance and rejection of data. These thresholds, chosen on the basis of previous research (Nelson 1983; Singh 1984), include all the data beyond one standard deviation of the mean reflectance in any one reflectance class. Other data are rejected.

Detection of geomorphological change in playa basins

The detection of geomorphological change has been approached in two ways. Firstly, after identification on sequential images, the areas of change were examined in the field. Whilst this allowed a certain level of understanding of the type of geomorphological process in operation, it was difficult to quantify the changes because of the lack of pre-change data. Secondly, control sites had been established in each process-domain and these were being monitored on the ground. Each was examined immediately after image acquisition. These sites provided a quantitative reference for changes that might have occurred at similar sites elsewhere in the study area.

Changes were noted between January 1983 and April 1986 using Thematic Mapper data. In order to extend the length of the monitoring period, Multispectral Scanner data that dates back to 1981 have been used. In addition SPOT High Resolution visible data has recently become available for the area, and this has been incorporated. The extent of detected changes has varied significantly between the process-domains (Table 1). Most change has occurred in the playas (65.3% by area) with much lower levels of change in the braided rivers, the alluvial fans and on the pediments.

Environmental setting of playas

Three playas, Chott el Djerid, C. el Fedjadj and C. el Guettar (Fig. 1), have been studied intensively. Chott el Djerid and C. el Fedjadj occur in a zone of subsidence to the S of the Atlas Mountains (Demaison 1965); they form part of a series of playas stretching from the Sebkhet Mechrecherma (25 km NW of Gabes) to Chott el Melhrir in central Algeria (Fig. 1). Chott el Djerid is the largest playa and covers about 5360 km^2. It has an elongate northeastern arm, the Chott el Fedjadj, which continues eastward into the Sebkhet el Hamma, covering about 770 km^2. The geomorphology and hydrology of these playas have been investigated by Coque (1969), Coque & Jauzien (1967) and Meckelein (1977).

Chott el Guettar is about 75 km^2 in area. It is situated in an enclosed basin bounded to the N by the Djebel Orbata and to the S by the Djebel Berba. Both of these mountain ranges have active fans encroaching on the playa. The junction between the mountain front and alluvial fans on the Djebel Orbata marks the location of the Saharan Boundary Fault. Low ground and smaller mountains and hills are found to the E and W of the chott.

Geomorphological changes on the playas

Geomorphologically significant change has been detected on these playas that can be attributed to both surface and subsurface factors, in particular the seasonal nature of the hydrological regime, salt dynamics, and aeolian activity. The playas can be zoned according to surface detail, salt content and surface moisture content. Change detection imagery provides information on the movement of zonal boundaries, as well as helping to explain the characteristics of each zone.

Seasonal changes

Seasonal changes on playas have been identified previously by Glennie (1970), Langer & Kerr (1966) and Neal & Motts (1967) and are readily

TABLE 1. *Change detected in different Earth surface process-domains between January 1983 and April 1986 from Thematic Mapper imagery for a 108 km² study area in south-central Tunisia*

Major process-domain	Area, on TM quarter scene km²	Area in each category with +ve or −ve change km²	Proportion showing change %
Alluvial fans	359.0	29.8	12.05
Pediments	4436.8	251.2	17.66
Playas	2150.0	1403.2	65.23
River channels*	262	66.1	25.32

* Measurements refer to lengths of channel, not area.

detected by remote sensing. Imagery corresponding to wet and dry seasons has been compared for all three playas and five types of change of geomorphic significance have been identified (Table 2):

Surface moisture and vegetation changes (Chott el Guettar). Seasonal patterns relating to surface moisture and vegetation are best illustrated by examining the change detection image for winter and summer in Chott el Guettar (Fig. 2), particularly the fan delta NW of the playa. Vegetation cover on the fan delta ranges from moderate (30–35%) to poor (<5%) and consists of halophytic succulents, in particular *Crassula* spp., *Limoniastium gyonianum*, *Limonium* spp. There are a number of spring pot and spring neck complexes (Reeves 1965) which are sparsely vegetated, subdued depressions that feed into rills (Fig. 3). They, in turn, drain into three channels varying between 50 and 100 m wide and 2 m deep, that debouch onto the playa. These channels form embayments in the fan delta.

TABLE 2. *Classification of geomorphological change on Tunisian chotts*

Change category	Characteristic types of change
Intra-annual	Surface moisture Surface texture Surface composition Vegetation cover Aeolian activity
Inter-annual	*Marginal chott facies* Sedimentation (sheetwash, alluvial fans) Erosion (runoff, cliff retreat) *Central chott facies* Surface water flow

Elsewhere the fan-delta/playa boundary is marked by a low cliff up to 0.5 m in height. This cliff cannot be detected on remotely sensed imagery, but the difference between the moderately well vegetated fan delta and the sparsely vegetated playa margin locates the boundary accurately (Fig. 2). The change detection imagery (Fig. 2) indicates higher absorption in the middle infrared over most of the fan delta in winter than summer. This is a result, inevitably, of higher levels of soil and/or foliar moisture in winter and has been noted for other playa margins in the area (Epema 1986). The channels and adjacent playas are characterized by higher middle-infrared absorption in summer than in winter. This implies higher moisture levels in these areas during the summer. Field observations of the sparsely vegetated depressions and the channels in both September 1985 and May 1986 showed that they are indeed very moist at the surface in both seasons. However, another factor comes to bear as well: an examination of the channels in May 1986 revealed fresh salt efflorescences. Such efflorescences last for about a week after rainfall or runoff and indicate the recency of surface flow. The higher winter reflectance values appear to be caused in part by the surface salt efflorescence.

The area influenced by surface runoff in winter and groundwater seepage throughout the year forms a semicircle 20 km² in area. Similar patterns can be seen on change detection imagery derived from Multispectral Scanner data for 1981 and Thematic Mapper imagery for 1985/86.

The other areas contributing surface water to the Chott el Guettar appear to be less important. The effect of winter flows along the Oued el Rahr and Oued es Sedd systems to the E and NE, respectively, are visible in Fig. 2. However, water that might be flowing onto, or through the alluvial fans to the N shows little effect on the imagery. This may be due to either a lack of surface flow in the years examined, or a constant groundwater seepage that reduced the chance of seasonal differences. Most of the fans to the S of the playa

FIG. 3. A spring pot, with rills in the spring neck in the foreground; scene from the fan delta to the SW of El Guettar.

show the same pattern. However, substantial changes in moisture levels to the SE of the playa were seen on the change detection imagery of 1981 (Fig. 4).

There are a few nebkhas on the playa (4% cover of 23 sample sites) and little evidence of wind-blown sand or winnowing (13% cover of 23 sample sites). Therefore, in the absence of significant aeolian transport, groundwater seepage and surface runoff must control the transport of salts and sediments onto the playa. However, there is little evidence, from the change detection images and from field observations on four transects across the playa, of any seasonal variation in groundwater levels. Nevertheless, the main groundwater source to the NW undoubtedly generates higher seepage rates throughout the year than other sources because of the intensive irrigation in the Gafsa Oasis, and because of its larger catchment.

Surface runoff acts as the main transporting mechanism for sediment moving onto the playa and it is particularly important in the areas adjacent to the active fans. Most sediment therefore moves into the playa from the S forming

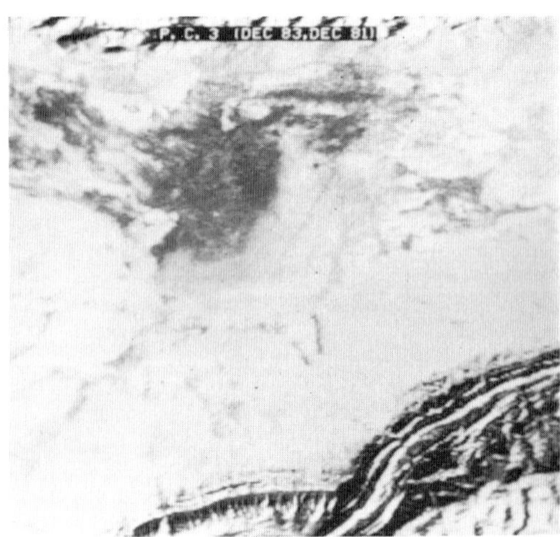

FIG. 4. A third principal component image of Chott el Guettar using Multispectral Scanner data from the wet and dry seasons of 1981. Areas with high levels of moisture change can be seen on one of the fans to the S of the playa and on the south-central playa. This area of change is related to deposition of sediments eroded from the fans.

a south-central depositional wedge which is recognizable in all change detection images as an area of lower summer reflectance (Figs 2 and 4). In actuality, the reflectance is due to the higher water-holding capacity of these sediments resulting in high moisture levels in winter. Salt transport, unlike sediment transport, is related to both surface water and groundwater fluxes and it is likely to mirror the hydrological regimes more strongly than the sedimentation patterns. The analysis of change detection imagery and ground observations has allowed the development of a provisional model of the relationships between surface-water and groundwater influences and sediment and salt fluxes on Chott el Guettar (Fig. 5).

Surface texture, surface composition and aeolian activity (Chott el Djerid). Seasonal changes in surface texture, surface composition and aeolian activity are best developed on Chott el Djerid. This is because salt types, which are strongly related to surface morphology (Langer & Kerr 1966), are well developed on this playa (Munday 1985). A transect of control sites was located to the N and S of the Fatnassa–Degache road on which the sampling interval was 1 km (Fig. 6). The control sites occur in four of the morphological zones defined by Mitchell (1983):

(1) Wind-sculpted areas of winnowing, nebkhas and small yardangs.
(2) Areas of thick salt crusts.

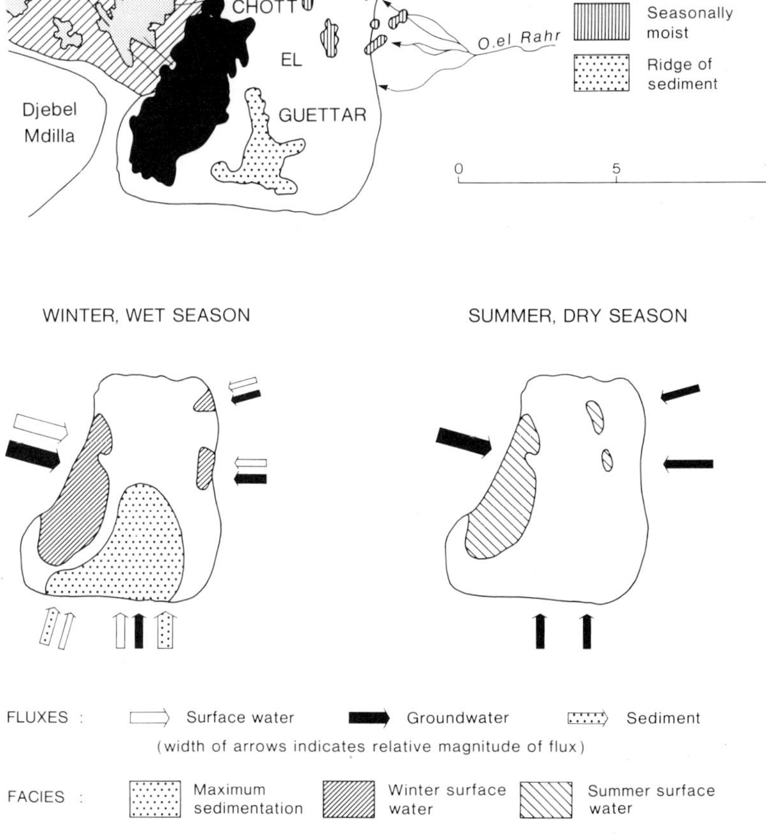

FIG. 5. Chott el Guettar: provisional model of seasonal water, salt and sediment dynamics.

FIG. 6. SPOT High Resolution Visible scanner image of the northern Chott el Djerid (copyright CNES, 1976). The Fatnassa–Degache road can be seen traversing the playa; the control sites are located at 1 km intervals to the N and S of the road. The four morphological zones (see text) can be identified: (A) The wind sculpted areas with winnowing, nebkhas and small yardangs; (B) areas dominated by *aioun*; (C) thick salt crust areas, and (D) areas of blistered thin salt crust with polygonal folding.

(3) Areas of blistered thin salt crust with polygonal folding.
(4) Areas dominated by groundwater upwellings known as *ain* (pl. aioun).

Uncorrected reflectance values at the control sites (Thematic Mapper data) have been analysed for 6 September 1985 and 6 April 1986 (Table 3). By far the smallest changes are found in the areas of the playa strongly affected by wind (the mean difference of the reflectance values is 11.6). The other three morphological zones in which control sites are located show far greater seasonal differences. The values for the areas of thick salt crust and *aioun* are similar (49.7 and 47.5, respectively), while that for the areas of thin salt crust are marginally higher (53.8).

Within one month of satellite overpass, ground site descriptions and data were collected at each control site. These data are summarized in four categories—aeolian and fluvial activity, hydrological regime, salt regime, and vegetation (Table 4). The changes in ground phenomena at the control sites parallel the changes in reflection and absorption.

The areas of the playa which are strongly affected by wind remained constant between September and May. The wind-affected area is

TABLE 3. *Summary of reflectance data for groups of control sites in the Chott el Djerid*

Control sites	Surface reflectance and seasonal differences, Digital Number (DN)			
	Blistered thin salt crust with polygonal ridging*	Thick salt crust†	*Aioun*‡	Wind-sculpted and hummocky§
N of Fatnassa–Degache road				
September 1985				
Max	134	124	74	98
Min	85	32	43	63
April 1986				
Max	90	82	131	92
Min	60	22	81	42
Seasonal difference				
range	25–74	5–82	30–78	(−7) −45
mean	52.1	46.1	55.6	13.7
S of Fatnassa–Degache road				
September 1985				
Max	145	145	60	93
Min	82	42	25	65
April 1986				
Max	75	85	100	82
Min	47	17	78	77
Seasonal difference				
range	26–84	(−7) −78	25–46	(−7) −34
mean	55.4	48.8	43.8	9.5
Mean difference in reflectance (all sites)	53.8	47.5	49.8	11.6

* $n=14$; † $n=42$; ‡ $n=10$; § $n=40$.

TABLE 4. *Chott el Djerid: geomorphic, hydrologic and vegetated character of sample sites spaced at 1 km intervals along the Fatnassa–Degache road from W to E. Observations made in September 1985 and May 1986*

Transect distance—each row represents an interval of 1 km

Vegetation, geomorphology & hydrology																			Mitchell (1983) Surface detail zones
Vegetation			Aeolian and fluvial activity					Salt regime							Hydrological regime				
Vegetated hummocks	Halophytes present	Other vegetation	Nebkha/Rabdu	Winnowing/fluting	Yardangs	Lag deposits (stony)	Sand accumulation	Salt efflorescences	Thin crust	Thick crust	Blistering	Polygonal ridging	Polygonal cracking	Thrust polygons	*Aioun*	Flow features	Moist surface	Standing water	
*			*			*		*									*	*	Blistered thin salt crust with polygonal ridging
		*						*											
	*							*			†			‡			†		
	*							*			†	‡					*		
	*							*			†	‡					*		
	*							*			†	†							
								*			†	†	‡						
								*	*			*	*			*			Thick salt crusts
								†	*			†	†			*			
								*	*			*	*			*			
										†			*	*					
										†			*	*					
				†						†			*	*					
				†						†			*	*					
				†						†			*	†					
										†			*	†					
										†			*	†		†			
										†			*	*		†			
										†			*			†			
										†			*	†		†			
								*			†		*						
								*			†		*	*					
							†	*					*				†	†	
							*						†			†		*	
													†					†	
													*	*			†		
													*	*	*				*Aioun*
													*	*	*				
													*	*	*				
				*									*		*				
*	*		*	*				†					†		*				Blistered thin salt crust with polygonal ridges
*	*		*	*									*						
*	*		*	*									*						
*	*		*	*				*											Wind sculpted & hummocky
*	*		*	*															
*	*		*	*															
*	*		*	*															
*	*		*	*			*												
*	*		*	*															
*	*		*	*															
*	*		*	*				*											
*	*		*	*				*											
*	*		*	*			*	*											
*	*		*	*				*											
*	*		*	*				*											

TABLE 4 (*continued*)

Vegetation, geomorphology & hydrology																			Mitchell (1983) Surface detail zones
Vegetation			Aeolian and fluvial activity					Salt regime							Hydrological regime				
Vegetated hummocks	Halophytes present	Other vegetation	Nebkha/Rabdu	Winnowing/fluting	Yardangs	Lag deposits (stony)	Sand accumulation	Salt efflorescences	Thin crust	Thick crust	Blistering	Polygonal ridging	Polygonal cracking	Thrust polygons	*Aioun*	Flow features	Moist surface	Standing water	
*	*		*	*				*											Wind sculpted & hummocky
*	*		*	*	*														
*	*		*	*	*														
*	*		*	*	*														
*	*		*	*	*														
*	*		*	*	*														
*	*		*	*		*													
*	*		*	*															
*	*		*	*															

* Feature present in winter wet and summer dry season.
† Feature present in summer dry season only.
‡ Feature present in winter wet season only.

topographically higher than that part of the playa in which salt crusts and *aioun* are found. Consequently, it was not extensively flooded in the 1985–86 winter and standing water was absent in May. Nevertheless, field evidence from May 1986 showed that standing water had accumulated in the depressions between nebkhas during the wet season.

The greatest changes in ground phenomena were noted at the control sites dominated by a thin salt crust. Neal & Motts (1967) have recognized the very dynamic nature of salt-crust surfaces in American playas. Here, changes were noted in 71.4% of the control sites (Table 4). The thin salt crusts on the Chott el Djerid are dominated by halite (41–94%), low amounts of gypsum (usually <4%) and variable amounts of wind-blown material, mainly quartz (1–33%). The soluble salt content ranges from 8–37% and the dominant size of the insoluble fraction is very fine sand, silt and clay (Munday 1985). The changes detected were almost all concerned with the degree of salt crust development (*ie* blistering, polygonal folding and cracking). Some of the control sites when examined in May 1986 had a highly reflective, thin and featureless salt crust varying from 0.1–1.0 cm in thickness. Other sites had standing water to a depth of a few centimetres (Table 4). The range of reflectance values was therefore bound to be quite high. As the water evaporated, the salts accumulated and the crust became thicker until, by the late summer, it ranged from 0.3–1.5 cm at the control sites. More importantly, the surface microrelief increased, blistering and polygonal patterns of folding and cracking had developed in response to the compressional stresses of drying (Christiansen 1963). This increase in surface relief could not cause such a dramatic increase in reflectance (Table 3) and the change in spectral response is therefore most likely attributable to the absence of standing water and to desiccation of the salt crust in the summer.

Changes in surface phenomena in the areas of thick salt crust and *aioun* were variable. The thick crusts have a similar chemistry to the thin ones. Halite varies from 45–97%, gypsum from 2–5%, and insoluble wind-blown material from 0–50%. The soluble salt content is higher, varying from 25–41% (Munday 1985). The texture of the thick crusts is coarser than the thin crusts with greater proportions of medium and fine sand.

Annual changes

Annual changes have been identified from images taken at similar times, but in different years. Satellite orbits and problems of image acquisition mean that it is almost impossible for imagery of the same day and week to be acquired. Imagery for the same month is feasible, especially in the dry season, but it is more difficult to obtain imagery for the wet season because of increased cloud cover.

Changes detected using these images are inevitably, longer term. They fall into two groups. Firstly, those due to processes that effect an annual increment *eg* of sediment. They occur either at the playa margin (Hardie *et al.* 1978; Reeves 1968), or in the centre of the playa. Secondly, there are changes due to processes which occur less frequently and which are the product of either long return-period storms or tectonic activity.

Changes at the playa margin. The marginal processes that have been detected in the field on the Chott el Guettar and C. el Fedjadj are mainly those which operate each wet season *eg* cliff retreat, rill erosion and sedimentation. The annual change that results from these processes is relatively small. Consequently, the detection of the effect of these processes is not possible with remotely sensed data unless the interval between images is very short.

However, the encroachment of alluvial fans onto playas is easily detected on remotely-sensed imagery (Fig. 7), providing that the dates of imagery bracket the sporadic events that cause the sedimentation.

Changes in the central playa. In the central playa, two phenomena related to salt and sediment redistribution have been identified. They occur on all the playas but are best developed on the Chott el Djerid and C. el Fedjadj. Winter runoff from surrounding higher ground gathers on the Sebkhet el Hamma (Fig. 8) and flows westward into the Chott el Fedjadj along very low gradients. At the outlet of the Sebkhet, the water is constricted and the 'channel' slope reaches levels as high as 0.007. Both factors produce relatively fast water flow for a playa environment and are areas of active erosion. On the thin salt crust of Chott el Djerid lines, relating to the surface flow from C. el Fedjadj, up to 0.5 km wide and with splays at their ends can be seen as patterns on the salt crust (Fig. 9). Darker 'flow lines' can be seen suggesting more than one flow event. The splays indicate flow into slight depressions. These are morphologically similar to the playa grooves which have been seen on other parts of the Chott el Djerid and which are also described by Reeves (1968) from Texas and Bonython & Mason (1943) from Australia. The imagery shows that at the splayed ends of the 'flow lines' the uppermost crusts overlap other crusts, or salt-rich zones (Fig. 9), again suggesting multiple flow events.

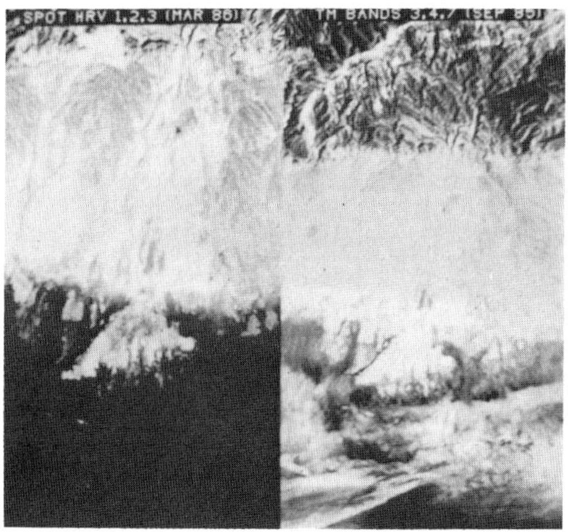

FIG. 7. *Landsat* Thematic Mapper (right) and SPOT High Resolution Visible scanner (left, copyright CNES, 1986) band composites from 1983 and 1985 respectively showing sedimentation of material onto the northern Chott el Djerid from a fan flanking Djebel el Asker.

Conclusions

Two categories of change have been detected on the three playas examined in this study using remotely-sensed data. They have seasonal and

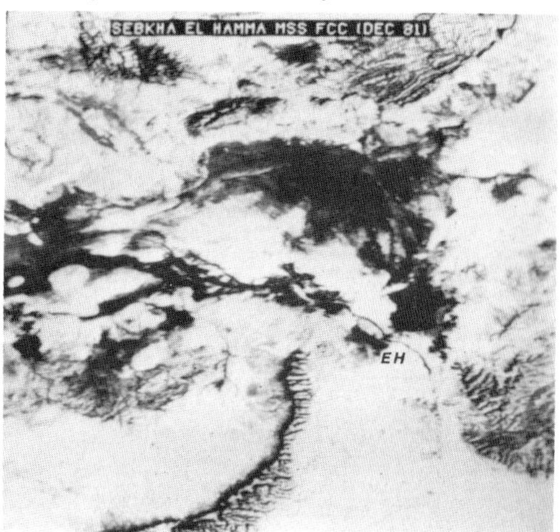

FIG. 8. *Landsat* Multispectral Scanner Band Composite image of the Sebkhet el Hamma from 1981. The detail in the chott has been increased by contrast stretching the playa at the expense of the surrounding areas, but the town and oases of El Hamma (EH) can still be seen. Vegetation and salt patterns relating to surface water flow from the surrounding uplands is seen funnelling through the neck of the sebkhet as it enters the Chott el Fedjadj to the W.

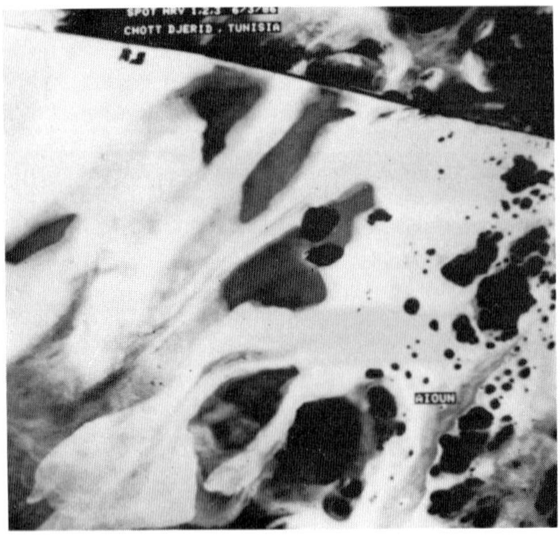

FIG. 9. SPOT High Resolution Visible Band Composite image of the central Chott el Djerid, March 1986. Two features are noticeable: (1) the variable size of the *aioun* in the E of the image and (2) the 'flow-lines' (playa grooves) formed in the thin salt crusts which channel water on the salt crusts to a depression terminating in splays.

annual dimensions. This suggests that the method is of use to geomorphologists, hydrologists and sedimentologists engaged in monitoring episodic geomorphological processes in deserts. Analysis shows that geomorphological change can be detected in three process-domains. However, only in one of these—the playa—is the method very successful. Some of the braided rivers in the study area have shifted their channels and this has been detected on remotely sensed imagery. The ability to detect activity on the alluvial fans has been restricted to only a few areas and most of the change detected has been directly attributed to seasonal vegetation differences rather than geomorphic processes.

Multispectral Scanner Band 7 and Thematic Mapper Bands 3 and 7 were found to contain the most useful spectral information when analysing changes on playas. This is due to their ability to provide data on soil and foliar moisture levels and, in the case of Thematic Mapper Band 7, on variations in sediment composition.

ACKNOWLEDGMENTS: This work has been carried out under the NASA Scientific Applications of Thematic Mapper Program by ACM & JRGT, and under NERC Contract F60/G6/12. ARJ acknowledges receipt of a NERC Research Training Award.

References

ASEM, A., KHALAF, F., ATTUSI, S. & PALOU, F. 1982. Classification of surface sediments in Kuwait using Landsat data. *Proceedings of the 1st Thematic Conference on Remote Sensing of Arid and Semi Arid Lands*, Cairo, 1057–64.

BIRD, A. C., WILLIAMS, T. H., BARRETT, M. E., MUNDAY, T. J. & TOWNSHEND, J. R. G. 1982. The Imperial College multichannel electronic image classifier and its applications to the classification of surface types of multispectral analysis. *Proceedings of the 1st Thematic Conference on Remote Sensing of Arid and Semi-Arid Lands*, Cairo, 665–675.

BODECHTEL, J. 1983. Requirements for spaceborne remote sensing in geology. *Remote Sensing: New satellite systems and potential applications*. European Space Agency Special Publication **205**, 53–57.

BONYTHON, C. W. & MASON, B. 1943. The filling and drifting of Lake Eyre. *Geographical Journal* **119**, 321–330.

BRUNSDEN, D. & THORNES, J. B. 1979. Landscape sensitivity and change. *Transactions of the Institute of British Geographers* **4**, 463–484.

BURROLLET, P. 1972. General geology of Tunisia. *In*: MARTIN, L. (ed.) *Guidebook to the Geology and History of Tunisia*. Society of Libyan Petroleum Geologists Handbook **3**, 51–58.

BYRNE, G. F., CRAPPER, P. F. & MAYO, J. K. 1980. Monitoring land cover by principal components analysis of multitemporal Landsat data. *Remote Sensing of the Environment* **10**, 175–184.

COQUE, R. 1969. *La Tunisie Pre-Saharienne: Etude Geomorphologique*. Armand Colin, Paris.

—— & JAUZIEN, A. 1967. The geomorphology and Quaternary geology of Tunisia. *In*: MARTIN, L. (ed.) *Guidebook to the Geology and History of Tunisia*. Society of Libyan Petroleum Geologists, Handbook **3**, 227–257.

DAVIES, P. A., GROLIER, M. J., SCHULTEJANN, P. A. & ELIASON, P. T. 1982. Discrimination of phosphate, gypsum, limestone, halite and quartz-sand deposits in south-central Tunisia by cluster analysis of Landsat Multispectral data. *Proceedings of the 1st Thematic Conference on Remote Sensing in Arid and Semi-Arid Lands*, Cairo, 337–360.

DEMAISON, G. J. 1965. The Triassic Salt in the Algerian Sahara. *In*: INSTITUTE OF PETROLEUM (ed.) *Salt Basins around Africa*. 91–100.

DOORNKAMP, J. C., BRUNSDEN, D. & JONES, D. K. C. 1980. *Geology, Geomorphology and Pedology of Bahrain*. Geobooks, Norwich.

DOTT, R. H. 1983. Episodic sedimentation—How normal is average? How rare is rare? Does it matter? *Journal of sedimentary Petrology* **53**, 5–23.

EPEMA, G. F. 1986. Multitemporal analysis of Thematic Mapper data for soil survey in southern Tunisia. *Proceedings of Symposium on Remote Sensing for Resource Development and Environmental Management*, Enschede, 245–249.

GLADWELL, D. R. 1982. Application of reflectance spectrometry to clay mineral determination in geological materials using portable radiometers. *Proceedings of the 2nd Thematic Conference of Remote Sensing for Exploration Geology*, Fort Worth, 29–38.

GLENNIE, K. W. 1970. *Desert sedimentary environments*. Elsevier, Amsterdam.

GRAETZ, R. D. & PECH, R. P. 1982. The utility of Landsat for monitoring the ephemeral water and herbage resources of arid lands: an example of rangeland management in the channel country of Australia. *Proceedings of the 1st Thematic Conference on Remote Sensing of Arid and Semi-Arid Lands*, Cairo, 1031–1046.

HARDIE, L. A., SMOOT, J. P. & EUGSTER, H. P. 1978. Saline lakes and their deposits: a sedimentological approach. *In*: MATTER, A. & TUCKER, M. E. (eds) Special Publication of the International Association of Sedimentologists **2**, 7–41.

HOWARTH, P. J. & BOASSON, E. 1983. Landsat digital enhancements for change detection in urban environments. *Remote Sensing of the Environment* **13**, 149–160.

HUNT, G. R. 1980. Electromagnetic radiation: the communication link in remote sensing. *In*: SIEGAL, B. S. & GILLESPIE, A. R. (eds) *Remote Sensing in Geology*. John Wiley & Sons, New York.

JONES, A. R. 1986*a*. The use of Thematic Mapper imagery for geomorphological mapping in arid and semi-arid environments. *Proceedings of Symposium on Remote Sensing for Resource Development and Environmental Management*, Enschede, 273–281.

—— 1986*b*. An evaluation of satellite Thematic Mapper imagery for geomorphological mapping in arid and semi-arid environments. *Proceedings of the 1st International Geomorphology Conference*, Manchester, John Wiley & Sons, 343–358.

KAHLE, A. 1984. Measuring spectra of arid lands. *In*: EL-BAZ, F. (ed.) *Deserts and Arid Lands*. Martinus Nijhoff, The Hague.

KLEMAS, V. & ABDEL-KADER, A. M. F. 1982. Remote sensing of coastal processes with emphasis on the Nile Delta. *Proceedings of the 1st Thematic Conference on Remote Sensing of Arid and Semi-Arid Lands*, Cairo, 389–416.

LANGER, A. M. & KERR, P. F. 1966. Mojave playa crusts: Physical properties and mineral content. *Journal of sedimentary Petrology* **36**, 377–396.

LODWICK, G. D. 1979. Measuring ecological changes in multitemporal Landsat data using principal components. *Proceedings of the 13th International Symposium Remote Sensing of the Environment*, Ann Arbor, Michigan, 1131–1142.

MECKELEIN, W. 1977. Zur Geomorphologie des Chott Djerid. *Stuttgarter geographische Studien* **91**, 247–301.

MILLINGTON, A. C. & TOWNSHEND, J. R. G. 1986. The potential of satellite remote sensing data for geomorphological investigations: an overview. *Proceedings of the 1st International Geomorphological Conference*, Manchester, John Wiley & Sons, 331–342.

MITCHELL, C. W. 1983. The soils of the Sahara with special reference to the Mahgreb. *Mahgreb Review* **8**, 29–37.

——, HOWARD, J. A. & MAINGUET, M. M. 1982. Soil degradation mapping from Landsat in North Africa and the Middle East. *Proceedings of the 1st Thematic Conference of Remote Sensing for Arid and Semi-Arid Lands*, Cairo, 899–908.

MUNDAY, T. J. 1985. *Multispectral remote sensing of surficial materials in an arid environment*. PhD Thesis, University of Reading.

NEAL, J. T. & MOTTS, W. S. 1967. Recent geomorphic changes in playas of western United States. *Journal of Geology* **75**, 511–525.

NELSON, R. F. 1983. Detecting forest canopy change due to insect activity using Landsat MSS. *Photogrammetric Engineering & Remote Sensing* **49**, 1303–1314.

REEVES, C. C. 1965. Spring pots and spring necks, new geomorphic features from Lynn and Terry Counties, Texas. *Compass* **43**, 41–45.

—— 1968. *Introduction to Paleolimnology*. Elsevier, Amsterdam.

RICHARDSON, J. A. & MILNE, A. K. 1983. Mapping fire burns and vegetation regeneration using principal components analysis. *Proceedings of the International Geoscience and Remote Sensing Symposium*, San Francisco, California, 51–56.

SCHUMM, S. A. 1979. Geomorphic thresholds: the concept and its applications. *Transactions of the Institute of British Geography* **4**, 485–515.

SINGH, A. 1984. *Tropical forest monitoring using digital Landsat data in northeastern India*. PhD thesis, University of Reading.

SUNHA, A. J. & VENKATACHALAM, P. 1982. Landsat spectral signatures studies with soil association and vegetation. *Proceedings of the 1st Thematic Conference of Remote Sensing of Arid and Semi-Arid Lands*, Cairo, 813–822.

TOWNSHEND, J. R. G. & HANCOCK, P. 1981. The use of remote sensing in mapping surficial materials. *In*: TOWNSHEND, J. R. G. (ed.) *Terrain Analysis and Remote Sensing*. George Allen & Unwin, London, 204–218.

A. C. MILLINGTON & A. R. JONES,* Department of Geography, University of Reading, Reading RG6 2AB, UK.

N. QUARMBY & J. R. G. TOWNSHEND, NERC Unit for Thematic Information & Services, University of Reading, Reading RG6 2AB, UK.

* *Present address:* NERC Institute of Terrestrial Ecology, Penrhos Road, Bangor, Gwynedd, Wales.

Index

Page numbers in *italics* refer to figures, those in **bold** refer to tables